Identities

1. $(\mathbf{A} \times \mathbf{B}) \cdot \mathbf{C} = \mathbf{A} \cdot (\mathbf{B} \times \mathbf{C})$
2. $\mathbf{A} \times (\mathbf{B} \times \mathbf{C}) = \mathbf{B}(\mathbf{A} \cdot \mathbf{C}) - \mathbf{C}(\mathbf{A} \cdot \mathbf{B})$
3. $\nabla(fg) = f\nabla g + g\nabla f$
4. $\nabla(a/b) = (1/b)\nabla a - (a/b^2)\nabla b$
5. $\nabla(\mathbf{A} \cdot \mathbf{B}) = (\mathbf{B} \cdot \nabla)\mathbf{A} + (\mathbf{A} \cdot \nabla)\mathbf{B} + \mathbf{B} \times (\nabla$
6. $\nabla \cdot (f\mathbf{A}) = (\nabla f) \cdot \mathbf{A} + f(\nabla \cdot \mathbf{A})$
7. $\nabla \cdot (\mathbf{A} \times \mathbf{B}) = \mathbf{B} \cdot (\nabla \times \mathbf{A}) - \mathbf{A} \cdot (\nabla \times \mathbf{B})$
8. $(\nabla \cdot \nabla)f = \nabla^2 f$
9. $\nabla \times (\nabla f) = 0$
10. $\nabla \cdot (\nabla \times \mathbf{A}) = 0$
11. $\nabla \times (f\mathbf{A}) = (\nabla f) \times \mathbf{A} + f(\nabla \times \mathbf{A})$
12. $\nabla \times (\mathbf{A} \times \mathbf{B}) = (\mathbf{B} \cdot \nabla)\mathbf{A} - (\mathbf{A} \cdot \nabla)\mathbf{B} + (\nabla \cdot \mathbf{B})\mathbf{A} - (\nabla \cdot \mathbf{A})\mathbf{B}\dagger$
13. $\nabla \times (\nabla \times \mathbf{A}) = \nabla(\nabla \cdot \mathbf{A}) - \nabla^2 \mathbf{A}$ (Sec. 1.10)
14. $$(\mathbf{A} \cdot \nabla)\mathbf{B} = \left[A_x \frac{\partial B_x}{\partial x} + A_y \frac{\partial B_x}{\partial y} + A_z \frac{\partial B_x}{\partial z}\right]\hat{\mathbf{x}} + \left[A_x \frac{\partial B_y}{\partial x} + A_y \frac{\partial B_y}{\partial y} + A_z \frac{\partial B_y}{\partial z}\right]\hat{\mathbf{y}} + \left[A_x \frac{\partial B_z}{\partial x} + A_y \frac{\partial B_z}{\partial y} + A_z \frac{\partial B_z}{\partial z}\right]\hat{\mathbf{z}}$$
15. $\nabla'(1/r) = \hat{\mathbf{r}}/r^2$. This is the gradient calculated at (x', y', z'), and $\mathbf{r}$ is the vector pointing from (x', y', z') to (x, y, z).
16. $\nabla(1/r) = -\hat{\mathbf{r}}/r^2$. This is the gradient calculated at (x, y, z) with the same vector $\mathbf{r}$.
17. $\mathcal{A} = \frac{1}{2}\oint_C \mathbf{r} \times d\mathbf{l}$, where the surface of area $\mathcal{A}$ is plane. The vector $\mathbf{r}$ extends from an arbitrary origin to a point on the curve C that bounds $\mathcal{A}$.
18. $\int_v \nabla f\, dv = \int_{\mathcal{A}} f\, d\mathcal{A}$
19. $\int_v (\nabla \times \mathbf{A})dv = -\int_{\mathcal{A}} \mathbf{A} \times d\mathcal{A}$, where $\mathcal{A}$ is the area of the closed surface that bounds the volume v.
20. $\oint_C f\, d\mathbf{l} = -\int_{\mathcal{A}} \nabla f \times d\mathcal{A}$, where C is the closed curve that bounds the open surface of area $\mathcal{A}$.

Theorems

1. The divergence theorem. $\int_{\mathcal{A}} \mathbf{A} \cdot d\mathcal{A} = \int_v \nabla \cdot \mathbf{A}\, dv$ where $\mathcal{A}$ is the area of the closed surface that bounds the volume v.
2. Stokes's theorem: $\oint_C \mathbf{A} \cdot d\mathbf{l} = \int_{\mathcal{A}} (\nabla \times \mathbf{A}) \cdot d\mathcal{A}$.

---

†Cartesian coordinates only.

RAMSEY LIBRARY - UNCA
3 0509 1307363 /

DATE DUE

OCT 29 2004
FE 09 '05
AP 07 '05

GAYLORD
PRINTED IN U.S.A.

QC760 .L65 2000
Lorrain, Paul
Fundamentals of
electromagnetic phenomena

9-6-01

Fundamentals of Electromagnetic Phenomena

Paul Lorrain
Université de Montréal and McGill University

Dale R. Corson
Cornell University

François Lorrain
Collège Jean-de-Brébeuf, Montréal

W.H. Freeman and Company
New York

Cover

Solar *coronal loops.* See the fourth Example in Sec. 14.2. The loops emerge from an *active region,* where there are intense magnetic fields. The curvature of the *limb* (edge) of the sun is an indication of the size of the loops: the sun has a radius of 700 megameters.

This is the superposition of three separate images taken in the ultraviolet at $\lambda = 17.1$, 19.5, and 28.4 nanometers. The false colors blue, green, and red were assigned to each image, respectively.

You can see this image, as well as many others, and films, too, at the following Web site: http://vestige.lmsal.com/TRACE/Public/AGU. Select image trace.425.jpg.

The letters "lmsal" stand for "Lockheed Martin Solar and Astrophysics Laboratory." "TRACE" stands for "Transition Region and Coronal Explorer," which is the name of a satellite.

Photo courtesy of Dr. Neal Hurlburt, Lockheed Martin and Astrophysics Laboratory.

Publisher: Susan F. Brennan
Cover Design: Blake Logan
Production Coordinator: Paul Rohloff
Manufacturing: Quebecor Printing, Martinsburg, West Virginia
Text Design and Composition: Publication Services, Champaign, Illinois
Marketing Director: John Britch

Library of Congress Cataloging-in-Publication Data
Lorrain, Paul.
Fundamentals of electromagnetic phenomena / Paul Lorrain, Dale R. Corson, François Lorrain.
p. cm.
Includes bibliographical references and index.
1. Electromagnetism. I. Corson, Dale R. II. Lorrain, François, 1945–

QC760 .L65 2000
537—dc21 00-039306

© 2000 by W.H. Freeman and Company. All rights reserved.

No part of this book may be reproduced by any mechanical, photographic, or electronic process, or in the form of a phonographic recording, nor may it be stored in a retrieval system, transmitted, or otherwise copied for public or private use, without written permission from the publisher.

Printed in the United State of America

First printing 2000

CONTENTS

†Asterisks indicate material that can be omitted without losing continuity.

GETTING STARTED

Dear Reader.

There is a long road ahead. Long? Well, an interesting road is never long enough, so you might not find the road too long, after all.

This book concerns *phenomena*. There is a lot of math, but the basic objective is to understand *phenomena*. Understand. That is an ambitious objective! Nature is *so* complicated that we often have the impression that no one really understands *anything!* Well, we all try!

This book aims to give you a *working knowledge* of electromagnetic phenomena. *Knowledge is Power,* but only if you know how to use it! That is why the more than 100 Examples and 350 Problems explore such a wide variety of applications of the theory.

The role of this book is to *open doors!*

Electromagnetic phenomena. There are electric, magnetic, and electromagnetic devices everywhere. Think of the immense fields of electronics, instrumentation, communications, computer science, and power technology. Think of cellular telephones, optical fibers, radio and TV receivers, remote controls for operating vehicles on the surface of Mars, and thousands and thousands of other devices. You will find many references to devices here.

There are also electromagnetic phenomena everywhere in nature. Again, to name just a few, there are lightning, sunsets, the earth's magnetic field, the van Allen belts circling the earth, sunspots, and the galaxies.

Of course we do not, and we cannot, discuss all that is known about electromagnetic phenomena. That would require a thousand scientists, who would write a thousand large books, over a period of many years.

So, even if you work through this book from cover to cover, you will not be proficient in *all* the aspects of electromagnetic phenomena. But you will have a start.

Problems. Solve as many Problems as you can; there is not much point in just reading the theory. What is important is what you can *do* with it!

There are three kinds of problems.

The problems that you will find here are of the *first kind:* problems that someone, somewhere, has already solved. That is where you have to start. But beware! *Solutions* (ours, for example!) are not necessarily satisfactory; they can be clumsy, or they can be half-truths, or they can be dead wrong. It has been said that it is less important to answer a question than it is to question the answer! Engineers and scientists are in a hurry, like everybody else, so *solutions* tend to become dogmas that nobody questions.

Problems of the *second kind* are more interesting: they are ones that are known, but that no person has ever solved. That is what research papers are about. Several of the Examples here are drawn from research papers written by the authors and colleagues. Here again, beware! All research papers are refereed, but neither author nor referee is infallible—and both are normally members of the same clique, with its own doctrine.

So be skeptical!

But there are also problems of a *third kind:* problems that no one has ever *thought* about! Undoubtedly, they are innumerable. They are the most interesting, but the main difficulty is to formulate them. It has also been said that "The kind of people we most need are not those who are good at answering well-posed old questions, but those who are capable of posing new ones."

The library. Browse through *Technology Review, Physics World, Physics Today,* and *The Industrial Physicist,* regularly. You will find them in your college library. They will open a *lot* of doors. Many of the papers there are written mostly for mature scientists, so just skip what you do not understand. Also, get to know engineering handbooks such as the *Electronics Engineers' Handbook* and *Reference Data for Radio Engineers*. There are many others.

On your marks!

Get set!

Go!

How far will *you* go?

PREFACE

This book is designed for one- or two-semester courses on Electromagnetism. It is an off-shoot of the bench-mark book *Electromagnetic Fields and Waves* by the same authors and the same publisher, which has been translated into Arabic, Chinese, French, German, Portuguese, and Spanish.

Here also the aim is to provide the reader with a *working knowledge* of Electromagnetism. We stress not only the basic principles, but also many applications, ranging from subsurface exploration to sunspots. There are over 100 Examples and over 350 Problems. According to one teacher, the problems are unsurpassable! There are both easy and "real-world" problems. Some of the answers are given at the end of the book.

A small part of the theory, and a number of the Examples, concern current work by the authors and colleagues, referred to in the References.

The first two chapters concern vectors and phasors. There are eight chapters on electric fields (four optional), three on Relativity (all optional), seven on magnetic fields (three optional), one that groups together the Maxwell equations, three on waves, and one on radiation, for a total of 25 chapters. Electric circuits are discussed briefly, where appropriate. Asterisks indicate sections and chapters that can be skipped without losing continuity. Each chapter is preceded by a Table of Contents, and followed by a Summary and a set of Problems classified by section number. Finally, there are four Appendixes, a Bibliography, and various Tables.

Teachers should read the Teaching Notes that follow this Preface.

The design of this book resulted from consultation with eleven teachers in three stages. At the very beginning I had an avalanche of suggestions from David Kaplan of Yeshiva University in New York, Paul Lafrance of the Université du Québec à Trois-Rivières, Eric T. Lane, of the University of Tennessee at Chattanooga, James McTavish, of the Liverpool John Moores University, and Hywel Morgan of the University of Glasgow.

At later stages I received further suggestions from Randolf R. Aldinger of Gettysburg College, Mario Belloni of Davidson College, Daniel L. Hatten of the Rose-Hulman Institute of Technology, R.A. Heelis of the University of Texas at Dallas, Steven Mellema of Gustavus Adolfus College, and Philip J. Siemens of Oregon State University.

Even though I expressed my gratitude individually to each of those teachers, I reiterate my thanks here. Their collaboration is deeply appreciated.

Daniel Hatten contributed a section in Chapter 9 on the use of spreadsheets for the calculation of electromagnetic fields, as well as the related problems. I thank him for sharing his extensive experience in this field.

Aside from making a multitude of suggestions, Paul Lafrance ably checked the complete text. That was a long and difficult task indeed because of the innumerable mathematical symbols and cross-references. His collaboration was deeply appreciated. Any remaining errors are my personal responsibility. Joseph Miskin, who plotted the curves and surfaces for *Electromagnetic Fields and Waves,* plotted many new ones for this book.

I am grateful to Serge Koutchmy, of the Institut d'astrophysique de Paris, to Jean-Louis Lemouël of the Institut de physique du Globe de Paris, to the late Neville Robinson of Saint Catherine's College of Oxford University, to Ernesto Martín and José Margineda of the Universidad de Murcia, and to Antonio Castellanos of the Universidad de Sevilla, for their hospitality.

I am particularly indebted to Prof. A. E. Williams-Jones, chair of the Department of Earth and Planetary Sciences of McGill University in Montréal, where most of this book was written.

Holly Hodder, Physics editor for Freeman, was extremely helpful in the early stages of the preparation of this book. Her successor Susan Brennan, together with her assistants Ayisha Day, Serena Tan, and Brian Donnellan, guided the operation most skillfully right up to final publication.

It is a pleasure to thank Rob Siedenburg for his careful copy editing, and to thank Kris Engberg and Rhonda Ries, all of Publication Services, Champaign, Illinois, who were most helpful at the typesetting stage.

Last, but not least, I owe much to my wife, Dorothée Sainte-Marie, who has tolerated my addiction to Physics for many years.

I shall be grateful to readers who will write to me, either to offer suggestions or about corrigenda.

PAUL LORRAIN
Department of Earth and Planetary Sciences
McGill University
Montréal, Québec, Canada H3A 2A7

May 2000

TEACHING NOTES

Electric circuits. We have followed the advice of several teachers and discussed circuits only briefly. Depending on the curriculum, you might wish to skip circuits entirely. If, on the contrary, you wish to stress circuits, then look up *Electromagnetic Fields and Waves* by the same authors and the same publisher (3rd edition, 1988).

Problems. Answers to some of the problems are given at the end of the book, to two significant figures. Of course teachers may ask for more accuracy. Teachers will find many solved problems in *Electromagnetism: Principles and Applications,* by Paul Lorrain and Dale Corson, also published by Freeman. That book complements the present one, at about the same level, and we have borowed many problems from it. There are more difficult problems in *Electromagnetic Fields and Waves.* Instructors who are pressed for time can obtain a *Solutions Manual* for the present book from the publisher (see below).

Difficult problems can be solved in class. Indeed, it is not a bad policy to do little more in class than solve problems, of course explaining along the way the principles involved.

Asterisks point out the sections and chapters that can be skipped without losing continuity. Teachers thus have much leeway to plan a one-semester course. Here are further ideas on what can be skipped without tears, and what is truly essential.

Chapter 1, *Vectors.* For some classes, this chapter will serve only as a quick review. But students *must* be proficient in the use of the four basic operators $\boldsymbol{\nabla}$, $\boldsymbol{\nabla}\cdot$, $\boldsymbol{\nabla}\times$, and $\boldsymbol{\nabla}^2$, at least in Cartesian coordinates. It is advisable to check by setting some problems.

Chapter 2, *Phasors.* Traditionally, phasors are never taught explicitly in Physics courses. That is unfortunate because they do require a fair amount of thought, and they are inescapable. Again, it is advisable to set problems.

Chapters 3 to 10, *Electric fields.* It is the custom to start books on Electromagnetism with a long discussion of electrostatics. That seems to be the place to start, but it might be advisable to spend less time on this and to spend more time on waves. Here, we have possibly discussed electrostatics too extensively.

If you are short of time, then you should preferably whittle down the beginning of the book, rather than the end.

Chapters 11 to 13, *Relativity.* If your students are wide-awake, then you will probably find that Relativity is particularly difficult to teach. There is no end to the absurd situations that you can get into because "If it makes sense, then it's wrong!" Again, if you are short of time, skip those chapters, even though they are clearly fundamental.

Chapters 14 to 20, *Magnetic fields.* These chapters are much more interesting and useful than the chapters on electrostatics.

Chapter 21, *The Maxwell Equations.* A must. Do read Appendix C on *The Story of the "Maxwell" Equations,* and possibly look up the references given there. Appendix C is, to a certain extent, a tribute to Heaviside.

Chapters 22 to 24, *Electromagnetic Waves.* These are important and interesting, even though they do not stress Optics enough. Try to not skip them. We have just about deleted rectangular metallic waveguides because they are now relatively uninteresting.

Chapter 25, *Radiation.* Coming at the end, this chapter is liable to be omitted. It would be wiser to omit some of the electrostatics.

Finally, teachers should browse through just the first few chapters of the biography of Edwin Land (McElheny, 1998), the inventor of Polaroid and of instant photography. He filed the first of his 535 patents when he was 20!

Note: a solutions manual is available for instructors, both in print and electronic format; adopters can contact W. H. Freeman to receive a copy: e-mail facultyservices @bfwpub.com.

LIST OF SYMBOLS

Space, Time, Mechanics

Length	l, L, s, r
Area	$\mathscr{A}$
Volume	v
Solid angle	Ω
Unit vector along $\boldsymbol{r}$	$\hat{\boldsymbol{r}}$
Unit vectors in Cartesian coordinates	$\hat{\boldsymbol{x}}, \hat{\boldsymbol{y}}, \hat{\boldsymbol{z}}$
Unit vectors in cylindrical coordinates	$\hat{\boldsymbol{\rho}}, \hat{\boldsymbol{\phi}}, \hat{\boldsymbol{z}}$
Unit vectors in spherical coordinates	$\hat{\boldsymbol{r}}, \hat{\boldsymbol{\theta}}, \hat{\boldsymbol{\phi}}$
Unit vector normal to a surface	$\hat{\boldsymbol{n}}$
Wavelength	λ
Wavelength in free space	λ_0
Wavelength in a waveguide	λ_g
Radian length	$ƛ = \lambda/2\pi$
Wave number	$2\pi/\lambda = 1/ƛ$
Time	t
Period	$T = 1/f$
Frequency	$f = 1/T$
Angular frequency	$\omega = 2\pi f$
Angular velocity	ω
Velocity, speed	$\boldsymbol{v}, v$
Mass	m
Mass per unit volume	m'
Force	$\boldsymbol{F}$
Force per unit volume	$\boldsymbol{F}'$
Pressure	p
Energy	$\mathscr{E}$
Energy per unit volume	$\mathscr{E}'$
Power	P
Power per unit volume	P'

Electricity and Magnetism

Speed of light	c
Electronic charge (absolute value of)	e
Boltzman constant	k
Planck's constant	h
Electric charge	Q
Electric charge per unit volume	ρ, $\tilde{Q}$
Electric charge per unit area	σ
Electric potential	V
Electric field strength	$\boldsymbol{E}$
Permittivity of vacuum	ϵ_0
Electric polarization	$\boldsymbol{P}$
Electric susceptibility	χ_e
Relative permittivity	ϵ_r
Electric current	I
Electric current per unit area	$\boldsymbol{J}$
Conductivity	σ
Mobility	$\mathcal{M}$
Resistance	R
Capacitance	C
Vector potential	$\boldsymbol{A}$
Magnetic flux density	$\boldsymbol{B}$
Magnetic flux	Φ
Magnetic susceptibility	χ_m
Permeability of vacuum	μ_0
Relative permeability	μ_r
Self-inductance	L
Mutual inductance	M
Impedance	Z
Poynting vector	$\mathcal{S}$

Mathematical Symbols

Approximately equal to	$\approx$
Proportional to	$\propto$
Exponential of	$\exp x$

$(-1)^{1/2}$	j
Vector	$\boldsymbol{F}$
Magnitude of $\boldsymbol{F}$	F
Peak value of F	F_m
Gradient of V	∇V
Divergence of $\boldsymbol{E}$	$\nabla \cdot \boldsymbol{E}$
Curl of $\boldsymbol{E}$	$\nabla \times \boldsymbol{E}$
Laplacian of V	$\nabla^2 V = \nabla \cdot (\nabla V)$

CHAPTER

1

VECTOR OPERATORS

This introductory chapter is meant to help readers who are not yet proficient in the use of vector *operators*.

We frequently refer to the fields of electric charges and currents. For example, we consider the force between two electric charges to arise from an interaction between either one of the charges and the *field* of the other.

Mathematically, a *field* is a function that describes a physical quantity at all points in space. In *scalar fields* this quantity is specified by a single number for each point. Temperature, density, and electric potential are examples of scalar quantities that can vary from one point to another in space. In *vector fields* the physical quantity is a vector, specified by both a number and a direction. Wind velocity and gravitational force are examples of vector fields.

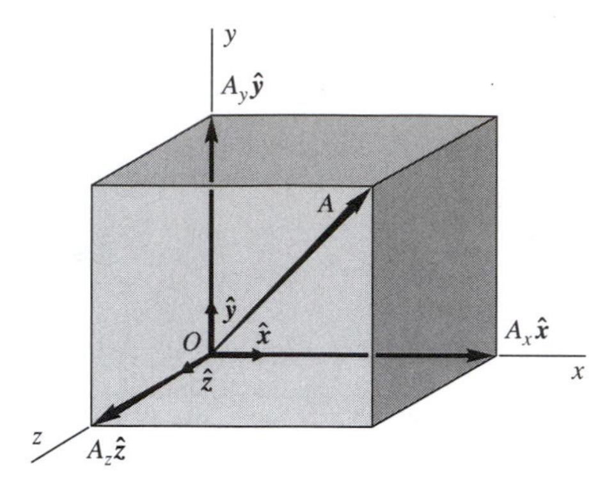

Figure 1-1 A vector $\boldsymbol{A}$ and its three component vectors $A_x\hat{\boldsymbol{x}}$, $A_y\hat{\boldsymbol{y}}$, $A_z\hat{\boldsymbol{z}}$, which, when they are placed end to end, are equivalent to $\boldsymbol{A}$. The unit vectors $\hat{\boldsymbol{x}}, \hat{\boldsymbol{y}}, \hat{\boldsymbol{z}}$ point in the positive directions of the coordinate axes and are of unit magnitude.

Vector quantities are designated by ***boldface italic type,*** and *unit vectors* carry a circumflex: $\hat{\boldsymbol{x}}, \hat{\boldsymbol{y}}, \hat{\boldsymbol{z}}$.

We follow the usual custom of using right-hand Cartesian coordinate systems, as in Fig. 1-1: the positive z-direction is the direction of advance of a right-hand screw rotated in the sense that turns the positive x-axis into the positive y-axis through the 90° angle.

Scalar quantities are designated by *lightface italic type*.

1.1 VECTOR ALGEBRA

Figure 1-1 shows a vector $\boldsymbol{A}$ and its three *components* A_x, A_y, A_z. If we define two vectors

$$\boldsymbol{A} = A_x\hat{\boldsymbol{x}} + A_y\hat{\boldsymbol{y}} + A_z\hat{\boldsymbol{z}}, \qquad \boldsymbol{B} = B_x\hat{\boldsymbol{x}} + B_y\hat{\boldsymbol{y}} + B_z\hat{\boldsymbol{z}}, \tag{1-1}$$

where $\hat{\boldsymbol{x}}, \hat{\boldsymbol{y}}, \hat{\boldsymbol{z}}$ are the *unit vectors* along the x-, y-, and z-axes, respectively, then

$$\boldsymbol{A} + \boldsymbol{B} = (A_x + B_x)\hat{\boldsymbol{x}} + (A_y + B_y)\hat{\boldsymbol{y}} + (A_z + B_z)\hat{\boldsymbol{z}}, \tag{1-2}$$

$$\boldsymbol{A} - \boldsymbol{B} = (A_x - B_x)\hat{\boldsymbol{x}} + (A_y - B_y)\hat{\boldsymbol{y}} + (A_z - B_z)\hat{\boldsymbol{z}}, \tag{1-3}$$

$$\boldsymbol{A} \cdot \boldsymbol{B} = A_xB_x + A_yB_y + A_zB_z = AB\cos\phi, \tag{1-4}$$

$$\boldsymbol{A} \times \boldsymbol{B} = \begin{vmatrix} \hat{\boldsymbol{x}} & \hat{\boldsymbol{y}} & \hat{\boldsymbol{z}} \\ A_x & A_y & A_z \\ B_x & B_y & B_z \end{vmatrix} = AB\sin\phi\hat{\boldsymbol{c}} = \boldsymbol{C}, \tag{1-5}$$

as in Fig. 1-2, where

$$A = (A_x^2 + A_y^2 + A_z^2)^{1/2} \tag{1-6}$$

is the *magnitude* of $\boldsymbol{A}$, and similarly for $\boldsymbol{B}$.

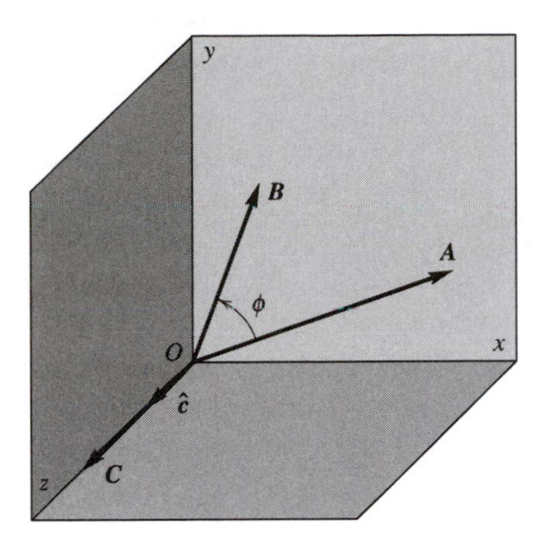

Figure 1-2 Two vectors $\boldsymbol{A}$ and $\boldsymbol{B}$ and the unit vector $\hat{\boldsymbol{c}}$, normal to the plane containing $\boldsymbol{A}$ and $\boldsymbol{B}$. The positive directions for ϕ and $\hat{\boldsymbol{c}}$ follow the right-hand screw rule. The vector product $\boldsymbol{A} \times \boldsymbol{B}$ is equal to $AB \sin \phi \, \hat{\boldsymbol{c}}$, and $\boldsymbol{B} \times \boldsymbol{A} = -\boldsymbol{A} \times \boldsymbol{B}$.

The quantity $\boldsymbol{A} \cdot \boldsymbol{B}$, which is read "A dot B," is the *scalar,* or *dot, product* of $\boldsymbol{A}$ and $\boldsymbol{B}$, and $\boldsymbol{A} \times \boldsymbol{B}$, read "A cross B," is their *vector,* or *cross, product.*

1.2 THE GRADIENT ∇f

A *scalar point-function* is a scalar quantity, say, temperature, that is a function of the coordinates. Consider a scalar point-function f that is continuous and differentiable. We want to know how f changes over the infinitesimal distance dl in Fig. 1-3. The differential

$$df = \frac{\partial f}{\partial x} dx + \frac{\partial f}{\partial y} dy + \frac{\partial f}{\partial z} dz \tag{1-7}$$

is the scalar product of the two vectors

$$\boldsymbol{dl} = dx\,\hat{\boldsymbol{x}} + dy\,\hat{\boldsymbol{y}} + dz\,\hat{\boldsymbol{z}} \tag{1-8}$$

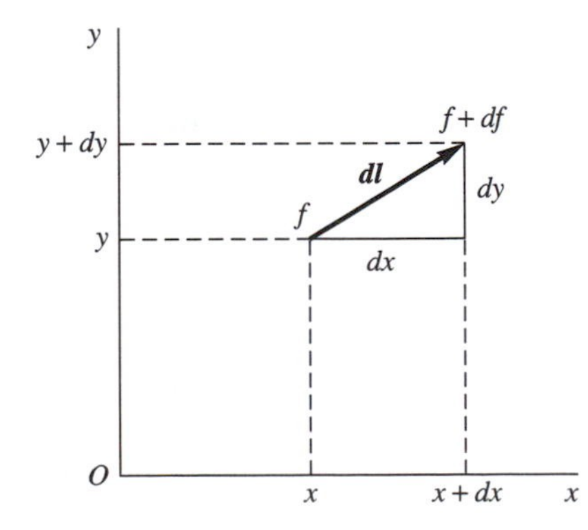

Figure 1-3 A scalar-point function changes from f to $f + df$ over the distance $\boldsymbol{dl}$.

and

$$\boldsymbol{\nabla} f = \frac{\partial f}{\partial x}\hat{\boldsymbol{x}} + \frac{\partial f}{\partial y}\hat{\boldsymbol{y}} + \frac{\partial f}{\partial z}\hat{\boldsymbol{z}}. \tag{1-9}$$

This second vector, whose components are the rates of change of f with distance along the coordinate axes, is called the *gradient* of f. The symbol $\boldsymbol{\nabla}$ is read "del,"

$$\boldsymbol{\nabla} = \hat{\boldsymbol{x}}\frac{\partial}{\partial x} + \hat{\boldsymbol{y}}\frac{\partial}{\partial y} + \hat{\boldsymbol{z}}\frac{\partial}{\partial z}. \tag{1-10}$$

Note the value of the magnitude of the gradient:

$$|\boldsymbol{\nabla} f| = \left[\left(\frac{\partial f}{\partial x}\right)^2 + \left(\frac{\partial f}{\partial y}\right)^2 + \left(\frac{\partial f}{\partial z}\right)^2\right]^{1/2}. \tag{1-11}$$

Thus

$$df = \boldsymbol{\nabla} f \cdot \boldsymbol{dl} = |\boldsymbol{\nabla} f|\,|\boldsymbol{dl}| \cos\theta, \tag{1-12}$$

where θ is the angle between the vectors $\boldsymbol{\nabla} f$ and $\boldsymbol{dl}$.

What direction should one choose for $\boldsymbol{dl}$ to maximize df? One should choose the direction for which $\cos\theta = 1$ or $\theta = 0$, that is, the direction of $\boldsymbol{\nabla} f$. Therefore the gradient of a scalar function at a given point is a vector having the following properties:

1. Its components are the rates of change of the function along the directions of the coordinate axes.
2. Its magnitude is the maximum rate of change with distance.
3. Its direction is that of the maximum rate of change with distance.
4. It points toward larger values of the function.

The gradient is a vector point-function that derives from a scalar point-function. Its magnitude and direction are those of the maximum space rate of change of f.

EXAMPLE The Elevation of a Point on the Surface of the Earth

As an example of the gradient, consider Fig. 1-4 in which E, the elevation above sea level, is a function of the x- and y-coordinates measured on a horizontal plane. Points at a given elevation define a *contour line*. The gradient of the elevation E at a given point then has the following properties:

1. It is perpendicular to the contour line at that point.
2. Its magnitude is equal to the maximum rate of change of elevation with displacement measured in a horizontal plane at that point.
3. It points toward an increase in elevation.

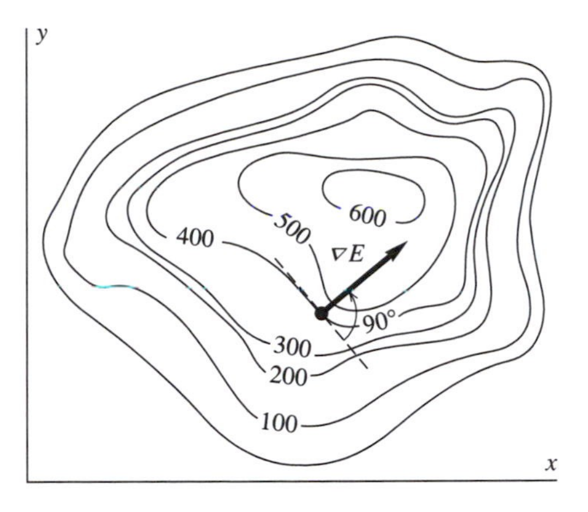

Figure 1-4 Topographic map of a hill. The numbers shown give the elevation E in meters. The gradient of E is the slope of the hill at the point considered, and it points toward an increase in elevation. The arrow shows ∇E at one point where the elevation is 400 meters.

1.3 FLUX

It is often necessary to calculate the flux of a vector quantity through a surface. By definition, the *flux* $d\Phi$ of $\boldsymbol{B}$ through an infinitesimal surface $d\boldsymbol{\mathcal{A}}$ is

$$d\Phi = \boldsymbol{B} \cdot d\boldsymbol{\mathcal{A}}, \tag{1-13}$$

where the vector $d\boldsymbol{\mathcal{A}}$ is normal to the surface. The flux $d\Phi$ is therefore the component of the vector normal to the surface, multiplied by $d\mathcal{A}$. For a surface of finite area $\mathcal{A}$,

$$\Phi = \int_{\mathcal{A}} \boldsymbol{B} \cdot d\boldsymbol{\mathcal{A}}. \tag{1-14}$$

If the surface is closed, the vector $d\boldsymbol{\mathcal{A}}$ points *outward* by convention.

If the surface is *not* closed, then the vector $d\boldsymbol{\mathcal{A}}$ is related to the direction of the bounding curve C by the right-hand screw rule.

EXAMPLE Fluid Flow

Consider fluid flow, and let ρ be the density, $\boldsymbol{v}$ the velocity, and $d\boldsymbol{\mathcal{A}}$ an element of area situated in the fluid. The scalar product $\rho\boldsymbol{v} \cdot d\boldsymbol{\mathcal{A}}$ is equal to the mass of fluid that crosses $d\boldsymbol{\mathcal{A}}$ in 1 second, in the direction of the vector $d\boldsymbol{\mathcal{A}}$. Thus the flux of $\rho\boldsymbol{v}$ through a closed surface, or the integral of $\rho\boldsymbol{v} \cdot d\boldsymbol{\mathcal{A}}$ over that surface, is equal to the net rate at which mass leaves the enclosed volume. In an incompressible fluid this flux would be equal to zero.

1.4 THE DIVERGENCE $\nabla \cdot \boldsymbol{B}$

The outward flux of a vector through a closed surface can be calculated either from Eq. 1-14 or as follows. Consider an infinitesimal volume $dx\,dy\,dz$ and a vector $\boldsymbol{B}$, as in Fig. 1-5, whose components B_x, B_y, B_z are functions of x, y, z. The value of B_x at the

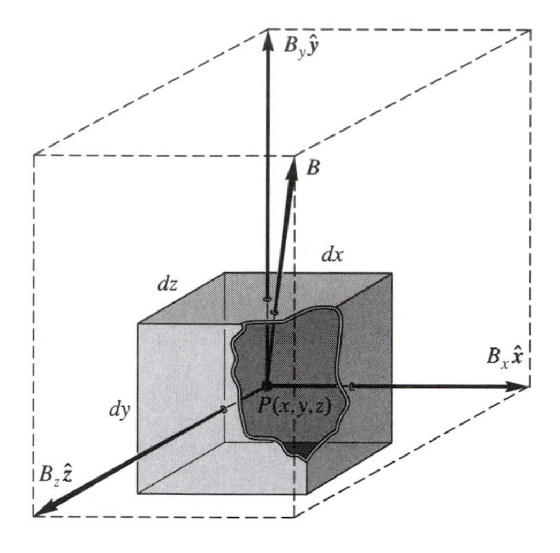

Figure 1-5 Element of volume $dx\,dy\,dz$ and the vector $\boldsymbol{B}$ at the point P. The dashed parallelepiped has one corner at P.

center of the right-hand face may be considered to be the average value over that face. Through the right-hand face of the volume element, the outgoing flux is

$$d\Phi_R = \left(B_x + \frac{\partial B_x}{\partial x}\frac{dx}{2}\right) dy\,dz, \tag{1-15}$$

since the normal component of $\boldsymbol{B}$ at the right-hand face is the x-component of $\boldsymbol{B}$ at that face. The volume being infinitesimal, we neglect higher-order derivatives of the components of $\boldsymbol{B}$.

At the left-hand face, the outgoing flux is

$$d\Phi_L = -\left(B_x - \frac{\partial B_x}{\partial x}\frac{dx}{2}\right) dy\,dz. \tag{1-16}$$

There is a minus sign before the parenthesis because $B_x\hat{\boldsymbol{x}}$ points inward at this face and $d\boldsymbol{\mathcal{A}}$ points outward.

Thus the outward flux through the two faces is

$$d\Phi_L + d\Phi_R = \frac{\partial B_x}{\partial x}\,dx\,dy\,dz = \frac{\partial B_x}{\partial x}\,dv, \tag{1-17}$$

where dv is the volume of the infinitesimal element.

If we calculate the net flux through the other pairs of faces in the same manner, we find that the total outward flux for the element of volume dv is

$$d\Phi_{\text{tot}} = \left(\frac{\partial B_x}{\partial x} + \frac{\partial B_y}{\partial y} + \frac{\partial B_z}{\partial z}\right) dv. \tag{1-18}$$

Suppose now that we have two adjoining infinitesimal volume elements and that we add the flux emerging through the bounding surface of the first volume to the flux emerging through the bounding surface of the second. At the common face, the fluxes are equal in magnitude but opposite in sign, and they cancel. The sum of the flux from the first volume and that from the second is therefore the flux emerging through the bounding surface of the combined volumes.

To extend this calculation to a finite volume, we sum the individual fluxes for each of the infinitesimal volume elements in the finite volume, and so the total outward flux is

$$\Phi_{\text{tot}} = \int_v \left(\frac{\partial B_x}{\partial x} + \frac{\partial B_y}{\partial y} + \frac{\partial B_z}{\partial z} \right) dv. \tag{1-19}$$

At any given point in the volume, the quantity

$$\frac{\partial B_x}{\partial x} + \frac{\partial B_y}{\partial y} + \frac{\partial B_z}{\partial z}$$

is thus the *outgoing* flux per unit volume. We call this quantity the *divergence* of $\boldsymbol{B}$ at the point. The divergence of a vector point-function is a scalar point-function.

According to the rule for the scalar product, we write the *divergence* of $\boldsymbol{B}$ as

$$\nabla \cdot \boldsymbol{B} = \frac{\partial B_x}{\partial x} + \frac{\partial B_y}{\partial y} + \frac{\partial B_z}{\partial z}. \tag{1-20}$$

See the Examples below.

1.5 THE DIVERGENCE THEOREM

Now the total outward flux of a vector $\boldsymbol{B}$ is equal to the surface integral of the normal outward component of $\boldsymbol{B}$. Thus, if we denote by $\mathcal{A}$ the area of the surface bounding v, the total outward flux is

$$\Phi_{\text{tot}} = \int_{\mathcal{A}} \boldsymbol{B} \cdot d\boldsymbol{\mathcal{A}} = \int_v \left(\frac{\partial B_x}{\partial x} + \frac{\partial B_y}{\partial y} + \frac{\partial B_z}{\partial z} \right) dv = \int_v \nabla \cdot \boldsymbol{B}\, dv. \tag{1-21}$$

These relations apply to any continuously differentiable† vector field $\boldsymbol{B}$. Thus

$$\int_{\mathcal{A}} \boldsymbol{B} \cdot d\boldsymbol{\mathcal{A}} = \int_v \nabla \cdot \boldsymbol{B}\, dv. \tag{1-22}$$

This is the *divergence theorem.* Note that the first integral involves only the values of $\boldsymbol{B}$ on the *surface* of area $\mathcal{A}$, whereas the second involves the values of $\boldsymbol{B}$ throughout the *volume* v.

---

†A function is *continuously differentiable* if its first derivatives are continuous.

EXAMPLES

A hermetic box contains oxygen.

1. The temperature, the pressure, and the mass density ρ are all constant and uniform. There is thermal agitation, but the time-averaged velocity $\boldsymbol{v}$ of a molecule in the region of an arbitrary point P is zero, and the net mass flux through the surface S of a small sphere surrounding P is zero. So the integral of $\rho\boldsymbol{v}$ over S is zero, and $\boldsymbol{\nabla}\cdot(\rho\boldsymbol{v}) = 0$. Also, $d\rho/dt = 0$.
2. Now pump oxygen in. The mass density ρ increases and there is an inward mass flux through S. The averaged radial velocity over S is negative, the integral of $\rho\boldsymbol{v}$ over S is negative, $\boldsymbol{\nabla}\cdot(\rho\boldsymbol{v})$ is negative, and $d\rho/dt$ is positive.
3. Now pump oxygen out. Gas flows out of S, the integral of $\rho\boldsymbol{v}$ over S is positive, $\boldsymbol{\nabla}\cdot(\rho\boldsymbol{v})$ is positive, and $d\rho/dt$ is negative.

In an incompressible fluid, $\boldsymbol{\nabla}\cdot(\rho\boldsymbol{v})$ is everywhere equal to zero, since the outward mass flux per unit volume is zero.

Within an explosion, $\boldsymbol{\nabla}\cdot(\rho\boldsymbol{v})$ is positive.

1.6 THE LINE INTEGRAL $\int_a^b \boldsymbol{B}\cdot\boldsymbol{dl}$. CONSERVATIVE FIELDS

The integrals

$$\int_a^b \boldsymbol{B}\cdot\boldsymbol{dl}, \qquad \int_a^b \boldsymbol{B}\times\boldsymbol{dl}, \qquad \text{and} \qquad \int_a^b f\,\boldsymbol{dl},$$

evaluated from the point a to the point b over some specified curve, are examples of *line integrals*.

In the first, which is especially important, the term under the integral sign is the product of an element of length $\boldsymbol{dl}$ on the curve and the local value of $\boldsymbol{B}$ according to the rule for the scalar product.

A vector field $\boldsymbol{B}$ is *conservative* if the line integral of $\boldsymbol{B}\cdot\boldsymbol{dl}$ around any closed curve is zero:

$$\oint \boldsymbol{B}\cdot\boldsymbol{dl} = 0\,. \qquad (1\text{-}23)$$

The circle on the integral sign indicates that the path of integration is closed.

1.7 THE CURL $\boldsymbol{\nabla}\times\boldsymbol{B}$

We now turn to the curl operator, which is somewhat more complex than the gradient and divergence operators. All three are essential for a proper discussion of electromagnetic fields. The Laplacian operator of Sec. 1.9 is also essential, although less commonly used.

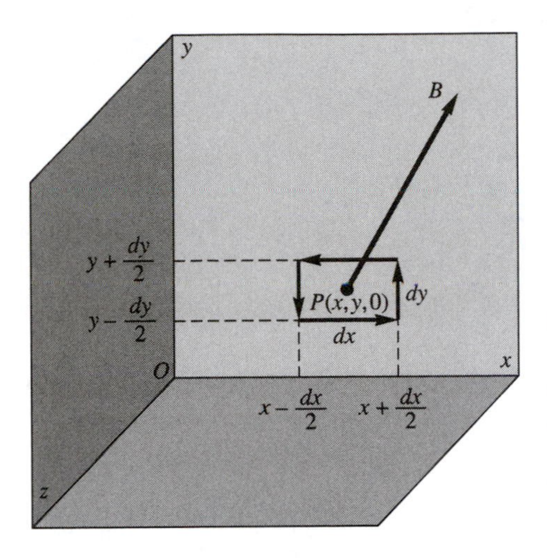

Figure 1-6 Closed, rectangular path in the xy-plane, centered on the point $P(x, y, 0)$, where the vector $\boldsymbol{B}$ has the value illustrated by the arrow. The integration around the path proceeds in the direction of the arrows, in accordance with the right-hand screw rule applied to the z-axis.

For any given field $\boldsymbol{B}$ and for a path $\boldsymbol{dl}$ situated in the xy-plane,

$$\boldsymbol{B}\cdot\boldsymbol{dl} = B_x\,dx + B_y\,dy \tag{1-24}$$

and

$$\oint \boldsymbol{B}\cdot\boldsymbol{dl} = \oint B_x\,dx + \oint B_y\,dy. \tag{1-25}$$

Now consider the infinitesimal path in Fig. 1-6. There are two contributions to the first integral on the right-hand side of Eq. 1-25, one at $y - dy/2$ and one at $y + dy/2$:

$$\oint B_x\,dx = \left(B_x - \frac{\partial B_x}{\partial y}\frac{dy}{2}\right)dx - \left(B_x + \frac{\partial B_x}{\partial y}\frac{dy}{2}\right)dx. \tag{1-26}$$

There is a minus sign before the second term because the path element at $y + dy/2$ points in the negative x-direction. Therefore, for this infinitesimal path,

$$\oint B_x\,dx = -\frac{\partial B_x}{\partial y}\,dy\,dx. \tag{1-27}$$

Similarly,

$$\oint B_y\,dy = \frac{\partial B_y}{\partial x}\,dx\,dy, \tag{1-28}$$

and

$$\oint \boldsymbol{B}\cdot\boldsymbol{dl} = \left(\frac{\partial B_y}{\partial x} - \frac{\partial B_x}{\partial y}\right)dx\,dy \tag{1-29}$$

for the infinitesimal path of Fig. 1-6.

If we set

$$g_3 = \frac{\partial B_y}{\partial x} - \frac{\partial B_x}{\partial y}, \tag{1-30}$$

then

$$\oint \boldsymbol{B} \cdot d\boldsymbol{l} = g_3 \, d\mathcal{A}, \tag{1-31}$$

where $d\mathcal{A} = dx\,dy$ is the area enclosed by the infinitesimal path. Note that this is correct only if the line integral runs in the positive direction in the xy-plane, that is, in the direction in which one would turn a right-hand screw to make it advance in the positive direction along the z-axis.

Consider now g_3 and the other two symmetric quantities as the components of a vector

$$\boldsymbol{\nabla} \times \boldsymbol{B} = \left(\frac{\partial B_z}{\partial y} - \frac{\partial B_y}{\partial z}\right)\hat{\boldsymbol{x}} + \left(\frac{\partial B_x}{\partial z} - \frac{\partial B_z}{\partial x}\right)\hat{\boldsymbol{y}} + \left(\frac{\partial B_y}{\partial x} - \frac{\partial B_x}{\partial y}\right)\hat{\boldsymbol{z}}, \tag{1-32}$$

which may be written as

$$\boldsymbol{\nabla} \times \boldsymbol{B} = \begin{vmatrix} \hat{\boldsymbol{x}} & \hat{\boldsymbol{y}} & \hat{\boldsymbol{z}} \\ \dfrac{\partial}{\partial x} & \dfrac{\partial}{\partial y} & \dfrac{\partial}{\partial z} \\ B_x & B_y & B_z \end{vmatrix}. \tag{1-33}$$

This is the *curl* of $\boldsymbol{B}$. The quantity g_3 is its z-component.

If we choose a vector $d\boldsymbol{\mathcal{A}}$ that points in the direction of advance of a right-hand screw turned in the direction chosen for the line integral, then

$$d\boldsymbol{\mathcal{A}} = d\mathcal{A}\,\hat{\boldsymbol{z}} \tag{1-34}$$

and

$$\oint \boldsymbol{B} \cdot d\boldsymbol{l} = (\boldsymbol{\nabla} \times \boldsymbol{B}) \cdot d\boldsymbol{\mathcal{A}}. \tag{1-35}$$

This means that the line integral of $\boldsymbol{B} \cdot d\boldsymbol{l}$ around the edge of the area $d\boldsymbol{\mathcal{A}}$ is equal to the scalar product of the curl of $\boldsymbol{B}$ by this element of area.

The sign convention is similar to the one above: the line integral runs in the positive direction around the element of area $d\boldsymbol{\mathcal{A}}$ if it is the direction in which one would turn a right-hand screw to make it advance in the direction of $\boldsymbol{\nabla} \times \boldsymbol{B}$.

In general, $\boldsymbol{\nabla} \times \boldsymbol{B}$ is *not* normal to $\boldsymbol{B}$. See Prob. 1-7.

The curl of a gradient is identically equal to zero:

$$\boldsymbol{\nabla} \times (\boldsymbol{\nabla} f) \equiv 0. \tag{1-36}$$

EXAMPLE Fluid Stream

Near the bottom of a fluid stream the velocity v is proportional to the distance from the bottom. Set the x-axis parallel to the direction of flow and the z-axis perpendicular to the stream bottom. Then

$$v_x = cz, \qquad v_y = 0, \qquad v_z = 0, \tag{1-37}$$

and the curl of the velocity vector is

$$\boldsymbol{\nabla} \times \boldsymbol{v} = \begin{vmatrix} \hat{\boldsymbol{x}} & \hat{\boldsymbol{y}} & \hat{\boldsymbol{z}} \\ \dfrac{\partial}{\partial x} & \dfrac{\partial}{\partial y} & \dfrac{\partial}{\partial z} \\ cz & 0 & 0 \end{vmatrix} = c\hat{\boldsymbol{y}}. \tag{1-38}$$

1.8 STOKES'S THEOREM

Equation 1-35 is true only for a path so small that $\boldsymbol{\nabla} \times \boldsymbol{B}$ is nearly constant over the surface $d\boldsymbol{\mathcal{A}}$ bounded by the path. What happens when the path is so large that this condition is not met? In such cases we divide the surface—any finite surface† bounded by the path of integration in question—into elements of area $d\boldsymbol{\mathcal{A}}_1$, $d\boldsymbol{\mathcal{A}}_2$, and so forth, as in Fig. 1-7. For any one of these small areas,

$$\oint \boldsymbol{B} \cdot d\boldsymbol{l} = (\boldsymbol{\nabla} \times \boldsymbol{B}) \cdot d\boldsymbol{\mathcal{A}}. \tag{1-39}$$

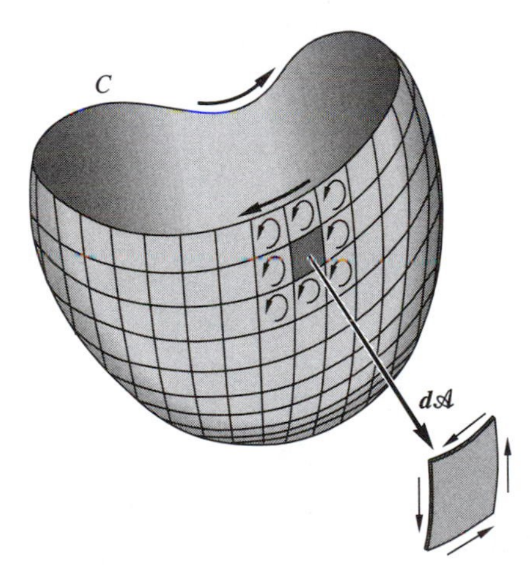

Figure 1-7 An arbitrary surface of area $\mathcal{A}$ bounded by the curve C. The sum of the line integrals around the curvilinear squares shown is equal to the line integral around C. The positive direction for the vector $d\boldsymbol{\mathcal{A}}$ follows the right-hand screw rule.

†This must be an *orientable* surface, that is, a surface with two sides. Not all surfaces have two sides; a Möbius strip, for instance, has only one side.

We add the left-hand sides of these equations for all the $d\mathcal{A}$'s and then all the right-hand sides. The sum of the left-hand sides is the line integral around the external boundary, since there are always two equal and opposite contributions to the sum along every common side between adjacent $d\mathcal{A}$'s. The sum of the right-hand sides is the integral of $(\nabla \times \boldsymbol{B}) \cdot d\boldsymbol{\mathcal{A}}$ over the finite surface. Thus

$$\oint_C \boldsymbol{B} \cdot d\boldsymbol{l} = \int_{\mathcal{A}} (\nabla \times \boldsymbol{B}) \cdot d\boldsymbol{\mathcal{A}}, \tag{1-40}$$

where $\mathcal{A}$ is the area of any open surface bounded by the closed curve C.

This is *Stokes's theorem.* It relates the line integral over a given path to a surface integral over any finite surface bounded by that path. Figure 1-7 illustrates the sign convention.

EXAMPLE Conservative Fields

Under what condition is a vector field conservative? In other words, under what condition is the line integral of $\boldsymbol{B} \cdot d\boldsymbol{l}$ around an arbitrary closed path equal to zero? From Stokes's theorem, the line integral of $\boldsymbol{B} \cdot d\boldsymbol{l}$ around an arbitrary closed path is zero if $\nabla \times \boldsymbol{B} = 0$ everywhere. This condition is met if

$$\boldsymbol{B} = \nabla f, \tag{1-41}$$

which is equivalent to

$$\nabla \times \boldsymbol{B} = 0. \tag{1-42}$$

A field $\boldsymbol{B}$ that is the gradient of some scalar point-function f is therefore *conservative.*

1.9 THE LAPLACIAN OPERATOR ∇^2

The divergence of the gradient of f is the *Laplacian* of f:

$$\nabla \cdot \nabla f = \nabla^2 f = \frac{\partial^2 f}{\partial x^2} + \frac{\partial^2 f}{\partial y^2} + \frac{\partial^2 f}{\partial z^2}, \tag{1-43}$$

where ∇^2 is the *Laplacian operator.*

We have defined the Laplacian of a scalar point-function f. It is also useful to define the Laplacian of a vector point-function $\boldsymbol{B}$:

$$\nabla^2 \boldsymbol{B} = \nabla^2 B_x \hat{x} + \nabla^2 B_y \hat{y} + \nabla^2 B_z \hat{z}. \tag{1-44}$$

This relation applies only to Cartesian coordinates. See Sec. 1.10.

1.10 SCALARS AND VECTORS IN SPHERICAL COORDINATES

It is always possible to use Cartesian coordinates, whatever the geometry of the field, but other coordinate systems are often much more convenient. For example, if there is cylindrical symmetry, it is highly preferable to use cylindrical coordinates; if there is spherical symmetry, then spherical coordinates are in order.

With *spherical coordinates,* the position of a point P in space is defined by the coordinates r, θ, ϕ, as in Fig. 1-8. The angle ϕ is undefined for points on the z-axis. The vector $\boldsymbol{r}$ that defines the position of P is simply $r\hat{\boldsymbol{r}}$, the coordinates θ and ϕ being given by the orientation of $\hat{\boldsymbol{r}}$.

Note that the unit vectors $\hat{\boldsymbol{r}}$, $\hat{\boldsymbol{\theta}}$, $\hat{\boldsymbol{\phi}}$ do not maintain the same orientations when P moves about.

The unit vectors are mutually orthogonal and

$$\hat{\boldsymbol{r}} \cdot \hat{\boldsymbol{r}} = 1, \ldots \qquad \hat{\boldsymbol{r}} \cdot \hat{\boldsymbol{\theta}} = 0, \ldots \tag{1-45}$$

$$\hat{\boldsymbol{r}} \times \hat{\boldsymbol{\theta}} = \hat{\boldsymbol{\phi}}, \ldots \qquad \hat{\boldsymbol{\theta}} \times \hat{\boldsymbol{r}} = -\hat{\boldsymbol{\phi}}, \ldots \tag{1-46}$$

See Prob. 1-8 for the relations between the unit vectors $\hat{\boldsymbol{r}}$, $\hat{\boldsymbol{\theta}}$, $\hat{\boldsymbol{\phi}}$, and the unit vectors $\hat{\boldsymbol{x}}$, $\hat{\boldsymbol{y}}$, $\hat{\boldsymbol{z}}$.

One calculates scalar and vector products in the same way as with Cartesian coordinates. For example, if

$$\boldsymbol{J} = J_r\hat{\boldsymbol{r}} + J_\theta\hat{\boldsymbol{\theta}} + J_\phi\hat{\boldsymbol{\phi}}, \tag{1-47}$$

$$\boldsymbol{B} = B_r\hat{\boldsymbol{r}} + B_\theta\hat{\boldsymbol{\theta}} + B_\phi\hat{\boldsymbol{\phi}}, \tag{1-48}$$

then

$$\boldsymbol{J} \cdot \boldsymbol{B} = J_rB_r + J_\theta B_\theta + J_\phi B_\phi, \tag{1-49}$$

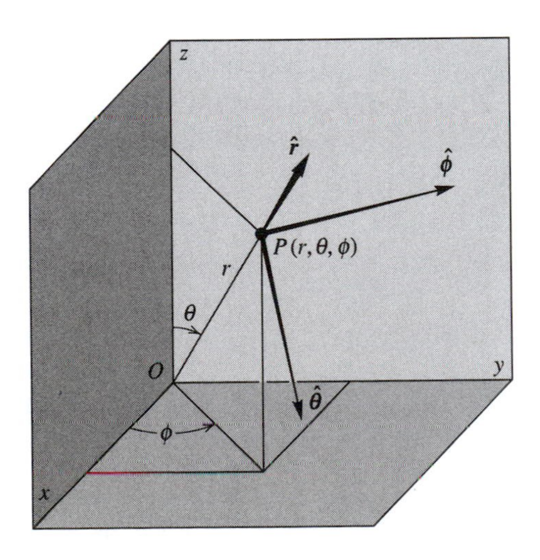

Figure 1-8 Spherical coordinates.

and

$$\boldsymbol{J} \times \boldsymbol{B} = (J_r \hat{\boldsymbol{r}} + J_\theta \hat{\boldsymbol{\theta}} + J_\phi \hat{\boldsymbol{\phi}}) \times (B_r \hat{\boldsymbol{r}} + B_\theta \hat{\boldsymbol{\theta}} + B_\phi \hat{\boldsymbol{\phi}}) \tag{1-50}$$

$$= \begin{vmatrix} \hat{\boldsymbol{r}} & \hat{\boldsymbol{\theta}} & \hat{\boldsymbol{\phi}} \\ J_r & J_\theta & J_\phi \\ B_r & B_\theta & B_\phi \end{vmatrix} \tag{1-51}$$

$$\begin{aligned} &= (J_\theta B_\phi - J_\phi B_\theta)\hat{\boldsymbol{r}} \\ &+ (J_\phi B_r - J_r B_\phi)\hat{\boldsymbol{\theta}} \\ &+ (J_r B_\theta - J_\theta B_r)\hat{\boldsymbol{\phi}}. \end{aligned} \tag{1-52}$$

Note the cyclic permutation of the subscripts. Note also that the terms are positive when the coordinates appear in the proper cyclic order, and are otherwise negative.

Here is a second important point. With Cartesian coordinates, a single operator $\boldsymbol{\nabla}$ serves for the gradient, divergence, curl, and Laplacian. But, with other coordinates, there does not exist an operator $\boldsymbol{\nabla}$.

Figure 1-9 shows that, if P moves by a distance $\boldsymbol{dr}$, then

$$\boldsymbol{dr} = dr\hat{\boldsymbol{r}} + r\,d\theta\,\hat{\boldsymbol{\theta}} + r\sin\theta\,d\phi\,\hat{\boldsymbol{\phi}}. \tag{1-53}$$

So the volume element

$$dv = (dr)(r\,d\theta)(r\sin\theta\,d\phi) = r^2\sin\theta\,dr\,d\theta\,d\phi. \tag{1-54}$$

From Fig. 1-9, the elements of area that are perpendicular to $\hat{\boldsymbol{r}}$, $\hat{\boldsymbol{\theta}}$, $\hat{\boldsymbol{\phi}}$ are

$$d\mathcal{A}_r = (r\,d\theta)(r\sin\theta\,d\phi) = r^2\sin\theta\,d\theta\,d\phi, \tag{1-55}$$

$$d\mathcal{A}_\theta = dr(r\sin\theta\,d\phi) = r\sin\theta\,dr\,d\phi, \tag{1-56}$$

$$d\mathcal{A}_\phi = dr(r\,d\theta) = r\,dr\,d\theta. \tag{1-57}$$

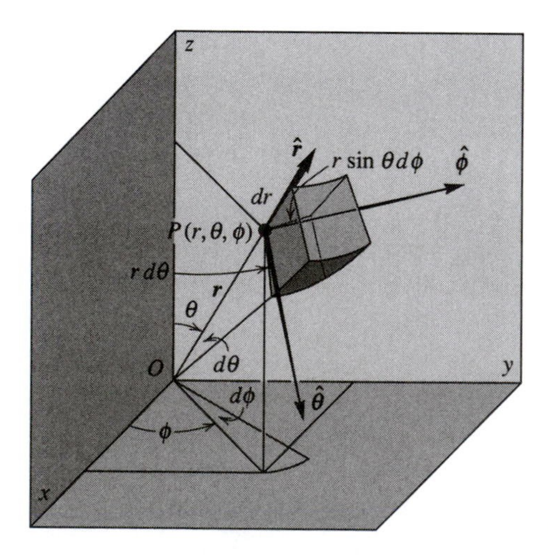

Figure 1-9 Element of volume in spherical coordinates.

It is easy to express the *gradient* in spherical coordinates. Since the gradient is the vector rate of change of a scalar function f,

$$\boldsymbol{\nabla} f = \frac{\partial f}{\partial r}\hat{\boldsymbol{r}} + \frac{1}{r}\frac{\partial f}{\partial \theta}\hat{\boldsymbol{\theta}} + \frac{1}{r\sin\theta}\frac{\partial f}{\partial \phi}\hat{\boldsymbol{\phi}}. \tag{1-58}$$

This definition is meaningless on the z-axis, where $\sin\theta = 0$.

The *divergence* is more difficult. The divergence is the outward flux of a vector quantity, say $\boldsymbol{B}$, per unit volume. Refer to the element of volume shown in Fig. 1-9. At P, the face of the element of volume that is normal to $\boldsymbol{r}$ has an area $(r\,d\theta)(r\sin\theta\,d\phi)$. It follows that the net outward flux through the faces that are normal to $\hat{\boldsymbol{r}}$ is

$$\frac{\partial}{\partial r}[B_r(r\,d\theta)(r\sin\theta\,d\phi)]\,dr,$$

and the contribution of that flux to the divergence is that flux, divided by the element of volume dv, or

$$\begin{aligned} &\frac{1}{r^2\sin\theta\,dr\,d\theta\,d\phi}\frac{\partial}{\partial r}[B_r(r\,d\theta)(r\sin\theta\,d\phi)]\,dr \\ &= \frac{1}{r^2\sin\theta}\frac{\partial}{\partial r}(B_r r^2\sin\theta). \end{aligned} \tag{1-59}$$

Proceeding similarly with the outward flux through the faces that are normal to $\hat{\boldsymbol{\theta}}$, and then with the outward flux through the faces that are normal to $\hat{\boldsymbol{\phi}}$, we find that

$$\boldsymbol{\nabla}\cdot\boldsymbol{B} = \frac{1}{r^2\sin\theta}\left[\frac{\partial}{\partial r}(B_r r^2\sin\theta) + \frac{\partial}{\partial\theta}(B_\theta r\sin\theta) + \frac{\partial}{\partial\phi}(B_\phi r)\right], \tag{1-60}$$

$$= \frac{2}{r}B_r + \frac{\partial B_r}{\partial r} + \frac{B_\theta}{r}\cot\theta + \frac{1}{r}\frac{\partial B_\theta}{\partial\theta} + \frac{1}{r\sin\theta}\frac{\partial B_\phi}{\partial\phi}. \tag{1-61}$$

This divergence is meaningless on the z-axis, where $\sin\theta = 0$.

Clearly, it is not possible to define a single operator $\boldsymbol{\nabla}$ that can fit both the gradient and the divergence; the same principle applies to the curl and to the Laplacian.

We now give without proof the expressions for the *curl* of a vector and for the *Laplacian* of a scalar in spherical coordinates:

$$\begin{aligned} (r^2\sin\theta)\,\boldsymbol{\nabla}\times\boldsymbol{B} = &\left[\frac{\partial}{\partial\theta}(r\sin\theta B_\phi) - \frac{\partial}{\partial\phi}(rB_\theta)\right]\hat{\boldsymbol{r}} \\ &+ \left[\frac{\partial}{\partial\phi}B_r - \frac{\partial}{\partial r}(r\sin\theta B_\phi)\right]r\,\hat{\boldsymbol{\theta}} \\ &+ \left[\frac{\partial}{\partial r}(rB_\theta) - \frac{\partial}{\partial\theta}B_r\right]r\sin\theta\,\hat{\boldsymbol{\phi}}. \end{aligned} \tag{1-62}$$

and

$$\nabla^2 f = \frac{2}{r}\frac{\partial f}{\partial r} + \frac{\partial^2 f}{\partial r^2} + \frac{\cot\theta}{r^2}\frac{\partial f}{\partial \theta} + \frac{1}{r^2}\frac{\partial^2 f}{\partial \theta^2} + \frac{1}{r^2 \sin^2\theta}\frac{\partial^2 f}{\partial \phi^2}, \tag{1-63}$$

except, again, on the z-axis.

It is *only* in Cartesian coordinates that the Laplacian of a vector is the sum of the Laplacians of its components. In all other coordinate systems,

$$\nabla^2 \boldsymbol{B} = \nabla(\nabla \cdot \boldsymbol{B}) - \nabla \times (\nabla \times \boldsymbol{B}). \tag{1-64}$$

Figures 1-10 and 1-11 show a point P and an element of volume dv in cylindrical coordinates. See the table on the back of the front cover for the vector operators in that system.

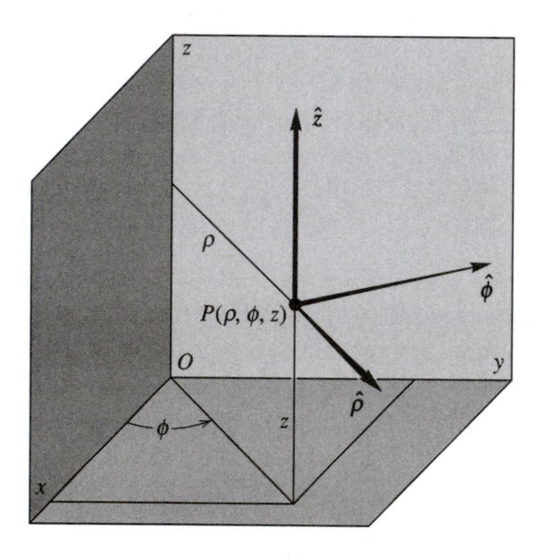

Figure 1-10 Cylindrical coordinate system.

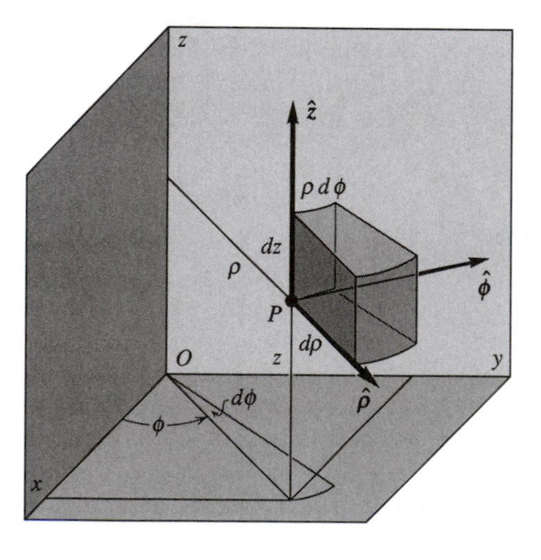

Figure 1-11 Element of volume in cylindrical coordinates.

1.11 SUMMARY

The *gradient* ∇f is a vector whose magnitude and direction are those of the maximum rate of increase of the scalar point-function f with distance at a point.

The *flux* Φ of a vector $\boldsymbol{B}$ through a surface of area $\mathcal{A}$ is the scalar

$$\Phi = \int_{\mathcal{A}} \boldsymbol{B} \cdot d\boldsymbol{\mathcal{A}}. \tag{1-14}$$

If the surface is closed, the vector $d\boldsymbol{\mathcal{A}}$ points outward, by convention.

The *divergence* of $\boldsymbol{B}$

$$\nabla \cdot \boldsymbol{B} = \frac{\partial B_x}{\partial x} + \frac{\partial B_y}{\partial y} + \frac{\partial B_z}{\partial z} \tag{1-20}$$

is the outward flux of $\boldsymbol{B}$ per unit volume at a point.

The *divergence theorem* states that

$$\int_v \nabla \cdot \boldsymbol{B}\, dv = \int_{\mathcal{A}} \boldsymbol{B} \cdot d\boldsymbol{\mathcal{A}}, \tag{1-22}$$

where $\mathcal{A}$ is the area of the closed surface bounding the volume v.

The *line integral*

$$\int_a^b \boldsymbol{B} \cdot d\boldsymbol{l}$$

over a specified curve is the sum of the $\boldsymbol{B} \cdot d\boldsymbol{l}$ terms for each element $d\boldsymbol{l}$ of the curve between points a and b. For a closed curve C that bounds a surface $\mathcal{A}$, we have *Stokes's theorem:*

$$\oint_C \boldsymbol{B} \cdot d\boldsymbol{l} = \int_{\mathcal{A}} (\nabla \times \boldsymbol{B}) \cdot d\boldsymbol{\mathcal{A}}, \tag{1-40}$$

where

$$\nabla \times \boldsymbol{B} = \begin{vmatrix} \hat{\boldsymbol{x}} & \hat{\boldsymbol{y}} & \hat{\boldsymbol{z}} \\ \dfrac{\partial}{\partial x} & \dfrac{\partial}{\partial y} & \dfrac{\partial}{\partial z} \\ B_x & B_y & B_z \end{vmatrix} \tag{1-33}$$

is the *curl* of the vector point-function $\boldsymbol{B}$. The above surface integral applies to a surface of area $\mathcal{A}$ bounded by the curve C.

The *Laplacian* is the divergence of the gradient:

$$\boldsymbol{\nabla} \cdot \boldsymbol{\nabla} f = \boldsymbol{\nabla}^2 f = \frac{\partial^2 f}{\partial x^2} + \frac{\partial^2 f}{\partial y^2} + \frac{\partial^2 f}{\partial z^2}. \tag{1-43}$$

The *Laplacian of a vector* in *Cartesian* coordinates is defined as

$$\boldsymbol{\nabla}^2 \boldsymbol{B} = \boldsymbol{\nabla}^2 B_x \hat{\boldsymbol{x}} + \boldsymbol{\nabla}^2 B_y \hat{\boldsymbol{y}} + \boldsymbol{\nabla}^2 B_z \hat{\boldsymbol{z}}. \tag{1-44}$$

In spherical coordinates (Figs. 1-8 and 1-9), the unit vectors $\hat{\boldsymbol{r}}$, $\hat{\boldsymbol{\theta}}$, $\hat{\boldsymbol{\phi}}$ are orthogonal. Also,

$$\boldsymbol{dr} = dr\,\hat{\boldsymbol{r}} + r\,d\theta\,\hat{\boldsymbol{\theta}} + r\sin\theta\,d\phi\,\hat{\boldsymbol{\phi}}, \tag{1-53}$$

$$dv = r^2 \sin\theta\,dr\,d\theta\,d\phi. \tag{1-54}$$

The vector operators for spherical coordinates appear on the back of the front cover. They are meaningless on the z-axis, where $\sin\theta = 0$.

In other than Cartesian coordinates,

$$\boldsymbol{\nabla}^2 \boldsymbol{B} = \boldsymbol{\nabla}(\boldsymbol{\nabla} \cdot \boldsymbol{B}) - \boldsymbol{\nabla} \times (\boldsymbol{\nabla} \times \boldsymbol{B}). \tag{1-64}$$

The page facing the front cover shows many useful vector identities and theorems.†

PROBLEMS

1-1. *(1.1)*‡ Show that the angle between $\boldsymbol{A} = 2\hat{\boldsymbol{x}} + 3\hat{\boldsymbol{y}} + \hat{\boldsymbol{z}}$ and $\boldsymbol{B} = \hat{\boldsymbol{x}} - 6\hat{\boldsymbol{y}} + \hat{\boldsymbol{z}}$ is 130.6°.

1-2. *(1.1)*

(a) Show that $(\boldsymbol{A} \times \boldsymbol{B}) \cdot \boldsymbol{C}$ is the volume of the parallelepiped whose edges are $\boldsymbol{A}$, $\boldsymbol{B}$, $\boldsymbol{C}$, when the vectors start from the same point.

(b) Show that $(\boldsymbol{A} \times \boldsymbol{C}) \cdot \boldsymbol{B} = -(\boldsymbol{A} \times \boldsymbol{B}) \cdot \boldsymbol{C}$. Observe that the sign changes when the cyclic order of the vectors changes.

1-3. *(1.1)* Let C be a simple plane closed curve. Prove that the area $\mathcal{A}$ enclosed by C is given by

$$\mathcal{A} = \frac{1}{2}\oint_C \boldsymbol{r} \times \boldsymbol{dl},$$

where the vector $\boldsymbol{r}$ goes from an *arbitrary* origin to the element $\boldsymbol{dl}$ on the curve, and where the positive directions for $\mathcal{A}$ and $\boldsymbol{dl}$ obey the right-hand screw rule.

†See Jean Van Bladel (1985), Appendices 1 and 2, for an extensive collection of vector identities and theorems.

‡Section numbers are provided in parentheses after problem numbers.

1-4. (*1.2*) The vector $\boldsymbol{r}$ points from $P'(x', y', z')$ to $P(x, y, z)$.

(a) Show that if P is fixed and P' is allowed to move, then $\boldsymbol{\nabla}'(1/r) = \hat{\boldsymbol{r}}/r^2$, where $\hat{\boldsymbol{r}}$ is the unit vector along $\boldsymbol{r}$.

(b) Show that, similarly, if P' is fixed and P is allowed to move, then $\boldsymbol{\nabla}(1/r) = -\hat{\boldsymbol{r}}/r^2$.

1-5. (*1.4*)

(a) Show that $\boldsymbol{\nabla} \cdot \boldsymbol{r} = 3$.

(b) What is the flux of $\boldsymbol{r}$ through a spherical surface of radius a?

1-6. (*1.5*) Show that

$$\int_v \boldsymbol{\nabla} f \, dv = \int_{\mathscr{A}} f \, d\boldsymbol{\mathscr{A}},$$

where $\mathscr{A}$ is the area of the closed surface bounding the volume v. You can prove this by multiplying both sides by $\boldsymbol{c}$, where $\boldsymbol{c}$ is any vector independent of the coordinates. Then use Identity 6 (from inside the front cover) and the divergence theorem.

1-7. (*1.7*) Since $\boldsymbol{A} \times \boldsymbol{B}$ is normal to $\boldsymbol{B}$, it seems, offhand, that $\boldsymbol{\nabla} \times \boldsymbol{B}$ must be normal to $\boldsymbol{B}$. That is *wrong*.

As a counterexample, show that $(\boldsymbol{\nabla} \times \boldsymbol{B}) \cdot \boldsymbol{B} = -1$ if $\boldsymbol{B} = y\hat{\boldsymbol{x}} + \hat{\boldsymbol{z}}$.

1-8. (*1.10*)

(a) Check, by inspection of Fig. 1-8, that the unit vectors in Cartesian and spherical coordinates are related as follows:

$$\hat{\boldsymbol{r}} = \sin\theta\cos\phi\,\hat{\boldsymbol{x}} + \sin\theta\sin\phi\,\hat{\boldsymbol{y}} + \cos\theta\,\hat{\boldsymbol{z}},$$

$$\hat{\boldsymbol{\theta}} = \cos\theta\cos\phi\,\hat{\boldsymbol{x}} + \cos\theta\sin\phi\,\hat{\boldsymbol{y}} - \sin\theta\,\hat{\boldsymbol{z}}, \qquad \hat{\boldsymbol{\phi}} = -\sin\phi\,\hat{\boldsymbol{x}} + \cos\phi\,\hat{\boldsymbol{y}}.$$

(b) Show that

$$\hat{\boldsymbol{x}} = \sin\theta\cos\phi\,\hat{\boldsymbol{r}} + \cos\theta\cos\phi\,\hat{\boldsymbol{\theta}} - \sin\phi\,\hat{\boldsymbol{\phi}},$$

$$\hat{\boldsymbol{y}} = \sin\theta\sin\phi\,\hat{\boldsymbol{r}} + \cos\theta\sin\phi\,\hat{\boldsymbol{\theta}} + \cos\phi\,\hat{\boldsymbol{\phi}}, \qquad \hat{\boldsymbol{z}} = \cos\theta\,\hat{\boldsymbol{r}} - \sin\theta\,\hat{\boldsymbol{\theta}}.$$

1-9. (*1.10*) A vector $\boldsymbol{F}$ has the same magnitude and direction at all points in space. Choose the z-axis parallel to $\boldsymbol{F}$. Then, in Cartesian coordinates, $\boldsymbol{F} = F\hat{\boldsymbol{z}}$.

Express $\boldsymbol{F}$ in spherical coordinates.

1-10. (*1.10*) Show, by differentiating the appropriate expressions for $\boldsymbol{r}$, that the velocity $\dot{\boldsymbol{r}}$ in spherical coordinates is $\dot{r}\hat{\boldsymbol{r}} + r\dot{\theta}\hat{\boldsymbol{\theta}} + r\sin\theta\,\dot{\phi}\hat{\boldsymbol{\phi}}$.†

1-11. (*1.10*) A force $\boldsymbol{F}$ is of the form $(K/r^3)\hat{\boldsymbol{r}}$ in spherical coordinates, where K is a constant. Is the field conservative?

1-12. (*1.7*) In the coordinate systems that we have used so far, vectors and the operator $\boldsymbol{\nabla}$ all have three components each. However, in relativity theory (Chaps. 11 to 13), it is often more convenient to consider only *two* components, one that is parallel to a given direction and one that is perpendicular. For example, one writes that $\boldsymbol{r} = \boldsymbol{r}_{\parallel} + \boldsymbol{r}_{\perp}$.

†It is the custom to set $\dot{x} \equiv \partial x/\partial t$.

If the chosen direction is the x-axis, then

$$\boldsymbol{r}_{\parallel} = x\hat{\boldsymbol{x}} \qquad \text{and} \qquad \boldsymbol{r}_{\perp} = y\hat{\boldsymbol{y}} + z\hat{\boldsymbol{z}}.$$

Also, $\boldsymbol{\nabla} = \boldsymbol{\nabla}_{\parallel} + \boldsymbol{\nabla}_{\perp}$, with

$$\boldsymbol{\nabla}_{\parallel} = \hat{\boldsymbol{x}}\frac{\partial}{\partial x}, \qquad \boldsymbol{\nabla}_{\perp} = \hat{\boldsymbol{y}}\frac{\partial}{\partial y} + \hat{\boldsymbol{z}}\frac{\partial}{\partial z}.$$

Then

$$\boldsymbol{\nabla} V = \boldsymbol{\nabla}_{\parallel} V + \boldsymbol{\nabla}_{\perp} V.$$

Show that

$$\boldsymbol{\nabla}\cdot\boldsymbol{A} = \boldsymbol{\nabla}_{\parallel}\cdot\boldsymbol{A}_{\parallel} + \boldsymbol{\nabla}_{\perp}\cdot\boldsymbol{A}_{\perp}, \qquad \boldsymbol{\nabla}\times\boldsymbol{A} = \boldsymbol{\nabla}_{\parallel}\times\boldsymbol{A}_{\perp} + \boldsymbol{\nabla}_{\perp}\times\boldsymbol{A}_{\parallel} + \boldsymbol{\nabla}_{\perp}\times\boldsymbol{A}_{\perp}.$$

CHAPTER

2

PHASORS

This short chapter discusses a second mathematical prerequisite for the study of electromagnetic fields, namely, phasors. We use phasors mostly, but not exclusively, in relation with electromagnetic waves.

One uses phasors to represent quantities that are sine or cosine functions of time, of space coordinates, or of both.

The functions $\sin \omega t$ and $\cos \omega t$ play a major role in modern technology, mostly because of the relative ease with which they can be generated. They are also relatively easy to manipulate mathematically. All other periodic functions, such as square waves, for example, are much more difficult to generate and much more difficult to manipulate mathematically.

One often has to solve linear differential equations involving sine and cosine functions with constant coefficients. As we shall see, the use of phasors then has the immense advantage of transforming these differential equations into simple algebraic equations.

But first, in order to understand how to use phasors, we must review complex numbers.

2.1 COMPLEX NUMBERS

A *complex number* is of the form

$$z = a + jb, \tag{2-1}$$

where $j = (-1)^{1/2}$ and where a and b are real numbers, such as 2.5, 3, or –10. Complex numbers can be plotted in the *complex plane,* as in Fig. 2-1. The quantity a is said to be the *real part,* and bj the *imaginary part,* of the complex number.

One can express complex numbers in *Cartesian form,* as above, or in *polar form,* as follows. First,

$$z = a + bj = r\cos\theta + jr\sin\theta = r(\cos\theta + j\sin\theta), \tag{2-2}$$

where

$$r = (a^2 + b^2)^{1/2} \tag{2-3}$$

is the *modulus* of the complex number z, and the angle θ is its *argument.*

With the angle θ expressed in *radians,*

$$\cos\theta = 1 - \frac{\theta^2}{2!} + \frac{\theta^4}{4!} - \frac{\theta^6}{6!} + \cdots, \tag{2-4}$$

$$\sin\theta = \theta - \frac{\theta^3}{3!} + \frac{\theta^5}{5!} - \frac{\theta^7}{7!} + \cdots, \tag{2-5}$$

$$\exp j\theta = 1 + j\theta - \frac{\theta^2}{2!} - \frac{j\theta^3}{3!} + \frac{\theta^4}{4!} + \frac{j\theta^5}{5!} - \cdots, \tag{2-6}$$

and

$$\cos\theta + j\sin\theta = \exp j\theta. \tag{2-7}$$

Of course, $j \times j = -1, j \times j \times j = -j$, etc.

Thus, from Eq. 2-2,

$$z = a + bj = r\exp j\theta. \tag{2-8}$$

We have here a complex number expressed both in Cartesian form and in polar form.

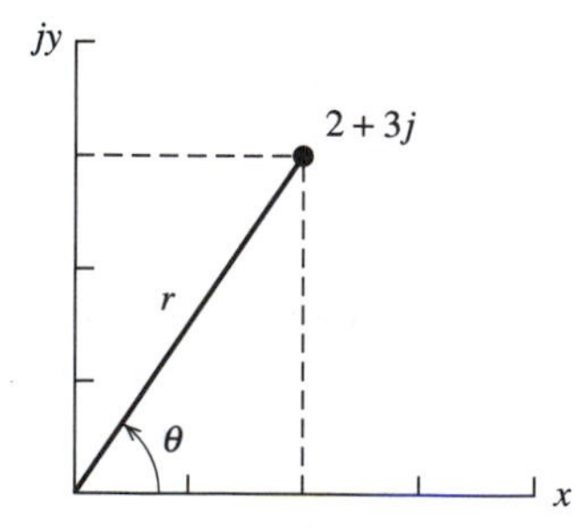

Figure 2-1 The complex number 2 + 3j plotted in the complex plane.

If a is positive, then θ is in either the first or the fourth quadrant. If a is negative, θ is in either the second or third quadrant. Use the proper angle! For example, the argument of $-1 + j$ is $3\pi/4$, not $-\pi/4$.

Note that

$$\exp j\frac{\pi}{2} = j, \quad \exp j\pi = -1, \quad \exp\left(j\frac{3\pi}{2}\right) = -j, \quad \exp j2\pi = 1. \tag{2-9}$$

If

$$z = a + bj = r\exp j\theta, \tag{2-10}$$

where $\theta = \tan^{-1}(b/a)$, then the *complex conjugate* of z is

$$z^* = a - bj = r\exp(-j\theta). \tag{2-11}$$

To obtain the complex conjugate of a complex expression, one changes the sign before j everywhere. For example, if

$$z = \frac{a + bj}{c + dj}, \qquad \text{then} \qquad z^* = \frac{a - bj}{c - dj}. \tag{2-12}$$

2.1.1 Addition and Subtraction of Complex Numbers

With complex numbers in Cartesian form, one simply adds or subtracts the real and imaginary parts separately:

$$(a + bj) + (c + dj) = (a + c) + (b + d)j. \tag{2-13}$$

If the numbers are in polar form, one first transforms them into Cartesian form.

2.1.2 Multiplication and Division of Complex Numbers

In Cartesian form, one proceeds as follows:

$$(a + bj)(c + dj) = (ac - bd) + (ad + bc)j, \tag{2-14}$$

$$\frac{a + bj}{c + dj} = \frac{(a + bj)(c - dj)}{(c + dj)(c - dj)} = \frac{(ac + bd) + (-ad + bc)j}{c^2 + d^2}. \tag{2-15}$$

In polar form,

$$(r_1\exp j\theta_1)(r_2\exp j\theta_2) = r_1 r_2 \exp j(\theta_1 + \theta_2), \tag{2-16}$$

$$\frac{r_1\exp j\theta_1}{r_2\exp j\theta_2} = \frac{r_1}{r_2}\exp j(\theta_1 - \theta_2). \tag{2-17}$$

Remember to express the angles in *radians*.

EXAMPLES

$$(4 + 5j) + (2 - 3j) = 6 + 2j, \tag{2-18}$$

$$(4 + 5j)^2 = (16 - 25) + 40j = -9 + 40j, \tag{2-19}$$

$$\frac{1}{4 + 5j} = \frac{4 - 5j}{(4 + 5j)(4 - 5j)} = \frac{4 - 5j}{16 + 25} = \frac{4 - 5j}{41} = 0.098 - 0.122j, \tag{2-20}$$

$$\left(5 \exp j\frac{\pi}{3}\right)\left(2 \exp j\frac{\pi}{2}\right) = 10 \exp j\frac{5\pi}{6}, \tag{2-21}$$

$$\frac{5 \exp(j\pi/3)}{2 \exp(j\pi/2)} = 2.5 \exp\left(-j\frac{\pi}{6}\right). \tag{2-22}$$

2.2 PHASORS

Electric currents and voltages, electric fields, and magnetic fields are often sinusoidal functions of time. For example, an *alternating current* is of the form

$$I = I_m \cos(\omega t + \alpha), \tag{2-23}$$

where I_m is the *maximum value* of the current, $\omega = 2\pi f$ is the *circular,* or *angular, frequency,* and f is the *frequency.* The quantity in parentheses is the *phase,* or *phase angle,* α being the phase at $t = 0$.

The point **I** of Fig. 2-2 rotates on a circle of radius I_m in the complex plane at an angular velocity ω. Then its projection on the real axis is

$$I = I_m \cos \omega t. \tag{2-24}$$

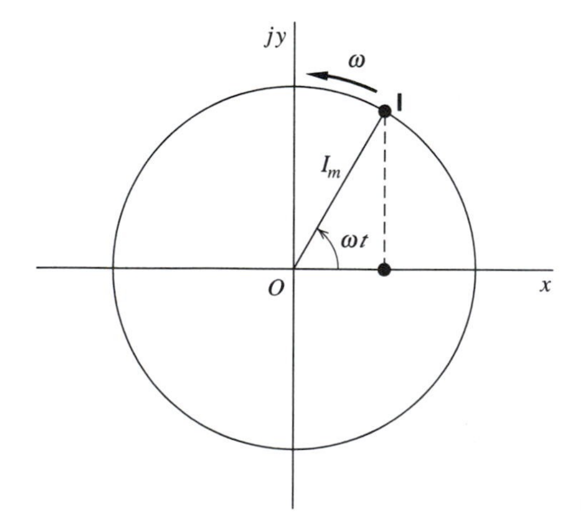

Figure 2-2 Point **I** in the complex plane describes a circle of radius I_m about the origin O. It represents the *phasor* $I_m \exp j\omega t$. Its projection on the x-axis is $I_m \cos \omega t$.

We have set $\alpha = 0$ for simplicity. Therefore

$$I = \text{Re}\,(I_m \exp j\omega t) = \text{Re}\,\mathbf{I}, \tag{2-25}$$

where the operator Re means "Real part of what follows." The quantity in parentheses is the *phasor* $\mathbf{I}$ of Fig. 2-2:

$$\mathbf{I} = I_m \exp j\omega t = I_m \cos \omega t + jI_m \sin \omega t. \tag{2-26}$$

So the phasor $\mathbf{I}$ is equal to the variable I *plus* the parasitic imaginary term $jI_m \sin \omega t$.

Then

$$\frac{d\mathbf{I}}{dt} = j\omega\mathbf{I}, \qquad \frac{d^2\mathbf{I}}{dt^2} = (j\omega)^2\mathbf{I} = -\omega^2\mathbf{I}, \qquad \frac{d^n\mathbf{I}}{dt^n} = (j\omega)^n\mathbf{I}, \qquad \text{etc.} \tag{2-27}$$

You can easily check that

$$\text{Re}\,(j\omega\mathbf{I}) = \frac{dI}{dt}, \qquad \text{Re}\,(-\omega^2\mathbf{I}) = \frac{d^2 I}{dt^2}, \qquad \text{etc.} \tag{2-28}$$

In other words,

$$\text{Re}\,\mathbf{I} = I, \tag{2-29}$$

$$\text{Re}\,\frac{d\mathbf{I}}{dt} = \text{Re}\,j\omega\mathbf{I} = \frac{dI}{dt}, \tag{2-30}$$

$$\text{Re}\,\frac{d^2\mathbf{I}}{dt^2} = \text{Re}\,(j\omega)^2\mathbf{I} = \frac{d^2 I}{dt^2}, \qquad \text{etc.} \tag{2-31}$$

Therefore, it one replaces the variable I by the phasor $\mathbf{I}$, then the *operator* d/dt becomes a *factor* $j\omega$, and a *differential* equation involving time derivatives becomes an *algebraic* equation!

A phasor can be multiplied by a complex number:

$$(a + bj)\mathbf{I} = (a + bj)I_m \exp j\omega t = rI_m \exp j(\omega t + \theta), \tag{2-32}$$

where r is the modulus of $a + bj$ and θ is its argument.

One can also divide a phasor by a complex number:

$$\frac{\mathbf{I}}{a + bj} = \frac{I_m}{r} \exp j(\omega t - \theta). \tag{2-33}$$

Instead of having a cosine function of the time t, one might have a cosine of a coordinate:

$$E = E_m \cos kx. \tag{2-34}$$

The corresponding phasor would then be

$$\mathbf{E} = E_m \exp jkx. \tag{2-35}$$

In a traveling wave, one has a cosine function of both t and, say, z:

$$E = E_m \cos(\omega t - kz), \tag{2-36}$$

where k is the *wave number.* This wave travels in the positive direction of the z-axis. In phasor form,

$$\mathbf{E} = E_m \exp j(\omega t - kz) \tag{2-37}$$

and

$$\frac{\partial \mathbf{E}}{\partial t} = j\omega \mathbf{E}, \qquad \frac{\partial \mathbf{E}}{\partial z} = -jk\mathbf{E}. \tag{2-38}$$

Vector quantities can also be expressed in phasor form. For example, a force could be a cosine function of time:

$$\boldsymbol{F} = (F_m \cos \omega t)\hat{\boldsymbol{x}}. \tag{2-39}$$

Then

$$\mathbf{F} = (F_m \exp j\omega t)\hat{\boldsymbol{x}} \tag{2-40}$$

is both a vector and a phasor.

2.3 USING PHASORS

To use phasors, one first expresses the sine or cosine functions in the form $x_m \cos(\omega t + \theta)$, and then one uses the phasor

$$\mathbf{x} = x_m \exp j(\omega t + \theta) \equiv x_m \cos(\omega t + \theta) + jx_m \sin(\omega t + \theta). \tag{2-41}$$

One then performs the calculation with the phasors. The result almost invariably stays in phasor form. However, if one requires a real function, one simply rejects the imaginary part.

EXAMPLE Solving a Second-Order Linear Differential Equation with Phasors

One of the most common types of differential equation is the following:

$$m\frac{d^2x}{dt^2} + b\frac{dx}{dt} + kx = F_m \cos \omega t. \tag{2-42}$$

Here all the terms are real. This equation describes the motion of a mass m subjected to the applied force $F_m \cos \omega t$, to a restoring force $-kx$, and to a damping force $-bv$, as in Fig. 2-3: the product of the mass m by the acceleration d^2x/dt^2 is equal to the sum of the applied forces.

The steady-state solution is of the form $x_m \cos(\omega t + \theta)$, where x_m is the amplitude of the motion. It is a simple matter to solve this equation with phasors. We use the phasor

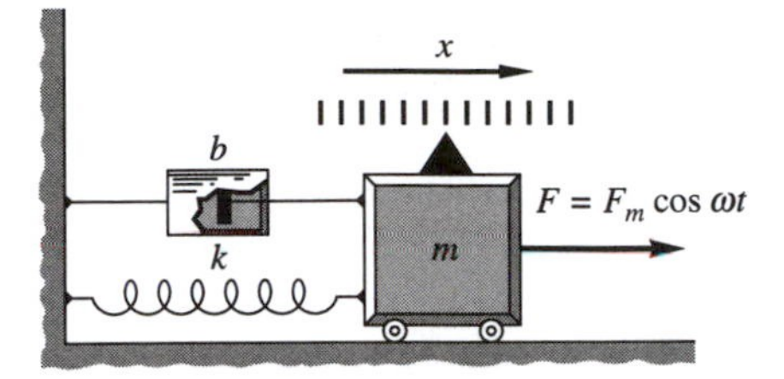

Figure 2-3 Mass m subjected to a force $F = F_m \cos \omega t$, to a restoring force $-kx$ exerted by a spring, and to a damping force $-b\, dx/dt$ exerted by a dashpot. At rest, $x = 0$.

$$\mathbf{x} = x_m \exp j(\omega t + \theta), \tag{2-43}$$

whose real part is the displacement x. We also set

$$\mathbf{F} = F_m \exp j\omega t. \tag{2-44}$$

Substitution into the differential equation is trivial:

$$-m\omega^2 \mathbf{x} + bj\omega \mathbf{x} + k\mathbf{x} = \mathbf{F} \tag{2-45}$$

and

$$\mathbf{x} = \frac{\mathbf{F}}{k - m\omega^2 + jb\omega}. \tag{2-46}$$

Thus, expressing the denominator in polar form,

$$x_m = \frac{F_m}{\left[(k - m\omega^2)^2 + b^2\omega^2\right]^{1/2}}, \qquad \theta = -\tan^{-1}\frac{b\omega}{k - m\omega^2}. \tag{2-47}$$

The actual displacement is the real part of the phasor $\mathbf{x}$, or

$$x = x_m \cos(\omega t + \theta). \tag{2-48}$$

2.4 PRODUCTS OF PHASORS

One often requires the *average* value of the product of two sinusoidal quantities. Now if one tries to multiply phasors, one runs into trouble. Consider a simple example. Suppose an alternating voltage $V = V_m \cos \omega t$ is applied across a resistance R. Then the current $I = (V_m \cos \omega t)/R$. The instantaneous power dissipated in the resistor is

$$P_{\text{inst}} = IV = \frac{V_m^2 \cos^2 \omega t}{R}, \tag{2-49}$$

and the average power is

$$P_{\text{av}} = \frac{V_m^2}{2R} = \frac{V_{\text{rms}}^2}{R}, \tag{2-50}$$

the average value of $\cos^2 \omega t$ being $\frac{1}{2}$. Here V_{rms} is the *root mean square* voltage, or the square root of the mean value of the square of V:

$$V_{\mathrm{rms}} = \frac{V_m}{2^{1/2}} = 0.707 V_m \tag{2-51}$$

for a sine or a cosine function.

If one uses the phasors

$$\mathbf{V} = V_m \exp j\omega t \qquad \text{and} \qquad \mathbf{I} = \frac{V_m \exp j\omega t}{R}, \tag{2-52}$$

then

$$\mathbf{IV} = \frac{V_m^2 \exp 2j\omega t}{R}, \tag{2-53}$$

whose real part $(V_m^2/R)\cos 2\omega t$ is *neither* the instantaneous *nor* the average power. So phasors must *not* be multiplied in this way.

Suppose one has two sinusoidal quantities of the *same frequency*

$$A = A_m \cos \omega t \qquad \text{and} \qquad B = B_m \cos(\omega t + \theta). \tag{2-54}$$

Then the time-averaged value of their product is

$$\langle A_m (\cos \omega t) B_m \cos(\omega t + \theta)\rangle$$

$$= \langle A_m B_m \cos \omega t (\cos \omega t \cos \theta - \sin \omega t \sin \theta)\rangle \tag{2-55}$$

$$= \langle A_m B_m (\cos^2 \omega t \cos \theta - \cos \omega t \sin \omega t \sin \theta)\rangle, \tag{2-56}$$

where $\langle x \rangle$ means "average value of x." Now the average value of $\cos^2 \omega t$ over one full cycle is $\frac{1}{2}$, as we saw above, while the average value of $\cos \omega t \sin \omega t$ is zero. Then

$$\langle A_m (\cos \omega t) B_m \cos(\omega t + \theta)\rangle = \tfrac{1}{2} A_m B_m \cos \theta = A_{\mathrm{rms}} B_{\mathrm{rms}} \cos \theta, \tag{2-57}$$

where *rms* values are defined as in Eq. 2-51.

If one uses the phasors

$$\mathbf{A} = A_m \exp j\omega t \qquad \text{and} \qquad \mathbf{B} = B_m \exp j(\omega t + \theta), \tag{2-58}$$

then the time-averaged value of the product of their real parts is given correctly by

$$\tfrac{1}{2}\mathrm{Re}(\mathbf{AB}^*) = \tfrac{1}{2}\mathrm{Re}\{A_m(\exp j\omega t) B_m \exp[-j(\omega t + \theta)]\}, \tag{2-59}$$

$$= \tfrac{1}{2}\mathrm{Re}[A_m B_m \exp(-j\theta)] = \tfrac{1}{2} A_m B_m \cos \theta, \tag{2-60}$$

$$= A_{\mathrm{rms}} B_{\mathrm{rms}} \cos \theta, \tag{2-61}$$

as above.

The time-averaged value of the product of two phasors **A** and **B** is thus equal to one half the real part of the product **AB***, where **B*** is the complex conjugate of **B**. Recall that the complex conjugate of **B** is the same as **B**, but with j replaced by $-j$.

2.5 QUOTIENTS OF PHASORS

Dividing one phasor by another *of the same frequency* yields a complex number:

$$\frac{r_1 \exp j(\omega t + \alpha)}{r_2 \exp j(\omega t + \beta)} = \frac{r_1}{r_2} \exp j(\alpha - \beta). \tag{2-62}$$

2.6 ROTATING VECTORS

If one expresses a rotating vector in phasor form, one runs into another kind of trouble. Suppose that

$$\boldsymbol{E} = \hat{\boldsymbol{x}} E_m \cos \omega t + \hat{\boldsymbol{y}} E_m \sin \omega t. \tag{2-63}$$

Then the vector $\boldsymbol{E}$ rotates in real space as in Prob. 2-10, and $d\boldsymbol{E}/dt$ is perpendicular to $\boldsymbol{E}$.

In phasor form,

$$\mathbf{E} = E_m (\exp j\omega t)\hat{\boldsymbol{x}} + E_m \exp j\left(\omega t - \frac{\pi}{2}\right)\hat{\boldsymbol{y}} \tag{2-64}$$

and

$$\frac{d\mathbf{E}}{dt} = j\omega \mathbf{E}. \tag{2-65}$$

This equation is correct. The trouble here is that it *appears* to say that $d\mathbf{E}/dt$ is collinear with **E**, which is wrong, as one can see from Fig. 2-4. *The factor j rotates a phasor by* $+\pi/2$ as in Eq. 2-9.

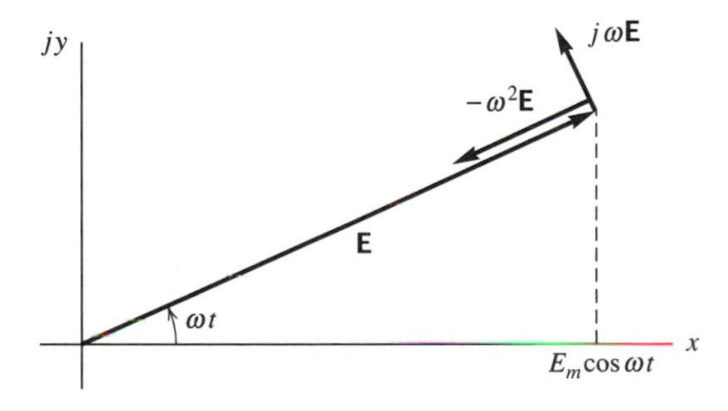

Figure 2-4 The phasor **E** has a constant magnitude, but rotates counterclockwise at an angular velocity of ω radians/second. Its time derivative is $d\mathbf{E}/dt$, or $j\omega\mathbf{E}$, in the direction shown. Its second time derivative is $-\omega^2\mathbf{E}$.

2.7 NOTATION

We have used boldface sans-serif type for phasors, and the usual lightface italic type for the other variables. However, it is customary to use lightface italic type for phasors, as for any other variable, and to omit the operator Re. So, in practice, we write

$$x = x_m \exp j\omega t, \qquad \frac{dx}{dt} = j\omega x, \qquad \frac{d^2x}{dt^2} = -\omega^2 x, \qquad \text{etc.,} \tag{2-66}$$

with the tacit understanding that the imaginary parts are parasitic.

2.8 SUMMARY

A *complex number* z is of the form $a + bj$, where $j = (-1)^{1/2}$ and a and b are real numbers. It is the custom to plot complex numbers in the complex plane as in Fig. 2-1, and thus

$$z = a + bj = r\exp(j\theta). \tag{2-8}$$

The *complex conjugate* of a complex number is its mirror image with respect to the real axis:

$$z^* = a - bj = r\exp(-j\theta). \tag{2-11}$$

Addition and *subtraction* of complex numbers are simpler with the Cartesian form:

$$(a + bj) + (c + dj) = (a + c) + j(b + d). \tag{2-13}$$

However, multiplication and division are simpler with the polar form:

$$(r_1 \exp j\theta_1)(r_2 \exp j\theta_2) = r_1 r_2 \exp j(\theta_1 + \theta_2), \tag{2-16}$$

$$\frac{r_1 \exp j\theta_1}{r_2 \exp j\theta_2} = \frac{r_1}{r_2} \exp j(\theta_1 - \theta_2). \tag{2-17}$$

If one has to deal with the time derivatives of a quantity of the form $I = I_m \cos \omega t$, it is usually advisable to substitute the *phasor*

$$\mathsf{I} = I_m \exp j\omega t = I_m \cos \omega t + jI_m \sin \omega t. \tag{2-26}$$

Then

$$\frac{d\mathsf{I}}{dt} = j\omega\mathsf{I}, \qquad \frac{d^2\mathsf{I}}{dt^2} = -\omega^2\mathsf{I}, \qquad \frac{d^n\mathsf{I}}{dt^n} = (j\omega)^n\mathsf{I}, \tag{2-27}$$

with the understanding that only the real parts are physically meaningful.

The phasor

$$\mathsf{E} = \boldsymbol{E}_m \exp j(\omega t - kz) \tag{2-37}$$

represents a *plane wave* traveling in the positive direction of the z-axis, where k is the *wave number*. The quantities $\mathbf{E}$ and E_m can be vectors. Then $\mathbf{E}$ is both a phasor and a vector.

One occasionally requires the time average of the product of two sinusoidal quantities *of the same frequency,* such as I and V in an alternating-current circuit. This is given by $\frac{1}{2}\operatorname{Re}\mathbf{IV}^*$, where $\mathbf{I}$ and $\mathbf{V}$ are phasors.

The ratio of two phasors, again of the *same frequency,* is a complex number.

PROBLEMS

2-1. (*2.1*) Complex numbers in polar form are often written as $r\angle\theta$, where r is the modulus and θ is the argument, expressed in radians.

Express $1 + 2j$ in this way.

2-2. (*2.1*)
(a) Express the complex numbers $1 + 2j, -1 + 2j, -1 - 2j$, and $1 - 2j$ in polar form.
(b) Simplify the following expressions, leaving them in Cartesian coordinates: $(1 + 2j)(1 - 2j)$, $(1 + 2j)^2$, $1/(1 + 2j)^2$, $(1 + 2j)/(1 - 2j)$.

2-3. (*2.1*) What happens to a complex number in the complex plane when it is (a) multiplied by j, (b) multiplied by j^2, (c) divided by j?

2-4. (*2.2*) Find the real parts of the phasors $(1 + 3j)\exp j(\omega t + 2)$ and $[\exp j(\omega t + 2)]/(1 + 3j)$.

2-5. (*2.3*) Solve the following differential equation by means of phasors:

$$2\frac{d^2x}{dt^2} + 3\frac{dx}{dt} + 4x = 5\cos 6t.$$

2-6. (*2.4*) Find the rms values for the waveforms shown in Fig. 2-5.

2-7. (*2.4*) A certain electric circuit draws a current of 1.00 ampere rms when it is fed at 120 volts rms, 60 hertz. The current lags the voltage by $\pi/4$ radian.
(a) Express V and I in the form of phasors, and calculate the time-averaged power dissipation.
(b) Now calculate the power $V_{\text{rms}}I_{\text{rms}}\cos\theta$, where θ is $\pi/4$.

2-8. (*2.5*) Find the value of

$$\frac{5.14\exp j(\omega t + 3)}{3.72\exp j(\omega t + 5)}$$

in Cartesian form.

2-9. (*2.5*) Show that

$$\ln(a + bj) = \frac{1}{2}\ln(a^2 + b^2) + j\tan^{-1}\frac{b}{a},$$

if a is positive. Then the point $a + bj$ lies in either the first or the fourth quadrant. If a is not positive, be careful to use the proper angle (Sec. 2.1).

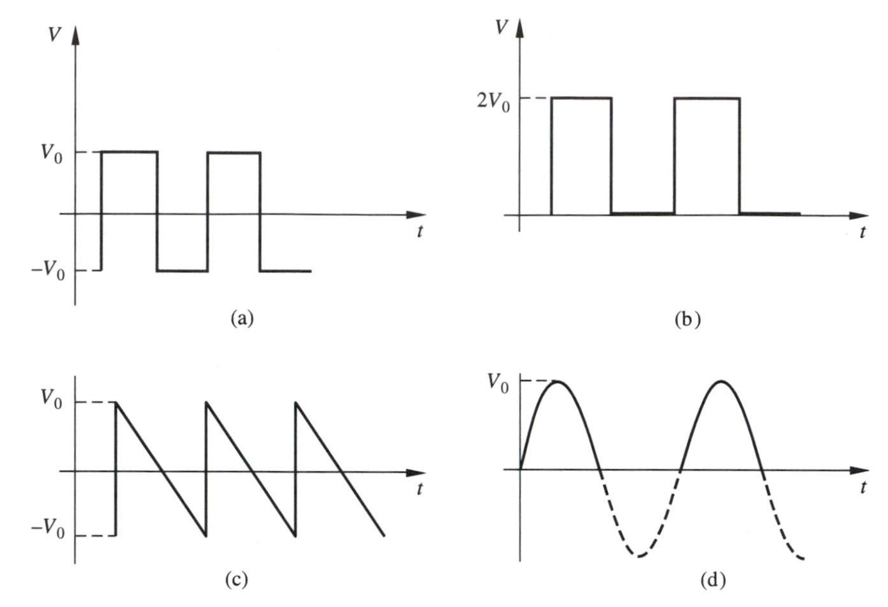

Figure 2-5

2-10. *(2.6)* Show that the phasor $\mathbf{V} = V_m[\hat{x} \exp j\omega t + \hat{y} \exp j(\omega t - \pi/2)]$ represents a vector of constant magnitude V_m that rotates in the positive direction in the xy-plane at the angular velocity ω.

2-11. *(2.1)* The square root of a complex number

Set $a + jb = (c + jd)^2$. Then $c + jd$ is the square root of $a + jb$. Show that

$$c^2 = \frac{a \pm (a^2 + b^2)^{1/2}}{2},$$

$$d^2 = \frac{b^2}{2\left[a \pm (a^2 + b^2)^{1/2}\right]}.$$

CHAPTER

3

ELECTRIC FIELDS I

Coulomb's Law and Gauss's Law

In Chapters 3 through 10 we study the electric fields that result from the accumulation of electric charges. We must simplify: the electric charges will either be stationary, or they will move slowly. Also, we shall neglect their magnetic fields.

Simplification is the very essence of Science: Nature is so exceedingly complex that it is impossible to understand a phenomenon without concentrating on specific aspects. According to legend, Newton discovered the law of gravity after watching an apple fall from a tree. Think of all the variables: the color of the apple, its orientation and its shape, the time of day, etc., etc. If he had tried to take everything into account, his successors would still be experimenting with lemons and plums, and with sticks and stones.

So we shall assume that the speed v of our moving charges is much smaller than the speed of light c, or more precisely that $v^2 \ll c^2$. It is only in Chapters 11 through

13 that we shall be able to discuss fast-moving charges. But even very slow-moving charges have appreciable magnetic fields, which we shall disregard for the moment.

This first chapter on electric fields concerns two fundamental and well-established laws, Coulomb's law and Gauss's law.

3.1 THE ELECTRIC FIELD STRENGTH *E*

An electrically charged body is surrounded by an electric field. For example, at a point P near an isolated conducting sphere that carries a charge Q, the *electric field strength* is

$$\boldsymbol{E} = \frac{Q}{4\pi\epsilon_0 r^2}\hat{\boldsymbol{r}}, \tag{3-1}$$

where r is the distance between the center of the sphere and P, $\hat{\boldsymbol{r}}$ is a unit vector that points outward, and the constant

$$\epsilon_0 = 8.854187817 \times 10^{-12} \text{ coulomb}^2/(\text{newton meter}^2) \tag{3-2}$$

is the *permittivity of free space*.†

Electric fields can be complex, depending on the complexity of the charge distribution.

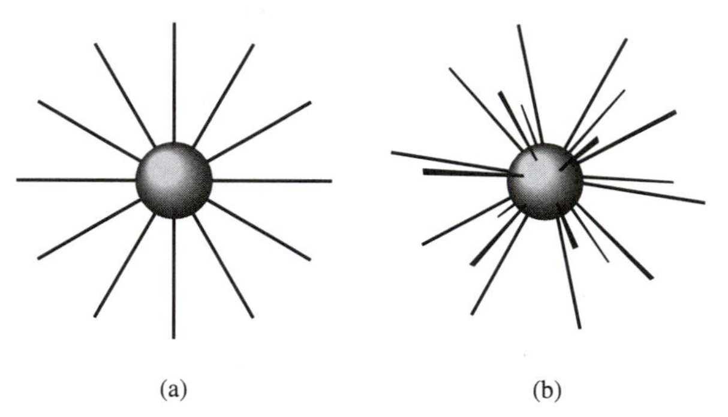

Figure 3-1 Electric field lines near a charged conducting sphere. (a) In this usual two-dimensional representation, the line density is proportional to $1/r$, giving the false impression that E is proportional to $1/r$. (b) This more realistic figure shows electric field lines that are all of the same length and equally spaced on the sphere, projected onto the plane of the paper. The field line density is seen to decrease much faster with increasing distance.

†The denominator in Eq. 3-1 is bothersome: why is there a factor of $4\pi\epsilon_0$ in the denominator? Could not the units of electric charge, or of electric field, be changed to eliminate that factor? Coulomb's law, Eq. 3-5, has the same denominator. The system of units that is based on Eq. 3-1 is said to be *rationalized*. With unrationalized units, Coulomb's law is written $F = Q_1Q_2/r^2$, without the factor of $1/(4\pi\epsilon_0)$. Then the coulomb is a *mechanical* quantity whose dimensions are $M^{1/2}L^{3/2}/T$! That is bad enough; indeed, it is going back to early 19th-century Physics. But also, the fact of "simplifying" Coulomb's law, which is *never* used, complicates most of the useful equations by adding factors of 4π or of $1/(4\pi)$. Rationalized units were discovered and proposed by Heaviside in about 1885. See Appendix C. Well over one century later, many authors still use unrationalized units.

To plot an *electric field line,* join end-to-end short vectors $\boldsymbol{ds} = dx\,\hat{\boldsymbol{x}} + dy\,\hat{\boldsymbol{y}}$ pointing in the direction of $\boldsymbol{E}$. First calculate $\boldsymbol{E}$ at an arbitrary starting point, say x_0, y_0. Then, on a field line,

$$\frac{dy}{dx} = \frac{E_y}{E_x}, \tag{3-3}$$

and the next point is at $x_0 + dx$, $y_0 + dy$, etc.

Keep in mind that electric field lines, like magnetic field lines, are no more than "thinking crutches;" their only function is to show the direction of the field at a given point in space. They are extremely useful for visualizing fields, even though individual field lines do not really exist. See Fig. 3-1.

3.2 COULOMB'S LAW

If you have two point charges Q_a and Q_b separated by a distance r, the electric fields interact as in Fig. 3-2, and the force exerted *by* Q_a *on* Q_b is

$$\boldsymbol{F}_{ab} = \boldsymbol{E}_a Q_b \tag{3-4}$$

$$= \frac{Q_a Q_b}{4\pi\epsilon_0 r^2}\hat{\boldsymbol{r}}_{ab} \tag{3-5}$$

$$\approx 9 \times 10^9 \frac{Q_a Q_b}{r^2}\hat{\boldsymbol{r}}_{ab}, \tag{3-6}$$

where $\hat{\boldsymbol{r}}_{ab}$ is a unit vector that points *from* Q_a *to* Q_b. This is *Coulomb's law.* The force is repulsive if the two charges are of the same sign, and attractive if the charges have different signs. The exponent of r is known to be equal to 2 within one part in 10^{16}.

Note that each charge has its own radial field, whether or not the second charge is present.

We cannot give a proper definition of the coulomb until Chapter 17. For the moment, we may either take the value of ϵ_0 as given and use Coulomb's law as a provisional definition of the unit of charge or say that the coulomb is equal to 6.24151×10^{18} electron charges.

Equation 3-4 provides a definition for the electric field strength $\boldsymbol{E}$ at a point: it is the force exerted on a unit charge at that point. So an electric field is expressed in newtons/coulomb, and

$$1\,\frac{\text{newton}}{\text{coulomb}} = 1\,\frac{\text{joule/meter}}{\text{coulomb}} = 1\,\frac{\text{joule/coulomb}}{\text{meter}} = 1\,\frac{\text{volt}}{\text{meter}},$$

1 volt being 1 joule/coulomb.

If charge flows at the rate of 1 coulomb per second, then the current is 1 *ampere.* What if either Q_a or Q_b moves? Does Coulomb's law remain valid?

1. If Q_a is stationary and Q_b is not, then Coulomb's law applies to the force on Q_b, whatever the velocity of Q_b. This is an experimental fact: the trajectories of charged particles in oscilloscopes, mass spectrographs, and ion accelerators are always calculated on that basis.

2. If Q_a is not stationary, then Coulomb's law is no longer valid. We return to this fact in Chapter 13.

Coulomb's law applies to a pair of charges situated in a vacuum. It also applies for dielectrics and conductors if F_{ab} is the direct force between Q_a and Q_b, irrespective of the forces arising from other charges within the medium.

With extended charges, "the distance between the charges" has no definite meaning. Moreover, the presence of Q_b can modify the charge distribution within Q_a, and vice versa, leading to a complicated variation of force with distance.

Electric forces in nature are enormous when compared to gravitational forces, for which

$$F = 6.67259 \times 10^{-11} \frac{m_a m_b}{r^2}. \tag{3-7}$$

For example, the gravitational force on a proton at the surface of the sun (mass $= 2 \times 10^{30}$ kilograms, radius $= 7 \times 10^8$ meters) is equal to the electric force between one proton and one *microgram* of electrons, separated by a distance equal to the sun's radius. Or the electric repulsion between two electrons (mass $= 9.1 \times 10^{-31}$ kilogram) is about 4.2×10^{42} times as large as their gravitational attraction.

There are two reasons why, fortunately, we are not normally conscious of the enormous electric forces. First, ordinary matter is truly neutral, or so it seems. Experiments have shown that no atom or molecule carries a charge greater than 10^{-20} times the electronic charge. Second, the mobility of some of the electrons in matter prevents the accumulation of any appreciable quantity of charge of either sign.

3.3 THE PRINCIPLE OF SUPERPOSITION

If there are several charges, each one imposes its own field irrespective of the others, and the resultant $\boldsymbol{E}$ is simply the vector sum of the individual $\boldsymbol{E}$'s. This is the *principle of superposition*. See Fig. 3-2a,b.

From now on we shall use *primed coordinates* for the source point P', where the charges are, and *unprimed coordinates* for the field point P, where we calculate the field.

For example, if the electric charge is distributed continuously over a finite volume as in Fig. 3-3, then at the field point $P(x,y,z)$,

$$\boldsymbol{E} = \frac{1}{4\pi\epsilon_0} \int_{v'} \frac{\rho \hat{\boldsymbol{r}}}{r^2} dv', \tag{3-8}$$

where ρ is the volume charge density at the source point $P'(x',y',z')$, $\hat{\boldsymbol{r}}$ is the unit vector pointing from P' to P, r is the distance from P' to P, and dv' is the element of volume $dx'\,dy'\,dz'$ at P'.

If there is also a surface charge of surface density σ coulombs/meter^2 over an area $\mathcal{A}'$, then one must add the surface integral

$$\frac{1}{4\pi\epsilon_0} \int_{\mathcal{A}'} \frac{\sigma \hat{\boldsymbol{r}}}{r^2} d\mathcal{A}'.$$

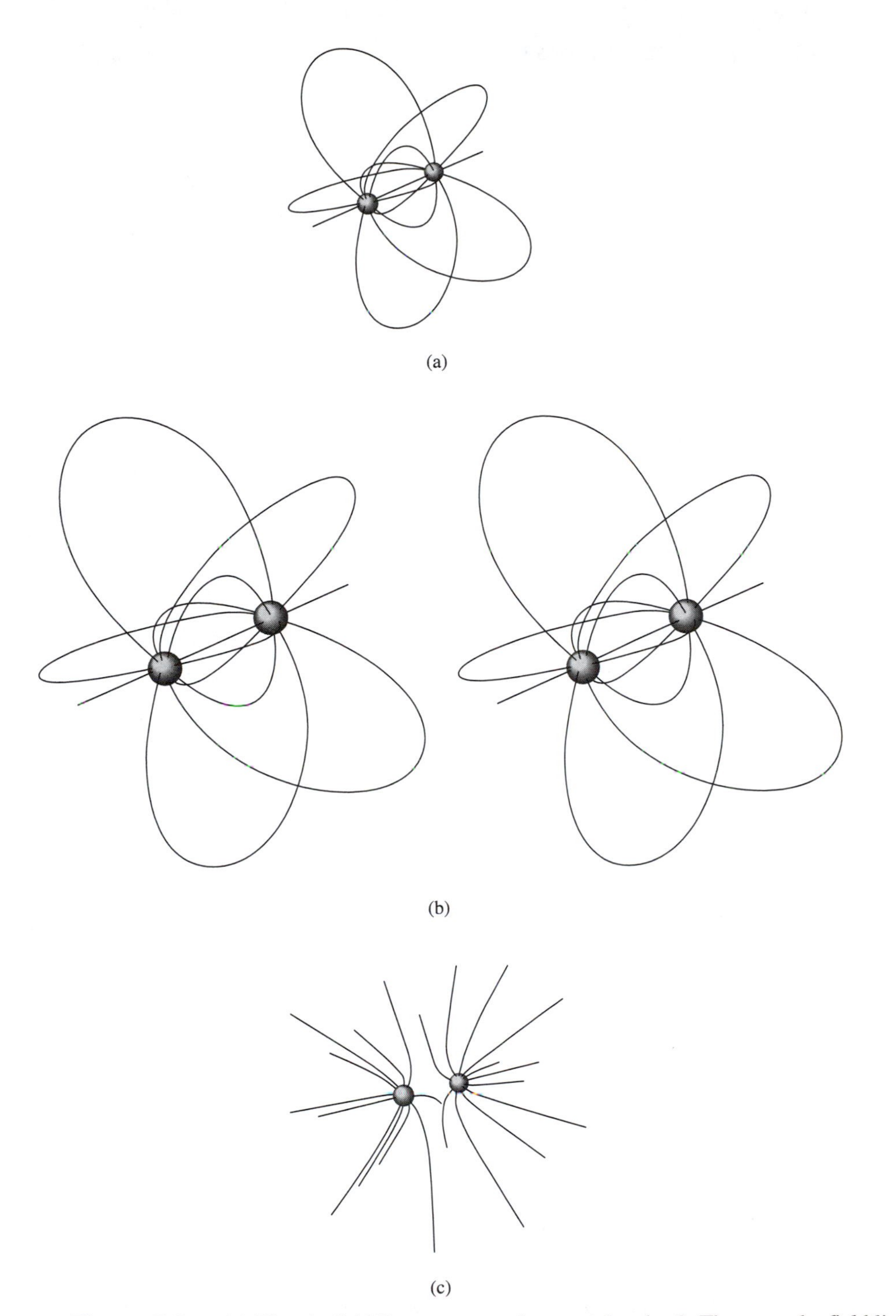

Figure 3-2a (a) Electric field lines near two charges $+Q$ and $-Q$. These are the field lines for the vector sum of the two fields. The force is attractive: electric field lines "are under tension." (b) The same field, shown in three dimensions.† (See footnote on p. 38.) (c) The field of two equal charges $+Q$ and $+Q$. The force is repulsive: electric field lines "repel laterally."

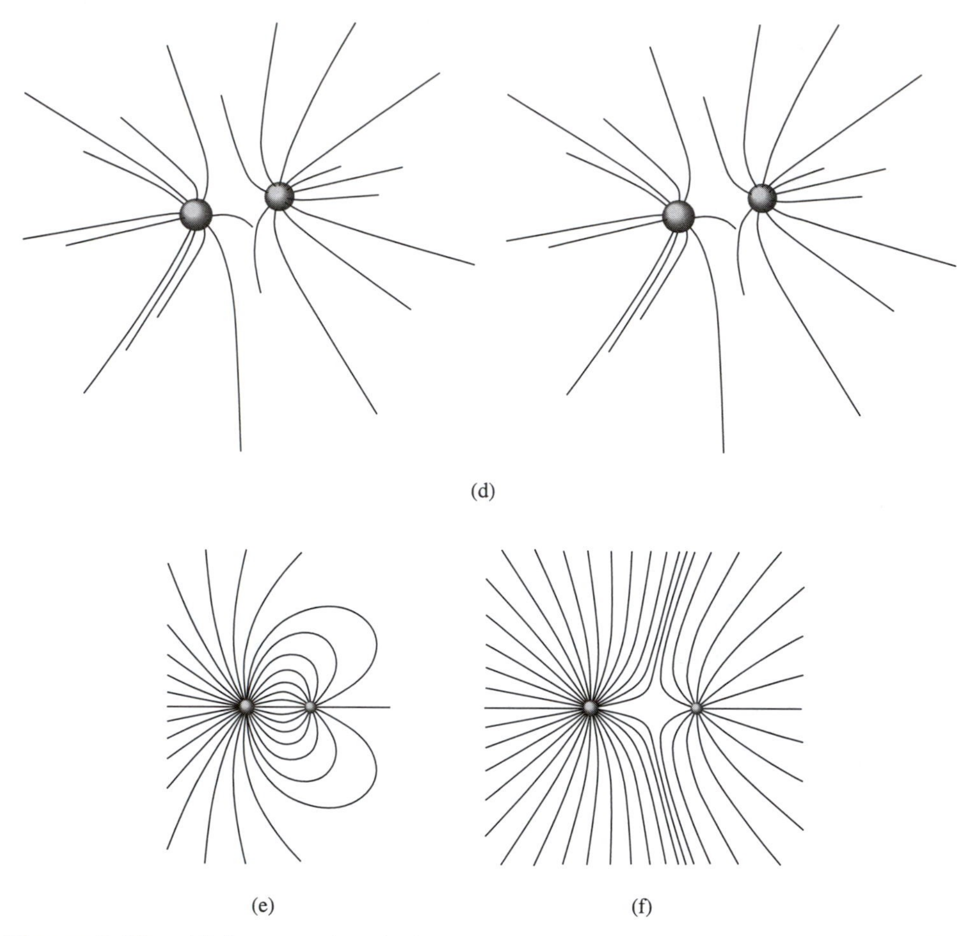

Figure 3-2b (d) Stereoscopic pair showing the field of two charges $+Q$ and $+Q$.† (e) The field of charges $+2Q$ and $-Q$. Some of the field lines of the charge $+2Q$ extend to infinity. (f) The field of charges $+2Q$ and $+Q$.

Or if there is a line charge of length L of line density λ coulombs/meter, then one adds the line integral

$$\frac{1}{4\pi\epsilon_0}\int_L \frac{\lambda\hat{\boldsymbol{r}}}{r^2}\,dl'.$$

†This is a stereoscopic pair of figures that show again the same field lines, seen from two neighboring points. You can see the field in 3-D in the following way. Look in the direction of the figures, a long distance away. You will then see four figures, instead of two. Now adjust your eyes so as to merge the two middle figures, leaving three figures instead of four, and concentrate on the middle figure. It might take you several tries, but it is worth the effort! Not all persons are able to adjust their eyes properly.

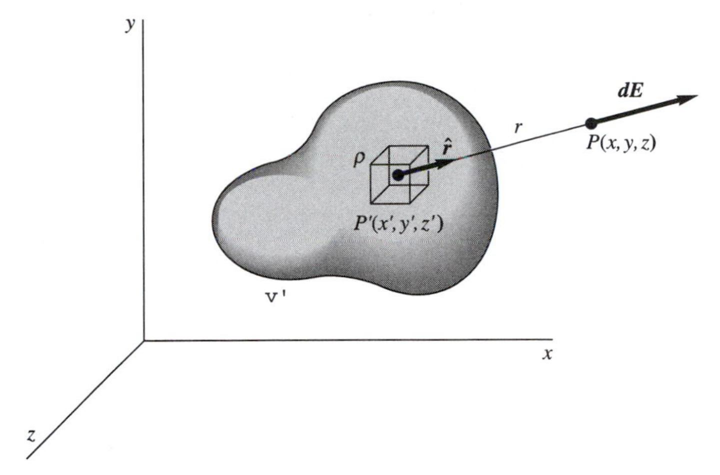

Figure 3-3 Charge distribution of volume density ρ occupying a volume v. The element of volume at $P'(x', y', z')$ has a field $\boldsymbol{dE}$ at $P\,(x, y, z)$.

EXAMPLE The Electric Field of a Pair of Point Charges

We shall return to this field in one of the Examples in Sec. 4.3.

Refer to Fig. 3-4. Let us calculate $\boldsymbol{E}$ at the point $P(x,y)$. At that point,

$$E_1 = \frac{Q_1}{4\pi\epsilon_0 d_1^2}, \tag{3-9}$$

where

$$d_1^2 = (a + x)^2 + y^2 \tag{3-10}$$

and

$$\boldsymbol{E}_1 = \frac{Q_1}{4\pi\epsilon_0 d_1^2}(\cos\theta_1\,\hat{\boldsymbol{x}} + \sin\theta_1\,\hat{\boldsymbol{y}}) \tag{3-11}$$

$$= \frac{Q_1}{4\pi\epsilon_0 d_1^2}\left(\frac{a+x}{d_1}\hat{\boldsymbol{x}} + \frac{y}{d_1}\hat{\boldsymbol{y}}\right) \tag{3-12}$$

$$= \frac{Q_1}{4\pi\epsilon_0 d_1^3}[(a+x)\,\hat{\boldsymbol{x}} + y\,\hat{\boldsymbol{y}}]. \tag{3-13}$$

Similarly,

$$E_2 = \frac{Q_2}{4\pi\epsilon_0 d_2^2}, \tag{3-14}$$

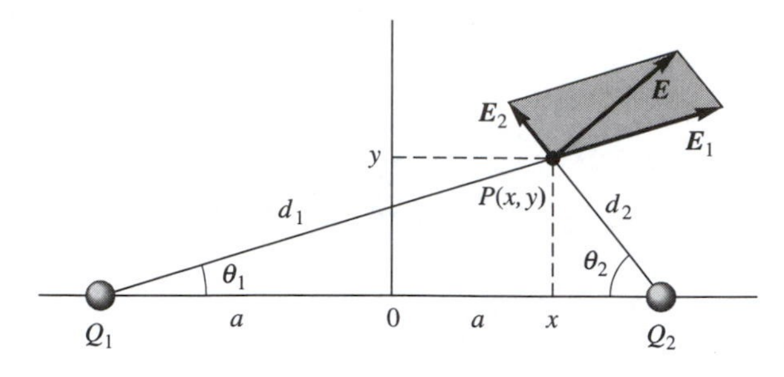

Figure 3-4 Point P in the field of two point charges Q_1 and Q_2.

where

$$d_2^2 = (a - x)^2 + y^2, \tag{3-15}$$

and

$$\boldsymbol{E}_2 = \frac{Q_2}{4\pi\epsilon_0 d_2^2}(-\cos\theta_2\,\hat{\boldsymbol{x}} + \sin\theta_2\,\hat{\boldsymbol{y}}). \tag{3-16}$$

Note the negative sign before the $\cos\theta_2$ term: the x-component of $\boldsymbol{E}_2$ is negative, on the figure. So

$$\boldsymbol{E}_2 = \frac{Q_2}{4\pi\epsilon_0 d_2^2}\left(-\frac{a-x}{d_2}\hat{\boldsymbol{x}} + \frac{y}{d_2}\hat{\boldsymbol{y}}\right) \tag{3-17}$$

$$= \frac{Q_2}{4\pi\epsilon_0 d_2^3}[(x-a)\hat{\boldsymbol{x}} + y\,\hat{\boldsymbol{y}}] \tag{3-18}$$

and, at $P(x,y)$,

$$\boldsymbol{E} = \boldsymbol{E}_1 + \boldsymbol{E}_2 \tag{3-19}$$

$$= \frac{Q_1}{4\pi\epsilon_0 d_1^3}[(a+x)\hat{\boldsymbol{x}} + y\,\hat{\boldsymbol{y}}] + \frac{Q_2}{4\pi\epsilon_0 d_2^3}[(x-a)\hat{\boldsymbol{x}} + y\,\hat{\boldsymbol{y}}]. \tag{3-20}$$

Let us check. Set $Q_1 = Q$, $Q_2 = -Q$. At $x = 0$, for all y, field lines must be parallel to the x-axis, so that the y-components must cancel. Do they? For $x = 0$, $d_1 = d_2 = d$ and, with $Q_1 = Q$, $Q_2 = -Q$, from Eq. 3-20,

$$\boldsymbol{E}_1 + \boldsymbol{E}_2 = \frac{Q}{4\pi\epsilon_0 d^3}(a\,\hat{\boldsymbol{x}} + y\,\hat{\boldsymbol{y}}) - \frac{Q}{4\pi\epsilon_0 d^3}(-a\,\hat{\boldsymbol{x}} + y\,\hat{\boldsymbol{y}}) \tag{3-21}$$

$$= \frac{Q}{4\pi\epsilon_0 d^3}2a\,\hat{\boldsymbol{x}} = 2\frac{Q}{4\pi\epsilon_0 d^2}\frac{a}{d}, \tag{3-22}$$

as expected.

Here is another simple check. At the origin, $d_1 = d_2 = a$ and, again with $Q_1 = Q$, $Q_2 = -Q$,

$$E_1 + E_2 = \frac{Q}{4\pi\epsilon_0 a^3}(2a\hat{x}) = \frac{Q}{4\pi\epsilon_0 a^2}2\hat{x}. \tag{3-23}$$

The field is twice as large as if there were only one charge at a distance a, and it points to the right on the figure, as expected.

3.4 THE ELECTRIC POTENTIAL V AND THE CURL OF E

Imagine a test charge Q' that is situated at point A and that can move about in an electric field $\boldsymbol{E}$. The field exerts on it a force $Q'\boldsymbol{E}$. To move the charge away from point A, you must exert a force $-Q'\boldsymbol{E}$. Then the energy required to move Q' at a constant velocity from A to a point B along a given path is equal to the line integral of the force $-Q'\boldsymbol{E}$ over the path going from A to B:

$$\mathscr{E} = -\int_A^B Q'\boldsymbol{E}\cdot d\boldsymbol{l}. \tag{3-24}$$

We assume that Q' is so small that it does not appreciably disturb the charges that generate the field.

If the path is closed, then the total work done *on* the charge Q' is

$$\mathscr{E} = -\oint Q'\boldsymbol{E}\cdot d\boldsymbol{l}. \tag{3-25}$$

To evaluate this integral we first consider the field of a single stationary point charge Q. Then

$$\oint Q'\boldsymbol{E}\cdot d\boldsymbol{l} = Q'\oint\left(\frac{Q}{4\pi\epsilon_0 r^2}\hat{\boldsymbol{r}}\right)\cdot d\boldsymbol{l} = \frac{Q'Q}{4\pi\epsilon_0}\oint\frac{\hat{\boldsymbol{r}}\cdot d\boldsymbol{l}}{r^2}. \tag{3-26}$$

Now $\hat{\boldsymbol{r}}\cdot d\boldsymbol{l}$ is the change dr in r over the path $d\boldsymbol{l}$, so that

$$\frac{\hat{\boldsymbol{r}}\cdot d\boldsymbol{l}}{r^2} = \frac{dr}{r^2} = -d\left(\frac{1}{r}\right).$$

But the sum of the increments of $1/r$ over a closed path is zero, since r has the same value at the beginning and at the end. So the integral is zero, and the net work done in moving Q' around any closed path in the field of the fixed charge Q is zero.

If the electric field is that of some fixed charge distribution, then the line integrals corresponding to each individual charge are all zero. Thus, for any distribution of fixed charges,

$$\oint \boldsymbol{E}\cdot d\boldsymbol{l} = 0. \tag{3-27}$$

An electrostatic field is therefore conservative (Sec. 1.6). This important property follows from the fact that the electric force in the field of a point charge is radial.

We can now show that the work done in moving a test charge Q' at a constant velocity from point A to point B is independent of the path followed. Let m and n be any two paths leading from A to B, as in Fig. 3-5. These two paths form a closed curve, and the work done in going from A to B along m, and then back to A along n is zero. Then the work done in going from A to B is the same along m as it is along n.

Now let us choose a reference point $P_0(x_0,y_0,z_0)$, and let us define a scalar function V of $P(x,y,z)$ such that

$$V_P - V_{P_0} = -\int_{P_0}^{P} \boldsymbol{E} \cdot \boldsymbol{dl} = \int_{P}^{P_0} \boldsymbol{E} \cdot \boldsymbol{dl}. \tag{3-28}$$

This is the energy required to move a unit charge from P_0 to P. This definition is unambiguous because the integral is the same for all paths leading from P to P_0. Then, for any pair of points A and B,

$$V_A - V_B \equiv -\int_{A}^{B} (\nabla V) \cdot \boldsymbol{dl} = \int_{A}^{B} \boldsymbol{E} \cdot \boldsymbol{dl}, \tag{3-29}$$

as in Fig. 3-5, and therefore

$$\boldsymbol{E} = -\nabla V. \tag{3-30}$$

The electric potential V describes the field completely; the negative sign makes $\boldsymbol{E}$ point toward a *decrease* of V.

But note that V is not uniquely defined, because the point P_0 is arbitrary. One can add to V any quantity that is independent of the coordinates without affecting $\boldsymbol{E}$.

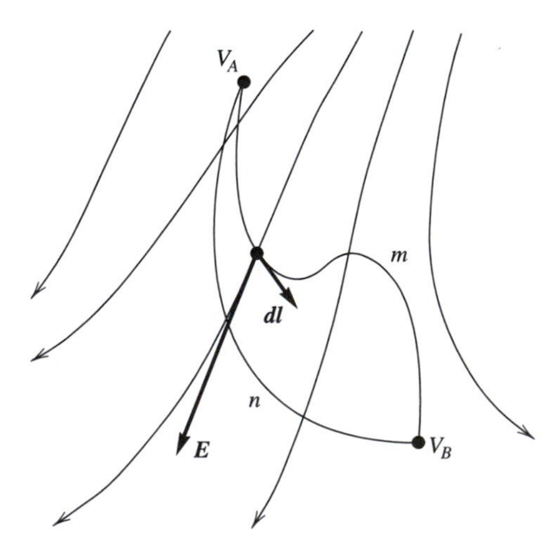

Figure 3-5 The potential difference $V_A - V_B$ between two points is equal to the line integral of $\boldsymbol{E} \cdot \boldsymbol{dl}$ from A to B, where $\boldsymbol{E}$ is the electric field strength and $\boldsymbol{dl}$ is an element of the path along which the integral runs. The light lines are lines of $\boldsymbol{E}$.

From Eq. 3-27 and from Stokes's theorem (Sec. 1.8),

$$\nabla \times \boldsymbol{E} = 0. \tag{3-31}$$

This is also obvious from the fact that $\nabla \times \boldsymbol{E} = -\nabla \times \nabla V \equiv 0$.

Remember that we are dealing here with *static* fields due to essentially stationary charges. If there were time-dependent currents, then $\nabla \times \boldsymbol{E}$ would not necessarily be zero, and ∇V would account for only part of $\boldsymbol{E}$. We shall investigate these more complicated phenomena starting in Chapter 18.

3.4.1 The Electric Potential V at a Point

Equation 3-29 shows that $\boldsymbol{E}$ is related to the *difference* in potential between two points. If one wishes to speak of the potential at a given point, one must define the potential at the *reference* point P_0. Then, at a point P,

$$V = \int_P^{P_0} \boldsymbol{E} \cdot d\boldsymbol{l}. \tag{3-32}$$

As a rule, P_0 is chosen to be at zero, or at *ground* potential. When the charges extend over a finite region, it is usually convenient to choose P_0 at infinity.

If the field is that of a single isolated point charge, then

$$V = \int_r^{\infty} \frac{Q}{4\pi\epsilon_0 r^2} dr = \frac{Q}{4\pi\epsilon_0 r}. \tag{3-33}$$

The sign of V is the same as that of Q. See Fig. 3-6.

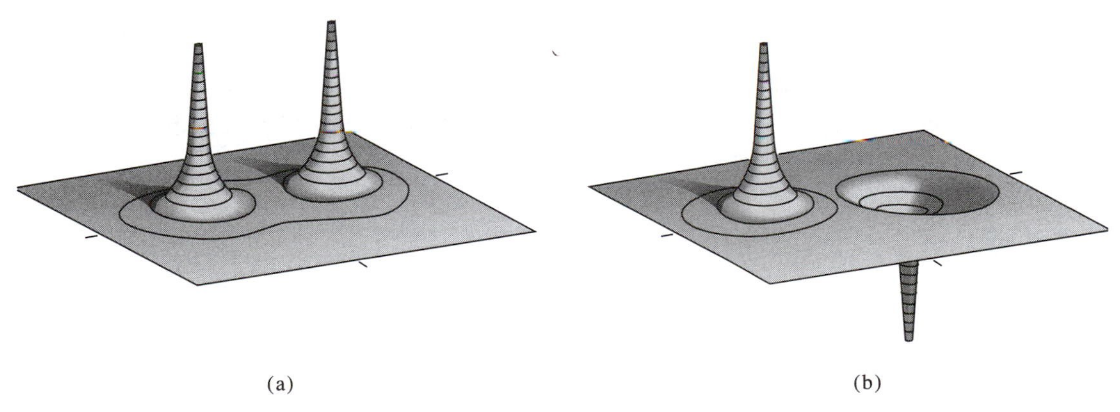

Figure 3-6 (a) The potential V, plotted vertically, in a plane containing two equal charges. (b) The potential V, again plotted vertically, in a plane containing two charges that are of equal magnitude, but of opposite signs.

Since the energy $\mathscr{E}$ required to bring a charge Q from a point P_0, where $V = 0$, to P is VQ, $V = \mathscr{E}/Q$, and the unit of V is the joule/coulomb, or *volt*.

The principle of superposition applies to V as well as to $\mathbf{E}$. For any charge distribution of volume charge density ρ,

$$V = \frac{1}{4\pi\epsilon_0}\int_{v'} \frac{\rho\, dv'}{r}, \tag{3-34}$$

with r as in Fig. 3-3. The volume v' encloses all the charges. If there are surface charges, one adds a surface integral. If there are line charges, one adds a line integral.

For some geometries, the above integral diverges. In those cases, one first calculates $\mathbf{E}$, and then integrates to find V.

For example, the integral for the potential V at a distance D from an infinite line charge of linear density λ that extends from $-\infty$ to $+\infty$ is proportional to $\ln(2/0)$! But $\mathbf{E}$ is easy to calculate, thanks to Gauss's law of Sec. 3.6.

EXAMPLE The Electrostatic Potential V at the Surface of the Sun

The surface of the sun† that one sees in white light is known as the *photosphere*. See Fig. 17-11. Its temperature is about 6,400 kelvins, and the pressure about 1.2×10^4 pascals, or about one-eighth of the atmospheric pressure at the surface of the earth. All of the elements are present, but hydrogen (90%) and helium (9%) predominate. Only a small fraction of the hydrogen, about 0.04%, is ionized, and the number densities of free electrons and protons are about equal.

What is the electrostatic potential V at the surface of the sun? The potential is not zero, because electrons tend to escape more easily than the much heavier ions. Must the net current be zero? Yes because, otherwise, the electrostatic charge on the sun would increase indefinitely—and the sun has been around for quite a while. What is your guess? Is V positive or negative?

It is this evaporation that causes the *solar wind*. A similar phenomenon occurs on other stars and in galaxies.

This evaporation does not seem to have any appreciable astrophysical significance. Even giant galaxies have center-to-surface potential differences that are only of the order of one kilovolt, as on the sun.

See Prob. 3-15, where you can calculate the electrostatic charge carried by the earth; in the case of the earth, we have an entirely different phenomenon.

The fraction F of the particles that possess enough energy to escape is given by

$$F = \exp\left(-\frac{\text{escape kinetic energy}}{kT}\right), \tag{3-35}$$

†For a general survey of solar phenomena, see Zirin (1988).

where k is the Boltzmann constant, and T the surface temperature. This is the *Maxwell velocity distribution function*. The exponent is $mv^2/(2kT)$.

1. First, what is the escape kinetic energy for an *uncharged* particle of mass m at the surface of a star of mass M and radius R?

   The energy required to pull m away from the radius R to infinity, against the gravitational attraction, is GMm/R. Calling the escape velocity v,

$$\frac{1}{2}mv^2 = G\frac{Mm}{R}. \tag{3-36}$$

   So the escape *kinetic energy* is proportional to the particle mass m, but the escape *velocity* v is independent of m.

   If electrons and protons were uncharged, then we would find that the fractions F for electrons and for protons would be

$$F_e = \exp\left(-\frac{GMm_e}{RkT}\right), \qquad F_p = \exp\left(-\frac{GMm_p}{RkT}\right), \tag{3-37}$$

   and then we would find that

$$\frac{F_e}{F_p} = \exp\left[\frac{GM}{RkT}(m_p - m_e)\right] \gg 1. \tag{3-38}$$

   This last exponent is a very large number, about 4×10^{14}, and the ratio on the left is an even larger number—very much larger.

2. Now what is the escape kinetic energy for a *charged* particle?

   The additional energy that is required to pull a particle of charge q out to infinity is $-qV$, where V is the electric potential at the radius R, or *minus* the electric potential energy.† The extra energy $-qV$ is positive if q and V have opposite signs, or if the electric force is attractive.

   The escape kinetic energy for a particle of mass m and charge q is thus

$$\frac{1}{2}mv^2 = G\frac{Mm}{R} - qV. \tag{3-39}$$

3. So what is the equilibrium potential V at the surface of the star? Assume free electrons and protons at the same temperature, and zero net current.

   The number densities of protons and electrons remain constant in a static situation. So $F_e = F_p$, and the escape *kinetic energies* for protons and electrons are

†The electric potential energy qV is proportional to $1/r$ and is equal to zero when q is at an infinite distance from the star. It is positive if q and V have the same sign but is otherwise negative. In both cases, the electric force tries to *decrease* the potential energy. If you sketch a graph of qV as a function of r, you will verify that the electric force is repulsive if q and V have the same sign, but is otherwise attractive.

equal. Set $e = +1.6 \times 10^{-19}$ coulomb. That is the magnitude of the proton and electron charges. Then

$$G\frac{Mm_p}{R} - eV = G\frac{Mm_e}{R} + eV, \tag{3-40}$$

$$G\frac{M(m_p - m_e)}{R} = 2eV, \tag{3-41}$$

$$G\frac{Mm_p}{R} \approx 2eV, \qquad V \approx G\frac{Mm_p}{2eR}, \tag{3-42}$$

and the electrostatic potential V at the surface of the star is *positive*.

4. For the case of the sun, see the Table facing the back cover for the values of the physical constants. In that case,

$$V \approx \frac{(6.7 \times 10^{-11})(2 \times 10^{30})(1.7 \times 10^{-27})}{2 \times (1.6 \times 10^{-19})(7 \times 10^{8})} \approx +1000 \text{ volts}. \tag{3-43}$$

Now, from Gauss's law of Sec. 3.6, $\quad V = Q/(4\pi\epsilon_0 R)$,

$$Q \approx 4 \times 3.14 \times (8.85 \times 10^{-12})(7 \times 10^{8}) \times 1000 \approx +80 \text{ coulombs}. \tag{3-44}$$

There are slightly more free protons than free electrons in the sun. The excess mass of protons is $80 \times (1.7 \times 10^{-27})/(1.6 \times 10^{-19}) \approx 1 \times 10^{-6}$ kilogram, or about 1 milligram, whereas the sun has a mass of 2×10^{30} kilograms!

Because the escape *kinetic energies* are equal, the escape *velocities* are different, the escaping protons are much slower than the escaping electrons, and there is a net positive space charge density in the atmosphere of the sun.

3.4.2 Equipotential Surfaces and Lines of *E*

The set of points in space that are at a given potential V defines an *equipotential surface*. For example, equipotential surfaces near an isolated point charge are concentric spheres, whereas equipotential surfaces near an infinite line charge are coaxial cylinders. These are the simplest cases. More complicated charge distributions have more complicated equipotential surfaces.

But how are equipotential surfaces related to $\boldsymbol{E}$? Recall from Sec. 3.4 that $\boldsymbol{E} = -\nabla V$. So $\boldsymbol{E}$ points in the direction of *maximum rate of decrease* of V. This means that the vector $\boldsymbol{E}$ is normal to the local equipotential surface, and that it points toward *lower* values of V.

The isolated point charge is a good example. Say the charge is positive. Then, in the neighborhood, V is positive and decreases with distance, ∇V points radially *inward,* and $\boldsymbol{E}$ points radially *outward.*

We return to lines of $\boldsymbol{E}$ in Sec. 6.7.

3.5 THE ELECTRIC FIELD INSIDE AND OUTSIDE MACROSCOPIC BODIES

Macroscopic bodies consist of positively charged nuclei and negative electrons. This brings up three questions.

1. Can one calculate the field *outside* an electrically charged body by assuming that the charge distribution inside the body is continuous? If so, then one can calculate the field by integrating over the charge distribution. Otherwise, one must find some other form of calculation.

   It is, in fact, usually appropriate to treat the discrete charges carried by nuclei and by electrons within macroscopic bodies as though they were continuous. Even the largest nuclei have diameters that are only of the order of 10^{-14} meter. Nuclei and electrons are so small and so closely packed, compared to the dimensions of ordinary macroscopic objects, that one may assume a smoothly varying electric charge density measured in coulombs per cubic meter or per square meter.

2. Now what about the electric field *inside* a charged body? Clearly the electric field strength in the immediate neighborhood of a nucleus or of an electron is enormous. Also at a given fixed point, this electric field changes erratically with time, because the charges are never perfectly stationary. It is not useful for our purposes to look at the electric field as closely as that. We shall be satisfied to calculate space- and time-averaged values of $\boldsymbol{E}$ and V inside a charged body by assuming a continuous distribution of charge.

3. Is it, then, really possible to define the electric field at a point P' inside a continuously distributed charge? It appears at first sight that the contribution to the potential, dV, due to the charge element $\rho\, dv'$ at P' is infinite, since r is zero. In fact, dV is not infinite.

   Consider a spherical shell of thickness dr and radius r centered on P'. The charge in this shell contributes at P' a dV of $(4\pi r^2\, dr\, \rho)/(4\pi\epsilon_0 r) = (1/\epsilon_0) r\, dr\, \rho$. Another shell of smaller radius contributes a smaller dV. The electric potential V therefore converges, and the integral is finite. A similar argument shows that $\boldsymbol{E}$ also converges.

   One can therefore calculate the electric fields of real charge distributions by the usual techniques of the integral calculus, both inside and outside the distributions.

3.6 GAUSS'S LAW

Gauss's law relates the flux of $\boldsymbol{E}$ through a closed surface to the total charge enclosed within that surface.

Consider Fig. 3-7, in which a finite volume v bounded by a surface $\mathcal{A}$ encloses a charge Q. We can calculate the outward flux of $\boldsymbol{E}$ through $\mathcal{A}$ as follows. The flux of $\boldsymbol{E}$ through the element of area $d\boldsymbol{\mathcal{A}}$ is

$$\boldsymbol{E} \cdot d\boldsymbol{\mathcal{A}} = \frac{Q}{4\pi\epsilon_0} \frac{\hat{\boldsymbol{r}} \cdot d\boldsymbol{\mathcal{A}}}{r^2}. \tag{3-45}$$

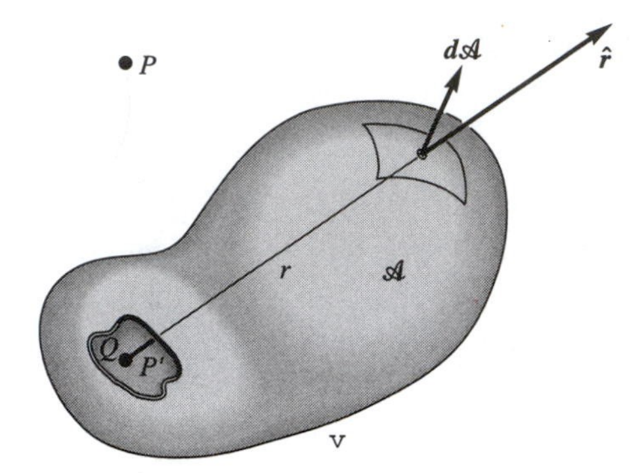

Figure 3-7 A point charge Q located inside a volume v bounded by the surface of area $\mathcal{A}$. Gauss's law states that the surface integral of $\boldsymbol{E} \cdot \boldsymbol{d\mathcal{A}}$ over $\mathcal{A}$ is equal to Q/ϵ_0. The vector $\boldsymbol{d\mathcal{A}}$ points *outward.*

Now $\hat{\boldsymbol{r}} \cdot \boldsymbol{d\mathcal{A}}$ is the projection of $\boldsymbol{d\mathcal{A}}$ on $\hat{\boldsymbol{r}}$. Then

$$\boldsymbol{E} \cdot \boldsymbol{d\mathcal{A}} = \frac{Q}{4\pi\epsilon_0} d\Omega, \tag{3-46}$$

where $d\Omega$ is the solid angle subtended by $\boldsymbol{d\mathcal{A}}$ at the point P': the solid angle $d\Omega$ is equal to the area $\boldsymbol{d\mathcal{A}}$ projected on a plane perpendicular to $\hat{\boldsymbol{r}}$, divided by r^2.

To find the outward flux of $\boldsymbol{E}$, we integrate over the closed area $\mathcal{A}$, or over a solid angle of 4π. Thus

$$\int_{\mathcal{A}} \boldsymbol{E} \cdot \boldsymbol{d\mathcal{A}} = \frac{Q}{\epsilon_0}. \tag{3-47}$$

If Q is outside v at P, the integral is equal to zero: the solid angle subtended by any closed surface is 4π at a point P' inside, and zero at a point P outside. Equation 3-47 is *Gauss's law in integral form.*†

If the charge occupies a finite volume, then

$$\int_{\mathcal{A}} \boldsymbol{E} \cdot \boldsymbol{d\mathcal{A}} = \frac{1}{\epsilon_0} \int_v \rho \, dv, \tag{3-48}$$

where $\mathcal{A}$ is the area of the surface bounding the volume v, and ρ is the electric charge density. We assume that there are no surface charges.

If we apply the divergence theorem to the left-hand side, we find that

$$\int_v \nabla \cdot \boldsymbol{E} \, dv = \frac{1}{\epsilon_0} \int_v \rho \, dv. \tag{3-49}$$

†We have followed the usual custom of starting out with Coulomb's law, and then deducing Gauss's law from it. This procedure seems rational enough, but the latter law is, in fact, more general. Gauss's law applies to moving charges, whatever their velocity or acceleration, though Coulomb's law, as stated in Sec. 3.2, is valid only if Q_a is stationary.

Because this equation applies to any finite volume v, the integrands are equal and

$$\nabla \cdot \boldsymbol{E} = \frac{\rho}{\epsilon_0} \tag{3-50}$$

at every point in space.

This is *Gauss's law in differential form.* Observe that it relates the local charge density to the *space derivatives* of $\boldsymbol{E}$, and not to $\boldsymbol{E}$ itself.

When it is expressed in differential form, Gauss's law is a *local* law in that it relates the behavior of $\boldsymbol{E}$ in the infinitesimal neighborhood of a given point to the value of the charge density at that point. However, when it is expressed in integral form, as in Eq. 3-47, Gauss's law is *nonlocal,* because it concerns a finite region and not a specific point in space.

Many laws of nature, in particular the fundamental laws of electromagnetism, can be formulated in two such equivalent forms, one local and the other nonlocal. With the local forms of physical laws, in the guise of differential equations, one views phenomena as the result of processes occurring in the immediate neighborhood of every point in space.

EXAMPLE The Field of a Uniform Spherical Charge

A spherical charge Q has a radius R and a uniform density ρ, as in Fig. 3-8. Let us find E and V as functions of the distance r from the center of the sphere. By symmetry, both E and V are independent of the spherical coordinates θ and ϕ.

By symmetry, $\boldsymbol{E}$ is radial. It points outward if Q is positive.

1. The electric field strength $\boldsymbol{E}$

   Outside the charge distribution, at P where $r > R$, we imagine a sphere of radius r and surface area $4\pi r^2$. The enclosed charge is

$$Q = \tfrac{4}{3}\pi R^3 \rho. \tag{3-51}$$

   From Gauss's law,

$$E_o = \frac{Q}{4\pi\epsilon_0 r^2} = \frac{R^3 \rho}{3\epsilon_0 r^2} \tag{3-52}$$

   as if all the charge were situated at the center O.

   Inside the sphere, at P'', the charge enclosed by the imaginary sphere of radius $r < R$ is $Q(r/R)^3$. Using Gauss's law again, we find that

$$E_i = \frac{Q(r/R)^3}{4\pi\epsilon_0 r^2} = \frac{Qr}{4\pi\epsilon_0 R^3} = \frac{\rho r}{3\epsilon_0}. \tag{3-53}$$

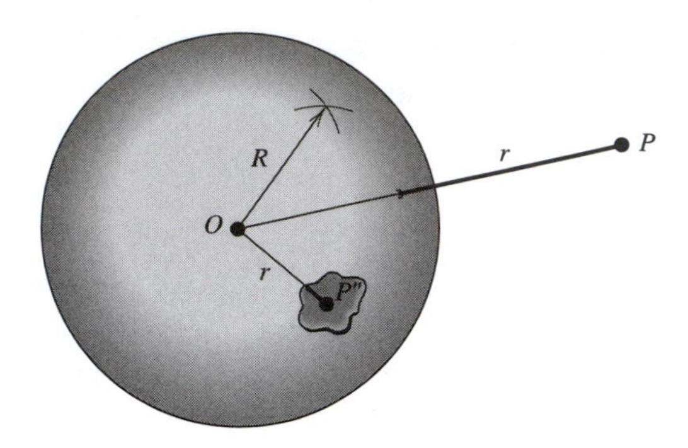

Figure 3-8 Spherical charge distribution.

Figure 3-9 shows E as a function of r.

Note that $E_i = E_o$ at $r = R$. This is in accordance with Gauss's law because a spherical shell of infinitesimal thickness just inside the surface of the sphere carries zero charge.

2. The electric potential V

At P, in Fig. 3-8,

$$V_o = \frac{Q}{4\pi\epsilon_0 r} \tag{3-54}$$

because the field is the same as that of a point charge Q at O.

To find the potential at a point P'' inside, we use Eq. 3-33:

$$V_i = \int_r^\infty E\,dr = \int_r^R E_i\,dr + \int_R^\infty E_o\,dr. \tag{3-55}$$

The last integral is simply the potential at $r = R$ of a point charge Q situated at O, or $Q/(4\pi\epsilon_0 R)$. Thus

$$V_i = \int_r^R \frac{Qr\,dr}{4\pi\epsilon_0 R^3} + \frac{Q}{4\pi\epsilon_0 R} = \frac{Q}{4\pi\epsilon_0 R}\left(\frac{3}{2} - \frac{r^2}{2R^2}\right). \tag{3-56}$$

See Fig. 3-9.

EXAMPLE The Electric Field of a Pair of Parallel, Oppositely Charged Wires

Electric power is distributed through parallel-wire lines. Let us calculate $\boldsymbol{E}$ and V near a pair of line charges of opposite polarity. We shall return to this field when discussing the parallel-wire transmission line in Chapter 24. Our discussion here will parallel that of the Example of Sec. 3.3

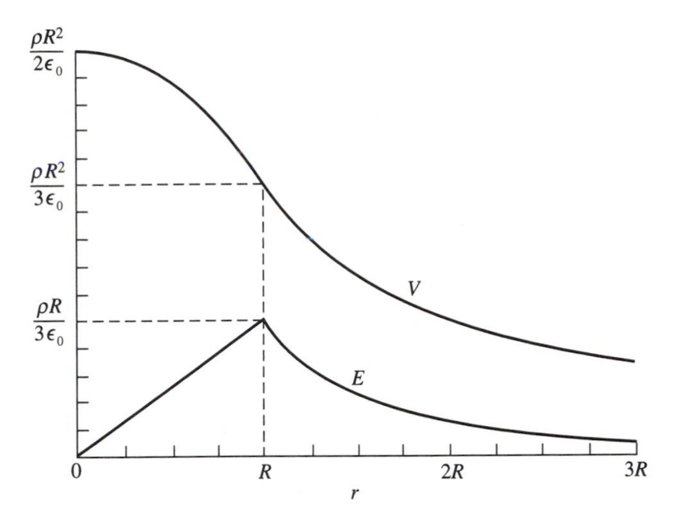

Figure 3-9 The potential V and the electric field strength E as functions of the radial distance r from a spherical charge distribution of radius R and volume charge density ρ.

Two long and parallel line charges, of charge densities

$$\lambda_A = +\lambda, \qquad \lambda_B = -\lambda \text{ coulombs/meter} \tag{3-57}$$

are separated by a distance $2a$ as in Fig. 3-10. The field is independent of the z-axis, which is perpendicular to the paper.

We wish to calculate both V and $\boldsymbol{E}$ but, as we pointed out in Sec. 3.4.1, in this specific case, we must first calculate $\boldsymbol{E}$. The potential V will follow.

1. $\boldsymbol{E}$. According to the principle of superposition of Sec. 3.3, the net field is the vector sum of the fields of the two wires. Let us start with the field $\boldsymbol{E}_A$ of the left-hand, positive, wire alone. Imagine a cylindrical surface of radius r_A centered on that wire. That field is radial and, from Gauss's law, over a unit length of the imaginary cylinder, the flux of $\boldsymbol{E}$, times ϵ_0, is equal to the enclosed charge:

$$\epsilon_0(2\pi r_A E_A) = \lambda, \qquad \text{so that} \quad E_A = \frac{\lambda}{2\pi\epsilon_0 r_A}, \tag{3-58}$$

where

$$r_A = [(a + x)^2 + y^2]^{1/2}. \tag{3-59}$$

That field points away from the positive wire, as in the figure.

Similarly, for the negative wire, the field

$$E_B = \frac{\lambda}{2\pi\epsilon_0 r_B} \tag{3-60}$$

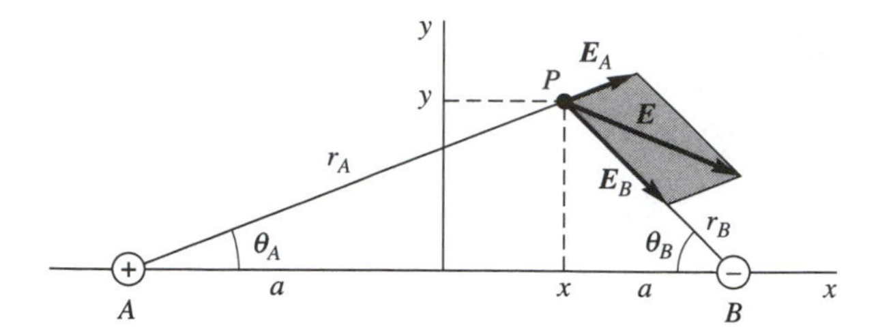

Figure 3-10 Pair of wires, A and B, perpendicular to the paper. Wire A carries a charge $+\lambda$ per meter, and wire B a charge $-\lambda$ per meter. The field $\boldsymbol{E}_A$ points away from A, $\boldsymbol{E}_B$ points toward B, and the net field $\boldsymbol{E}$ points in the direction shown.

points toward the wire, again as in the figure, with

$$r_B = [(a - x)^2 + y^2]^{1/2}. \tag{3-61}$$

To find the net $\boldsymbol{E}$ at the point $P(x,y)$, we need to find the vector sum of the two $\boldsymbol{E}$'s. So we require their x- and y-components. First,

$$\boldsymbol{E}_A = E_A(\cos\theta_A\,\hat{\boldsymbol{x}} + \sin\theta_A\,\hat{\boldsymbol{y}}) = E_A\left[\frac{(a + x)\,\hat{\boldsymbol{x}} + y\,\hat{\boldsymbol{y}}}{r_A}\right] \tag{3-62}$$

and, similarly,

$$\boldsymbol{E}_B = E_B(\cos\theta_B\,\hat{\boldsymbol{x}} - \sin\theta_B\,\hat{\boldsymbol{y}}) = E_B\left[\frac{(a - x)\,\hat{\boldsymbol{x}} - y\,\hat{\boldsymbol{y}}}{r_B}\right], \tag{3-63}$$

so that

$$\boldsymbol{E} = \boldsymbol{E}_A + \boldsymbol{E}_B \tag{3-64}$$

$$= \frac{\lambda}{2\pi\epsilon_0}\left[\left(\frac{a + x}{r_A^2} + \frac{a - x}{r_B^2}\right)\hat{\boldsymbol{x}} + \left(\frac{1}{r_A^2} - \frac{1}{r_B^2}\right)y\,\hat{\boldsymbol{y}}\right]. \tag{3-65}$$

Let us check. On the y-axis, $x = 0$ and $r_A = r_B$. Then $E_y = 0$, as expected: the field line is horizontal on the figure. Also, at the origin $x = 0$, $y = 0$, $r_A = r_B = a$. Then $E_y = 0$ and

$$E_x = \frac{\lambda}{2\pi\epsilon_0}\frac{2a}{a^2} = 2\frac{\lambda}{2\pi\epsilon_0 a}, \tag{3-66}$$

or twice the E of a single wire carrying a charge density of λ, at a distance a, again as expected.

If you plot a few field lines, you should find lines like those of Fig. 3-11. The field lines can be seen as *circles* passing through the wires. Intriguing!

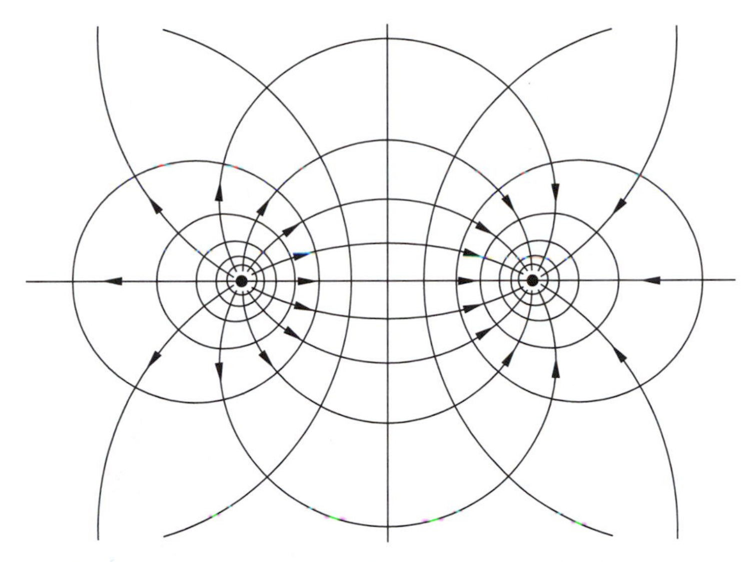

Figure 3-11 Electric field lines (arrowheads) and equipotential lines for the pair of charged wires of Fig. 3-10.

2. *V*. To find the potential *V* at the point *P*, we refer to Sec. 3.4.1:

$$V = \int_P^{P_0} \boldsymbol{E} \cdot \boldsymbol{dl} = \int_P^{P_0} E_x \, dx + E_y \, dy, \tag{3-67}$$

where P_0 is some point where we set $V = 0$.

Let us take it easy: set $y = 0$. On the x-axis, $E_y = 0$,

$$V_{y=0} = \int_x^{P_0} E_x \, dx, \tag{3-68}$$

and

$$r_A = a + x, \qquad r_B = a - x. \tag{3-69}$$

Our value of V will apply only on the x-axis. Also, for the reference point P_0 where $V = 0$, choose the origin $x = 0$. Then

$$V_{y=0} = \frac{\lambda}{2\pi\epsilon_0} \int_x^0 \left[\frac{a + x}{(a + x)^2} + \frac{a - x}{(a - x)^2} \right] dx \tag{3-70}$$

$$= \frac{\lambda}{2\pi\epsilon_0}\int_x^0 \left[\frac{1}{a+x} + \frac{1}{a-x}\right] dx \tag{3-71}$$

$$= \frac{\lambda}{2\pi\epsilon_0}[\ln(a+x) - \ln(a-x)]_x^0 \tag{3-72}$$

$$= \frac{\lambda}{2\pi\epsilon_0}\left[\ln\frac{a+x}{a-x}\right]_x^0 \tag{3-73}$$

$$= \frac{\lambda}{2\pi\epsilon_0}\left(\ln 1 - \ln\frac{a+x}{a-x}\right). \tag{3-74}$$

Now $\ln 1 = 0$ and

$$V_{y=0} = -\frac{\lambda}{2\pi\epsilon_0}\ln\frac{a+x}{a-x} = \frac{\lambda}{2\pi\epsilon_0}\ln\frac{a-x}{a+x}. \tag{3-75}$$

Again, let us check. At $x = 0$, V is proportional to $\ln 1 = 0$, so $V = 0$. Correct. At $x = -a$, V is proportional to $\ln\ \infty$, or to $+\infty$. That is also correct, because the potential on the supposedly infinitely thin, positive, left-hand wire is infinite, and positive. At $x = a$, V is proportional to $\ln\ 0$, or to $-\infty$, which is again correct.

We could also write that

$$V = \frac{\lambda}{2\pi\epsilon_0}\ln\frac{r_B}{r_A}. \tag{3-76}$$

We state without proof that this expression applies to an arbitrary point $P(x,y)$.

Fine, but what about *real* wires? In household wiring, both the wire diameter and a are of the order of one or two millimeters and the above approximation is of no use. Why? Because the charges crowd on the inside surfaces of the wires, and the distances r_A and r_B are not definable.

In the case of the high-voltage transmission lines that you see in the countryside, the wire diameter is of the order of centimeters, a is of the order of meters, and the approximation is satisfactory.†

---

†These lines almost invariably comprise either three or six wires (plus one or two ground wires), and carry three-phase current, with

$$V_1 = V_0 \cos\omega t, \qquad V_2 = V_0\cos[\omega t + (2\pi/3)], \qquad V_3 = V_0\cos[\omega t + (4\pi/3)].$$

Why three-phase current? Because, for a given transmitted power, line losses are lower. See Lorrain and Corson (1990), page 344. Here we assume steady voltages.

Let us set the wire diameter equal to 2 centimeters and $a = 1$ meter. Then the above approximation is satisfactory. On the x-axis, on the right-hand side of the positive wire, in Fig. 3-10,

$$r_A = 1 \times 10^{-2} \text{ meter}, \qquad r_B = 2 - (1 \times 10^{-2}) = 1.99 \text{ meter} \tag{3-77}$$

and

$$V_A = \frac{\lambda}{2\pi\epsilon_0} \ln\left(\frac{1.99}{1 \times 10^{-2}}\right) = 1.80 \times 10^{10} \lambda \ln 199, \tag{3-78}$$

$$= 1.80 \times 10^{10} \times 5.29\lambda = +9.53 \times 10^{10} \lambda \text{ volts.} \tag{3-79}$$

But what about λ? Well, set $V_A = +10{,}000$ volts and $V_B = -10{,}000$ volts. Then

$$\lambda = \frac{10{,}000}{9.53 \times 10^{10}} = 1.05 \times 10^{-7} \text{ coulomb/meter.} \tag{3-80}$$

How many electrons is that? Since the electron charge is -1.60×10^{-19} coulomb,

$$\lambda = \frac{1.05 \times 10^{-7}}{-1.60 \times 10^{-19}} = 6.56 \times 10^{11} \text{ electrons/meter.} \tag{3-81}$$

So the generator that maintains the voltage difference of 20,000 volts between the wires has transferred that many electrons from the positive wire to the negative wire. Is that an appreciable fraction of the conduction electrons in the wires? No! As we shall see in Sec. 4.5.1, the density of conduction electrons in copper is 8.1×10^{28} meter^{-3}.

Figure 3-11 shows a set of equipotential lines. They are *circles* again. Sliding the figure along the axis perpendicular to the paper generates equipotential surfaces.

The potential is large and positive near the positive wire, large and negative near the negative wire, and zero along the y-axis.

Note how the equipotentials are perpendicular to the lines of $\boldsymbol{E}$. That is because $\boldsymbol{E} = -\nabla V$ (Eq. 3-30), so that $\boldsymbol{E}$ points in the direction opposite from the maximum rate of increase of V (Sec. 1.2) everywhere.

EXAMPLE The Average Potential over a Spherical Surface. Earnshaw's Theorem

As another illustration of the use of Gauss's law, we shall prove that the average potential $\langle V \rangle$ over any spherical surface of radius R and center O has the following two

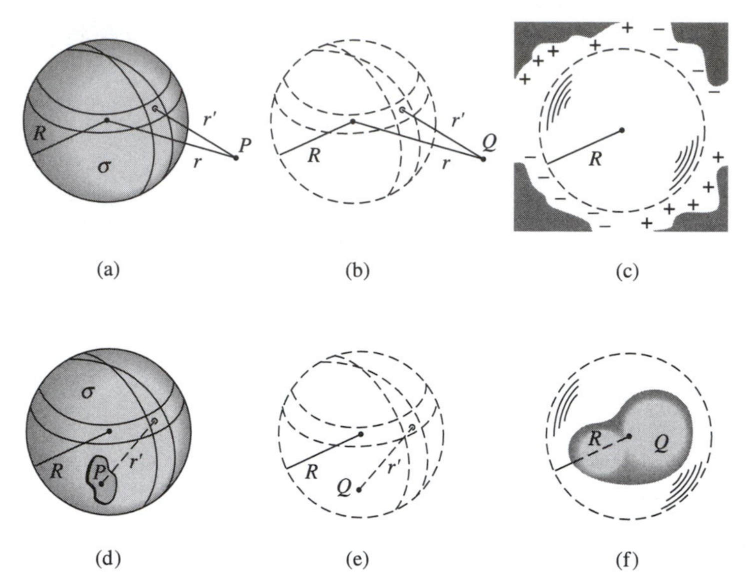

Figure 3-12 (a) Spherical surface carrying a uniform charge distribution σ. (b) Point charge Q outside an imaginary spherical surface of radius R. (c) Imaginary spherical surface in a charge-free region. The average potential over the surface is equal to the potential at the center. (d) Spherical surface carrying a uniform charge distribution σ. Point P is situated inside. (e) Point charge Q inside an imaginary spherical surface of radius R. (f) Imaginary spherical surface enclosing a charge Q. The average potential over the surface is equal to $Q/(4\pi\epsilon_0 R)$.

properties: if there are no charges inside, then $\langle V \rangle$ is equal to the value of V at the center O. If there is a net charge Q inside and no charge outside, then $\langle V \rangle$ is equal to $Q/(4\pi\epsilon_0 R)$.

1. Let us first think of an isolated spherical shell of radius R carrying a uniform surface charge density σ and a total charge $Q = 4\pi R^2 \sigma$, as in Fig. 3-12(a). Then, from Gauss's law, at some point P *outside,* at a distance r from O, the field is the same as if we had a point charge Q at the origin:

$$E = \frac{Q}{4\pi\epsilon_0 r^2} \quad \text{and} \quad V = \frac{Q}{4\pi\epsilon_0 r}. \tag{3-82}$$

But

$$V = \int_{\mathcal{A}} \frac{\sigma \, d\mathcal{A}}{4\pi\epsilon_0 r'} = \frac{Q}{4\pi R^2} \int_{\mathcal{A}} \frac{d\mathcal{A}}{4\pi\epsilon_0 r'}, \tag{3-83}$$

where $\mathcal{A}$ is the area of the shell.

Equating these two values of V gives a purely geometric relation concerning a sphere of radius R and a point P at a distance $r > R$ from its center:

$$\frac{Q}{4\pi\epsilon_0 r} = \frac{Q}{4\pi R^2}\int_{\mathcal{A}} \frac{d\mathcal{A}}{4\pi\epsilon_0 r'}. \tag{3-84}$$

Look at Fig. 3-12(b). We now have an imaginary sphere of radius R and a charge Q situated *outside* at a distance r. The left-hand side of Eq. 3-84 is equal to the potential at the center of the imaginary sphere, and the right-hand side is the average potential $\langle V \rangle$ on the spherical surface. So we have demonstrated that, in Fig. 3-12(b), the average V on the sphere is just the value of V at the center.

This result applies to any charge distribution situated *outside* the sphere, as in Fig. 3-12(c), because of the principle of superposition (Sec. 3.3).

Now imagine for a moment that there is a potential maximum at some point O in a region where $\rho = 0$. Then the average potential over some sphere centered on O must be lower than the potential at O, which is contrary to the above result. We see that *there can never be a potential maximum in a charge-free region*. For the same reason, *there can never be a potential minimum* either.

This is *Earnshaw's theorem.*

It is occasionally desirable to create a potential well in space so as to trap either ions or electrons. Earnshaw's theorem shows that doing so is impossible with electrostatic fields.

2. We can proceed in a similar fashion to find the average potential over a spherical surface when the charges are *inside.*

We start again with a charge Q spread uniformly over the surface, as in Fig. 3-12(d). At any point P inside, $\boldsymbol{E}$ is zero for the following reason: by symmetry, E_θ and E_ϕ are zero. To find E_r, we apply Gauss's law to a concentric spherical surface having a radius smaller than R and thus enclosing zero charge. We find that $E_r = 0$. Then $\boldsymbol{E} = 0$ inside, and the V at a point P inside is equal to the V at the surface, namely, $Q/(4\pi\epsilon_0 R)$. So, at P in Fig. 3-12(d),

$$V = \frac{Q}{4\pi\epsilon_0 R} = \int_{\mathcal{A}} \frac{\sigma\, d\mathcal{A}}{4\pi\epsilon_0 r'} = \frac{Q}{4\pi R^2}\int_{\mathcal{A}} \frac{d\mathcal{A}}{4\pi\epsilon_0 r'}. \tag{3-85}$$

The last term is just the average potential over the imaginary sphere of Fig. 3-12(e). So the average potential over a spherical surface of radius R containing a point charge Q is $Q/(4\pi\epsilon_0 R)$, regardless of the position of Q inside. The same applies to a charge distribution Q of finite volume inside the sphere, as in Fig. 3-12(f).

3.7 SUMMARY

The force exerted *by* a stationary point charge Q_a *on* a point charge Q_b, either stationary or in motion, is given by

$$\boldsymbol{F}_{ab} = \frac{Q_a Q_b}{4\pi\epsilon_0 r^2}\hat{\boldsymbol{r}}_{ab} \approx 9 \times 10^9 \frac{Q_a Q_b}{r^2}\hat{\boldsymbol{r}}_{ab}, \qquad (3\text{-}5, 3\text{-}6)$$

where $\epsilon_0 = 8.85 \times 10^{-12}$ coulomb²/(newton meter²), r is the distance between the charges, and $\hat{\boldsymbol{r}}_{ab}$ is the unit vector pointing from Q_a to Q_b. This is *Coulomb's law.*

A current flowing at the rate of 1 coulomb/second has a magnitude of 1 *ampere.*

According to the *principle of superposition,* two or more $\boldsymbol{E}$'s acting at the same point add vectorially.

An electrostatic field is conservative:

$$\oint_C \boldsymbol{E} \cdot \boldsymbol{dl} = 0, \qquad (3\text{-}27)$$

where C is any closed curve. It follows that

$$\nabla \times \boldsymbol{E} = 0 \qquad (3\text{-}31)$$

and that

$$\boldsymbol{E} = -\nabla V, \qquad (3\text{-}30)$$

where

$$V = \frac{1}{4\pi\epsilon_0}\int_{v'} \frac{\rho\, dv'}{r} \qquad (3\text{-}34)$$

is the *electric potential* at a point P. Here ρ is the volume charge density, r is the distance between $P(x,y,z)$ and the element of volume dv' at $P'(x',y',z')$, and v' encloses all the charges. This integral applies to finite charge distributions, and it assumes that $V = 0$ at infinity.

Gauss's law follows from Coulomb's law. In integral form,

$$\int_{\mathcal{A}} \boldsymbol{E} \cdot d\boldsymbol{\mathcal{A}} = \frac{Q}{\epsilon_0}, \qquad (3\text{-}47)$$

where Q is the net charge contained inside the closed surface of area $\mathcal{A}$. In differential form,

$$\nabla \cdot \boldsymbol{E} = \frac{\rho}{\epsilon_0}. \qquad (3\text{-}50)$$

PROBLEMS

3-1. (*3.1*) Electric field strength

A charge $+Q$ is situated at $x = a$, $y = 0$, and a charge $-Q$ at $x = -a$, $y = 0$. Calculate the vector $\boldsymbol{E}$ at the point $x = a$, $y = 2a$.

3-2. (*3.2*) Coulomb's law

The force of attraction between two charges of 1 coulomb and of opposite signs, separated by a distance of 1 meter, is about 9×10^9 newtons.

How large is a cube of lead that has a weight of 9×10^9 newtons? Lead has a density of 1.13×10^4 kilograms/meter^3.

3-3. (*3.2*) Coulomb's law

(a) Calculate the electric field strength that would be just sufficient to balance the gravitational force of the earth on an electron.

(b) If this electric field were produced by a second electron located below the first one, what would be the distance between the two electrons?

The charge of the electron is -1.6×10^{-19} coulomb and its mass is 9.1×10^{-31} kilogram.

3-4. (*3.2*) Cathode-ray tube

The electron beam in the cathode-ray tube of an oscilloscope is deflected vertically and horizontally by two pairs of parallel deflecting plates that are maintained at appropriate voltages. As the electrons pass between one pair, they are accelerated by the electric field, and their kinetic energy increases. Thus, under steady-state conditions, we achieve an increase in the kinetic energy of the electrons without any expenditure of power in the deflecting plates, as long as the beam does not touch the plates.

Could this phenomenon be used for a perpetual motion machine?

3-5. (*3.2*) Cathode-ray tube

In a certain cathode-ray tube, the electrons are accelerated under a difference of potential of 5 kilovolts. After being accelerated, they travel horizontally over a distance of 200 millimeters.

Calculate the downward deflection over this distance caused by the gravitational force.

An electron carries a charge of -1.6×10^{-19} coulomb and has a mass of 9.1×10^{-31} kilogram.

3-6. (*3.4*) Macroscopic particle gun

It is possible to obtain a beam of fine particles in the following manner: Figure 3-13 shows a parallel-plate capacitor with a hole in the upper plate. If a particle of dust, say, is introduced into the space between the plates, it sooner or later comes into contact with one of the plates and acquires a charge of the same polarity. It then flies over to the opposite plate, and the process repeats itself. The particle oscillates back and forth, until it is lost either at the edges or through the central hole. One can thus obtain a beam emerging from the hole by admitting particles steadily into the capacitor. In order to achieve high velocities, the gun must, of course, operate in a vacuum.

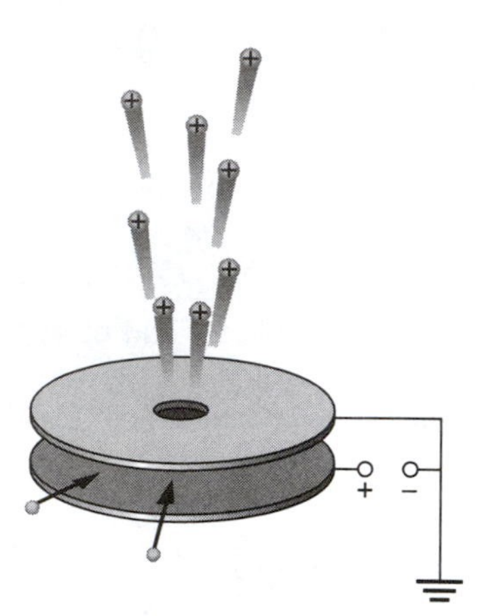

Figure 3-13

Beams of macroscopic particles are used for studying the impact of micrometeorites.

Now it has been found that a spherical particle of radius R lying on a charged plate acquires a charge

$$Q = 1.65 \times 4\pi\epsilon_r\epsilon_0 R^2 E_0 \text{ coulomb},$$

where E_0 is the electric field strength in the absence of the particle, and ϵ_r is the relative permittivity of the particle (Sec. 7.9). We assume that the particle is at least slightly conducting.

Assuming that the plates are 10 millimeters apart and that a voltage difference of 15 kilovolts is applied between them, find the velocity of a spherical particle 1 micrometer in diameter as it emerges from the hole in the upper plate. Assume that the particle has the density of water and that $\epsilon_r = 2$.

3-7. (*3.4*) Electrostatic spraying

When painting is done with an ordinary spray gun, part of the paint escapes deposition. The fraction lost depends on the shape of the surface, on drafts, etc., and can be as high as 80%. The use of the ordinary spray gun in large-scale industrial processes would therefore result in intolerable waste and pollution.

The efficiency of spray painting can be increased to nearly 100%, and the pollution reduced by a large factor, by charging the droplets of paint, electrically, and applying a voltage difference between the gun and the object to be coated.

It is found that, in such devices, the droplets carry a specific charge of roughly one coulomb per kilogram.

Assuming that the electric field strength in the region between the gun and the part is at least 10 kilovolts per meter, what is the minimum ratio of the electric force to the gravitational force?

3-8. (*3.4*) Cylindrical electrostatic analyzer

Figure 3-14 shows a cylindrical electrostatic analyzer or velocity selector. It consists of a pair of cylindrical conductors separated by a radial distance of a few millimeters.

Show that, if the radial distance between the cylindrical surfaces of average radius R is a, and if the voltage between them is V, then the particles collected at I have a velocity of

$$v = \left(\frac{Q}{m}\frac{VR}{a}\right)^{1/2}.$$

3-9. (*3.4*) Electrostatic seed-sorting device

It is possible to separate normal seeds from discolored ones and from foreign objects by means of device that operates as follows. The seeds drop one by one between a pair of photocells. If the color is not right, voltage is applied to a needle that deposits a charge on the seed. The seeds then fall between a pair of electrically charged plates that deflect the undesired ones into a separate bin. One such machine can sort peas at the rate of 100 per second, or about 2 metric tons per 24-hour day.

(a) If the seeds fall at the rate of 100 per second, over what distance must they fall if they must be spaced vertically by 20 millimeters when they pass between the photocells? Neglect air resistance.

(b) Assume that the seeds acquire a charge of 1.5×10^{-9} coulomb, that the deflecting plates are parallel and 50 millimeters apart, and that the potential difference between them is 25,000 volts. How far should the plates extend below the charging needle if the charged seeds must deflect by 40 millimeters on leaving the plates? Assume that the charging needle and the top of the deflecting plates are close to the photocell.

3-10. (*3.4*) Rutherford discovers the nucleus

In 1906, in the course of a historic experiment that demonstrated the small size of the atomic nucleus, Rutherford observed that an alpha particle ($Q_1 = 2 \times 1.6 \times 10^{-19}$

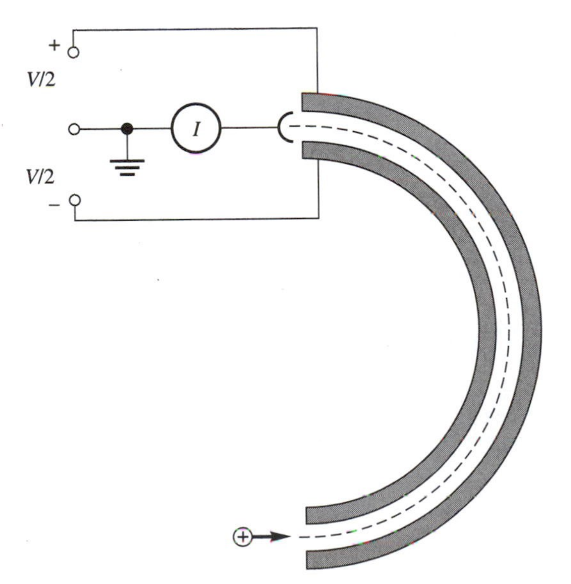

Figure 3-14

coulomb) having a kinetic energy of 7.68×10^6 electron volts ($7.68 \times 10^6 \times 1.6 \times 10^{-19}$ joule) rebounds backward in a head-on collision with a gold nucleus ($Q_2 = 79 \times 1.6 \times 10^{-19}$ coulomb).

(a) What is the distance of closest approach where the electrostatic potential energy is equal to the initial kinetic energy? Express your result in femtometers (10^{-15} meter).
(b) What is the maximum force of repulsion?
(c) What is the maximum acceleration in g's? The mass of the alpha particle is about 4 times that of a proton, or $4 \times 1.7 \times 10^{-27}$ kilogram.

3-11. (*3.4*) Electrostatic ion thruster

Ion thrusters correct either the attitude or the trajectory of satellites.

The force exerted by a thruster is equal to $m'v$, where m' is the mass of propellant ejected per second and v is the exhaust velocity with respect to the thruster.

Figure 3-15 shows a schematic diagram of a thruster that ejects a beam of charged particles. The propellant enters at P and is ionized in S. Electrodes A and B form a lens that accelerates the positive ions. A beam of positive ions exits on the right at a velocity determined by the accelerating voltage V. The ions of mass m carry charges ne, where e is the magnitude of the electronic charge. The current is I. Electrons emitted by the filament F neutralize the beam so as to prevent the satellite from charging up.

(a) Show that the thrust is given by $F = I[2Vm/(ne)]^{1/2}$.
(b) What is the value of F for a 0.1-ampere beam of protons when $V = 50$ kilovolts?
(c) If P is the power IV spent in accelerating the particles, show that

$$F = (2Pm')^{1/2} = \frac{2P}{v} = P\left(\frac{2m}{neV}\right)^{1/2}.$$

Thus, for given values of P and m', the thrust is independent of the charge-to-mass ratio of the ions. Or, for a given P, F is *inversely* proportional to v. The last expression shows that, *for a given power expenditure P*, it is preferable to use heavy ions carrying a single charge ($n = 1$) and to use as *low* an accelerating voltage V as possible.

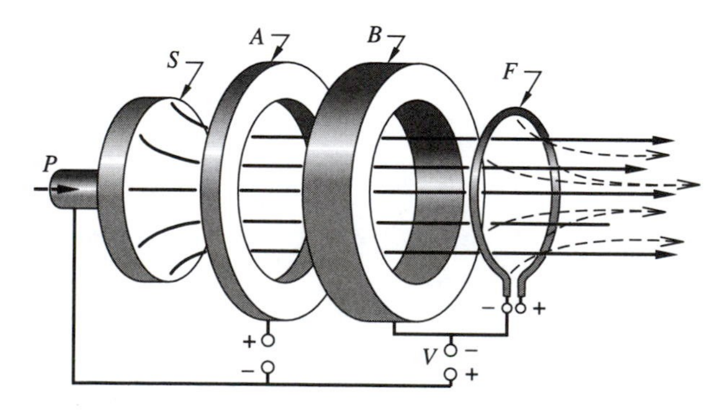

Figure 3-15

(d) If the electron source is turned off, and if the beam current I is 1 ampere, how long will it take the body of the satellite to attain a voltage equal to the accelerating voltage, if V is 50 kilovolts? Assume that the satellite is spherical and that it has a radius of 1 meter. At that point the thruster ceases to operate because the ions follow the satellite.

3-12. *(3.6)* The space derivatives of E_x, E_y, E_z

List all the relations between the various space derivatives of the three components of $\boldsymbol{E}$, using Gauss's law and the fact that an electrostatic field is conservative.

3-13. *(3.6)* Possible and impossible fields

An electric field points everywhere in the z-direction.

(a) What can you conclude about the values of the partial derivatives of $\boldsymbol{E}$ with respect to x, y, z (i) if the space charge density ρ is zero and (ii) if ρ is not zero?

(b) Sketch lines of $\boldsymbol{E}$ for one possible and for one impossible field, on the assumption that $\nabla \times \boldsymbol{E} = 0$.

3-14. *(3.6)* The conduction electron density at the surface of electrically charged copper

A copper atom has a diameter of about 0.3 nanometer.

(a) Calculate (i) the approximate number of atoms per square meter, (ii) the approximate surface charge density that would result if each atom gained one free electron, and (iii) the corresponding electric field strength.

(b) The maximum possible electric field strength in air is 3×10^6 volts/meter. How far apart are the excess electrons at that value of $\boldsymbol{E}$?

3-15. *(3.6)* The earth's electric charge

The electric field strength in the atmosphere near the surface of the earth is about 100 volts/meter and points downward. The potential increases with increasing height, up to about 300,000 volts. This field is maintained by thunderstorms, which deposit negative charge on the earth at the average rate of about 10^3 amperes.

Calculate the electric charge carried by the earth. It is possible to solve this problem in two equivalent ways.

When you are standing up, there is a potential difference of 200 volts between the top of your head and the ground. That is more than you need to be electrocuted! What is wrong? The explanation is that the current density in the atmosphere is only of the order of 10^{-12} ampere/meter^2, and the conductivity of a living organism is high. So the voltage difference across your body is less than what you could feel by very many orders of magnitude. See *Physics Today,* May 1999, pp. 15, 68.

3-16. *(3.6)* The coaxial line

Figure 24-2 shows a portion of a coaxial line.

Show that, at a distance r from the axis in the region between the two conductors, $E = \lambda/(2\pi\epsilon_0 r)$, where λ is the charge per unit length on the inner conductor. The vector $\boldsymbol{E}$ points outward if λ is positive.

3-17. *(3.6)* The force between a point charge and a line charge

A uniform linear distribution of a charge of λ coulombs/meter is situated at a distance r from a point charge Q of opposite sign.

(a) Calculate the force of attraction.

(b) Show that the force is the same as if the linear distribution were replaced by a single charge $Q' = 2\lambda r$ situated at the foot of the perpendicular drawn from Q.

3-18. *(3.6)* Proton beam

A 1.00-microampere beam of protons is accelerated through a difference of potential of 10,000 volts.

(a) Calculate the charge density in the beam, once the protons have been accelerated, assuming that the current density is uniform over a diameter of 2.00 millimeters and is zero outside.

(b) Calculate the radial E both inside and outside the beam.

(c) Draw a graph of the radial E for values of r ranging from 0 to 10.0 millimeters.

(d) The beam is situated on the axis of a grounded cylindrical conducting tube with an inside radius of 10.0 millimeters. Draw a graph of V inside the tube.

(e) Calculate the electric charge density per unit length on the inside of the tube.

3-19. *(3.6)* The field of an atomic nucleus

The radial dependence of the electric charge density inside a certain atomic nucleus of radius a is roughly described by $\rho = \rho_0(1 - r^2/a^2)$, for $r \leq a$, where $\rho_0 = 5.0 \times 10^{25}$ coulombs/meter^3 and $a = 3.4$ femtometers.

(a) What is the total charge Q?

(b) Find E and V outside the nucleus. What are the values of E and V at the surface?

(c) Find E and V inside the nucleus. What is the value of V at the center?

(d) Show that E is maximum at $r/a = 0.745$.

(e) Draw graphs showing $E/[2\rho_0/(15\epsilon_0)]$ and $V/[2\rho_0/(15\epsilon_0)]$ as functions of r for $r/a = 0$ to 5.

3-20. *(3.6)* Ion accelerator

A certain particle accelerator has a high-voltage electrode maintained under pressure in gaseous SF_6 in a metal tank. It is possible to maintain much higher voltages in this way than if the electrode were in air.

Assume that the electrode is spherical and that its radius is r_1. Its voltage is V. The tank has a radius r_2 and is grounded. The electric field strength is highest at the surface of the electrode. You are required to find values of r_1 and r_2 that will minimize this E.

For a given value of r_1, the optimum value of r_2 is infinite, which is absurd. Of course, cost, weight, and space limit r_2. So you must optimize r_1 for a given r_2, which is 483 millimeters in one specific case.

(a) Show that E at the surface of the high-voltage electrode ($r = r_1$) has a minimum value of $2V/r_1$ when $r_1 = r_2/2$.

(b) Explain qualitatively why there is an optimum radius r_1.

(c) Identifying an optimum condition is not sufficient. You must also evaluate how critical the condition is. Plot E/V at $r = r_1$ for $r_2 = 0.483$ meter and for values of r_1 ranging from 100 to 400 millimeters.

(d) What range of values of r_1 is permissible if E can be 10% larger than $2V/r_1$?

(e) Calculate $2V/r_1$ for $V = 5 \times 10^5$ volts and for the optimum r_1.

3-21. *(3.6)* Electrostatic precipitation

Electrostatic precipitation serves to eliminate dust particles from industrial gases, for example, to eliminate fly ash from the smoke of coal-fired electric power plants. A corona discharge ionizes the gas, and the ions charge the dust particles, which drift in the electric field to the electrodes, where they collect. Periodically, the electrodes are shaken, and the dust falls into a container.

In one type of precipitator, the anode is a grounded cylinder having an inside radius R of 150 millimeters, and the cathode is an axial wire maintained at a potential V of –50 kilovolts. The gas ionizes, and ions of both signs form in the corona discharge near the wire. The positive ions quickly reach the center wire, while the negative ions move out radially to the cylinder. The space charge is thus negative over most of the volume of the cylinder.

Under those conditions, experiments show that E is approximately uniform and equal to V/R for all values of r. If the dust particles are at least slightly conducting, they acquire a negative charge Q of $12\pi\epsilon_0 Ea^2$, where a is their radius. The charge is somewhat smaller if they are nonconducting.

(a) Let I be the radial electric current per meter and $\mathcal{M}$ the mobility (speed$/E$) of the negative ions.

Show that, for any r, $I = 2\pi r\rho\mathcal{M}E$, where ρ is the space charge density $\epsilon_0 E/r$.

(b) The drift velocity of the dust particles is given by *Stokes's law*: it is the force EQ divided by $6\pi\eta a$, where η is the viscosity of the gas.

Show that their drift velocity v is $2\epsilon_0 E^2 a/\eta$.

(c) Calculate I, ρ, v, and the time required for a dust particle to drift from the cathode to the anode when $\mathcal{M} = 2 \times 10^{-4}$ meter^2/(volt-second), $a = 5$ micrometers, and $\eta = 2 \times 10^{-5}$ kilogram/(meter-second).

This simplified theory neglects turbulence, which is important in practice.

3-22. *(3.6)* The expansion of the universe

In 1959 Lyttleton and Bondi suggested that the expansion of the universe could be explained on the basis of Newtonian mechanics if matter carries a net electric charge.

Imagine a spherical volume V of astronomical size containing un-ionized atomic hydrogen of uniform density N atoms per cubic meter, and assume that the proton charge is equal to $(1 + y)e$, where e is the magnitude of the electron charge.

(a) Find E at the radius R.

(b) Show that, for $y > 10^{-18}$, the electronic repulsion becomes greater than the gravitational attraction, so the gas expands.

(c) Show that the force of repulsion on an atom is then proportional to its distance R from the center and that, as a consequence, the radial velocity of an atom at R is proportional to R. Assume that the density is maintained constant by the continuous creation of matter in space.

(d) Show that the velocity v is R/T, where T is the time required for the radial distance R of a given atom to increase by a factor of e. This time T can be taken to be the age of the universe.

(e) In the Millikan oil-drop experiment, an electrically charged droplet of oil is suspended in the electric field between two plane horizontal electrodes. It is observed that the charge carried by the droplet changes by integral amounts within an accuracy of about 1 part in 10^5.

Show that the Millikan oil-drop experiment leads us to believe that y is less than about 10^{-16}.

3-23. (*3.6*) The average $\boldsymbol{E}$ over a spherical volume is equal to the value of $\boldsymbol{E}$ at the center.

We know that the force exerted by a uniform spherical charge distribution on an outside charge is the same as if the spherical charge were concentrated at its center.

(a) Use this fact to show that the field of a point charge is such that its volume average over a sphere is equal to its value at the center.

(b) Show that the same applies to any electrostatic field in a charge-free region.

CHAPTER

4

ELECTRIC FIELDS II

The Equations of Poisson and of Laplace. Electric circuits. Conduction

We now turn to Poisson's equation, which relates the local volume charge density ρ to the spatial rates of change of the potential V. This is a fundamental relation that follows immediately from Gauss's law. Laplace's equation is Poisson's, with ρ equal to zero. Both equations serve to calculate electric fields.

Charge conservation is an experimental fact. Whatever the circumstances, the net electric charge carried by a closed system is constant. We shall frame that law in a simple mathematical form, apply it a few sections later, and return to it on several occasions.

†Astericks indicate material that can be omitted without losing continuity.

The major part of this chapter pertains to conductors. Ordinary electric conductors contain conduction electrons that drift in the direction opposite to the applied $\boldsymbol{E}$. We shall find, among other things, that this drift velocity is surprisingly low and that the net volume charge density is normally zero.

4.1 THE EQUATIONS OF POISSON AND OF LAPLACE

Let us replace $\boldsymbol{E}$ by $-\boldsymbol{\nabla}V$ in Eq. 3-50. Then

$$\boldsymbol{\nabla}^2 V = -\frac{\rho}{\epsilon_0}. \tag{4-1}$$

This is *Poisson's equation.* It relates the space charge density ρ *at a given point* to the second space derivatives of V *in the region of that point.*

In a region where the charge density ρ is zero,

$$\boldsymbol{\nabla}^2 V = 0, \tag{4-2}$$

which is *Laplace's equation.*

The general problem of finding V in the field of a given charge distribution amounts to finding a solution to either Laplace's or Poisson's equation that will satisfy the given boundary conditions. Chapters 9 and 10 describe methods for solving both equations.

EXAMPLE The Field of a Uniform Spherical Charge

Consider again a spherical charge of uniform volume density ρ and radius R as in Fig. 3-8.

Outside the sphere, $\rho = 0$ and

$$\boldsymbol{\nabla}^2 V_o = 0. \tag{4-3}$$

Now, by symmetry, V_o is independent of both θ and ϕ. Therefore, from Sec. 1.10,

$$\frac{1}{r^2}\frac{\partial}{\partial r}\left(r^2\frac{\partial V_o}{\partial r}\right) = 0, \qquad \frac{\partial}{\partial r}\left(r^2\frac{\partial V_o}{\partial r}\right) = 0, \qquad r^2\frac{\partial V_o}{\partial r} = A, \tag{4-4}$$

$$\frac{\partial V_o}{\partial r} = \frac{A}{r^2}, \qquad E_o = \frac{A}{r^2}, \tag{4-5}$$

where A is a constant of integration. This is in agreement with Eq. 3-52 with $A = -Q/(4\pi\epsilon_0)$.

Inside the sphere,

$$\boldsymbol{\nabla}^2 V_i = -\frac{\rho}{\epsilon_0}, \tag{4-6}$$

$$\frac{1}{r^2}\frac{\partial}{\partial r}\left(r^2\frac{\partial V_i}{\partial r}\right) = -\frac{\rho}{\epsilon_0}, \tag{4-7}$$

$$\frac{\partial}{\partial r}\left(r^2\frac{\partial V_i}{\partial r}\right) = -\frac{\rho r^2}{\epsilon_0}, \tag{4-8}$$

$$r^2\frac{\partial V_i}{\partial r} = -\rho\frac{r^3}{3\epsilon_0} + B, \tag{4-9}$$

$$E_i = \frac{\rho r}{3\epsilon_0} - \frac{B}{r^2}, \tag{4-10}$$

where B is another constant of integration.

It is intuitively obvious that E_i cannot become infinite at $r = 0$; so B is zero and

$$E_i = \frac{\rho r}{3\epsilon_0} = \frac{Qr}{4\pi\epsilon_0 R^3}, \tag{4-11}$$

as in the first example in Sec. 3.6.

EXAMPLE The Thermionic Power Generator and the Thermionic Diode

The industrial revolution was a complex social phenomenon that was triggered largely by the 17th century invention of the steam engine, which could convert heat into mechanical work. The modern equivalent of the steam engine is of course the internal-combustion engine.

But how can you convert heat, not into mechanical work, but into *electricity?* You can do that with a gasoline engine or a nuclear reactor that drives an electric generator. You could also use a magnetohydrodynamic generator as in Sec. 17.1, or the Peltier effect. (You can look that up in an encyclopedia.) Here is another way.

Thermionic power generators convert heat into electricity without any moving parts except free electrons. They are used in Space with radioisotopes as heat sources, and also as topping units in steam power plants for improving the overall efficiency.

These converters consist of two plates, a hot cathode and a cold anode, as in Fig. 4-1. The gap can be evacuated, but it usually contains cesium vapor. Why cesium? For two reasons. First, cesium is adsorbed in the plates and reduces the work function.† Second, cesium vapor ionizes easily and supplies slow positive ions that help neutralize the electron space charge.

The hot cathode emits electrons that cross the gap, reach the cold anode, and flow back through an external load resistance R.

The current through R flows *from* the cathode *to* the anode, so that the "anode" is *negative* with respect to the cathode. It is only the electrons that have a sufficiently high kinetic energy on leaving the cathode that can reach the "anode".

†The energy required to remove an electron from inside a metal to a point just outside is $e\phi$, where e is the magnitude of the electron charge and ϕ is the *work function* characteristic of the metal.

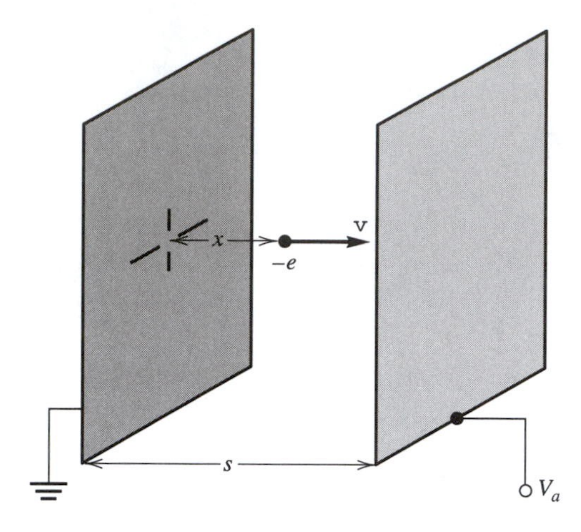

Figure 4-1 Schematic diagram of a thermionic diode. The cathode, on the left, emits electrons. An electron of charge $-e$ drifts in the direction of the anode at a velocity v. We have shown widely separated electrodes for clarity.

Since the analysis of the thermionic generator is relatively complex, we discuss here a different but related device, namely the *thermionic diode.*

The cathode and anode are plane and parallel, and they are separated by a distance s, as in Fig. 4-1, with s small compared to their linear extent. The cathode is at zero potential. We assume that the electrons have zero initial velocity, and that the current is not limited by the cathode temperature, but can be increased at will.

Since V depends only on x, by hypothesis, Poisson's equation 4-1 reduces to

$$\frac{d^2V}{dx^2} = -\frac{\rho}{\epsilon_0}, \tag{4-12}$$

where the electron space charge density ρ is negative. Thus d^2V/dx^2 is positive, but we have not yet found ρ.

Now the current density J is equal to the space charge density ρ multiplied by the electron velocity: $J = \rho v$. So

$$\frac{d^2V}{dx^2} = \frac{J}{\epsilon_0 v}, \tag{4-13}$$

where J is the *magnitude* of the current density.

By conservation of energy,

$$\frac{mv^2}{2} = eV, \tag{4-14}$$

where m is the mass of an electron and $-e$ is its electric charge: $e = +1.6 \times 10^{-19}$ coulomb. Then

$$\frac{d^2V}{dx^2} = \frac{J}{\epsilon_0(2eV/m)^{1/2}}. \tag{4-15}$$

To integrate, we multiply the left-hand side by $2(dV/dx)dx$ and the right-hand side by $2dV$. Then

$$\left(\frac{dV}{dx}\right)^2 = \frac{4J(mV/2e)^{1/2}}{\epsilon_0} + A, \tag{4-16}$$

where A is a constant of integration.

We now find the value of A. At the cathode, $V = 0$ and $A = (dV/dx)^2$. But dV/dx is zero at the cathode for the following reason. If one applies a voltage to the anode when the cathode is cold, dV/dx is positive and equal to V_a/s. If one now heats the cathode, it emits electrons, there is a negative space charge, and dV/dx near the cathode decreases. As long as dV/dx is positive at the cathode, the emitted electrons accelerate toward the anode and cannot return to the cathode. The current is then limited by the thermionic emission and not by V_a. This is contrary to what we assumed at the beginning. However, if dV/dx were negative, electrons could never leave the cathode, and there would be zero space charge, which is absurd. So dV/dx can be neither positive nor negative: it is zero, and A is also zero. (However, see below.) Then

$$\frac{dV}{dx} = 2\left(\frac{J}{\epsilon_0}\right)^{1/2}\left(\frac{m}{2e}\right)^{1/4} V^{1/4}, \tag{4-17}$$

$$V^{3/4} = 1.5\left(\frac{J}{\epsilon_0}\right)^{1/2}\left(\frac{m}{2e}\right)^{1/4} x + B. \tag{4-18}$$

The constant of integration B is zero because V is zero at $x = 0$. So

$$V = \left(\frac{9J}{4\epsilon_0}\right)^{2/3}\left(\frac{m}{2e}\right)^{1/3} s^{4/3}\left(\frac{x}{s}\right)^{4/3}. \tag{4-19}$$

When $x = s$, $V = V_a$. Therefore

$$V = V_a\left(\frac{x}{s}\right)^{4/3}. \tag{4-20}$$

Also, disregarding the sign of E,

$$E = \frac{4}{3}\frac{V_a}{s}\left(\frac{x}{s}\right)^{1/3}, \tag{4-21}$$

$$J = \frac{4\epsilon_0(2e/m)^{1/2}V_a^{3/2}}{9s^2} = 2.335 \times 10^{-6}\frac{V_a^{3/2}}{s^2} \qquad \text{amperes/meter}^2, \tag{4-22}$$

$$\rho = \frac{4\epsilon_0 V_a}{9s^2(x/s)^{2/3}}. \tag{4-23}$$

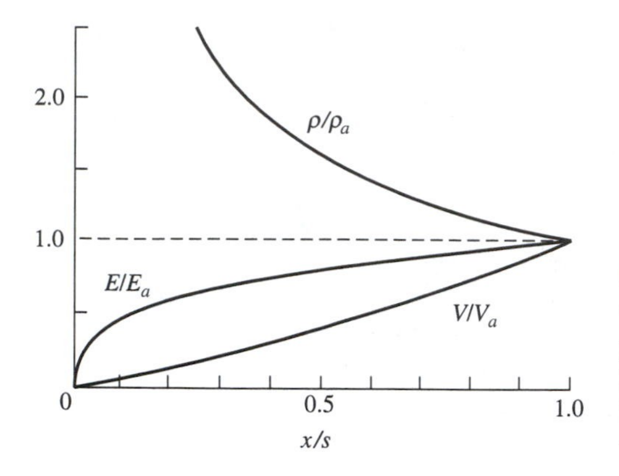

Figure 4-2 The space charge density ρ, the electric field strength E, and the potential V as functions of the distance x from the cathode of a plane-parallel diode. The subscript a refers to the value of the variable at the anode.

Equation 4-22 is known as the *Child-Langmuir law.* This law is valid only for a plane-parallel diode, with negligible edge effects and for electrons emitted with zero velocity. However, for any geometry, J is proportional to $V_a^{3/2}$.

Figure 4-2 shows V, E, and ρ as functions of x/s.

In an actual thermionic diode, the emission velocity of the electrons is finite and there is a potential minimum immediately in front of the cathode. Only electrons with an initial velocity larger than a certain value get past this minimum.

4.1.1 The Uniqueness Theorem for Electrostatic Fields

According to the *uniqueness theorem* for electrostatic fields, *there cannot exist more than one potential* $V(x, y, z)$ *that satisfies both Poisson's equation and a given set of boundary conditions.* This theorem is important because it leaves us free to use any method, even intuition, to find V. If we can, somehow, discover a function $V(x, y, z)$ that meets these two requirements, then it is the only possible potential function.

Several proofs of this theorem are known, each one based on its own particular set of assumptions. We simply assume that the uniqueness theorem always applies.

4.2 CONSERVATION OF ELECTRIC CHARGE

Consider a closed surface of area $\mathcal{A}$ enclosing a volume v. The volume charge density inside is ρ. Charges flow in and out, and the current density at a given point on the surface is $\boldsymbol{J}$ amperes/meter^2.

It is a well-established experimental fact that there is never any net creation of electric charge. Then any net outflow depletes the enclosed charge Q: at any given instant,

$$\int_{\mathcal{A}} \boldsymbol{J} \cdot d\boldsymbol{\mathcal{A}} = -\frac{d}{dt}\int_v \rho \, dv = -\frac{dQ}{dt}, \tag{4-24}$$

where the vector $d\boldsymbol{\mathcal{A}}$ points *outward,* according to the usual sign convention.

Applying the divergence theorem on the left, we find that

$$\int_v \nabla \cdot \boldsymbol{J}\, dv = -\int_v \frac{\partial \rho}{\partial t}\, dv. \tag{4-25}$$

We have transferred the time derivative under the integral sign, but then we must use a partial derivative because ρ can be a function of x, y, z, as well as of t.

Now the volume v is of any shape or size. Therefore

$$\nabla \cdot \boldsymbol{J} = -\frac{\partial \rho}{\partial t}. \tag{4-26}$$

Equations 4-24 and 4-26 are, respectively, the integral and differential forms of the *law of conservation of electric charge.*

4.3 OHM'S LAW

In good conductors, such as copper or aluminum, each atom possesses one or two conduction electrons that are free to roam about in the material.

Semiconductors may contain two types of mobile charges: conduction electrons and positive holes. A *hole* is a vacancy left by an electron liberated from the valence bond structure in the material. A hole behaves as a free particle of charge $+e$, and it moves through the semiconductor much as an air bubble rises through water.

In most good conductors and semiconductors, the current density $\boldsymbol{J}$ is proportional to $\boldsymbol{E}$:

$$\boldsymbol{J} = \sigma \boldsymbol{E}, \tag{4-27}$$

where σ is the *electric conductivity* of the material expressed in siemens/meter, where 1 *siemens*† is 1 ampere/volt. This relation was first stated by Ohm (1787–1854) and is known as *Ohm's law.* We find a more general form of this law in Chapter 19.

Table 4-1 shows the conductivities of some common materials.

Many conductors have temperature-dependent conductivities and can serve as temperature sensors. For example, platinum resistance thermometers have been in use for many years.

Ohm's law applies rigorously to metals, but it does not always apply to most other conductors. For example, in a certain type of semiconductor, J is proportional to the fifth power of E. Also, some conductors are not isotropic.

If Ohm's law applies, the *resistance* between two electrodes fixed to a sample of material is

$$R = \frac{V}{I}, \tag{4-28}$$

where V is the potential difference between the two electrodes and I is the current.

---

†After Ernst Werner von Siemens (1816–1892). The word therefore takes a terminal *s* in the singular: one siemens. The siemens was formerly called a "mho."

Table 4-1

Conductor	Conductivity σ, siemens/meter
Aluminum	3.77×10^7
Brass (65.8 Cu, 34.2 Zn)	1.59×10^7
Chromium	3.8×10^7
Copper	5.80×10^7
Gold	4.50×10^7
Graphite	7.1×10^4
Iron	1.0×10^7
Mumetal (75 Ni, 2 Cr, 5 Cu, 18 Fe)	0.16×10^7
Nickel	1.3×10^7
Seawater	~5
Silver	6.15×10^7
Tin	0.870×10^7
Zinc	1.86×10^7

EXAMPLES

For a cylinder of cross section $\mathscr{A}$, length L, uniform conductivity σ, and with electrodes on the ends as in Fig. 4-3*a*,

$$I = \mathscr{A}J = \mathscr{A}\sigma E = \frac{\sigma \mathscr{A} V}{L}, \tag{4-29}$$

$$R = \frac{L}{\sigma \mathscr{A}}. \tag{4-30}$$

The tube of Fig. 4-3*b* has inner and outer radii R_1 and R_2, respectively, a length L, and a uniform conductivity σ. There are copper electrodes on the inner and outer cylindrical surfaces. The resistance of a cylindrical element of thickness dr is $dr/(\sigma 2\pi r L)$. Then

$$R = \int_{R_1}^{R_2} \frac{dr}{\sigma 2\pi r L} = \frac{1}{2\pi\sigma L} \ln \frac{R_2}{R_1}. \tag{4-31}$$

EXAMPLE Geophysical Prospecting by the Resistivity Method

This is an application of the Example given in Sec. 3.1.

One can locate resistivity anomalies in the ground with the set-up shown in Fig. 4-4. The method is used for locating conducting ores and prehistoric artifacts. The current flowing between electrodes C_1 and C_2 establishes an electric field in the ground, and one measures the voltage V between electrodes P_1 and P_2 that are moved along the ground, but maintained at a fixed spacing b. With $b \ll a$, V/b is equal to

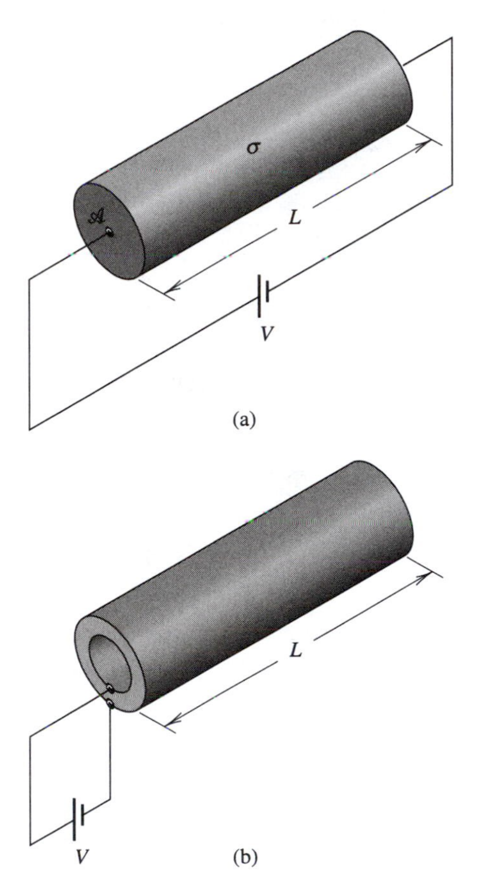

Figure 4-3 (a) Cylinder of conducting material with electrodes at both ends. (b) Tube of conducting material with electrodes on the inner and outer surfaces.

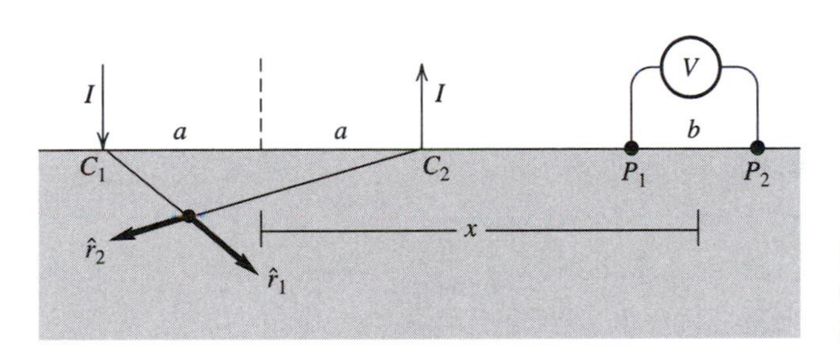

Figure 4-4 Locating conductivity anomalies in the ground.

E at the position x. Anomalies in ground conductivity show up in curves of E as a function of x. Let us see how that comes about.

Assume that the ground conductivity σ is uniform. The field $\boldsymbol{E}$, here, is the field in the ground, at the surface. So set $y = 0$. Then, from Fig. 4-4,

$$d_1 = x + a, \qquad d_2 = x - a. \tag{4-32}$$

Set also $Q_1 = Q$, $Q_2 = -Q$ at the positions of C_1 and C_2. Then

$$E = \frac{Q}{4\pi\epsilon_0(x+a)^3}(x+a) - \frac{Q}{4\pi\epsilon_0(x-a)^3}(x-a) \tag{4-33}$$

$$= \frac{Q}{4\pi\epsilon_0}\left[\frac{1}{(x+a)^2} - \frac{1}{(x-a)^2}\right]. \tag{4-34}$$

But the charge Q is unknown! It is easy to measure a voltage or a current, but charges are impossible to measure, except under very special circumstances. So let us look for a relation between the charge Q and the current I. Imagine a small *hemisphere* of radius $c \ll a$ centered on electrode C_1. On that surface,

$$E = \frac{Q}{4\pi\epsilon_0 c^2} \tag{4-35}$$

and

$$I = (2\pi c^2)\sigma E = 2\pi c^2 \sigma \frac{Q}{4\pi\epsilon_0 c^2} = \frac{Q\sigma}{2\epsilon_0}, \tag{4-36}$$

so that

$$Q = \frac{2\epsilon_0 I}{\sigma}. \tag{4-37}$$

Substituting into Eq. 4-34,

$$E = \frac{V}{b} = \frac{I}{2\pi\sigma}\left[\frac{1}{(x+a)^2} - \frac{1}{(x-a)^2}\right]. \tag{4-38}$$

The value of the chosen distance b depends on the expected size of the conductivity anomaly, and the current I required depends on the sensitivity of the voltmeter that is available for measuring V.

*4.4 ELECTRIC CIRCUITS. THE KIRCHHOFF LAWS

Kirchhoff's two laws are the fundamental laws of circuit theory. They are general: they apply even if the components are non-linear, active, time-dependent, and even hysteretic. (*Hysteresis* is a property of certain components whose parameters depend on their previous history.) Kirchhoff (1824–1887) discovered these laws when he was 21.

We first define key terms, illustrated by the circuit of Figure 4-5. Points A and B are called *nodes,* while $ABCDA$ and $AEFBA$ are *meshes.* A *branch,* such as AB or $BCDA$, connects two nodes together. Currents I_a and I_b are *mesh currents.* The downward *branch current* I_3 in AB is equal to $I_a - I_b$.

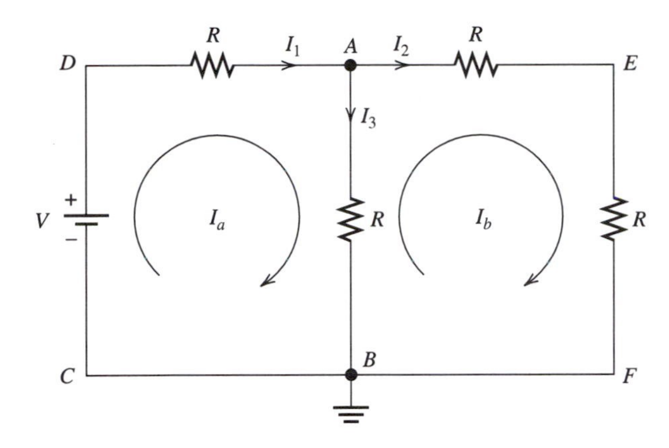

Figure 4-5 Two-mesh resistive circuit fed by a source supplying a steady voltage V.

The *Kirchhoff current law* is obvious: the algebraic sum of the currents entering a node is equal to zero. In the figure, $I_1 = I_2 + I_3$. If one uses mesh currents, rather than branch currents, this law is automatically satisfied.

The *Kirchhoff voltage law* is also obvious: the sum of the voltage drops and rises around a mesh is equal to zero.

These laws serve to calculate the voltages at the nodes and the currents in the branches.

EXAMPLE

Let us find the mesh currents I_a, I_b and the voltage V_A for the circuit of Fig. 4-5.

For the two meshes,

$$V - I_a R - (I_a - I_b)R = 0, \qquad -(I_b - I_a)R - 2I_b R = 0. \tag{4-39}$$

We have two equations and two unknowns, I_a and I_b. Solving,

$$I_a = 0.6V/R, \qquad I_b = 0.2V/R, \tag{4-40}$$

$$V_A = (I_a - I_b)R = 0.4V. \tag{4-41}$$

4.5 CONDUCTION

4.5.1 Conduction in a Steady Electric Field

For simplicity, we assume that the charge carriers are conduction electrons.

The detailed motion of an individual conduction electron is exceedingly complex because, every now and then, it collides with an atom and rebounds. The atoms, of course, vibrate about their equilibrium positions, because of thermal agitation, and exchange energy with the conduction electrons.

On average, each electron has a kinetic energy of $\frac{3}{2}kT$, where k is Boltzmann's constant and T is the temperature in kelvins. Thus, at room temperature, the speed v_{th} associated with thermal agitation is given by

$$\frac{mv_{th}^2}{2} = \frac{3}{2}kT = \frac{3}{2}(1.38 \times 10^{-23} \times 300) \approx 6 \times 10^{-21} \text{ joule,} \tag{4-42}$$

and

$$v_{th} \approx \left(\frac{12 \times 10^{-21}}{9.1 \times 10^{-31}}\right)^{1/2} \approx 10^5 \text{ meters/second.} \tag{4-43}$$

Under the action of a steady electric field, the cloud of conduction electrons drifts at a constant velocity $\boldsymbol{v}_d$ such that

$$\boldsymbol{J} = \sigma \boldsymbol{E} = -Ne\boldsymbol{v}_d, \tag{4-44}$$

where $\boldsymbol{v}_d$ points in the direction opposite to $\boldsymbol{J}$ and to $\boldsymbol{E}$, and N is the number of conduction electrons per cubic meter.

The drift speed is low. In copper, $N = 8.1 \times 10^{28}$ electrons/meter^3. If a current of 1 ampere flows through a wire having a cross section of 1 millimeter^2, $J = 10^6$ amperes/meter^2 and v_d works out to about 10^{-4} meter/second, or about 400 millimeters/*hour.* Then the drift speed is smaller than the thermal agitation speed by *nine* orders of magnitude!

In Eq. 4-44 v_d is small, but Ne is very large. In copper,

$$Ne = 8.1 \times 10^{28} \times 1.6 \times 10^{-19} \approx 10^{10} \text{ coulombs/meter}^3. \tag{4-45}$$

The low drift speed of conduction electrons is the source of many paradoxes. For example, a radio transmitting antenna is about 75 meters high and operates at about 1 megahertz. How can conduction electrons go from one end to the other and back in 1 microsecond? The answer is that they do not: they oscillate back and forth by a distance of the order of one atomic diameter, and that is enough to generate the required current.

Also, with such a low drift speed, how can you turn on a light, say one kilometer away, instantaneously? The explanation is that a wave propagates in the air, just outside the copper wires, at about the speed of light: the pair of wires acts as a waveguide. There is a longitudinal component of $\boldsymbol{E}$, both in the air and in the copper, that prods the conduction electrons one way or the other. The cloud of conduction electrons drifts very slowly, but it takes only a few microseconds before the cloud starts moving at the other end, one kilometer away. If the applied voltage is steady, the wave soon dies away and there remains inside the wire a constant electric field equal to $\boldsymbol{J}/\sigma$. If the applied voltage has a frequency of, say, 60 hertz, then the wave remains until you turn the voltage off.

The mean-free-path of a conduction electron between collisions is short: the time interval between collisions is of the order of 10^{-13} second, while the thermal speed is about 10^5 meters/second, as we saw above, so that the mean-free-path is of the order of 10^{-8} meter, or only about 100 atomic diameters. So the conduction electrons are "locked" to the crystal lattice. That fact is important because magnetic forces exerted on conduction electrons, as in an electric motor for example, transfer to the material.

4.5.2 The Mobility $\mathcal{M}$ of Conduction Electrons

The mobility of conduction electrons

$$\mathcal{M} = \frac{v_d}{E} = \frac{\sigma}{Ne} \tag{4-46}$$

is, by definition, a positive quantity.† We have used Eq. 4-44. The mobility is independent of $\boldsymbol{E}$ if Ohm's law applies. Thus

$$\sigma = Ne\mathcal{M} \tag{4-47}$$

where, as usual, we have taken e to be the *magnitude* of the electronic charge.

If the driving electric field is constant, then the drift velocity is constant. This means that the time-averaged net force on a conduction electron is zero, or that the average braking force due to the collisions just cancels the $-eE$ force exerted by the field.

What is the magnitude of this braking force? It is

$$-(-eE) = eE = \frac{ev_d}{\mathcal{M}} = -\frac{e}{\mathcal{M}}v_d, \tag{4-48}$$

from the definition of the mobility $\mathcal{M}$. The braking force and v_d point in opposite directions.

This situation is analogous to that of a body falling through water; after a while, the viscous force exactly cancels the gravitational force, the net force is zero, and the speed is constant.

The quantities N, $\mathcal{M}$, and σ for good conductors (gc) and for semiconductors (sc) are related as follows:

$$N_{\text{gc}} \gg N_{\text{sc}}, \qquad \sigma_{gc} \gg \sigma_{\text{sc}}, \qquad \mathcal{M}_{gc} \ll \mathcal{M}_{\text{sc}}. \tag{4-49}$$

4.5.3 Conduction in an Alternating Electric Field

If we disregard thermal agitation, there are two forces acting on a conduction electron: the driving force eE and the braking force of Eq. 4-48.

In an alternating electric field these two forces are unequal, and the equation of motion is

$$m^* \frac{dv_d}{dt} = -eE_m \exp j\omega t - \frac{e}{\mathcal{M}}v_d, \tag{4-50}$$

†Some authors assign to the mobility the sign of the charge carrier.

where m^* is the *effective mass*. This quantity takes collisions into account. As a rule, m^* is smaller than the mass of an isolated electron.

In silicon $m^* = 0.97m$, but in gallium arsenide (GaAs) the effective mass is only $0.07m$. Electron drift speeds *in solid-state devices* are of the order of 10^5 meters/second in silicon, and about 4 times larger in GaAs.

So electron *drift* speeds in solid-state devices are roughly *nine* orders of magnitude larger than in copper, or about equal to the *thermal* speeds at room temperature.

Replacing the time derivative by $j\omega$ and simplifying,

$$v_d = -\frac{\mathcal{M}}{1 + j\omega m^* \mathcal{M}/e} E_m \exp j\omega t. \tag{4-51}$$

Since $J = \sigma E = -Nev_d$,

$$\sigma = \frac{Ne\mathcal{M}}{1 + j\omega m^* \mathcal{M}/e}. \tag{4-52}$$

The conductivity is a complex number. This means that the alternating current is not in phase with the alternating voltage applied to the device.

If $\omega = 0$, we revert to Eq. 4-47. The above relation does *not* apply at frequencies of the order of 1 gigahertz or higher, where atomic phenomena become prominent.

For copper at room temperature,

$$\frac{\omega m^* \mathcal{M}}{e} = \frac{\omega m \sigma}{Ne^2} \tag{4-53}$$

$$= \frac{2\pi f \times 9.1 \times 10^{-31} \times 5.8 \times 10^7}{8.5 \times 10^{28} \times (1.6 \times 10^{-19})^2} \tag{4-54}$$

$$= 1.5 \times 10^{-13} f. \tag{4-55}$$

The imaginary term in Eq. 4-52 is negligible for $f \ll 7 \times 10^{12}$ hertz up to about 1 gigahertz the cloud of conduction electrons moves in phase with E, and $\sigma = Ne\mathcal{M}$.

Decreasing the temperature increases the mean-free-path of the carriers, which increases the mobility $\mathcal{M}$. At very low temperatures the conductivity of pure metals is imaginary:

$$\sigma = -j\frac{Ne^2}{m^* \omega}. \tag{4-56}$$

4.5.4 The Volume Charge Density ρ in a Conductor

1. Assume steady-state conditions and a homogeneous conductor. Then $\partial\rho/\partial t = 0$ and, from Sec. 4.2, $\nabla \cdot \boldsymbol{J} = 0$. If $\boldsymbol{J}$ is the conduction current density in a homogeneous conductor that satisfies Ohm's law $\boldsymbol{J} = \sigma \boldsymbol{E}$, then

$$\nabla \cdot \boldsymbol{J} = \nabla \cdot \sigma \boldsymbol{E} = \sigma \nabla \cdot \boldsymbol{E} = 0, \qquad \nabla \cdot \boldsymbol{E} = 0. \tag{4-57}$$

But the divergence of $\boldsymbol{E}$ is proportional to the volume charge density ρ, from Sec. 3.6. Thus, under steady-state conditions and in homogeneous conductors (σ independent of the coordinates), ρ is zero.

As a rule, the *surface* charge density on a conducting body carrying a current is not zero.

2. Now suppose that one injects a charge into a piece of copper by bombarding it with electrons. What happens to the charge density? In that case, from Sec. 4.2,

$$\boldsymbol{\nabla} \cdot \boldsymbol{J} = -\frac{\partial \rho}{\partial t}. \tag{4-58}$$

But, from Sec. 3.6,

$$\boldsymbol{\nabla} \cdot \boldsymbol{J} = \sigma \boldsymbol{\nabla} \cdot \boldsymbol{E} = \frac{\sigma \rho}{\epsilon_r \epsilon_0}, \tag{4-59}$$

where ϵ_r is relative permittivity of the material (Sec. 7.9). Thus

$$\frac{\partial \rho}{\partial t} = -\frac{\sigma \rho}{\epsilon_r \epsilon_0}, \qquad \rho = \rho_0 \exp\left(-\frac{\sigma t}{\epsilon_r \epsilon_0}\right), \tag{4-60}$$

and ρ decreases exponentially with time.

The relative permittivity ϵ_r of a good conductor is not measurable because conduction completely overshadows polarization (Sec. 7.2). One may presume that ϵ_r is of the order of 3, as in common dielectrics.

The inverse of the coefficient of t in the above exponent is the *relaxation time.*

We have neglected the fact that σ is frequency-dependent and is thus itself a function of the relaxation time. Relaxation times in good conductors are, in fact, short; in practice ρ may be set equal to zero. For example, the relaxation time for copper at room temperature is about 4×10^{-14} second, instead of $\approx 10^{-19}$ second, according to the above calculation.

3. In a homogeneous conductor carrying an alternating current, ρ is zero because Eq. 4-57 applies.

4. In a nonhomogeneous conductor carrying a current, ρ is not zero. For example, under steady-state conditions,

$$\boldsymbol{\nabla} \cdot \boldsymbol{J} = \boldsymbol{\nabla} \cdot (\sigma \boldsymbol{E}) = (\boldsymbol{\nabla} \sigma) \cdot \boldsymbol{E} + \sigma \boldsymbol{\nabla} \cdot \boldsymbol{E} = 0 \tag{4-61}$$

and

$$\boldsymbol{\nabla} \cdot \boldsymbol{E} = \frac{\rho}{\epsilon_r \epsilon_0} = -\frac{(\boldsymbol{\nabla} \sigma) \cdot \boldsymbol{E}}{\sigma}. \tag{4-62}$$

We have assumed that ϵ_r is uniform.

5. If the conductor moves in a magnetic field, then $\boldsymbol{J} = \sigma \boldsymbol{E}$ does not apply and a volume charge density can exist. See Secs. 17.5 and 17.6.

4.5.5 The Joule Effect

In the absence of an electric field, the cloud of conduction electrons remains in thermal equilibrium with the lattice of the host conductor. Upon application of an electric field, the electrons gain kinetic energy between collisions, and they share this extra energy with the lattice. The conductor thus heats up. This is the *Joule effect.*

What is the power gained by the conduction electrons? Consider a cube of the conductor, with side a. Apply a voltage V between opposite faces. The current is I. Then the power gained is VI, and the power dissipated as heat per cubic meter is

$$P' = \frac{VI}{a^3} = \left(\frac{V}{a}\right)\left(\frac{I}{a^2}\right) = EJ \tag{4-63}$$

$$= \sigma E^2 = \frac{J^2}{\sigma} \quad \text{watts/meter}^3. \tag{4-64}$$

If E and J are sinusoidal functions of the time,

$$P'_{av} = E_{rms} J_{rms} = \sigma E_{rms}^2 = \frac{J_{rms}^2}{\sigma}. \tag{4-65}$$

4.6 ISOLATED CONDUCTORS IN STATIC FIELDS

If one charges an isolated homogeneous conductor, the conduction electrons move about until they have reached their equilibrium positions and then, inside the conductor, there is zero $\boldsymbol{E}$.

It follows that (1) all points inside the conductor are at the same potential; (2) the volume charge density is zero, from Eq. 4-57; (3) any net static charge resides on the surface of the conductor; (4) $\boldsymbol{E}$ is normal at the surface of the conductor, for otherwise charges would flow along the surface; (5) just outside the surface, $E = \sigma_{ch}/\epsilon_0$, where σ_{ch} is the surface charge density, from Gauss's law.

Note the paradox: thanks to Gauss's law, one can express $\boldsymbol{E}$ at the surface of a conductor in terms of the *local* surface charge density alone, in spite of the fact that $\boldsymbol{E}$ depends on the magnitudes and positions of *all* the charges, whether they reside on the conductor or elsewhere.

What if the conductor is not homogeneous? For example, one might have a copper wire pressed onto a gold-plated terminal. Then, because of thermal agitation, some conduction electrons drift across the interface and establish a *contact potential,* usually of a fraction of a volt. The magnitude and sign of the contact potential depend on the nature of the materials.

EXAMPLE Hollow Conductor Enclosing a Charged Body

Figure 4-6 shows a cross section of a hollow conductor with a net electric charge Q within the cavity. The Gaussian surface lying within the conductor in a zero $\boldsymbol{E}$ encloses a zero net charge, because of Gauss's law. Then the surface charge on the *inside* surface of the conductor is $-Q$.

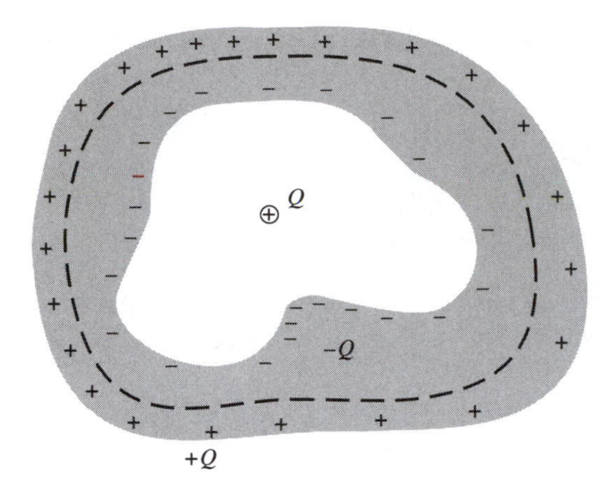

Figure 4-6 Section through a hollow conductor enclosing a body that carries a net charge Q. The dashed line is a section through a Gaussian surface lying entirely in the conducting material, where $E = 0$.

If the conductor carries a zero net charge, then the total charge on the *outside* surface is Q.

The surface charge density at a given point on the *outside* surface of the conductor is independent of the distribution of Q in the cavity. It is the same as if the conductor were solid and carried a net charge Q.

Inversely, the field inside the cavity is independent of the field outside the conductor. The conductor then acts as an *electrostatic shield.*

4.7 SUMMARY

Poisson's equation follows from the differential form of Gauss's law and from the relation $\boldsymbol{E} = -\boldsymbol{\nabla} V$:

$$\nabla^2 V = -\frac{\rho}{\epsilon_0}. \tag{4-1}$$

Setting the volume charge density ρ equal to zero yields *Laplace's equation:*

$$\nabla^2 V = 0. \tag{4-2}$$

The *law of conservation of electric charge* states that, whatever the circumstances, the net electric charge of a closed system is constant. Mathematically,

$$\boldsymbol{\nabla} \cdot \boldsymbol{J} = -\frac{\partial \rho}{\partial t}, \tag{4-26}$$

where $\boldsymbol{J}$ and ρ are, respectively, the current and charge densities.

Most conductors obey *Ohm's law*:

$$\boldsymbol{J} = \sigma \boldsymbol{E}, \tag{4-27}$$

where σ is the *conductivity,* expressed in siemens per meter.

The *Kirchhoff current law* states that the sum of the currents that enter a node, in an electric circuit, is equal to zero.

The *Kirchhoff voltage law* states that the sum of the voltage drops and rises around any closed path in a circuit is equal to zero.

In a time-independent electric field,

$$\boldsymbol{J} = \sigma \boldsymbol{E} = -Ne\boldsymbol{v}_d, \tag{4-44}$$

where N is the number of charge carriers per cubic meter, $-e$ is the charge on one of them, and $\boldsymbol{v}_d$ is their *drift velocity*. The *mobility* is defined by the equation

$$\mathcal{M} = \frac{|\boldsymbol{v}_d|}{E} = \frac{\sigma}{Ne} \tag{4-46}$$

and

$$\sigma = Ne\mathcal{M}. \tag{4-47}$$

In an alternating electric field the conductivity σ is complex:

$$\sigma = \frac{Ne\mathcal{M}}{1 + j\omega m^* \mathcal{M}/e}, \tag{4-52}$$

where m^* is the *effective mass* of a charge carrier.

If the current density is J and the electric field strength is E, then the time-averaged power dissipated per cubic meter in the form of heat is

$$P'_{av} = E_{rms} J_{rms} = \sigma E_{rms}^2 = \frac{J_{rms}^2}{\sigma} \qquad \text{watts/meter}^3. \tag{4-65}$$

This is the *Joule effect.*

Under static conditions there exists a net electric charge solely at the surface of a conductor, and $\boldsymbol{E}$ is zero inside. Just outside, $\boldsymbol{E}$ is normal to the surface, and its magnitude is σ_{ch}/ϵ_0, where σ_{ch} is the surface charge density.

PROBLEMS

4-1. (*4.4*) The potentiometer

Figure 4-7 shows a potentiometer circuit. Show that, when $I = 0$,

$$V_o = \frac{R_2}{R_1 + R_2} V_i.$$

This is a common type of circuit. It serves to measure a voltage, in this case V_o, without drawing current. In curve plotters the current I, after amplification, actuates a motor that displaces the pen and simultaneously moves the tap in the direction that decreases I. The resistances R_1 and R_2 act as a *potential divider.*

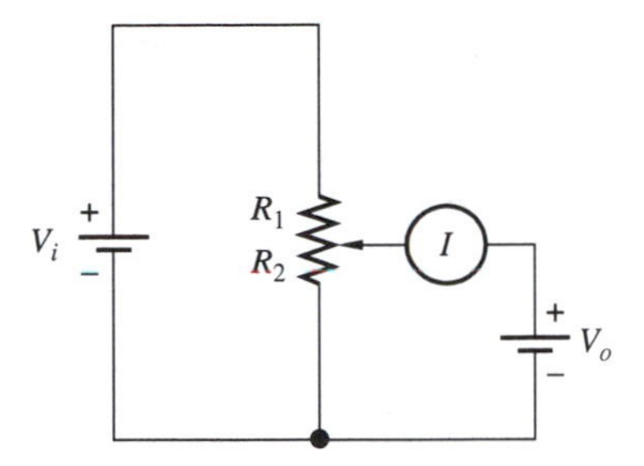

Figure 4-7

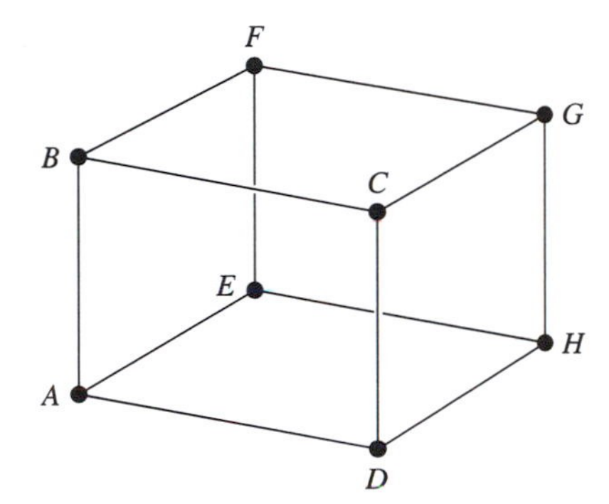

Figure 4-8

4-2. (*4.4*) Cube

Twelve equal resistances R form the edges of a cube, as in Fig. 4-8.

(a) A battery is connected between A and G.
Can you find one set of three points that are at the same potential?
Can you find another set?

(b) Now imagine that you have a sheet of copper connecting the first set, and another sheet of copper connecting the second set. This does not affect the currents in the resistors.
You can now show that

$$R_{AG} = (5/6)R.$$

4-3. (*4.5*) The conduction electron density in copper

The object of this problem is to illustrate the enormous magnitude of the electric charge densities in matter.

We take the example of the conduction electrons in copper. A copper atom contains 29 electrons, one of which is a conduction electron. Copper has an atomic weight of 64 and a density of 8.9×10^3 kilograms/meter^3. Suppose that you have two copper spheres, each one having a volume of 1 centimeter^3. The spheres are depleted of their conduction electrons and separated by a distance of 100 millimeters.

(a) Calculate the force of repulsion. Show that this force is equal to about 0.5% of the force of attraction between the sun and the earth. See the Table of physical constants on the page facing the back cover.

(b) Now calculate the energy stored in one sphere. That is of course the energy required to remove the conduction electrons.
(c) How much would this energy cost, at 10 cents per kilowatt-hour?
(d) To obtain this energy, you build a large nuclear reactor that can supply one gigawatt of electric power. How many years will it take to generate that energy?

4-4. (*4.5*) The drift speed of conduction electrons

Copper has an atomic weight of 64 and a density of 8.9×10^3 kilograms/meter^3.

(a) Calculate the number of atoms per cubic meter and the approximate diameter of an atom.
(b) Calculate the charge λ carried by the conduction electrons in 1 meter of copper wire 1 millimeter in diameter. There is one conduction electron per atom.
(c) Calculate the drift speed of the conduction electrons in meters per hour when the wire carries a current of 1 ampere.

4-5. (*4.5*) Conduction in a non-uniform medium

Two plane parallel copper electrodes are separated by a plate of thickness s whose conductivity σ varies linearly from σ_0, near the positive electrode, to $\sigma_0 + a$ near the negative electrode. Neglect edge effects. The current density is J.

Find the electric field strength in the conducting plate, as a function of the distance x from the positive electrode.

4-6. (*4.5*) Joule losses

A resistor has a resistance of 100 kilohms and a power rating of one-quarter watt. What is the maximum voltage that can be applied across it?

4-7. (*4.5*) Refraction of lines of E at the interface between media of different conductivities

We shall see in Sec. 8.2.3 that the tangential component of $\boldsymbol{E}$ is continuous at the interface between two media.

Show that, at the boundary between two media of conductivities σ_1 and σ_2, a line of $\boldsymbol{E}$, or a line of $\boldsymbol{J}$, is "refracted" in such a way that $\tan \theta_1/\sigma_1 = \tan \theta_2/\sigma_2$, where θ_1 and θ_2 are the angles formed by a line of $\boldsymbol{E}$ with the normal to the interface.

4-8. (*4.5*) The surface charge density at the interface between media of different conductivities

A current of density $\boldsymbol{J}$ flows in the direction normal to the interface between two media of conductivities σ_{co1} and σ_{co2}. The current flows from medium 1 to medium 2.

Show that the surface charge density σ_{ch} is $\epsilon_r\epsilon_0 J(1/\sigma_{co2} - 1/\sigma_{co1})$. Assume that ϵ_r has the same value on both sides. If the current is not normal to the interface, then the above J is the normal component of the current density.

4-9. (*4.5*) Conduction in a nonhomogeneous medium

In a nonhomogeneous medium, the conductivity σ is a function of the coordinates

Show that, under static conditions, or when $\boldsymbol{E} = -\boldsymbol{\nabla} V$, and if σ is nowhere equal to zero,

$$\nabla^2 V + \boldsymbol{\nabla} V \cdot \boldsymbol{\nabla}\tau = 0,$$

where $\tau = \ln \sigma$.

4-10. (*4.5*) The resistance of a spherical shell

A spherical shell of uniform conductivity σ has inner and outer radii R_1 and R_2, respectively. It has copper electrodes plated on the inner and outer surfaces.

Show that the resistance is $(1/R_1 - 1/R_2)/(4\pi\sigma)$.

4-11. (*4.5*) Resistive film

A square film of Nichrome, an alloy of nickel and chromium, has copper electrodes deposited on two opposite edges.

Show that the resistance between the electrodes depends only on the thickness of the film and on its conductivity, as long as the film is square. This *surface resistance* is expressed in *ohms per square.*

4-12. (*4.5*) A theorem on the resistance of a plate

A rectangular plate *ABCD* has a thickness s and a conductivity σ. With conducting electrodes on edges *AB* and *CD,* the resistance is R_1. With electrodes on *BC* and *DA,* the resistance is R_2.

Show that $R_1 R_2 = 1/(\sigma^2 s^2)$.

This equation also applies to any region bounded by equipotentials and lines of current flow.

4-13. (*4.5*) E and J inside a battery

A battery feeds a resistance R as in Fig. 4-9. The battery acts as a pump, forcing conduction electrons toward the negative electrode. The battery is cylindrical, of length s and cross-sectional area $\mathcal{A}$, with electrodes at each end. Then $|\nabla V| = V/s$ and, inside the battery, $J = \sigma(E_p - |\nabla V|)$, where E_p is the "pumping field."

Find the output voltage as a function of the current. Set $R = s/(\sigma \mathcal{A})$ as the output resistance of the battery.

4-14. (*4.5*) Mobility and electron drift

A simple model for the drift of a conduction electron is the following: the electron describes a ballistic trajectory for a while, under the action of the ambient electric field, and then the electron suffers an impact. Its velocity just after the impact is unrelated to its velocity before the impact, and we set it equal to zero. The electron then starts out on another ballistic trajectory, and the process repeats itself. Let the mean time between the collisions be Δt and the effective mass (Sec. 4.5.3) be m^*.

Find the mobility in terms of Δt and m^*.

See Prob. 4-4.

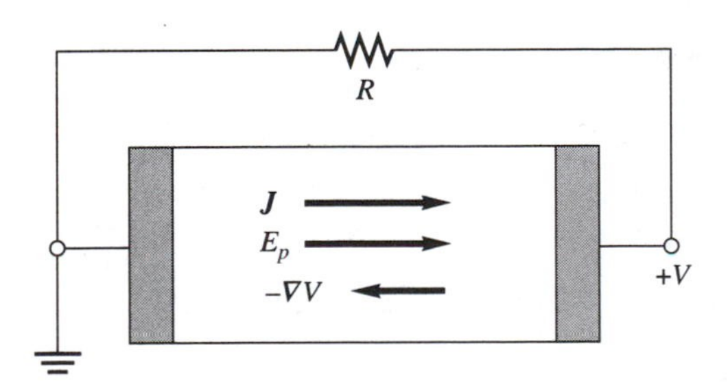

Figure 4-9

4-15. (*4.5*) Conduction by holes

We derived Eq. 4-50 on the assumption that the charge carriers are electrons. Suppose the carriers are holes. Then the charge changes sign, and both terms on the right are positive. If that is so, the complementary function, which one obtains on disregarding the forcing term $eE_m \exp j\omega t$, is

$$v_d = v_{d0} \exp\left(\frac{e}{m^* \mathcal{M}}\right) t.$$

Then v_d increases exponentially with time, which is absurd.

Show that, if the charge carriers are holes, then

$$m^* \frac{dv_d}{dt} = +eE_m \exp j\omega t - \frac{e}{\mathcal{M}} v_d,$$

and that

$$v_d = +\frac{\mathcal{M}}{1 + j\omega m^* \mathcal{M}/e} E_m \exp j\omega t, \qquad \sigma = \frac{Ne\mathcal{M}}{1 + j\omega m^* \mathcal{M}/e}.$$

4-16. (*4.5*) The resistojet

Figure 4-10 shows the principle of operation of a *resistojet* used as a thruster for correcting the trajectory or the attitude of a satellite. Hydrogen enters at *P*.

Assuming complete conversion of the electric energy to kinetic energy, calculate the thrust for a power input of 3 kilowatts and a flow of 0.6 gram of hydrogen per second.

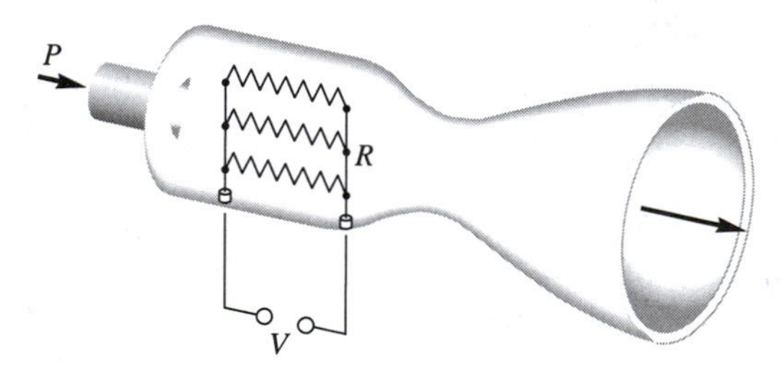

Figure 4-10

CHAPTER

5

ELECTRIC FIELDS III

Electric Multipoles

Electric multipoles are sets of point charges possessing certain symmetries. Their interest lies in the fact that real charged objects, such as antennas and atomic nuclei, possess electric fields that may be expressed as sums of multipole fields.

Aside from the monopole, which is a single point charge, the most useful type of multipole is the dipole, which consists of two charges of equal magnitudes and opposite signs, some distance apart. Most molecules act as small dipoles. Also many antennas radiate like oscillating electric dipoles.

5.1 THE ELECTRIC DIPOLE

The *electric dipole* is a common type of charge distribution. We return to it later in this chapter and in Chap. 25.

†Asterisks indicate material that can be omitted without losing continuity. However, the Legendre polynomials of Sec. 5.4.1 are required for Sec. 10.1.

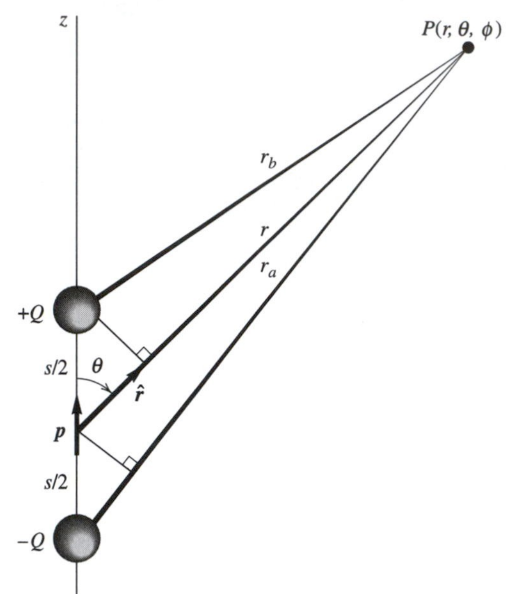

Figure 5-1 Two charges $+Q$ and $-Q$ separated by a distance s and forming a dipole. The dipole moment is $\boldsymbol{p}$. We calculate the potential at point P by summing the potentials of the two charges.

The electric dipole consists of two charges, one positive and one negative, of the same magnitude, separated by a distance s. We find V and $\boldsymbol{E}$ at a point P situated at a distance $r \gg s$, as in Fig. 5-1. At P,

$$V = \frac{Q}{4\pi\epsilon_0}\left(\frac{1}{r_b} - \frac{1}{r_a}\right), \tag{5-1}$$

where, if $r \gg s$,

$$r_a = r + \frac{s}{2}\cos\theta, \qquad \frac{r_a}{r} = 1 + \frac{s}{2r}\cos\theta. \tag{5-2}$$

To find the inverse of this ratio, recall the series

$$(1 \pm x)^n = 1 \pm nx + \frac{n(n-1)}{2 \times 1}x^2 \pm \frac{n(n-1)(n-2)}{3 \times 2 \times 1}x^3 + \cdots. \tag{5-3}$$

Setting $n = -1$, with $x = s\cos\theta/(2r) \ll 1$, we can keep only the first two terms and

$$\frac{r}{r_a} \approx 1 - \frac{s}{2r}\cos\theta. \tag{5-4}$$

Similarly, since

$$\frac{r_b}{r} = 1 - \frac{s}{2r}\cos\theta, \tag{5-5}$$

$$\frac{r}{r_b} = 1 + \frac{s}{2r}\cos\theta, \tag{5-6}$$

and

$$V = \frac{Qs}{4\pi\epsilon_0 r^2}\cos\theta \qquad (r^3 \gg s^3). \tag{5-7}$$

Note that the potential in the field of a dipole falls off as $1/r^2$, whereas the potential of a single point charge varies only as $1/r$. This is because the charges of a dipole appear close together for an observer some distance away, so their fields cancel more and more as the distance r increases.

The *dipole moment* $\boldsymbol{p} = Q\boldsymbol{s}$ is a vector that is directed from the negative to the positive charge. Then

$$V = \frac{\boldsymbol{p}\cdot\hat{\boldsymbol{r}}}{4\pi\epsilon_0 r^2}. \tag{5-8}$$

We can now find the electric field strength $\boldsymbol{E}$. In spherical coordinates,

$$E_r = -\frac{\partial V}{\partial r} = \frac{2p\cos\theta}{4\pi\epsilon_0 r^3}, \tag{5-9}$$

$$E_\theta = -\frac{1}{r}\frac{\partial V}{\partial\theta} = \frac{p\sin\theta}{4\pi\epsilon_0 r^3}, \tag{5-10}$$

$$E_\phi = -\frac{1}{r\sin\theta}\frac{\partial V}{\partial\phi} = 0, \tag{5-11}$$

$$\boldsymbol{E} = \frac{p}{4\pi\epsilon_0 r^3}(2\cos\theta\hat{\boldsymbol{r}} + \sin\theta\hat{\boldsymbol{\theta}}). \tag{5-12}$$

Thus E for a dipole falls off as the *cube* of the distance r.

Figure 5-2 shows lines of $\boldsymbol{E}$ and *equipotential lines* for an electric dipole. Rotating equipotential lines about the vertical axis generates *equipotential surfaces.*

More generally, the dipole moment of a charge distribution is

$$\boldsymbol{p} = \int_{v'} \boldsymbol{r}'\rho\,dv' = Q\left(\int_{v'} \boldsymbol{r}'\rho\,dv' \Big/ \int_{v'} \rho\,dv'\right) = Q\bar{\boldsymbol{r}}', \tag{5-13}$$

where Q is the net total charge occupying the volume v', and $\bar{\boldsymbol{r}}'$ defines the position of the *center of charge,* by analogy with the center of mass in mechanics.

If $Q = 0$, then $\bar{\boldsymbol{r}}' \to \infty$ and $Q\bar{\boldsymbol{r}}'$ is indeterminate. However, the integral of $\boldsymbol{r}'\rho\,dv'$ still provides the correct value of $\boldsymbol{p}$. If $Q = 0$, the dipole moment is independent of the choice of the origin (Prob. 5-1).

If $Q \neq 0$, the dipole moment of the distribution is zero when the origin is at the center of charge, for then $\bar{\boldsymbol{r}}' = 0$.

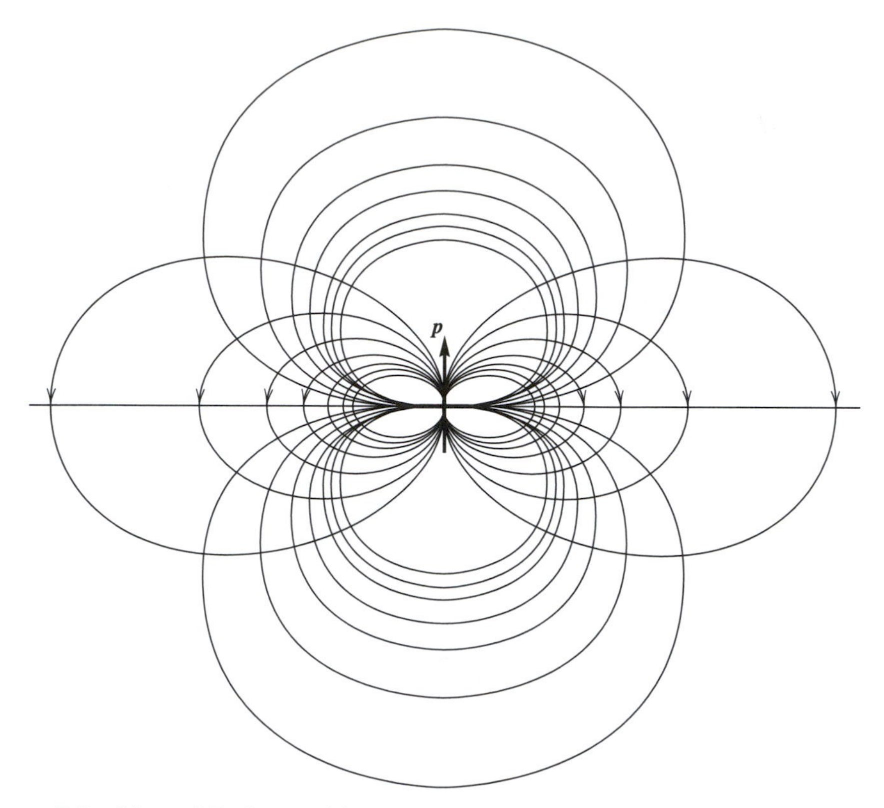

Figure 5-2 Lines of $\boldsymbol{E}$, shown with arrows, and equipotential lines for the electric dipole of Fig. 5-1. In the central region, the lines become too close together to be shown.

*5.2 THE LINEAR ELECTRIC QUADRUPOLE

The *linear electric quadrupole* is a set of three charges, as in Fig. 5-3. The separation s is again small compared to the distance r to the point P. At P,

$$V = \frac{1}{4\pi\epsilon_0}\left(\frac{Q}{r_a} - \frac{2Q}{r} + \frac{Q}{r_b}\right) = \frac{Q}{4\pi\epsilon_0 r}\left(\frac{r}{r_a} + \frac{r}{r_b} - 2\right). \tag{5-14}$$

We are running into trouble here: if we substitute Eqs. 5-4 and 5-6, after replacing $s/2$ by s, we find that $V = 0$. The explanation is that those approximations, which are based on Eqs. 5-2 and 5-5, are not good enough for use with the quadrupole. Here is a better approximation.

For the lower triangle in Fig. 5-3,

$$r_a^2 = r^2 + s^2 + 2rs\cos\theta. \tag{5-15}$$

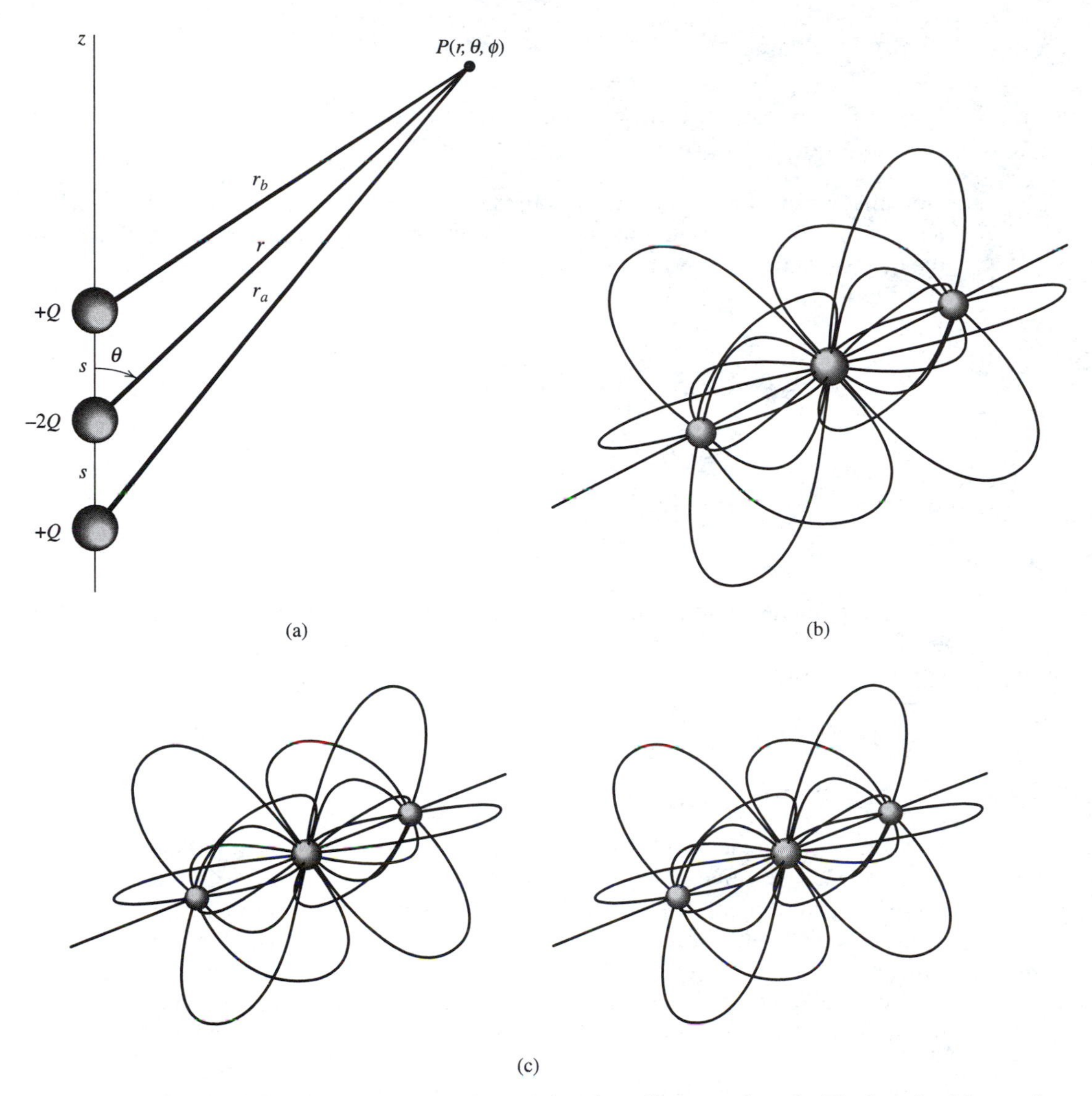

Figure 5-3 (a) Charges $+Q$, $-2Q$, $+Q$ forming a linear quadrupole. The length of the quadrupole is $2s$. (b) Electric field lines. (c) Stereo view of the field lines. See the footnote to Fig. 3-2.

Divide both sides by r^2, and take the inverse:

$$\frac{r}{r_a} = \left(1 + \frac{s^2}{r^2} + \frac{2s}{r}\cos\theta\right)^{-1/2} \tag{5-16}$$

$$= 1 - \frac{1}{2}\left(\frac{s^2}{r^2} + \frac{2s}{r}\cos\theta\right) + \frac{3}{8}\left(\frac{s^2}{r^2} + \frac{2s}{r}\cos\theta\right)^2 - \cdots. \tag{5-17}$$

Neglecting terms of the order of $(s/r)^3$ and higher,

$$\frac{r}{r_a} = 1 - \frac{s}{r}\cos\theta + \frac{s^2}{r^2}\left(\frac{3\cos^2\theta - 1}{2}\right). \tag{5-18}$$

Note that we are now taking into account terms of the order of $(s/r)^2$. So this is a better approximation. Indeed, it is only the third term of the series that is meaningful for the quadrupole.

Similarly,

$$\frac{r}{r_b} = 1 + \frac{s}{r}\cos\theta + \frac{s^2}{r^2}\frac{(3\cos^2\theta - 1)}{2}, \tag{5-19}$$

and

$$V = \frac{2Qs^2}{4\pi\epsilon_0 r^3}\frac{(3\cos^2\theta - 1)}{2} \qquad (r^3 \gg s^3). \tag{5-20}$$

The potential V of a linear electric quadrupole varies as $1/r^3$, whereas E, calculated as for the dipole, varies as $1/r^4$. The fields of the three charges cancel almost completely for $r \gg s$.

*5.3 ELECTRIC MULTIPOLES

It is possible to extend the concept of dipole and quadrupole to larger numbers of positive and negative charges. Such charge arrangements are known as *multipoles.* A single point charge is a *monopole.* A *dipole* is obtained by displacing a monopole through a small distance s_1 and replacing the original monopole by another of the same magnitude but of opposite sign. Likewise, a *quadrupole* is obtained by displacing a dipole by a small distance s_2 and then replacing the original dipole by one of equal magnitude but of opposite sign. For the linear quadrupole, $s_2 = s_1$.

The multipole concept can extend indefinitely. For example, the quadrupole can be displaced by a small distance s_3, and the original quadrupole replaced by one in which the signs of all the charges have been changed. This gives an *octupole.* A 2^l-pole requires l displacements $s_1, s_2, \ldots, s_l$.

We have seen that the dipole potential varies as $1/r^2$ and that the quadrupole potential varies as $1/r^3$. For the 2^l-pole, V varies as $1/r^{l+1}$ and E as $1/r^{l+2}$.

We have calculated the potential V in the field of a dipole and of a quadrupole in the following way. We found the sum of the potentials of the individual charges, expanded the sum as a power series, and then truncated. This approach is straightforward, and it has the advantage of showing exactly what approximations are involved. Problems 5-5 and 5-6 explore a different method that is more elegant, but that does not reveal the exact nature of the approximations.

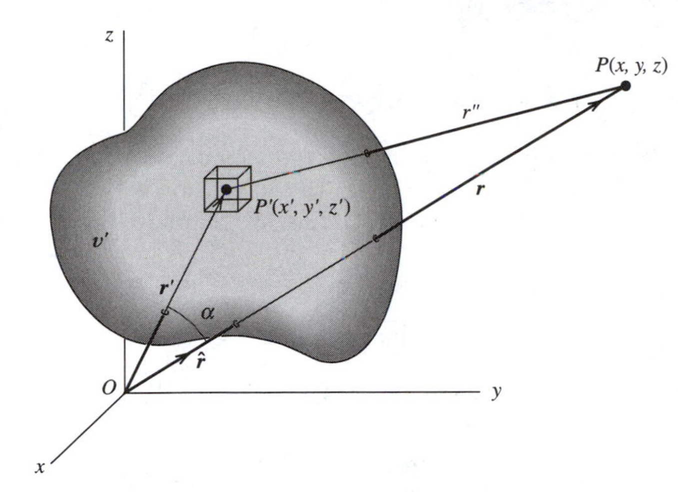

Figure 5-4 Arbitrary charge distribution enclosed within a volume v'. The potential at P is the same as if one had, at the origin, a monopole, plus a dipole, plus a quadrupole, plus an octupole, etc. However, r must be larger than the maximum value of r'.

*5.4 THE ELECTRIC FIELD OUTSIDE A CHARGE DISTRIBUTION, EXPANDED IN TERMS OF MULTIPOLES

A charge distribution of density $\rho(x', y', z')$ occupies a volume v' and extends to a maximum distance r'_{max} from an arbitrary origin O, as in Fig. 5-4. We select O either within the volume or close to it.

We shall see that the potential V at a point P *outside* the charge distribution such that $r > r'_{max}$ is the same as (1) the potential V_1 of a point charge, or monopole, equal to the net charge of the distribution, plus (2) the potential V_2 of a point dipole with a dipole moment equal to that of the charge distribution, plus (3) the potential V_3 of a point quadrupole with a quadrupole moment equal to that of the charge distribution, and so on, the monopole, dipole, quadrupole, etc., being all located at the arbitrary origin.

Similarly, the $\boldsymbol{E}$ at point P is the sum of the $\boldsymbol{E}$'s of the above monopole, dipole, quadrupole, etc.

If $Q = 0$, then V_2 is independent of the choice of origin. More generally, V_l is independent of the choice of origin if all the multipole moments up to the 2^{l-1}-pole are zero.

Since the V of a monopole decreases as $1/r$, that of a dipole as $1/r^2$, that of a quadrupole as $1/r^3$, etc., then a long distance away, where $r \gg r'_{max}$, the field of any charge distribution is simple. It is that of a point charge at the origin and $V \approx V_1$. Closer in, $V \approx V_1 + V_2$. Still closer in, V_3 becomes discernible, then V_4, etc., and the field becomes more and more complex.

*5.4.1 The Value of V. The Legendre Polynomials†

We wish to find V at some point P such that $r > r'_{max}$. From Fig. 5-4, this is

$$V = \int_{v'} \frac{\rho \, dv'}{4\pi\epsilon_o r''}, \tag{5-21}$$

†The Legendre polynomials are required for Sec. 10.1.

where

$$r'' = |\mathbf{r} - \mathbf{r}'| = [(x - x')^2 + (y - y')^2 + (z - z')^2]^{1/2}. \tag{5-22}$$

The point $P(x, y, z)$ is fixed. Thus r'' is a function of x', y', z', and we can expand $1/r''$ as a Taylor series near the origin. This expansion will lead us to a powerful method for the analysis of axisymmetric fields. We return to it in Chapter 10.

Let

$$w(u) = \frac{1}{|u\mathbf{r}' - \mathbf{r}|}. \tag{5-23}$$

The factor u is dimensionless. We require $w(1)$, which is $1/r''$:

$$w(1) = \frac{1}{r''} = w(0) + \left(\frac{\partial w}{\partial u}\right)_{u=0} + \frac{1}{2!}\left(\frac{\partial^2 w}{\partial u^2}\right)_{u=0} + \cdots, \tag{5-24}$$

where $w(0) = 1/r$. To calculate the derivatives on the right-hand side, we require the partial derivative of $|u\mathbf{r}' - \mathbf{r}|$ with respect to u:

$$\frac{\partial}{\partial u}|u\mathbf{r}' - \mathbf{r}| = \frac{\partial}{\partial u}[(ux' - x)^2 + (uy' - y)^2 + (uz' - z)^2]^{1/2} \tag{5-25}$$

$$= \frac{1}{2|u\mathbf{r}' - \mathbf{r}|}(2ux'^2 - 2x'x + 2uy'^2 - 2y'y + 2uz'^2 - 2z'z) \tag{5-26}$$

$$= \frac{ur'^2 - \mathbf{r}' \cdot \mathbf{r}}{|u\mathbf{r}' - \mathbf{r}|} = \frac{(u\mathbf{r}' - \mathbf{r}) \cdot \mathbf{r}'}{|u\mathbf{r}' - \mathbf{r}|}. \tag{5-27}$$

Thus

$$\frac{\partial w}{\partial u} = -\frac{1}{|u\mathbf{r}' - \mathbf{r}|^2}\frac{(u\mathbf{r}' - \mathbf{r}) \cdot \mathbf{r}'}{|u\mathbf{r}' - \mathbf{r}|} = -\frac{(u\mathbf{r}' - \mathbf{r}) \cdot \mathbf{r}'}{|u\mathbf{r}' - \mathbf{r}|^3} \tag{5-28}$$

and

$$\left(\frac{\partial w}{\partial u}\right)_{u=0} = \frac{\mathbf{r} \cdot \mathbf{r}'}{r^3} = \frac{\hat{\mathbf{r}} \cdot \mathbf{r}'}{r^2} = \frac{r' \cos\alpha}{r^2}, \tag{5-29}$$

where $\hat{\mathbf{r}}$ and α are as in Fig. 5-4. Also,

$$\frac{\partial^2 w}{\partial u^2} = \frac{3[(u\mathbf{r}' - \mathbf{r}) \cdot \mathbf{r}']^2}{|u\mathbf{r}' - \mathbf{r}|^5} - \frac{r'^2}{|u\mathbf{r}' - \mathbf{r}|^3} \tag{5-30}$$

and

$$\left(\frac{\partial^2 w}{\partial u^2}\right)_{u=0} = \frac{3(\mathbf{r}' \cdot \mathbf{r})^2}{r^5} - \frac{r'^2}{r^3} = \frac{r'^2(3\cos^2\alpha - 1)}{r^3}. \tag{5-31}$$

Table 5-1 Legendre polynomials

n	$P_n(\cos\alpha)$
0	1
1	$\cos\alpha$
2	$\dfrac{3\cos^2\alpha - 1}{2}$
3	$\dfrac{5\cos^3\alpha - 3\cos\alpha}{2}$
4	$\dfrac{35\cos^4\alpha - 30\cos^2\alpha + 3}{8}$
5	$\dfrac{63\cos^5\alpha - 70\cos^3\alpha + 15\cos\alpha}{8}$

In general,

$$\left(\frac{\partial^n w}{\partial u^n}\right)_{u=0} = \frac{r'^n}{r^{n+1}} n!\, P_n(\cos\alpha), \tag{5-32}$$

where

$$P_n(\cos\alpha) = \frac{1}{2^n n!} \frac{d^n}{d(\cos\alpha)^n} (\cos^2\alpha - 1)^n \tag{5-33}$$

is a *Legendre polynomial.* Table 5-1 gives the first five *Legendre polynomials,* while Fig. 5-5 shows the first four as functions of the angle α. For any α,

$$|P_n(\cos\alpha)| \le 1. \tag{5-34}$$

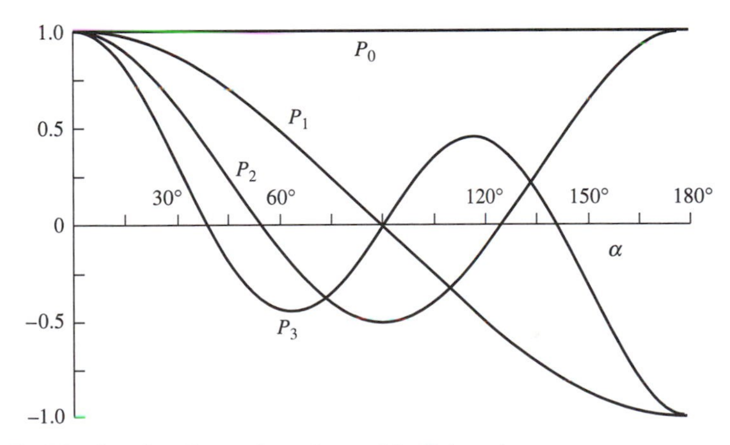

Figure 5-5 The first four Legendre polynomials $P_n(\cos\alpha)$.

Thus from Eq. 5-24,

$$\frac{1}{r''} = \frac{1}{r} + \frac{r'\cos\alpha}{r^2} + \frac{r'^2(3\cos^2\alpha - 1)}{2r^3} + \cdots = \sum_{n=0}^{\infty} \frac{1}{r^{n+1}} r'^n P_n(\cos\alpha). \tag{5-35}$$

Since $|P_n(\cos\alpha)| \leq 1$ as above, this series converges for $r'_{max} < r$.

Substituting in Eq. 5-21, we finally have that

$$V = \frac{1}{4\pi\epsilon_0 r}\int_{v'} \rho\, dv' + \frac{1}{4\pi\epsilon_0 r^2}\int_{v'} r'\cos\alpha\,\rho\, dv'$$
$$+ \frac{1}{4\pi\epsilon_0 r^3}\int_{v'} r'^2\,\frac{3\cos^2\alpha - 1}{2}\,\rho\, dv'$$
$$+ \frac{1}{4\pi\epsilon_0 r^4}\int_{v'} r'^3\,\frac{5\cos^3\alpha - 3\cos\alpha}{2}\,\rho\, dv' \tag{5-36}$$
$$+ \frac{1}{4\pi\epsilon_0 r^5}\int_{v'} r'^4\,\frac{35\cos^4\alpha - 30\cos^2\alpha + 3}{8}\,\rho\, dv' + \cdots$$
$$= V_1 + V_2 + V_3 + V_4 + V_5 + \cdots. \tag{5-37}$$

Let us examine the first three terms in succession.

*5.4.2 The Monopole Term

The first term is the V that one would have at P if the whole charge were concentrated at the arbitrary origin:

$$V_1 = \frac{Q}{4\pi\epsilon_0 r}, \tag{5-38}$$

where Q is the net charge in the distribution. This is the *monopole* term. It is zero if the net charge is zero. Its value depends on the position chosen for the origin.

*5.4.3 The Dipole Term

The second term varies as $1/r^2$, like the electric potential of a dipole. From Eq. 5-36

$$V_2 = \frac{1}{4\pi\epsilon_0 r^2}\int_{v'} r'\cos\alpha\,\rho\, dv' = \frac{\hat{\boldsymbol{r}}}{4\pi\epsilon_0 r^2}\cdot\int_{v'} \boldsymbol{r}'\rho\, dv', \tag{5-39}$$

where the integral on the right is the *dipole moment* of the charge distribution:

$$\boldsymbol{p} = \int_{v'} \boldsymbol{r}'\rho\, dv'. \tag{5-40}$$

Thus

$$V_2 = \frac{\boldsymbol{p} \cdot \hat{\boldsymbol{r}}}{4\pi\epsilon_0 r^2}, \tag{5-41}$$

as in Eq. 5-8.

*5.4.4 The Quadrupole Term

Now consider the term V_3 of Eq. 5-36. It involves a $1/r^3$ factor, like the V of the linear quadrupole of Sec. 5.2. If we calculate V_3 for the linear quadrupole with charges Q, $-2Q$, and Q at $z = s$, 0, and $-s$, respectively, we find that it is equal to the V of Eq. 5-20. Then V_3 is the potential of a small quadrupole at the origin. Now

$$V_3 = \frac{1}{4\pi\epsilon_0 r^3}\int_{v'} \frac{1}{2} r'^2 (3\cos^2\alpha - 1)\,\rho\, dv'$$

$$= \frac{1}{4\pi\epsilon_0 r^3}\int_{v'} \frac{1}{2}[3(\hat{\boldsymbol{r}} \cdot \boldsymbol{r}')^2 - r'^2]\,\rho\, dv'. \tag{5-42}$$

Set

$$\hat{\boldsymbol{r}} = l\hat{\boldsymbol{x}} + m\hat{\boldsymbol{y}} + n\hat{\boldsymbol{z}}, \tag{5-43}$$

where l, m, n are the *direction cosines* of $\hat{\boldsymbol{r}}$ and where

$$l^2 + m^2 + n^2 = 1. \tag{5-44}$$

Then expand and group terms. This yields

$$V_3 = \frac{1}{4\pi\epsilon_0 r^3}\Bigg(3mn\int_{v'} y'z'\rho\, dv' + 3nl\int_{v'} z'x'\rho\, dv' + 3lm\int_{v'} x'y'\rho\, dv' + \frac{3l^2 - 1}{2}\int_{v'} x'^2\rho\, dv' + \frac{3m^2 - 1}{2}\int_{v'} y'^2\rho\, dv' + \frac{3n^2 - 1}{2}\int_{v'} z'^2\rho\, dv'\Bigg). \tag{5-45}$$

These integrals, like the integral of Eq. 5-40, depend solely on the distribution of electric charge within v', and not on the coordinates x, y, z of the field point P. They specify the nine components of the quadrupole moment of the charge distribution:

$$p_{xx} = \int_{v'} x'^2 \rho\, dv' = Q\overline{x'^2}, \tag{5-46}$$

$$p_{yy} = \int_{v'} y'^2 \rho \, dv' = Q\overline{y'^2}, \tag{5-47}$$

$$p_{zz} = \int_{v'} z'^2 \rho \, dv' = Q\overline{z'^2}, \tag{5-48}$$

$$p_{yz} = p_{zy} = \int_{v'} y'z' \rho \, dv' = Q\overline{y'z'}, \tag{5-49}$$

$$p_{zx} = p_{xz} = \int_{v'} z'x' \rho \, dv' = Q\overline{z'x'}, \tag{5-50}$$

$$p_{xy} = p_{yx} = \int_{v'} x'y' \rho \, dv' = Q\overline{x'y'}. \tag{5-51}$$

The bar indicates, as usual, an average value. Thus

$$V_3 = \frac{1}{4\pi\epsilon_0 r^3}\left(3mnp_{yz} + 3nlp_{zx} + 3lmp_{xy} + \frac{3l^2-1}{2}p_{xx} + \frac{3m^2-1}{2}p_{yy} + \frac{3n^2-1}{2}p_{zz}\right). \tag{5-52}$$

If the charge distribution is symmetrical about the z-axis,

$$p_{yz} = p_{zx} = p_{xy} = 0, \qquad p_{xx} = p_{yy}. \tag{5-53}$$

It is then convenient to define a single quantity

$$\mathcal{Q} = 2(p_{zz} - p_{xx}) \tag{5-54}$$

that is also called the *quadrupole moment* of the charge distribution.

Because $l^2 + m^2 + n^2 = 1$,

$$V_3 = \frac{\mathcal{Q}}{16\pi\epsilon_0 r^3}(3n^2 - 1) = \frac{\mathcal{Q}}{16\pi\epsilon_0 r^3}(3\cos^2\theta - 1) \tag{5-55}$$

at the point r, θ, ϕ in spherical coordinates.

We can, of course, deduce the electric field strength from the relation $\boldsymbol{E} = -\nabla V$.

EXAMPLE The Quadrupole Moment of the Deuteron

The charge distribution in a deuteron (the nucleus of an atom of deuterium, comprising one proton and one neutron) has the shape of a prolate spheroid. (Rotating an ellipse about its *major* axis generates a *prolate* spheroid, while rotation about its *minor* axis generates an *oblate* spheroid.) The deuteron is therefore slightly elongated along one axis. The "radius" of the deuteron is 1.963 femtometer, or 1.963×10^{-15} meter, and its quadrupole moment is +0.28590 femtometer squared times the charge of the electron. See Eq. 5-54: the plus sign means that the spheroid is prolate; a negative sign would mean that the spheroid is oblate.

EXAMPLE The Field of a Set of Six Point Charges Set Symmetrically About the Origin

Figure 5-6 shows six charges. A point P in space is at a distance $r > a$ from the origin. The potential at P is given by the series of Eq. 5-37. Let us calculate the first three terms. From Eq. 5-38

$$V_1 = \frac{12Q}{4\pi\epsilon_0 r}. \tag{5-56}$$

From Eqs. 5-40 and 5-41,

$$\boldsymbol{p} = \sum \boldsymbol{r}' Q = 0, \qquad V_2 = 0. \tag{5-57}$$

From Eqs. 5-46 to 5-51,

$$p_{xx} = 2a^2 Q, \qquad p_{yy} = 2a^2 Q, \qquad p_{zz} = 8a^2 Q, \tag{5-58}$$

$$p_{yz} = p_{zx} = p_{xy} = 0. \tag{5-59}$$

Finally, from Eq. 5-52,

$$V_3 = \frac{1}{4\pi\epsilon_0 r^3}[(3l^2 - 1)a^2 Q + (3m^2 - 1)a^2 Q + (3n^2 - 1)4a^2 Q] \tag{5-60}$$

$$= \frac{a^2 Q}{4\pi\epsilon_0 r^3}(3l^2 + 3m^2 + 12n^2 - 6) \tag{5-61}$$

$$= \frac{a^2 Q}{4\pi\epsilon_0 r^3}(9n^2 - 3) = \frac{3a^2 Q}{4\pi\epsilon_0 r^3}(3n^2 - 1), \tag{5-62}$$

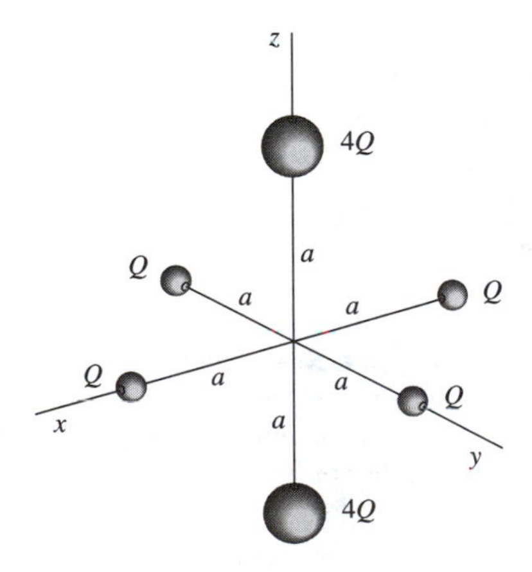

Figure 5-6 Six charges arranged symmetrically about the origin.

and

$$\mathcal{Q} = 12a^2Q. \tag{5-63}$$

It turns out that $V_4 = 0$, and that

$$V_5 = \frac{Qa^4}{16\pi\epsilon_0 r^5}[35(l^4 + m^4 + 4n^4) - 90n^2 - 12]. \tag{5-64}$$

At P,

$$V = V_1 + V_3 + V_5 + \cdots \tag{5-65}$$

$$= \frac{6Q}{4\pi\epsilon_0 r}\left\{2 + \frac{3n^2 - 1}{2}\frac{a^2}{r^2} + \frac{1}{24}[35(l^4 + m^4 + 4n^4) - 90n^2 - 12]\frac{a^4}{r^4} + \cdots\right\}. \tag{5-66}$$

Since the direction cosines l, m, n are each at most of the order of unity, with $l^2 + m^2 + n^2 = 1$, the coefficient of a^2/r^2 is also of the order of unity, and at $r \geq 10a$ the series reduces to its first term with an error of at most 1%. The terms V_3, V_5, . . . become progressively more prominent as r/a decreases.

Figure 5-7 shows an equipotential surface for this charge distribution.

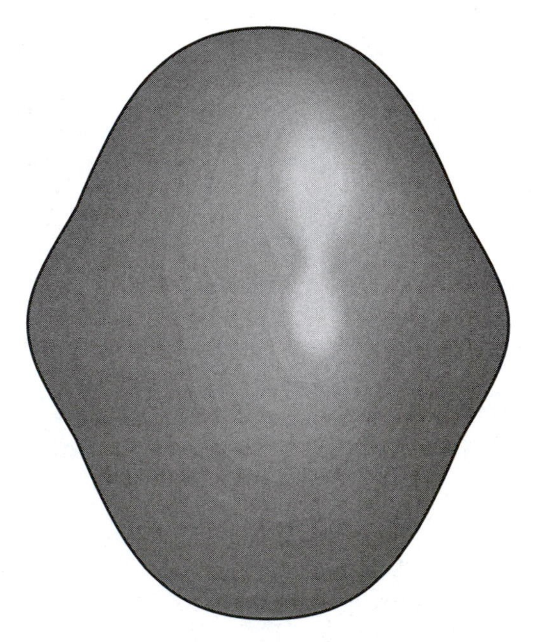

Figure 5-7 Equipotential surface for the charge distribution of Fig. 5-6. The value of V is the potential at the point $(0, 0, 1.8a)$. Further out, the equipotential surface is nearly spherical.

*5.5 SUMMARY

The *electric dipole* consists of a pair of charges of equal magnitude Q but of opposite signs, separated by a distance $\boldsymbol{s}$. Its *dipole moment* $\boldsymbol{p}$ is $Q\boldsymbol{s}$, a vector directed from the negative to the positive charge.

At a distance r from the dipole,

$$V = \frac{\boldsymbol{p} \cdot \hat{\boldsymbol{r}}}{4\pi\epsilon_0 r^2}, \qquad (r^3 \gg s^3). \tag{5-8}$$

The *linear electric quadrupole* consists of four charges Q, $-2Q$, Q, as in Fig. 5-3. Then

$$V = \frac{2Qs^2}{4\pi\epsilon_0 r^3} \frac{(3\cos^2\theta - 1)}{2}, \qquad (r^3 \gg s^3). \tag{5-8}$$

where $\cos\theta = \hat{\boldsymbol{z}} \cdot \hat{\boldsymbol{r}}$.

Some distance outside a charge distribution the potential can be written as a series

$$V = V_1 + V_2 + V_3 + V_4 + V_5 + \cdots. \tag{5-8}$$

Here V_1 is the potential at P due to a single charge, called a *monopole,* equal to the net charge of the distribution and situated at the position of the arbitrary origin. Similarly, V_2 is the potential at P due to a *dipole* whose dipole moment is equal to that of the distribution and also situated at the origin, etc.:

$$V_1 = \frac{Q}{4\pi\epsilon_0 r}, \tag{5-8}$$

$$V_2 = \frac{\boldsymbol{p} \cdot \hat{\boldsymbol{r}}}{4\pi\epsilon_0 r^2}, \tag{5-8}$$

$$V_3 = \frac{1}{4\pi\epsilon_0 r^3}\left(3mnp_{yz} + 3nlp_{zx} + 3lmp_{xy} + \frac{3l^2 - 1}{2}p_{xx} + \frac{3m^2 - 1}{2}p_{yy} + \frac{3n^2 - 1}{2}p_{zz}\right), \tag{5-52}$$

where l, m, n are the direction cosines of the vector $\boldsymbol{r}$ defining the position of P and where

$$p_{xx} = Q\overline{x'^2}, \qquad p_{yy} = Q\overline{y'^2}, \qquad p_{zz} = Q\overline{z'^2}, \tag{5-8}$$

$$p_{yz} = p_{zy} = Q\overline{y'z'}, \qquad p_{zx} = p_{xz} = Q\overline{z'x'}, \qquad p_{xy} = p_{yx} = Q\overline{x'y'}.$$

(5-49 to 5-51)

PROBLEMS

5-1. *(5.1)* The dipole moment of a charge distribution whose net charge is zero

Show that, if the net charge Q is zero, then the dipole moment of a charge distribution is independent of the choice of origin.

5-2. *(5.1)* The dipole moment of parallel line charges

Two line charges $+Q$ and $-Q$ extend, respectively, from $(-a, 0, c)$ to $(a, 0, c)$ and from $(-a, 0, -c)$ to $(a, 0, -c)$. Calculate their dipole moment.

5-3. *(5.1)* The dipole moment of a spherical shell of charge

Calculate the dipole moment of a spherical shell of radius R bearing a surface charge density $\sigma = \sigma_0 \cos\theta$.

5-4. *(5.1)* The dipole moment of a spherical shell of charge

(a) Calculate the dipole moment of a spherical shell of radius R whose surface charge density is $\sigma_0(1 + \cos\theta)$.

(b) What is the dipole moment if the center of the sphere is at $Z\hat{\mathbf{z}}$?

(c) What is the dipole moment if the center of the sphere is at $X\hat{\mathbf{x}} + Y\hat{\mathbf{y}} + Z\hat{\mathbf{z}}$?

5-5. *(5.1)* An alternate expression for the potential in the field of an electric dipole

We found that, in the field of an electric dipole,

$$V = \frac{Q}{4\pi\epsilon_0}\left(\frac{1}{r_b} - \frac{1}{r_a}\right).$$

Refer to Fig. 5-8. Show that, if the length of the dipole is small, then

$$V = \frac{Qs}{4\pi\epsilon_0}\left[\frac{d}{dz'}\left(\frac{1}{r'}\right)\right]_{z'=0},$$

where z' is the position of a point charge on the z-axis and $\mathbf{r}' = x\hat{\mathbf{x}} + y\hat{\mathbf{y}} + (z - z')\hat{\mathbf{z}}$.

5-6. *(5.2)* An alternate expression for the potential in the field of an electric quadrupole

See Prob. 5-5 and refer to Fig. 5-9. Show that the potential in the field of a linear electric quadrupole is

$$V = \frac{ps}{4\pi\epsilon_0}\left[\frac{d}{dz'}\left(\frac{\cos\theta}{r'^2}\right)\right]_{z'=0}.$$

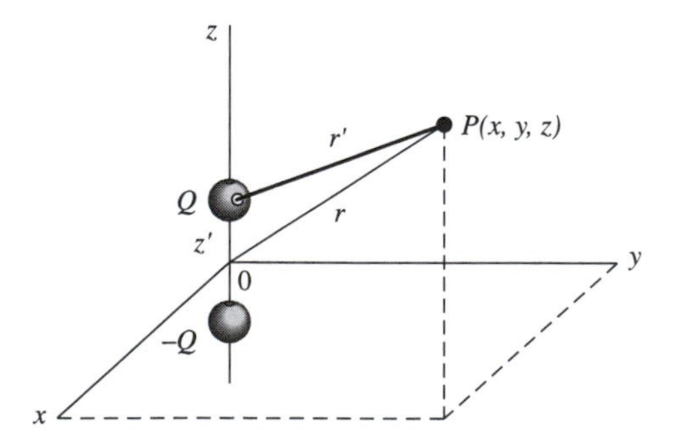

Figure 5-8

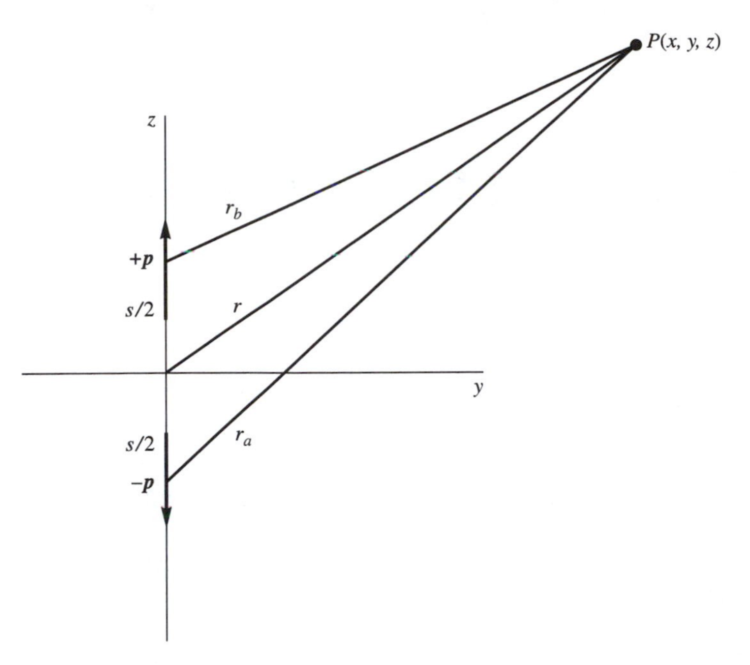

Figure 5-9

5-7. (*5.4*) Multipolar expansion of the field of a single point charge

A single point charge Q is situated at $P'(0, 0, s)$.

First expand its potential at point P in terms of multipoles. The vector $\boldsymbol{r}$ that defines the position of P forms an angle θ with the z-axis, and $r \gg s$. The distance from Q to P is r'. Disregard terms of the order of $(s/r)^4$ and higher. Then write out the values of V_1, V_2, V_3.

5-8. (*5.4.4*) The potential close to a dipole

Calculate V for a dipole exactly, and identify the quadrupole and octupole terms. The octupole term varies as $(s/r)^4$. You can therefore disregard terms in $(s/r)^5$, $(s/r)^6$, etc.

5-9. (*5.4.4*) The field of a charged cube

A cube of side $2a$ carries a uniform volume charge density ρ. The origin of coordinates is at the center. Calculate V_1, V_2, V_3 outside the cube.

5-10. (*5.4.4*) The field of a line charge

A line charge Q extends from $z = -a/2$ to $z = a/2$.

(a) Calculate the monopole, dipole, and quadrupole terms in the expansion for V.

(b) For what value of the distance r to the center of the charge is the quadrupole term less than 1% of the monopole term, if $3n^2 - 1$ is of the order of unity?

5-11. (*5.4.4*) The field of a set of six equal point charges

In Fig. 5-6, let all the charges be Q. Calculate V_4 and V_5 for $r > a$.

5-12. (*5.4.4*) The field of a quadrupole

Draw electric field lines for a quadrupole formed of charges $+Q$, $-Q$, $+Q$, $-Q$ at the corners of a square. Charges on a diagonal have the same sign.

Remember that, on a field line, $dy/dx = E_y/E_x$.

CHAPTER

6

ELECTRIC FIELDS IV

Energy, Capacitance, and Forces

This chapter concerns the energy stored in an electric field and the resulting forces exerted on charged conductors. It also deals with capacitors, devices designed to store electric energy‡.

---

†Asterisks indicate material that can be omitted without losing continuity.

‡See Booker (1982).

6.1 THE POTENTIAL ENERGY $\mathscr{E}$ OF A CHARGE DISTRIBUTION EXPRESSED IN TERMS OF CHARGES AND POTENTIALS

6.1.1 The Potential Energy of a Set of Point Charges

Imagine a set of N point charges distributed in space as in Fig. 6-1. There are no other charges in the neighborhood. A given charge occupies a point where the potential due to the *other* charges is V. That particular charge therefore possesses a potential energy, which is either positive or negative. The system as a whole possesses a potential energy $\mathscr{E}$ that we shall calculate.

Assume that the charges remain in equilibrium under the action of both the electric forces and restraining mechanical forces.

The potential energy of the system is equal to the work performed by the electric forces in the process of dispersing the charges out to infinity. After dispersal, the charges are infinitely remote from each other, and there is zero potential energy.

First, let Q_1 recede to infinity slowly, keeping the electric and the mechanical forces in equilibrium. There is zero acceleration and zero kinetic energy. The other charges remain fixed. The decrease in potential energy $\mathscr{E}_1$ is equal to Q_1 multiplied by the potential V_1 due to the other charges at the original position of Q_1:

$$\mathscr{E}_1 = \frac{Q_1}{4\pi\epsilon_0}\left(\frac{Q_2}{r_{12}} + \frac{Q_3}{r_{13}} + \cdots + \frac{Q_N}{r_{1N}}\right). \quad (6\text{-}1)$$

All the charges except Q_1 appear in the series between parentheses.

With Q_1 removed, let Q_2 recede to infinity, to some point infinitely distant from Q_1. The decrease in potential energy is now

$$\mathscr{E}_2 = \frac{Q_2}{4\pi\epsilon_0}\left(\frac{Q_3}{r_{23}} + \frac{Q_4}{r_{24}} + \cdots + \frac{Q_N}{r_{2N}}\right). \quad (6\text{-}2)$$

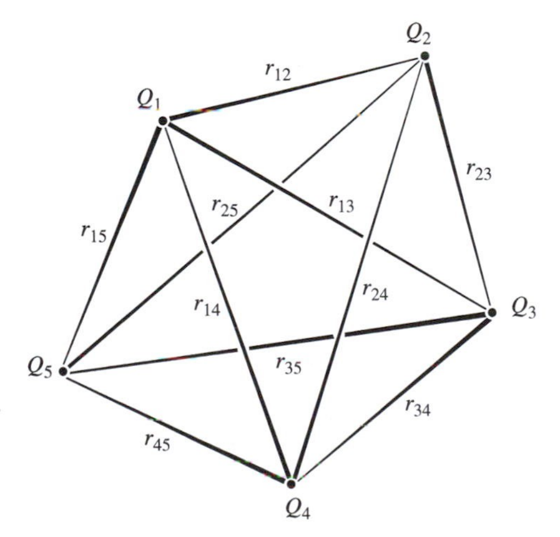

Figure 6-1 Set of point charges Q_1, Q_2, Q_3, . . . separated by distances r_{12}, r_{13}, r_{14}, etc.

The series for $\mathscr{E}_2$ has $N - 2$ terms. We continue the process for all the remaining charges, until finally the Nth charge can stay in position, since it lies in a zero field.

The total potential energy of the original charge distribution is then

$$\mathscr{E} = \mathscr{E}_1 + \mathscr{E}_2 + \mathscr{E}_3 + \cdots + \mathscr{E}_N \tag{6-3}$$

$$\begin{aligned} &= \frac{Q_1}{4\pi\epsilon_0}\left(0 + \frac{Q_2}{r_{12}} + \frac{Q_3}{r_{13}} + \frac{Q_4}{r_{14}} + \cdots + \frac{Q_N}{r_{1N}}\right) \\ &+ \frac{Q_2}{4\pi\epsilon_0}\left(0 + 0 + \frac{Q_3}{r_{23}} + \frac{Q_4}{r_{24}} + \cdots + \frac{Q_N}{r_{2N}}\right) \\ &+ \frac{Q_3}{4\pi\epsilon_0}\left(0 + 0 + 0 + \frac{Q_4}{r_{34}} + \cdots + \frac{Q_N}{r_{3N}}\right) + \cdots \\ &+ \frac{Q_N}{4\pi\epsilon_0}(0 + 0 + 0 + 0 + \cdots + 0). \end{aligned} \tag{6-4}$$

We now rewrite this array, adding, to the left of and below the diagonal line of zeros, terms that are equal to their counterparts to the right of and above the diagonal. Then every term of the series appears twice and

$$\begin{aligned} 2\mathscr{E} = &\frac{Q_1}{4\pi\epsilon_0}\left(0 + \frac{Q_2}{r_{12}} + \frac{Q_3}{r_{13}} + \frac{Q_4}{r_{14}} + \cdots + \frac{Q_N}{r_{1N}}\right) \\ &+ \frac{Q_2}{4\pi\epsilon_0}\left(\frac{Q_1}{r_{21}} + 0 + \frac{Q_3}{r_{23}} + \frac{Q_4}{r_{24}} + \cdots + \frac{Q_N}{r_{2N}}\right) \\ &+ \frac{Q_3}{4\pi\epsilon_0}\left(\frac{Q_1}{r_{31}} + \frac{Q_2}{r_{32}} + 0 + \frac{Q_4}{r_{34}} + \cdots + \frac{Q_N}{r_{3N}}\right) + \cdots \\ &+ \frac{Q_N}{4\pi\epsilon_0}\left(\frac{Q_1}{r_{N1}} + \frac{Q_2}{r_{N2}} + \frac{Q_3}{r_{N3}} + \frac{Q_4}{r_{N4}} + \cdots + 0\right). \end{aligned} \tag{6-5}$$

On the right, the first line is Q_1V_1, the second line is Q_2V_2, and so forth, where V_i is the potential in the undisturbed system due to all the charges except Q_i at the point occupied by Q_i. Thus

$$2\mathscr{E} = Q_1V_1 + Q_2V_2 + Q_3V_3 + \cdots + Q_NV_N, \tag{6-6}$$

and the potential energy of the initial charge configuration is

$$\mathscr{E} = \frac{1}{2}\sum_{i=1}^{N} Q_iV_i. \tag{6-7}$$

The reason for the factor of $\frac{1}{2}$ follows from the above calculation. Let all the charges be positive. Then the potential at the position of a given charge, just before it moves out to infinity, is less (except for Q_1) than the potential at the same point in the original charge distribution. On the average, the potential just before removal is one-half the potential in the original charge distribution.

This energy $\mathscr{E}$, which does *not* include the energy required to assemble the individual charges themselves, can be positive, negative, or zero. For example, for two charges of the same sign, $\mathscr{E}$ is positive. For charges of opposite signs, $\mathscr{E}$ is negative. For a single charge, $\mathscr{E}$ is zero.

But what is the energy required to simply modify a charge distribution, without dispersing it to infinity? This energy is clearly equal to the final potential energy minus the initial potential energy, whatever method one may choose to effect the change.

6.1.2 The Potential Energy of a Continuous Charge Distribution

For a continuous electric charge distribution, we replace Q_i by $\rho\, dv$ and the summation by an integration over any volume v that contains all the charge:

$$\mathscr{E} = \tfrac{1}{2}\int_v V\rho\, dv. \tag{6-8}$$

This integral is equal to the work performed by the electric forces in going from the given charge distribution to the situation where $\rho = 0$ everywhere, by dispersing all the charge to infinity, or by letting positive and negative charges coalesce, or by both processes combined.

At first sight, this is an obvious extension of the previous equation. It is not, because we have now included the energies required to assemble the individual macroscopic charges. In fact, as we shall see in Sec. 6.2, the above integral is always positive.

Observe that the potential V under the integral sign does not include the part that originates in the element of charge $\rho\, dv$ itself. We saw in Sec. 3.5 that the infinitesimal element of charge at a given point contributes nothing to V.

If there are surface charge densities σ, then their stored energy is

$$\mathscr{E} = \tfrac{1}{2}\int_{\mathscr{A}} \sigma V\, d\mathscr{A}, \tag{6-9}$$

where $\mathscr{A}$ includes all the surfaces carrying charge.

6.1.3 True Point and Line Charges

Suppose we have a spherical charge Q of uniform volume density and radius R. Then, using the value of the potential inside, V_i, that we calculated in the first Example in Sec. 3.6,

$$\mathscr{E} = \frac{1}{2}\int_0^R \left(\frac{Q}{4\pi R^3/3}\right)\left[\frac{Q}{4\pi\epsilon_0 R}\left(\frac{3}{2} - \frac{r^2}{2R^2}\right)\right] 4\pi r^2\, dr = \frac{3Q^2}{20\pi\epsilon_0 R}. \tag{6-10}$$

If R is zero, then $\mathscr{E}$ is infinite, which is nonsense. Electrons are presumably true point charges, and dealing with this absurd result poses difficult problems whose solutions are well beyond the scope of this book. With a true line charge, $\mathscr{E}$ is similarly infinite.

True point and line charges are therefore not allowed in the present context. Nonetheless, we follow the usual custom of speaking loosely of point and line charges when the radius is negligibly small. True surface charges cause no problems.

6.2 THE POTENTIAL ENERGY $\mathscr{E}$ OF AN ELECTRIC CHARGE DISTRIBUTION EXPRESSED IN TERMS OF $\mathbf{E}$

We have expressed the potential energy $\mathscr{E}$ of a charge distribution in terms of the charge density ρ and the potential V. Now both ρ and V are related to $\mathbf{E}$, so it should be possible to express $\mathscr{E}$ solely in terms of $\mathbf{E}$. That is what we shall do here. We shall find that

$$\mathscr{E} = \int_v \frac{\epsilon_0 E^2}{2}\, dv, \tag{6-11}$$

where the volume v includes all the regions where $\mathbf{E}$ exists. Thus we can calculate $\mathscr{E}$ by assigning to each point in space an *electric energy density* of $\epsilon_0 E^2/2$.

Since the above $\mathscr{E}$ is positive, that of Eq. 6-8 is also always positive.

These two equivalent expressions for $\mathscr{E}$, one in terms of ρ and V and the other in terms of $\mathbf{E}$, are both important. They emphasize different, but complementary, aspects of electrical phenomena. With the first expression, $\mathscr{E}$ is the potential energy of a system of *charges;* with the second, $\mathscr{E}$ is the energy stored in a *field.*

First we apply the above formula to the field of a charged spherical conductor, and then we give a general proof.

EXAMPLE The Potential Energy $\mathscr{E}$ of a Charged Conducting Sphere

We find the potential energy $\mathscr{E}$ of a conducting sphere of radius R carrying a charge Q in three different ways.

First method

The whole charge Q is at the potential $Q/(4\pi\epsilon_0 R)$. Then

$$\mathscr{E} = \frac{1}{2}Q\frac{Q}{4\pi\epsilon_0 R} = \frac{Q^2}{8\pi\epsilon_0 R}. \tag{6-12}$$

Second method

Imagine that the radius of the charged spherical conductor increases slowly and eventually becomes infinite. The total mechanical work performed by the charges is equal to the initial potential energy.

During the expansion, the field outside remains unaffected (Gauss's law again!). An element of charge $\sigma\, d\mathscr{A} = \epsilon_0 E\, d\mathscr{A}$ is subjected to an electric field strength equal to $E/2$ (Sec. 6.8), and the work performed by this element, when the radius of the sphere increases by dR, is

$$d\mathscr{E} = (\epsilon_0 E\, d\mathscr{A})\left(\frac{E}{2}\right)dR = \frac{\epsilon_0 E^2}{2}\, dv, \tag{6-13}$$

where dv is the element of volume swept by the element of area $d\mathscr{A}$. After the radius has expanded to infinity, the total work performed by the charges is

$$\mathscr{E} = \int_R^\infty \frac{\epsilon_0 E^2}{2}\, dv = \int_R^\infty \frac{\epsilon_0}{2}\left(\frac{Q}{4\pi\epsilon_0 r^2}\right)^2 4\pi r^2\, dr = \frac{Q^2}{8\pi\epsilon_0 R} \tag{6-14}$$

as above.

Third method

Let $\mathscr{A}(r)$ be the area of any surface of radius r, concentric with the conducting sphere and outside it, as in Fig. 6-2. Then, from Gauss's law (Sec. 3.6)

$$\mathscr{E} = \tfrac{1}{2}QV = \tfrac{1}{2}\int_{\mathscr{A}(r)} \epsilon_0 E\, d\mathscr{A} \int_R^\infty E\, dr. \tag{6-15}$$

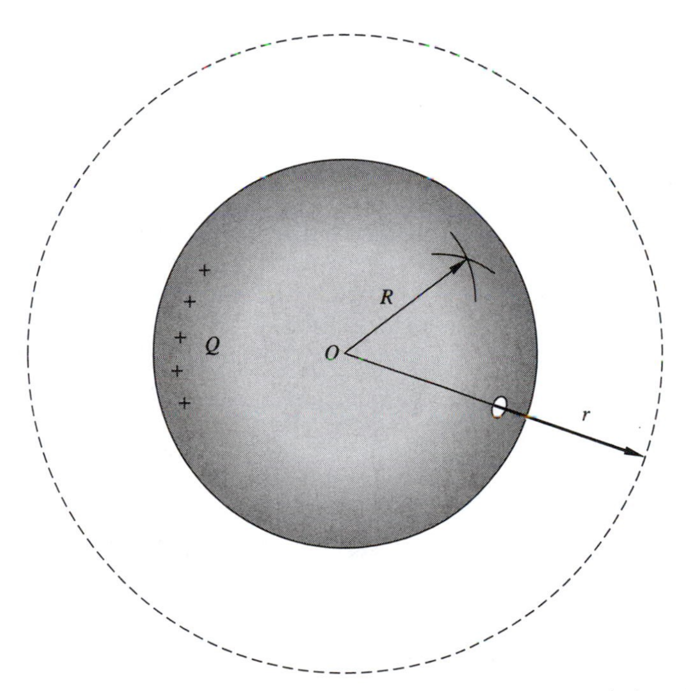

Figure 6-2 Spherical conductor carrying a charge Q, and a concentric imaginary spherical surface of area $\mathscr{A}(r)$.

Since the two integrals are independent of each other,

$$\mathscr{E} = \frac{1}{2}\int_R^\infty \left(\int_{\mathscr{A}(r)} \epsilon_0 E \, d\mathscr{A}\right) E \, dr \tag{6-16}$$

$$= \frac{1}{2}\int_R^\infty (4\pi r^2 \epsilon_0 E) E \, dr = \int_R^\infty \frac{\epsilon_0 E^2}{2} 4\pi r^2 \, dr \tag{6-17}$$

$$= \int_R^\infty \frac{\epsilon_0 E^2}{2} dv = \frac{Q^2}{8\pi\epsilon_0 R}, \tag{6-18}$$

as previously.

---

*6.2.1 The Potential Energy $\mathscr{E}$ of a Charge Distribution Expressed in Terms of $\boldsymbol{E}$: General Proof

We exclude unrealistic cases where V would be discontinuous, for that would require an infinite $\boldsymbol{E}$. We also set $V = 0$ at infinity, which excludes charges of infinite extent. Then V has a finite maximum V_{max} and a finite minimum V_{min} with $V_{max} \geq 0$, $V_{min} \leq 0$.

Let V_{min} be negative, and imagine a conductor that occupies all points where $V = V_{min}$. The conductor expands out to the equipotential $V_{min} + dV$. This does not affect the rest of the field. Eventually, the conductor reaches the equipotential $V = 0$. In the course of the expansion, any charge encountered accumulates on the surface of the conductor.

According to the second method above, the work performed by the charges is equal to the integral of $\epsilon_0 E^2/2$ over the volume swept out.

Now let V_{max} be greater than zero, and imagine another conductor occupying the region where $V = V_{max}$. It expands as above until it reaches the equipotential $V = 0$. Again the work performed is the integral of $\epsilon_0 E^2/2$ over the volume swept out.

If the two conductors meet, then, immediately before contact, the surface charge densities are equal in magnitude (same $\boldsymbol{E}$), opposite in sign, and at the same potential. They cancel. The charge density is now zero everywhere, either because the charges are dispersed to infinity or because positive and negative charges have neutralized.

The initial stored energy is thus given correctly by the integral of $\epsilon_0 E^2/2$.

6.3 THE CAPACITANCE OF AN ISOLATED CONDUCTOR

Imagine a finite conductor situated a long distance from any other body and carrying a charge Q. If Q changes, the conductor's potential also changes. As we show below, the ratio Q/V is a constant. The *capacitance* of the isolated conductor is

$$C = \frac{Q}{V}. \tag{6-19}$$

Thus the capacitance of an isolated conductor is equal to the extra charge required to increase its potential by 1 volt. The unit of capacitance is the *farad,* or coulomb per volt.

The energy stored in the field of an isolated conductor is

$$\mathscr{E} = \frac{QV}{2} = \frac{CV^2}{2} = \frac{Q^2}{2C}. \tag{6-20}$$

We now show that the capacitance C of an isolated conductor depends solely on its size and shape. The conductor is in air. The potential in the region surrounding the conductor is $V(x, y, z)$. It obeys Laplace's equation, and it is zero at infinity because the conductor is of finite size, by hypothesis. At the surface of the conductor the charge density σ is $\epsilon_0 E$ (Sec. 4.6), or ϵ_0 times the rate of change of $V(x, y, z)$ in the direction normal to the surface. An immediate consequence is that the value of the conductor potential determines the surface charge density $\sigma = -\epsilon_0 \boldsymbol{\nabla} V$ on the conductor. Therefore the conductor potential also determines the total charge Q on the conductor.

Observe that the equation $\nabla^2 V = 0$ is linear, so that any multiple of $V(x, y, z)$ is also a solution. If V increases everywhere by some factor a, this new V obeys Laplace's equation outside the conductor, is zero at infinity, and is equal to aV_c on the conductor. Furthermore, it is the only continuous function of x, y, z that satisfies these three conditions. We conclude that if the conductor's potential increases by the factor a, then V increases everywhere by the same factor a, and both σ and Q likewise increase by the same factor.

The charge Q on an isolated conductor is thus proportional to its voltage V, and its capacitance $C = Q/V$ depends solely on the size and shape of the conductor.

EXAMPLE The Capacitance of a Conducting Sphere

If an isolated conducting sphere of radius R carries a charge Q, the potential at its surface is $Q/(4\pi\epsilon_0 R)$ and

$$\begin{aligned} C = 4\pi\epsilon_0 R &= 1.11 \times 10^{-10} R \text{ farad} \\ &= 111 R \text{ picofarads.} \end{aligned} \tag{6-21}$$

6.4 THE CAPACITANCE BETWEEN TWO CONDUCTORS

We now have two uncharged isolated conductors. Transferring a charge Q from one to the other establishes a potential *difference* V between them. By definition, the *capacitance* between the conductors is Q/V. The capacitance depends solely on the geometry of the conductors and on their relative positions.

Pairs of conductors arranged specifically to possess capacitance are called *capacitors.*

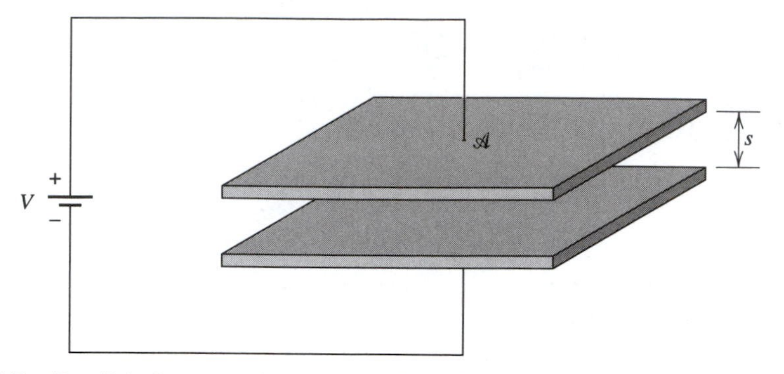

Figure 6-3 Parallel-plate capacitor connected to a battery.

EXAMPLE The Parallel-Plate Capacitor

A parallel-plate capacitor (Fig. 6-3) consists of two conducting plates of area $\mathcal{A}$, separated by a distance s. The plates carry charges Q and $-Q$. We neglect edge effects. From Gauss's law,

$$E = \frac{\sigma}{\epsilon_0} = \frac{Q}{\mathcal{A}\epsilon_0}, \qquad V = \frac{Qs}{\mathcal{A}\epsilon_0}. \tag{6-22}$$

Then

$$C = \frac{\epsilon_0 \mathcal{A}}{s}. \tag{6-23}$$

Also, the stored energy is

$$\mathscr{E} = \frac{QV}{2} = \frac{CV^2}{2} = \frac{Q^2}{2C} = \frac{Q^2 s}{2\epsilon_0 \mathcal{A}}, \tag{6-24}$$

or

$$\mathscr{E} = \int \frac{\epsilon_0 E^2}{2} dv = \frac{\epsilon_0}{2}\left(\frac{Q}{\epsilon_0 \mathcal{A}}\right)^2 \mathcal{A}s = \frac{Q^2 s}{2\epsilon_0 \mathcal{A}}. \tag{6-25}$$

EXAMPLE Capacitors in Parallel and in Series

Capacitors connected in parallel share the same voltage. Thus, for two capacitors, as in Fig. 6-4(a),

$$C = \frac{Q_1 + Q_2}{V} = \frac{Q_1}{V} + \frac{Q_2}{V} = C_1 + C_2. \tag{6-26}$$

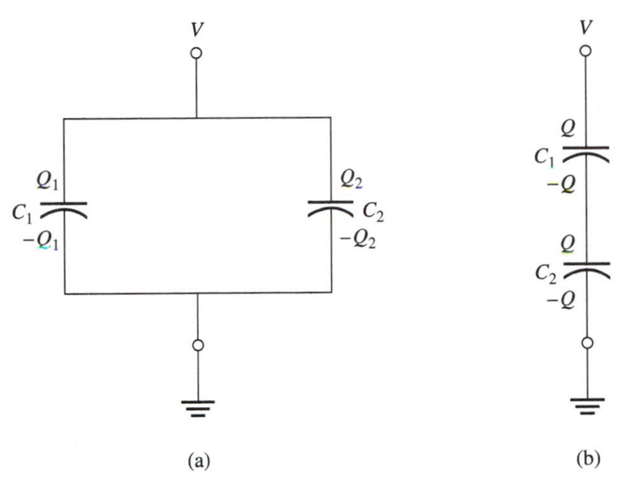

Figure 6-4 (a) Two capacitors connected in parallel. (b) Two capacitors connected in series.

Capacitors connected in series carry the same charges, as in Fig. 6-4(b). Then

$$V = \frac{Q}{C_1} + \frac{Q}{C_2} = \frac{Q}{C}, \tag{6-27}$$

$$\frac{1}{C} = \frac{1}{C_1} + \frac{1}{C_2}, \qquad C = \frac{C_1 C_2}{C_1 + C_2}. \tag{6-28}$$

So capacitors connected in parallel add as resistors in series, and capacitors connected in series add as resistors in parallel.

*6.5 TIME CONSTANT†

One connects a resistor across the terminals of a capacitor as in Fig. 6-5(a). The initial voltage on the capacitor is V_0. What happens? At the time t,

$$I = -\frac{dQ}{dt}, \quad \text{and } V = IR. \tag{6-29}$$

So

$$V = \frac{Q}{C} = IR = -\frac{dQ}{dt}R, \quad \text{and } \frac{dQ}{dt} = -\frac{Q}{RC}. \tag{6-30}$$

In practice, one measures voltages, not charges. So we divide the numerators in the last equation by C, with $Q/C = V$:

$$\frac{dV}{dt} = -\frac{V}{RC}, \quad \text{and } V = V_0 \exp\left(-\frac{t}{RC}\right). \tag{6-31}$$

†For a more extensive discussion of electric circuits, see Lorrain, Corson, and Lorrain (1988).

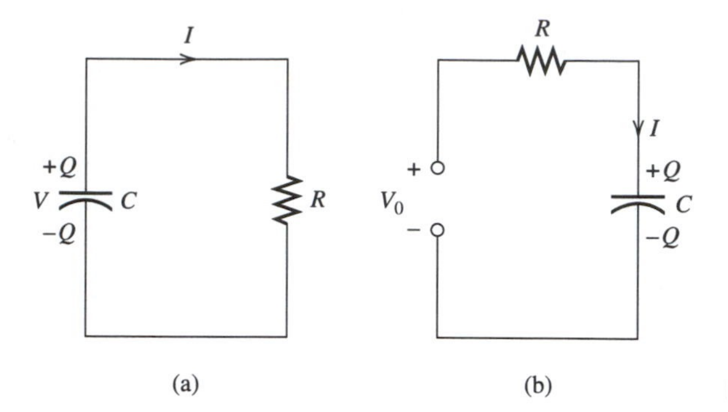

Figure 6-5

The product RC is called the *time constant* of the circuit because, after a time RC, the voltage on the capacitor has decreased by a factor of e. For $t \gg RC$, $V \approx 0$.

But what if you apply a constant voltage V_0 to an initially uncharged capacitor through a resistor, as in Fig. 6-5(b)? Then $I = +dQ/dt$,

$$V_0 = IR + \frac{Q}{C} = \frac{dQ}{dt}R + \frac{Q}{C}, \quad \text{and} \quad \frac{dQ}{dt} = \frac{V_0}{R} - \frac{Q}{RC}. \tag{6-32}$$

Dividing again by C,

$$\frac{dV}{dt} = \frac{V_0}{RC} - \frac{V}{RC}, \quad \text{and} \quad V = V_0\left[1 - \exp\left(-\frac{t}{RC}\right)\right]. \tag{6-33}$$

The voltage V on the capacitor builds up from zero with the same time constant RC, and eventually reaches the asymptotic value V_0.

*6.6 ALTERNATING CURRENTS IN CAPACITORS

Figure 6-6(a) shows a source of alternating current connected to a capacitor. Then

$$Q = CV_m \cos \omega t, \tag{6-34}$$

$$I = \frac{dQ}{dt} = -\omega C V_m \sin \omega t = \omega C V_m \cos\left(\omega t + \frac{\pi}{2}\right). \tag{6-35}$$

The current *leads* the voltage by $\pi/2$ radians, as in Fig. 6-6(b).

In phasor notation (Chap. 2),

$$I = j\omega Q = j\omega C V, \tag{6-36}$$

where I, Q, V are all phasors. At a given instant the energy stored in the capacitor is

$$\mathscr{E}_{\text{inst}} = \frac{CV^2}{2} = \frac{CV_m^2 \cos^2 \omega t}{2} \tag{6-37}$$

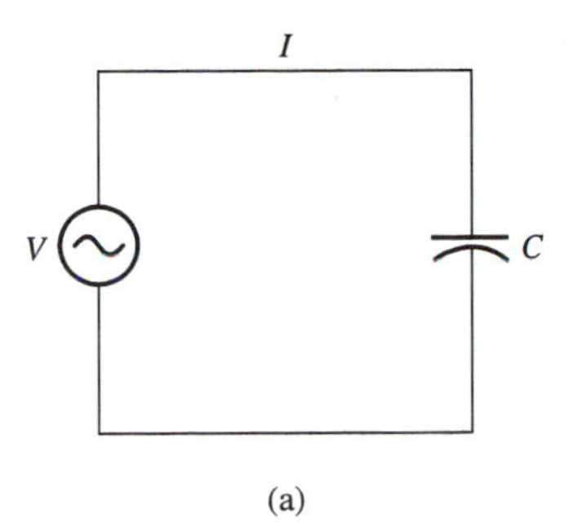

(a)

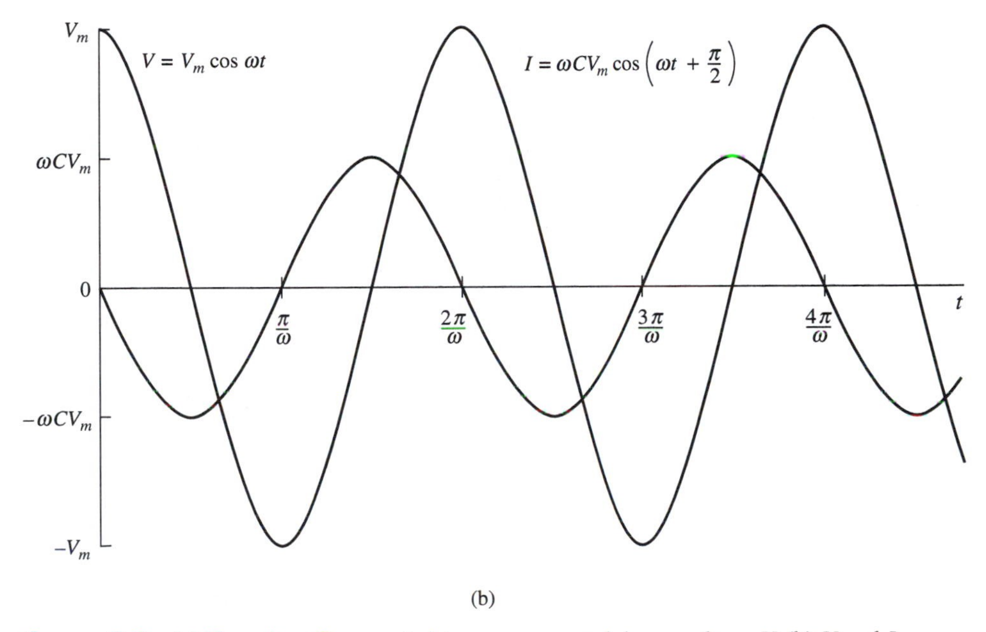

(b)

Figure 6-6 (a) Capacitor C connected to a source supplying a voltage V. (b) V and I as functions of the time for a capacitor. Here, the current leads the voltage by 90° and $\omega C = 0.5$.

and the average stored energy is

$$\mathscr{E}_{av} = \frac{CV_m^2/2}{2} = \frac{CV_{rms}^2}{2}. \tag{6-38}$$

*6.6.1 Impedance Z

By analogy with Ohm's law for DC circuits, we have *Ohm's law for AC circuits:*

$$I = \frac{V}{Z}, \tag{6-39}$$

where both I and V are phasors and Z is the *impedance* of the circuit, expressed in ohms. The impedance of a resistance is R.

For a capacitor, from Eq. 6-36,

$$Z = \frac{1}{j\omega C}. \tag{6-40}$$

Impedances in series and in parallel operate like resistances in series and in parallel.

Impedances are in general complex and thus of the form

$$Z = R + jX, \tag{6-41}$$

where X is the *reactance*. Given a two-terminal passive and linear circuit that comprises only resistors and capacitors, however complex, the impedance between its terminals is a complex number $R + jX$.

The reactance X of a capacitance C is negative:

$$jX = \frac{1}{j\omega C}, \qquad X = -\frac{1}{\omega C}. \tag{6-42}$$

The current in a capacitor leads the applied voltage; intuitively, you can guess that, whatever the arrangement of resistors and capacitors, the current at the input terminals will lead the voltage. Thus the reactance for the complete circuit will be negative, or capacitive.

Both the Kirchhoff current law and the Kirchhoff voltage law of Sec. 4.4 apply to alternating-current circuits, if one uses phasors and impedances.

EXAMPLE

At a frequency of 1 kilohertz, a one-microfarad capacitor has an impedance Z of -159j ohms. If you apply a voltage of 10 volts at that frequency between the terminals of the capacitor, then the current has a magnitude of 62.9 milliamperes, and it leads the voltage by 90 degrees.

6.7 ELECTRIC FORCES ON CONDUCTORS

An element of charge $\sigma\, d\mathcal{A}$ on the surface of a conductor experiences the electric field of all the *other* charges and is therefore subjected to an electric force. Under static conditions this force is perpendicular to the surface, for otherwise there would be a tangential field and a tangential current. The force also acts on the conductor, to which $\sigma\, d\mathcal{A}$ is bound by internal electric forces.

To calculate the magnitude of the electric force, consider a conductor carrying a surface charge density σ with an electric field strength $\boldsymbol{E}$ near the surface. From Gauss's law, this E is σ/ϵ_0. Now the force on $\sigma\, d\mathcal{A}$ is *not* $E\sigma\, d\mathcal{A}$, because the field that acts on $\sigma\, d\mathcal{A}$ is only that of the *other* charges in the system.

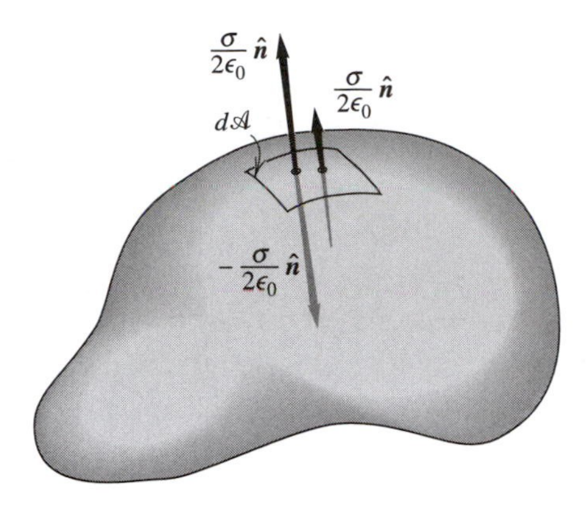

Figure 6-7 The local surface charge density σ on a conductor gives two oppositely directed electric fields, as shown by the two arrows on the left. All the other charges together give the field shown on the right. The net result is a field strength of σ/ϵ_0 outside and zero inside. The unit vector $\hat{\boldsymbol{n}}$ points *outward,* as usual.

We can find the field of $\sigma d\mathcal{A}$ itself from Gauss's law. The flux of $\boldsymbol{E}$ emerging from $\sigma d\mathcal{A}$ is $\sigma d\mathcal{A}/\epsilon_0$, half of it inward and half outward, as in Fig. 6-7. Then $\sigma d\mathcal{A}$ provides exactly half the total $\boldsymbol{E}$ at a point outside, close to the surface and cancels the field of the other charges, inside.

Therefore the E acting on $\sigma d\mathcal{A}$ is $\sigma/2\epsilon_0$, and the force on the element of area $d\mathcal{A}$ of the conductor is

$$dF = \frac{\sigma}{2\epsilon_0}\sigma d\mathcal{A} = \frac{\sigma^2}{2\epsilon_0}d\mathcal{A}. \tag{6-43}$$

The surface force density is

$$F' = \frac{dF}{d\mathcal{A}} = \frac{\sigma^2}{2\epsilon_0} = \frac{\epsilon_0 E^2}{2} \qquad \text{newtons/meter}^2. \tag{6-44}$$

The force per unit area on a conductor is equal to the energy density in the field.

The net electrostatic force on a conductor of area $\mathcal{A}$ is

$$\boldsymbol{F} = \frac{\epsilon_0}{2}\oint_{\mathcal{A}} E^2 \boldsymbol{d\mathcal{A}}, \tag{6-45}$$

where the vector $\boldsymbol{d\mathcal{A}}$ points outward. The local electric force tends to pull the conductor into the field. In other words, an electrostatic field exerts a negative pressure on a conductor. The net force on a conductor depends on the way σ and $\boldsymbol{E}$ vary along its surface. In air, or in a vacuum, electric forces are usually negligible. However, they can be appreciable in dielectrics.

Figure 6-8 shows lines of $\boldsymbol{E}$ for four pairs of *line* charges. See also Fig. 3-2.

In Fig. 6-8(a) and (c), the force is attractive and the lines of $\boldsymbol{E}$ are clearly "under tension." Indeed, the tensile force per square meter is $\epsilon_0 E^2/2$, as we can infer from Eq. 6-44.

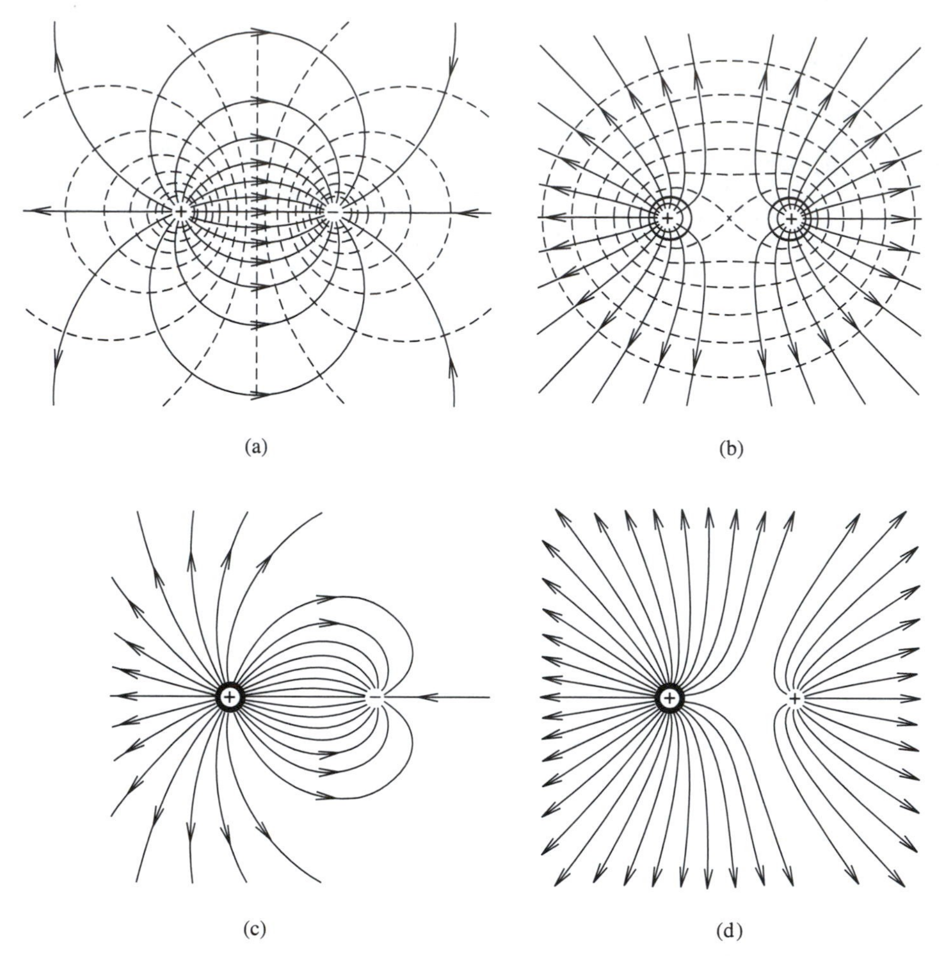

Figure 6-8 Lines of $\boldsymbol{E}$ between pairs of *line* charges. It is useful to visualize attractive electric forces as being caused by a tension in the lines of $\boldsymbol{E}$, as in (a) and (c). Similarly, repulsive electric forces may be thought of as being caused by a lateral repulsion between lines of $\boldsymbol{E}$, as in (b) and (d). In (c) and (d) the charge on the left is twice as large as the charge on the right. In (a) and (b), the dashed lines are equipotentials.

In Fig. 6-8(b) and (d), on the other hand, we can see lines of $\boldsymbol{E}$ "repelling" each other laterally. The repulsive surface force density in the region where the lines of $\boldsymbol{E}$ are parallel is also $\epsilon_0 E^2/2$.

Later we shall see that magnetic fields behave similarly.

6.8 CALCULATING ELECTRIC FORCES BY THE METHOD OF VIRTUAL WORK

We can also calculate electric forces by the *method of virtual work.* This method consists in postulating an infinitesimal displacement and then applying the principle of conservation of energy. We first define a system, and then we calculate the energy fed into it in the course of the displacement. This energy is equal to the increase in the internal energy of the system.

The method of virtual work is a general and reliable method for calculating forces, but only on two conditions: (1) one must be perfectly clear about exactly what system one is talking about, and (2) one must be particularly careful to use the proper signs.

EXAMPLE The Parallel-Plate Capacitor

A parallel-plate capacitor is connected to a battery supplying a fixed voltage V. See Fig. 6-9. We assume that the distance s between the plates *increases* by ds, and we apply the principle of virtual work. We are going to calculate energies related to the capacitor. Thus the net energy fed into the capacitor in the course of the displacement ds of the top plate will be equal to the increase in the electric energy stored in the capacitor.

Let $\mathscr{E}_F$ be the mechanical energy fed *into* the capacitor by the force $\boldsymbol{F}$, let $\mathscr{E}_B$ be the electric energy fed *into* the battery, and let $\mathscr{E}_E$ be the increase in the electric energy of the capacitor. Then

$$\mathscr{E}_F + \mathscr{E}_B = \mathscr{E}_E, \tag{6-46}$$

with

$$\mathscr{E}_F = F\,ds, \qquad \mathscr{E}_B = V\,dQ, \tag{6-47}$$

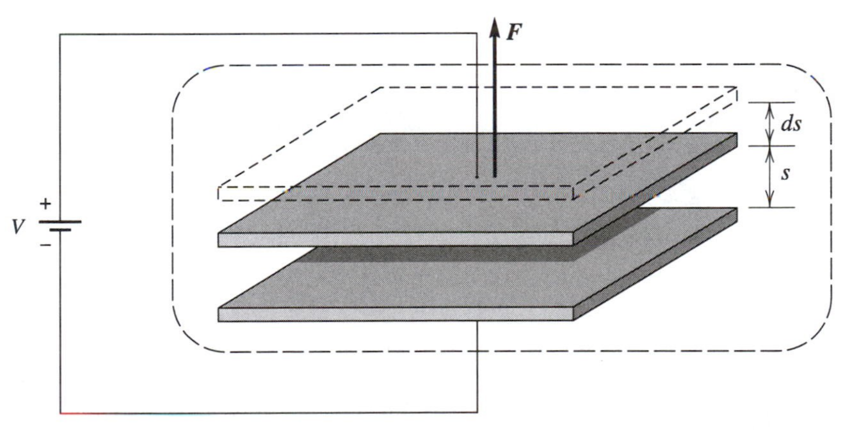

Figure 6-9 Parallel-plate capacitor connected to a battery supplying a fixed voltage V. We can calculate the force of attraction between the plates by imagining an equal but opposite force $\boldsymbol{F}$ that pulls one of the plates away by an infinitesimal distance ds. The dashed line around the plates reminds us that we apply the principle of the conservation of energy to the capacitor alone.

where dQ is the extra charge fed into the capacitor by the battery. Here

$$\mathscr{E}_B = V(VdC) = V^2 d\left(\frac{\epsilon_0 \mathscr{A}}{s}\right) = -V^2 \epsilon_0 \mathscr{A} \frac{ds}{s^2} = -(\epsilon_0 E^2)(\mathscr{A}\, ds). \tag{6-48}$$

If ds is positive, dC is negative, and energy flows from the capacitor into the battery.

Finally, since $E = V/s$,

$$\mathscr{E}_E = d\left(\frac{\epsilon_0 E^2}{2} \mathscr{A} s\right) = d\left(\frac{\epsilon_0 V^2 \mathscr{A} s}{2s^2}\right) = \epsilon_0 \frac{\mathscr{A} V^2}{2}\left(-\frac{ds}{s^2}\right) = -\frac{\epsilon_0 E^2}{2} \mathscr{A}\, ds. \tag{6-49}$$

Note the negative sign! The energy density $\epsilon_0 E^2/2$ decreases faster than the volume $\mathscr{A}s$ increases.

Thus

$$F\, ds - \epsilon_0 E^2 \mathscr{A}\, ds = -\frac{\epsilon_0 E^2}{2} \mathscr{A}\, ds, \tag{6-50}$$

$$F = \frac{\epsilon_0 E^2}{2} \mathscr{A}, \tag{6-51}$$

and the force per unit area is equal to the energy per unit volume, as previously.

Now try a negative ds. You will find that half the energy supplied by the battery becomes mechanical energy, and the other half becomes electric energy.

6.9 SUMMARY

The *potential energy* of a charge distribution is given by either one of two integrals:

$$\mathscr{E} = \frac{1}{2}\int_v V\rho\, dv \tag{6-8}$$

or

$$\mathscr{E} = \frac{1}{2}\int_v \epsilon_0 E^2\, dv. \tag{6-11}$$

In the first integral the volume v contains all the charges, while in the second it includes all the field. The assignment of an *electric energy density* $\epsilon_0 E^2/2$ to every point in space leads to the correct potential energy of a charge distribution.

If the potential of an isolated conductor is V when its charge is Q, then its *capacitance* C is Q/V *farads*. This quantity depends solely on the geometry of the conductor. The capacitance between two conductors is again Q/V, where Q is now the charge transferred from one to the other and V is the potential between them. In both cases the stored energy is

$$\mathscr{E} = \frac{QV}{2} = \frac{CV^2}{2} = \frac{Q^2}{2C}. \qquad (6\text{-}20)$$

If a capacitor C discharges through a resistor R, then the voltage across the capacitor decreases by a factor of e in one *time constant* RC.

Similarly, if a source V_0 charges a capacitor through a resistor R, then the difference between V_0 and the voltage V on the capacitor decreases by a factor of e in the same time constant RC.

If an alternating voltage $V_m \cos \omega t$ is applied to the terminals of a capacitor C, then the current leads the voltage by $\pi/2$ radians:

$$I = \omega C V_m \cos\left(\omega t + \frac{\pi}{2}\right) \qquad (6\text{-}35)$$

or, in phasor notation,

$$I = \frac{V}{Z} = j\omega CV, \qquad (6\text{-}36)$$

where the impedance Z is $1/(j\omega C)$.

The force per unit area exerted on a conductor situated in an electric field is equal to the electric energy density at the surface:

$$F' = \frac{\epsilon_0 E^2}{2} = \frac{\sigma^2}{2\epsilon_0}, \qquad (6\text{-}44)$$

where σ is the surface charge density.

It is useful to visualize electric forces as being caused by lines of $\mathbf{E}$ that are under tension and that repel each other laterally.

We can also calculate the electric force on a conductor by the principle of virtual work, which is simply the principle of conservation of energy applied to an infinitesimal disturbance of a system.

PROBLEMS

6-1. (*6.1*) The dipole and the quadrupole

Calculate the potential energies of an electric dipole and of a linear electric quadrupole.

6-2. (*6.2*) The potential energy of a sphere of charge

This problem again illustrates the immensity of electric forces and energies.

(a) Calculate the electric potential energy of a sphere of radius R carrying a total charge Q uniformly distributed throughout its volume.

(b) Calculate the gravitational potential energy of a sphere of radius R' and total mass M.

(c) Calculate the gravitational potential energy of the moon. See The Table of Physical Constants at the end of the book.

(d) Imagine that you can assemble a sphere of protons with a density equal to that of water. What would be the radius of this sphere if its electric potential energy were sufficient to blow up the moon?

(e) What is the voltage at the surface of the sphere of protons?

6-3. *(6.2)* The energy in the field of an electric dipole

Show that the energy in the field of an electric dipole of moment p outside a sphere of radius R is $p^2/(12\pi\epsilon_0 R^3)$.

6-4. *(6.2)* Electrostatic energy in a coaxial line

Coaxial lines comprise two coaxial cylindrical conductors. They serve mostly for the transmission of high-frequency signals (Chapter 24).

The outer radius of the inner conductor is 3.00 millimeters, and the inner radius of the outer conductor is 5.00 millimeters. Calculate the electric energy in the field, per meter, when the voltage difference is 5.00 volts.

6-5. *(6.2)* The energy in the field of a charged conducting sphere

A conducting sphere of radius R carries a charge Q.

(a) Calculate the stored energy.

(b) Calculate the radius $R_{0.5}$ within which the field energy is one half the total stored energy.

(c) Calculate the radius $R_{0.9}$.

6-6. *(6.3)* The earth's electric field

(a) Calculate the capacitance of the earth. Its radius is 6.4×10^6 meters.

(b) The earth carries a negative charge, and the vertical field, at the surface, is about 100 volts per meter. What is the total charge?

(c) Calculate the potential at the surface.

6-7. *(6.4)* A capacitor consisting of two concentric spheres is arranged so that the outer sphere can be separated and removed without disturbing the charges on either.

The radius of the inner sphere is a, that of the outer sphere is b, and the charges are Q and $-Q$, respectively.

(a) If the outer sphere is removed and restored to its original form, find the increase in energy when the two spheres are separated by a large distance.

(b) Where does this extra energy come from?

6-8. *(6.4)* Parallel-plate capacitor

Show that the capacitance of a parallel-plate capacitor having N plates is

$$C = 8.85(N-1)A/t \text{ picofarads,}$$

where A is the area of one side of one plate, and t is the distance between the plates. (One picofarad is 10^{-12} farad.)

6-9. *(6.4)* Parallel-plate capacitor

Can you suggest reasonable dimensions for a 1 picofarad (10^{-12} farad) air-insulated parallel-plate capacitor that is to operate at a potential difference of 500 volts?

6-10. (*6.4*) Electrostatic energy

(a) Two capacitors are connected in series as in Fig. 6-4(b). Calculate the ratio of the stored energies.

(b) Calculate this ratio for capacitors connected in parallel as in Fig. 6-4(a).

6-11. (*6.4*) Electrostatic energy

Two capacitors of capacitances C_1 and C_2 carry charges Q_1 and Q_2, respectively.

(a) Calculate the amount of energy dissipated when they are connected in parallel.

(b) How is this energy dissipated?

6-12. (*6.4*) Charging a capacitor through a resistor

A source supplying a voltage V charges a capacitor C through a resistance R.

Calculate the energy supplied by the source, that dissipated by the resistor, and that stored in the capacitor, after an infinite time.

You should find that one half of the energy is dissipated in R, and that the other half is stored in C.

6-13. (*6.4*) Cylindrical capacitor

(a) Show that the capacitance per unit length of a cylindrical capacitor is $C' = 2\pi\epsilon_0/\ln(R_2/R_1)$, where R_1 and R_2 are the inner and outer radii.

(b) Calculate the capacitance per meter when $R_2/R_1 = e = 2.718$.

See also Problem 6-4.

6-14. (*6.4*) Spherical capacitor

(a) Show that the capacitance of a spherical capacitor of inner and outer radii R_1 and R_2 is $C = 4\pi\epsilon_0 R_1 R_2/(R_2 - R_1)$.

(b) Calculate the capacitance when $R_1 = 100$ millimeters and $R_2 = 200$ millimeters.

6-15. (*6.4*) Connecting charged capacitors in parallel

Two capacitors C_1 and C_2 are charged to voltages V_1 and V_2, respectively and then connected in parallel, positive terminal to positive terminal and negative to negative.

(a) What is the final voltage?

(b) What happens to the stored energy?

6-16. (*6.6*) Capacitor and resistor in parallel

A capacitor is connected in parallel with a resistor. Show that

$$Z = \frac{R - jR^2\omega C}{R^2\omega^2C^2 + 1}.$$

Show that, with $C = 1 \times 10^{-9}$ farad and $R = 1000$ ohms, at 10 kilohertz, $Z = 1000 - 62.8j$ ohms.

6-17. (*6.6*) Low-pass *RC* filter

The circuit of Fig. 6-10 is commonly used to reduce the AC component at the output of a power supply.

As a rule, the term *power supply* applies to a circuit that transforms 120-volt 60-hertz alternating current into direct current. One simple type is the battery charger.

A power supply usually comprises three parts: a transformer, a rectifying circuit, and a filter. The output of the rectifying circuit is a pulsating direct current that can be considered to be a steady direct current, plus an alternating current. The function of

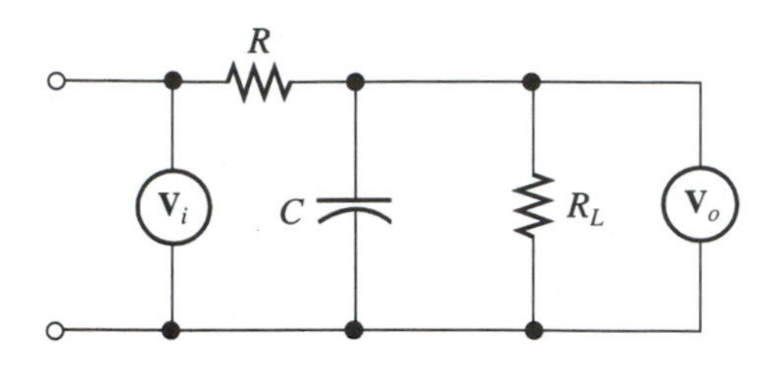

Figure 6-10 Simple low-pass *RC* filter for attenuating the alternating component at the output of a power supply, with $R_L \gg R \gg 1/(\omega C)$.

the filter is to attenuate the alternating part. Filters are usually dispensed with in battery chargers.

The filter is often followed, or even replaced, by an electronic circuit that maintains either the output voltage or the output current at a set value.

The circuit shown is a simple form of low-pass filter. The load resistance R_L is assumed to be large compared with $1/(\omega C)$.

(a) Show that the alternating component is attenuated by the factor

$$V_o/V_i \approx 1/(R\omega C)$$

if $R\omega C \gg 1$. The voltage drop across the resistance R should be small compared with that across R_L, so

$$R_L \gg R \gg 1/(\omega C).$$

(b) Plot a curve of V_o/V_i as a function of frequency from 0 to 150 hertz for $R = 10^4$ ohms and $C = 100$ microfarads.

(c) Plot the corresponding curve of the attenuation in *decibels* as a function of frequency, where the number of decibels is $20 \log V_o/V_i$.

6-18. (*6.7*) Electric forces and lines of $\boldsymbol{E}$

(a) Use the method of Sec. 6.7 to calculate the force of attraction between two charges Q and $-Q$ separated by a distance $2D$.

(b) Repeat the calculation for two charges of equal sign.

6-19. (*6.7*) High-voltage generator

Imagine the following mechanism for generating high voltages. One plate of a parallel-plate capacitor is fixed and connected to ground. The other plate is movable. When the plates are close together at a distance s, a contact closes and the movable plate charges to the voltage V. Then the contact opens, the movable plate moves out to a distance ns, and its voltage increases to nV, disregarding edge effects. At this point another contact closes, and the movable plate discharges to ground through a load resistance R.

(a) Verify that there is conservation of energy.

(b) Can you suggest a more convenient geometry for such a high-voltage generator?

6-20. (*6.7*) The surface force on a balloon carrying an electric charge

It is suggested that a balloon made of light conducting material could be kept approximately spherical by connecting it to a high-voltage supply. The balloon has a diameter of 100 millimeters, and the maximum breakdown field in air is 3 megavolts/meter.

(a) What is the maximum permissible voltage?
(b) What gas pressure, in atmospheres, inside the balloon would have the same effect?
(c) How large could the surface mass density of the balloon be?

6-21. (*6.8*) Stored energy

Four charges $+Q$, $-Q$, $+Q$, $-Q$ occupy the corners of a square of side a, with the positive charges on one diagonal and the negative charges on the other.

(a) Calculate the stored energy $\mathscr{E}$, and sketch a curve of $\mathscr{E}$ as a function of a.
(b) A mechanism constrains the charges to stay at the corners of a square but allows a to vary. What will happen?
(c) Calculate the forces on the charges by the method of virtual work.
(d) Compare the virtual work values with the values deduced from Coulomb's law.

6-22. (*6.8*) The forces on the plates of a parallel-plate capacitor

Show that the force of attraction between the plates of a parallel-plate capacitor that is not connected to a battery is $\epsilon_0 E^2 \mathscr{A}/2$, as in Eq. 6-51.

CHAPTER

7

ELECTRIC FIELDS V

Dielectric Materials A: Bound Charges and the Electric Flux Density $\boldsymbol{D}$

Dielectrics differ from conductors in that they possess no carriers of free charge that can drift about under the control of an electric field. In a true dielectric, the charges are all bound to their atoms or molecules, and they can be forced to move by only minute distances, positive charges going one way and negative charges the other way. A dielectric in which this displacement has taken place is said to be *polarized*.

Some molecules possess a permanent dipole moment and are said to be *polar.* Other molecules are *nonpolar.*

A given substance can be a dielectric under normal circumstances but become a conductor under appropriate conditions. For example, *photoconductors* are normally nonconducting, but become conducting when exposed to light.

In this first chapter on dielectric media we are concerned with the basic concepts of polarization, susceptibility, and relative permittivity, mostly in static fields.

7.1 THE THREE BASIC POLARIZATION PROCESSES

1. Under the action of an applied electric field, the center of charge† of the electron cloud in a molecule moves slightly with respect to the center of charge of the nuclei. This is *electronic polarization.* The displacement is minute, even on the atomic scale, typically 10^{-8} times the diameter of the atom.
2. Polar molecules align themselves and become further polarized in an applied electric field. This is *orientational polarization.* Collisions arising from thermal agitation partly disrupt this alignment.
3. The third basic process is *atomic polarization,* in which *ions* of opposite signs in a solid, such as NaCl, move in opposite directions when subjected to an electric field. The ferroelectric dielectrics of Sec. 8.1.4 exhibit atomic polarization.

A polarized dielectric possesses its own field, which adds to that of the other charges. The two fields can be comparable in magnitude.

7.2 THE ELECTRIC POLARIZATION $\boldsymbol{P}$

If, in the neighborhood of a given point, the average vector dipole moment per molecule in a given direction is $\boldsymbol{p}$, and if there are N molecules per cubic meter, then

$$\boldsymbol{P} = N\boldsymbol{p} \tag{7-1}$$

is the *electric polarization* at that point. So $\boldsymbol{P}$ is the *dipole moment per unit volume* at a given point, expressed in coulombs/meter^2. Recall from Sec. 5.4 that the dipole moment of a charge distribution is independent of the choice of origin when the net charge of the distribution is zero, as it is for a normal, neutral molecule.

7.3 FREE AND BOUND CHARGES

Polarization causes charges to accumulate, either within the dielectric or at its surface. We refer to such charges as *bound.* Other charges are said to be *free.* Examples of free charges are the conduction electrons in good conductors, the charge carriers in semi-

†The center of charge is analogous to the center of mass in mechanics.

conductors, and electrons injected into a dielectric by means of a high-energy electron beam.†

7.3.1 The Bound Surface Charge Density σ_b

Imagine an element of area $d\mathcal{A}$ inside a dielectric as in Fig. 7-1. Say the dielectric is nonpolar. When the dielectric is polarized, the center of positive charge $+Q$ of a molecule lies at a distance $\boldsymbol{s}$ from the center of negative charge $-Q$. This $\boldsymbol{s}$ is the same for all the molecules over an infinitesimal region.

Upon application of an electric field, n_+ positive charges cross the element of area by moving in the direction of $\boldsymbol{s}$, and n_- negative charges cross it by moving in the opposite direction. The net charge that crosses $d\mathcal{A}$ in the direction of $\boldsymbol{s}$ is therefore

$$dQ = n_+Q - n_-(-Q) = (n_+ + n_-)Q. \tag{7-2}$$

Now $n_+ + n_-$ is simply the number of molecules within the imaginary parallelepiped of Fig. 7-1, whose volume is $\boldsymbol{s} \cdot \boldsymbol{d\mathcal{A}}$. Then

$$dQ = NQ\boldsymbol{s} \cdot \boldsymbol{d\mathcal{A}} = N\boldsymbol{p} \cdot \boldsymbol{d\mathcal{A}} = \boldsymbol{P} \cdot \boldsymbol{d\mathcal{A}}, \tag{7-3}$$

where $Q\boldsymbol{s}$ is the dipole moment $\boldsymbol{p}$ of a single molecule.

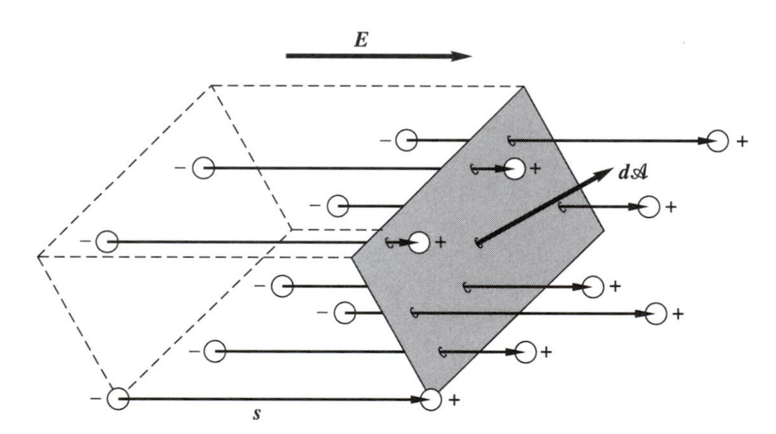

Figure 7-1 Element of area $\boldsymbol{d\mathcal{A}}$ situated inside a dielectric. The vector $\boldsymbol{d\mathcal{A}}$ is normal to the element. Here $\boldsymbol{E}$ is the externally applied field plus the field of the dipoles within the dielectric. Under the action of this $\boldsymbol{E}$, the molecules either stretch, or rotate and stretch, and a net charge of $NQ\boldsymbol{s} \cdot \boldsymbol{d\mathcal{A}}$ crosses the element of the area.

†The concepts of free and bound charges, and even the concept of polarization, are not as clear-cut as one might wish. The distinction between free and bound charges rests on the assumption that the dielectric consists of electrically neutral molecules. If there are no well-defined molecules, as in a crystal of NaCl, for example, the values of the free and bound surface charge densities are arbitrary. We confine most of our discussion here to dielectrics composed of molecules.

If $d\mathcal{A}$ lies on the surface of a dielectric, then dQ accumulates there and the *bound surface charge density* is

$$\sigma_b = \frac{dQ}{d\mathcal{A}} = \boldsymbol{P} \cdot \hat{\boldsymbol{n}}, \tag{7-4}$$

where $\hat{\boldsymbol{n}}$ is the unit vector normal to the surface and pointing *outward*. Thus σ_b is equal in magnitude to the normal component of $\boldsymbol{P}$, pointing outward.

7.3.2 The Bound Volume Charge Density ρ_b

We now demonstrate that inside a dielectric the bound volume charge density ρ_b is equal to $-\boldsymbol{\nabla} \cdot \boldsymbol{P}$. The net bound charge that flows out of a volume v across an element $d\mathcal{A}$ of its surface is $\boldsymbol{P} \cdot \boldsymbol{d\mathcal{A}}$, as we found above. The net bound charge that flows out of the closed surface of area $\mathcal{A}$ delimiting a volume v entirely situated within the dielectric (so as to exclude surface charges) is thus

$$Q_{\text{out}} = \int_{\mathcal{A}} \boldsymbol{P} \cdot \boldsymbol{d\mathcal{A}}, \tag{7-5}$$

and the net charge that remains within v must be $-Q_{\text{out}}$. If ρ_b is the volume density of the charge remaining within v, then

$$\int_v \rho_b \, dv = -Q_{\text{out}} = -\int_{\mathcal{A}} \boldsymbol{P} \cdot \boldsymbol{d\mathcal{A}} = -\int_v \boldsymbol{\nabla} \cdot \boldsymbol{P} \, dv. \tag{7-6}$$

We have used the divergence theorem to transform the second integral into the third. Since this equation applies to any volume chosen as above, the integrands are equal at every point in the dielectric and the *bound volume charge density* is

$$\rho_b = -\boldsymbol{\nabla} \cdot \boldsymbol{P}. \tag{7-7}$$

7.3.3 The Current Densities $\boldsymbol{J}_p$ and $\boldsymbol{J}_f$

The motion of bound charges under the action of a time-dependent electric field generates a *polarization current*. Consider a small surface $\boldsymbol{d\mathcal{A}}$ situated inside a dielectric as in Fig. 7-1, but normal to $\boldsymbol{P}$. As the polarization increases from zero to $\boldsymbol{P}$, a net charge $dQ_b = P\,d\mathcal{A}$ crosses $d\mathcal{A}$ in the direction of $\boldsymbol{P}$. More generally, if $\boldsymbol{P}$ increases by $\boldsymbol{dP}$ in a time interval dt and if $\boldsymbol{d\mathcal{A}}$ is normal to $\boldsymbol{dP}$, then a current

$$I = \frac{dQ_b}{dt} = \frac{dP\,d\mathcal{A}}{dt} \tag{7-8}$$

flows through $\boldsymbol{d\mathcal{A}}$ in the direction of $\boldsymbol{P}$.

Thus if, at a given point in space, $\boldsymbol{P}$ is a function of time, the motion of bound charge results in a *polarization current density*

$$\boldsymbol{J}_p = \frac{\partial \boldsymbol{P}}{\partial t}. \tag{7-9}$$

Setting $\boldsymbol{J}_f$ equal to the current density of free charges, the law of conservation of charge of Sec. 4.2 can be rewritten as

$$\boldsymbol{\nabla} \cdot \boldsymbol{J}_f = -\frac{\partial \rho_f}{\partial t}. \tag{7-10}$$

7.4 THE ELECTRIC FIELD OF A POLARIZED DIELECTRIC

We have seen that polarization causes charges to accumulate, either at the surface of a dielectric or inside, with charge densities σ_b and ρ_b, respectively. Now Coulomb's law applies to *any* net charge density, regardless of any matter that may be present. The potential V ascribable to a polarized dielectric that occupies a volume v' bounded by a surface of area $\mathscr{A}'$ is therefore the same as if the bound charges were located in a vacuum:

$$V = \frac{1}{4\pi\epsilon_0}\int_{v'} \frac{\rho_b dv'}{r} + \frac{1}{4\pi\epsilon_0}\int_{\mathscr{A}'} \frac{\sigma_b d\mathscr{A}'}{r}, \tag{7-11}$$

where r is the distance between the element of bound charge at $P'(x',y',z')$ and the point $P(x, y, z)$ where one calculates V.

If there are also free charges present, then one adds similar integrals for the free charges. Therefore

$$\boldsymbol{E} = \frac{1}{4\pi\epsilon_0}\int_{v'} \frac{\rho\hat{\boldsymbol{r}}}{r^2}\, dv' + \frac{1}{4\pi\epsilon_0}\int_{\mathscr{A}'} \frac{\sigma\hat{\boldsymbol{r}}}{r^2}\, d\mathscr{A}', \tag{7-12}$$

where $\hat{\boldsymbol{r}}$ points from P' to P and where ρ and σ are total charge densities, free plus bound.

It does *not* follow that one can calculate $\boldsymbol{E}$ from ρ and σ because the free and bound charge densities depend on $\boldsymbol{E}$.

EXAMPLE The Field of an Infinite-Sheet Electret

An *electret* is the electrical equivalent of a permanent magnet. In most dielectrics the polarization vanishes immediately upon removal of the electric field, but some dielectrics retain their polarization for long periods. For example, certain polymers have extrapolated lifetimes of several thousand years at room temperature.

As a rule, electrets have the form of sheets, with the polarization normal to the surface. One way of *poling* an electret is to place the material in an electric field of about 10^8 volts/meter at about 100°C. A bound surface charge builds up as the molecules orient themselves, and the sample is then cooled down to room temperature without removing the electric field.

One commonly used material is polyvinylidene fluoride (PVF_2). This is a polymer composed of a chain of CH_2–CF_2 units. Its remnant polarization is typically 50 to 70 millicoulombs/meter^2. This material is used in various types of transducers, microphones for example, because it also has the property of going into a metastable polarized state when stretched.

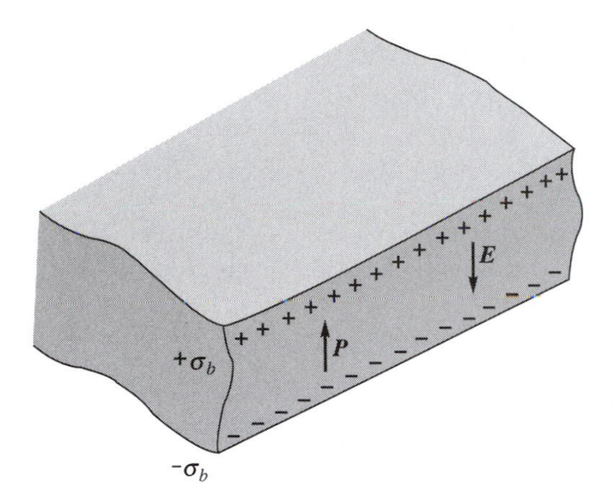

Figure 7-2 Section through a sheet electret. The vectors $\boldsymbol{E}$ and $\boldsymbol{P}$ point in opposite directions.

For a sheet electret, neglecting edge effects, the polarization P is uniform and equal to σ_b as in Fig. 7-2. Then ρ_b is zero, from Eq. 7-7. Inside the sheet,

$$E = \frac{\sigma_b}{\epsilon_0}, \quad P = \sigma_b. \tag{7-13}$$

7.5 GAUSS'S LAW IN DIELECTRICS

Say a given volume v contains various dielectrics, some of which may be partly inside and partly outside. The total free and bound charge within v is $Q_f + Q_b = Q$. There are no surface charges on the surface of v. Then *Gauss's law* (Section 3.6) relates the outward flux of $\boldsymbol{E}$ through the surface of area $\mathcal{A}$ to the net enclosed charge Q:

$$\boxed{\int_{\mathcal{A}} \boldsymbol{E} \cdot d\boldsymbol{\mathcal{A}} = \frac{Q_f + Q_b}{\epsilon_0} = \frac{Q}{\epsilon_0}.} \tag{7-14}$$

This is Gauss's law in its more general form.

If the volume v lies entirely inside a dielectric, there are no surface charges and

$$\int_{\mathcal{A}} \boldsymbol{E} \cdot d\boldsymbol{\mathcal{A}} = \frac{1}{\epsilon_0}\int_v (\rho_f + \rho_b)\, dv = \frac{1}{\epsilon_0}\int_v \rho\, dv, \tag{7-15}$$

where $\rho = \rho_f + \rho_b$ is the *total charge density*. Applying the divergence theorem to the surface integral of $\boldsymbol{E}$ gives the volume integral of $\boldsymbol{\nabla} \cdot \boldsymbol{E}$. Equating the integrands then yields

$$\boxed{\boldsymbol{\nabla} \cdot \boldsymbol{E} = \frac{\rho_f + \rho_b}{\epsilon_0} = \frac{\rho}{\epsilon_0},} \tag{7-16}$$

which is again Gauss's law, expressed in differential form.

This is one of *Maxwell's four fundamental equations of electromagnetism.*

7.6 POISSON'S AND LAPLACE'S EQUATIONS FOR V IN DIELECTRICS

Since $\boldsymbol{E} = -\boldsymbol{\nabla} V$, from Sec. 3.4, it follows that

$$\nabla^2 V = -\rho/\epsilon_0. \tag{7-17}$$

This is *Poisson's equation* for V in dielectrics.

Laplace's equation applies to regions where the total electric charge density is zero:

$$\nabla^2 V = 0, \tag{7-18}$$

as in Sec. 4.1.

7.7 THE ELECTRIC FLUX DENSITY $\boldsymbol{D}$. THE DIVERGENCE OF $\boldsymbol{D}$

According to Eq. 7-16,

$$\boldsymbol{\nabla} \cdot \boldsymbol{E} = \frac{\rho_f + \rho_b}{\epsilon_0}. \tag{7-19}$$

But we found in Sec. 7.3 that $\rho_b = -\boldsymbol{\nabla} \cdot \boldsymbol{P}$. Therefore

$$\boldsymbol{\nabla} \cdot (\epsilon_0 \boldsymbol{E} + \boldsymbol{P}) = \rho_f. \tag{7-20}$$

We conclude that the vector

$$\boldsymbol{D} = \epsilon_0 \boldsymbol{E} + \boldsymbol{P} \tag{7-21}$$

is such that its divergence is equal to the *free* volume charge density. This quantity is called the *electric flux density*. Thus

$$\boldsymbol{\nabla} \cdot \boldsymbol{D} = \rho_f. \tag{7-22}$$

It follows that, for any volume v that lies entirely inside a dielectric and that encloses a free charge Q_f with no surface charges,

$$\int_v \boldsymbol{\nabla} \cdot \boldsymbol{D}\, dv = \int_{\mathcal{A}} \boldsymbol{D} \cdot d\boldsymbol{\mathcal{A}} = \int_v \rho_f\, dv = Q_f. \tag{7-23}$$

This concurs with the relation $\boldsymbol{D} = \epsilon_0 \boldsymbol{E} + \boldsymbol{P}$:

$$\int_{\mathcal{A}} \boldsymbol{D} \cdot d\boldsymbol{\mathcal{A}} = \epsilon_0 \int_{\mathcal{A}} \boldsymbol{E} \cdot d\boldsymbol{\mathcal{A}} + \int_{\mathcal{A}} \boldsymbol{P} \cdot d\boldsymbol{\mathcal{A}} = Q + \int_v \boldsymbol{\nabla} \cdot \boldsymbol{P}\, dv \tag{7-24}$$

$$= Q - \int_v \rho_b\, dv = Q - Q_b = Q_f. \tag{7-25}$$

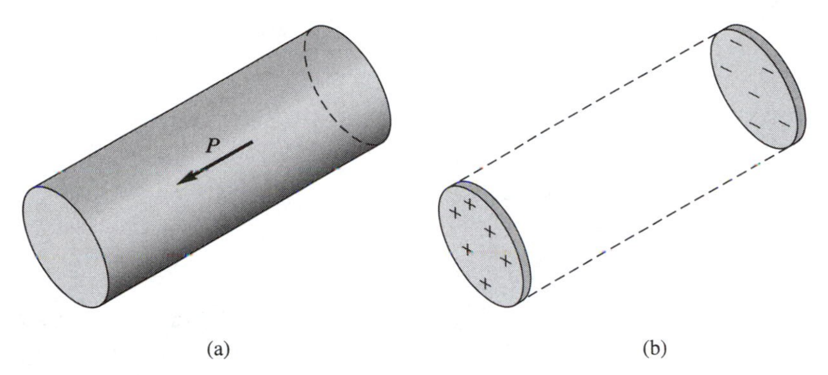

Figure 7-3 (a) Idealized bar electret polarized uniformly parallel to its axis. (b) The $\boldsymbol{E}$ field of the bar electret is the same as that of a pair of circular plates carrying uniform surface charge densities P of opposite polarities.

The fact that the divergence of $\boldsymbol{D}$ depends solely on the free charge density does *not* mean that $\boldsymbol{D}$ itself depends only on ρ_f. To find $\boldsymbol{D}$, we must integrate Eq. 7-22 subject to whatever boundary conditions apply.

In using the divergences of $\boldsymbol{E}$, $\boldsymbol{P}$, and $\boldsymbol{D}$, we assume implicitly the existence of the space derivatives. If we have to deal with the interface between two media, where these derivatives do not exist, then we must use the integral form of Gauss's law, which is therefore more general.

EXAMPLE The Infinite-Sheet Electret

Inside the electret of the example in Sec. 7.4, $\boldsymbol{E} = (\sigma_b/\epsilon_0)\hat{\boldsymbol{n}}$ and points downward. Also, $\boldsymbol{P}$ points upward and its magnitude is σ_b.Therefore $\boldsymbol{E}$ and $\boldsymbol{P}/\epsilon_0$ are equal in magnitude, but point in opposite directions, as in Fig. 7-2, so that $\boldsymbol{D} = \epsilon_0 \boldsymbol{E} + \boldsymbol{P} = 0$.

EXAMPLE The Bar Electret

Assume that, inside the bar electret of Fig. 7-3, the polarization $\boldsymbol{P}$ is uniform. (This is an interesting but unrealistic example because $\boldsymbol{P}$ is in fact nonuniform.) Then $\boldsymbol{\nabla} \cdot \boldsymbol{P}$ is zero and, from Eq. 7-7, ρ_b is zero. There are bound charges solely on the end faces. Then the $\boldsymbol{E}$-field, both inside and outside the bar electret, is that of a pair of circular sheets of charge of uniform charge densities $+P$ and $-P$, respectively, as in the figure. (The surface charge density on a conducting plate is greater near the edges than near the center.)

Outside, the lines of $\boldsymbol{D}$ are identical to the lines of $\boldsymbol{E}$ because $\boldsymbol{D} = \epsilon_0 \boldsymbol{E}$.

Inside, $\boldsymbol{D} = \epsilon_0 \boldsymbol{E} + \boldsymbol{P}$, with $\boldsymbol{E}$ and $\boldsymbol{P}$ pointing in roughly opposite directions. The divergence of $\boldsymbol{D}$ is everywhere zero. See Fig. 7-4.

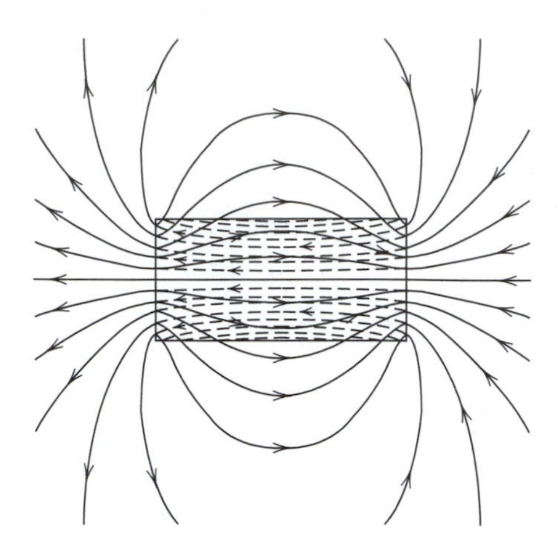

Figure 7-4 Lines of $\boldsymbol{E}$ (solid) for the idealized bar electret of Fig. 7-3. Lines of $\boldsymbol{D}$ are shown dashed inside the electret; outside they follow the lines of $\boldsymbol{E}$. The polarization is uniform and points to the left.

7.8 THE ELECTRIC SUSCEPTIBILITY χ_e

In most dielectrics $\boldsymbol{P}$ is proportional to $\boldsymbol{E}$ and points in the same direction. (This is *not* true of electrets, as we just saw!) Such dielectrics are *linear* and *isotropic*. So, in linear and isotropic dielectrics,

$$\boldsymbol{P} = \epsilon_0 \chi_e \boldsymbol{E}, \tag{7-26}$$

where χ_e is the *electric susceptibility* of the medium.

The dielectric is *homogeneous* if its electric susceptibility is independent of the coordinates.

7.9 THE RELATIVE PERMITTIVITY ϵ_r

In linear and isotropic dielectrics, Eq. 7-26 applies and

$$\boldsymbol{D} = \epsilon_0 \boldsymbol{E} + \boldsymbol{P} = \epsilon_0 (1 + \chi_e) \boldsymbol{E} = \epsilon_0 \epsilon_r \boldsymbol{E} = \epsilon \boldsymbol{E}, \tag{7-27}$$

where

$$\epsilon_r = 1 + \chi_e = \frac{\epsilon}{\epsilon_0} \tag{7-28}$$

is the *relative permittivity*. This quantity is dimensionless and larger than unity. The quantity ϵ is the *permittivity*. Its dimensions are the same as those of ϵ_0, namely, farads per meter. Thus, for linear and isotropic dielectrics,

$$\boldsymbol{P} = \epsilon_0 \chi_e \boldsymbol{E} = \epsilon_0 (\epsilon_r - 1) \boldsymbol{E}. \tag{7-29}$$

Table 7-1 shows the value of ϵ_r for some common dielectrics at three widely spaced frequencies.

The relative permittivities of gases at normal temperature and pressure are only slightly larger than unity. For example, $\epsilon_r = 1.000536$ for air at normal temperature and pressure.

Table 7-1 Relative permittivities of various materials near 25°C

	Frequency (hertz)		
Type	**100**	**10^6**	**10^{10}**
Barium titanate	1250	1140	100
Benzene	2.28	2.28	2.28
Birch (yellow)		2.7	1.95
Butyl rubber	2.43	2.40	2.38
Carbon tetrachloride	2.17	2.17	2.17
Fused silica	3.78	3.78	3.78
Glass (soda borosilicate)	5.0	4.84	4.82
Ice		4.15	3.20
Lucite	3.20	2.63	2.57
Neoprene	6.70	6.26	4.0
Polyethylene	2.26	2.26	2.26
Polystyrene	2.56	2.56	2.54
Sodium chloride		5.90	5.90
Soil (dry loam)		2.59	2.55
Steatite	6.55	6.53	6.51
Styrofoam	1.03	1.03	1.03
Teflon	2.1	2.1	2.08
Water	81	78.2	34
Wheat (red, winter)		4.3	2.6

The relative permittivities of good conductors are unknown because their conduction currents are so much larger than their polarization currents. One may assume that their permittivities are of the order of three to five, as for ordinary dielectrics.

EXAMPLE Free and Bound Volume Charge Densities

The presence of a free volume charge density ρ_f polarizes a dielectric and gives rise to a ρ_b. If the dielectric is homogeneous, isotropic, and linear, then

$$\boldsymbol{P} = \boldsymbol{D} - \epsilon_0 \boldsymbol{E} = \left(1 - \frac{1}{\epsilon_r}\right)\boldsymbol{D}, \tag{7-30}$$

$$\boldsymbol{\nabla} \cdot \boldsymbol{P} = \left(1 - \frac{1}{\epsilon_r}\right)\boldsymbol{\nabla} \cdot \boldsymbol{D}, \tag{7-31}$$

$$\rho_b = -\left(1 - \frac{1}{\epsilon_r}\right)\rho_f. \tag{7-32}$$

If ρ_f is zero, then ρ_b is zero and $\boldsymbol{\nabla} \cdot \boldsymbol{P}$ is also zero. If ρ_f is not zero, which is a rare occurrence, the total charge density ρ has the same sign as ρ_f, but it is smaller:

$$\rho = \rho_f + \rho_b = \rho_f - \boldsymbol{\nabla} \cdot \boldsymbol{P} = \frac{\rho_f}{\epsilon_r}. \tag{7-33}$$

As a consequence, *Gauss's law* (Eq. 7-19) can also be written as

$$\nabla \cdot \boldsymbol{E} = \frac{\rho_f}{\epsilon_r \epsilon_0}. \tag{7-34}$$

EXAMPLE Free and Bound Surface Charge Densities at the Interface Between a Dielectric and a Conductor

At the interface between a conductor and a linear and isotropic dielectric, the conductor carries a σ_f and the dielectric a σ_b. If the field is static, $\boldsymbol{E} = 0$ and $\boldsymbol{D} = 0$ inside the conductor. Since the tangential component of $\boldsymbol{E}$ is continuous across the interface, as we shall see in Sec. 8.2.3, both $\boldsymbol{D}$ and $\boldsymbol{E}$ are normal to the interface in the dielectric, close to the conductor.

Now imagine a thin Gaussian volume G straddling an area $\mathcal{A}$ of the interface as in Fig 7-5. Gauss's law, as stated in Sec. 7.5, tells us that the outward flux of $\boldsymbol{D}$ in Fig. 7-5 is equal to the enclosed free charge $\sigma_f \mathcal{A}$. Take the upward direction as positive. Then, since $|P| = |\sigma_b|$ in this example, from Sec. 7.3,

$$\sigma_f = D = \epsilon_r \epsilon_0 E = \epsilon_0 E + P = \epsilon_0 E - \sigma_b \tag{7-35}$$

and

$$E = \frac{\sigma_f + \sigma_b}{\epsilon_0}. \tag{7-36}$$

Observe that Gauss's law for $\boldsymbol{E}$, as stated in Eq. 7-14, leads to the same value of E.

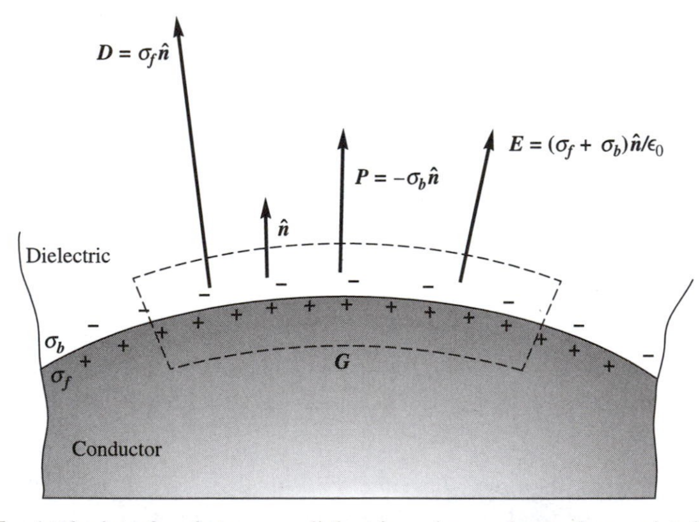

Figure 7-5 At the interface between a dielectric and a conductor there exists both a free surface charge density σ_f on the conductor and a bound surface charge density σ_b on the dielectric. If σ_f is positive, as it is here, then σ_b is negative and the vectors $\boldsymbol{D}, \boldsymbol{P}, \boldsymbol{E}$ point in the directions shown. With $\hat{\boldsymbol{n}}$ pointing into the dielectric as above, $\boldsymbol{P} = -\sigma_b \hat{\boldsymbol{n}}$. See Eq. 7-4.

Since $D = \epsilon_r \epsilon_0 E$,

$$\sigma_f = D = \epsilon_r(\sigma_f + \sigma_b), \qquad \sigma_b = -\left(1 - \frac{1}{\epsilon_r}\right)\sigma_f, \tag{7-37}$$

$$\sigma = \sigma_f + \sigma_b = \frac{\sigma_f}{\epsilon_r}. \tag{7-38}$$

Compare with Eqs. 7-32 and 7-33.

EXAMPLE Dielectric-Insulated Parallel-Plate Capacitor

Figure 7-6 shows a cross-section of a parallel-plate capacitor. We have shown air spaces on either side of the dielectric sheet so as to render our discussion more instructive, but as a rule the plates are in contact with the dielectric.

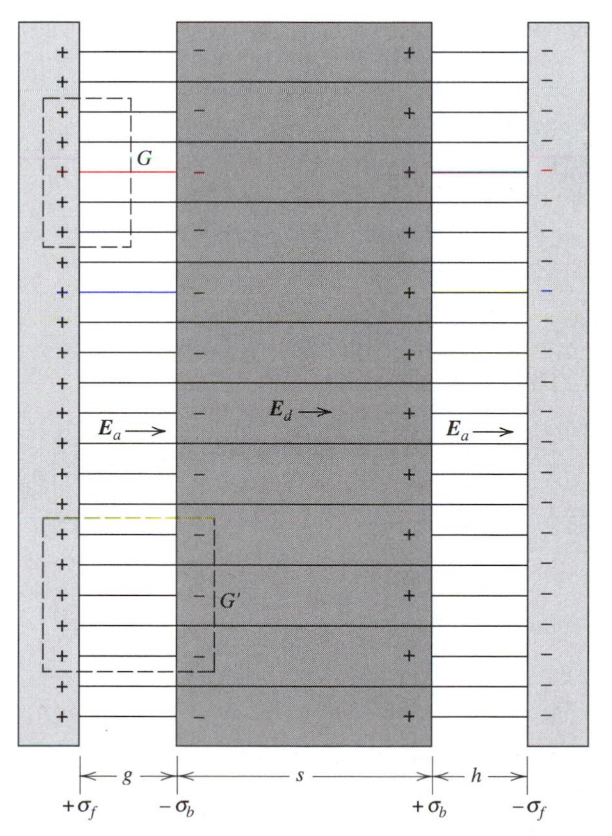

Figure 7-6 Pictorial representation of $\boldsymbol{E}$ inside a parallel-plate capacitor containing a sheet of dielectric with $\epsilon_r = 2$. Of course, individual lines of $\boldsymbol{E}$ do not exist. Here $|\sigma_b| = |\sigma_f|/2$.

We assume static charges and fields. Also, almost invariably, $\rho_f = 0$ in the dielectric and thus $\rho_b = 0$ and $\boldsymbol{\nabla} \cdot \boldsymbol{P} = 0$, from the first example above. Neglecting edge effects, the surface charge densities σ_f and σ_b are uniform. Similarly, the vectors $\boldsymbol{P}$, $\boldsymbol{E}$, and $\boldsymbol{D}$ are uniform within the dielectric and uniform within the air spaces.

The Gaussian volume G straddles the air-conductor interface. Within the conductor, $\boldsymbol{E} = 0$ and $\boldsymbol{D} = 0$. Therefore, in the air space,

$$E_a = \frac{\sigma_f}{\epsilon_0}, \qquad D_a = \sigma_f. \tag{7-39}$$

Similarly, for the Gaussian volume G',

$$E_d = \frac{\sigma_f + \sigma_b}{\epsilon_0}, \qquad D_d = \sigma_f = D_a. \tag{7-40}$$

But

$$E_d = \frac{D_d}{\epsilon_r \epsilon_0} = \frac{\sigma_f}{\epsilon_r \epsilon_0}. \tag{7-41}$$

It follows that

$$\sigma_b = -\left(1 - \frac{1}{\epsilon_r}\right)\sigma_f. \tag{7-42}$$

Of course, $|\sigma_b| = P$, from Sec. 7.3.

The voltage V across the capacitor is thus

$$V = \frac{\sigma_f}{\epsilon_0}(g + h) + \frac{\sigma_f}{\epsilon_r \epsilon_0} s = \frac{Q}{\epsilon_0 \mathscr{A}}\left(g + h + \frac{s}{\epsilon_r}\right), \tag{7-43}$$

where $\sigma_f = Q / \mathscr{A}$, Q being the magnitude of the charge and $\mathscr{A}$ the area of one plate.

The capacitance is

$$C = \frac{Q}{V} = \frac{\sigma_f \mathscr{A}}{V} = \frac{\epsilon_0 \mathscr{A}}{g + h + s/\epsilon_r}. \tag{7-44}$$

Without air spaces on either side of the dielectric, $g + h = 0$ and

$$V = \frac{Q}{\epsilon_r \epsilon_0 \mathscr{A}} s, \tag{7-45}$$

$$C = \frac{\epsilon_r \epsilon_0 \mathscr{A}}{s}. \tag{7-46}$$

This is ϵ_r times larger than the capacitance that we found in Sec. 6.4 for an air-insulated parallel-plate capacitor.

The measurement of a capacitance, with and without a dielectric, provides a convenient way of measuring a relative permittivity.

EXAMPLE The Field of Free Charges Embedded in a Dielectric

Imagine a small conducting sphere of radius a carrying a free charge Q_f and embedded in an infinite, homogeneous, isotropic, linear, and stationary (HILS) dielectric. The surface of the dielectric in contact with the charged sphere carries a bound charge of the opposite sign. At each point on the surface of the sphere Eq. 7-38 applies, and the net charge inside a sphere of radius $r > a$ is Q_f/ϵ_r. Thus, at any point inside the dielectric at a distance r from the charged conducting sphere,

$$E = \frac{Q_f/\epsilon_r}{4\pi\epsilon_0 r^2} = \frac{Q_f}{4\pi\epsilon_r\epsilon_0 r^2}. \tag{7-47}$$

Similarly, if a given distribution of free charges ρ_f is located inside a dielectric as above, the net charge density is ρ_f/ϵ_r and the E at any point is E_0/ϵ_r, where E_0 is the field that would obtain if the free charge distribution were situated in a vacuum.

If the dielectric is not infinite, then we must take into account the field of the bound charges at the surface.

7.10 THE DISPLACEMENT CURRENT DENSITY $\partial \mathbf{D}/\partial t$

Figure 7-7 shows a parallel-plate capacitor connected to a source of alternating voltage of frequency $\omega/2\pi$. The plates have an area $\mathcal{A}$ and are spaced by a distance s. We assume that edge effects are negligible. Then

$$I = \frac{V}{Z} = \frac{Es}{1/j\omega C} = j\omega Es\frac{\epsilon_r\epsilon_0\mathcal{A}}{s} \tag{7-48}$$

$$= \mathcal{A}j\omega\epsilon_r\epsilon_0 E = \mathcal{A}j\omega D = \mathcal{A}\frac{\partial D}{\partial t} = \frac{dQ}{dt}. \tag{7-49}$$

The current density in the capacitor is therefore $j\omega D$ or, more generally, $\partial \mathbf{D}/\partial t$. This quantity is called the *displacement current density* because $\mathbf{D}$ was formerly called the *electric displacement*.

The displacement current density consists of two parts:

$$\frac{\partial \mathbf{D}}{\partial t} = \frac{\partial}{\partial t}(\epsilon_0\mathbf{E} + \mathbf{P}) = \epsilon_0\frac{\partial \mathbf{E}}{\partial t} + \frac{\partial \mathbf{P}}{\partial t}. \tag{7-50}$$

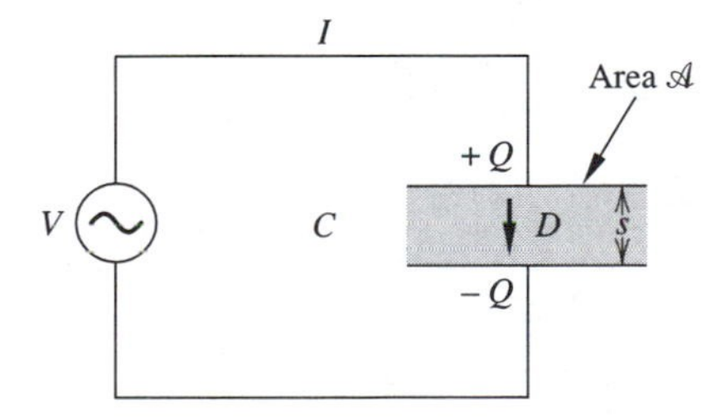

Figure 7-7 Parallel-plate capacitor connected to a source of alternating voltage. At a given instant, Q and $\mathbf{D}$ have the polarities shown. The displacement current density in the dielectric is $\partial \mathbf{D}/\partial t$.

The second term on the right is the polarization current density of Sec. 7.3, which results from the motion of bound charges. The first term, $\epsilon_0 \partial \boldsymbol{E}/\partial t$, is nameless. It can exist even in a vacuum.

We shall have many occasions to refer to the displacement current density.

7.11 SUMMARY

In a *polar dielectric*, each molecule possesses a permanent dipole moment. Upon application of an electric field, the molecules orient themselves more or less, and their dipole moments increase. The molecules of a *nonpolar dielectric* acquire a dipole moment when subjected to an electric field.

The dipole moment per cubic meter at a given point is called the *electric polarization* $\boldsymbol{P}$.

Polarization causes bound charges to accumulate, with *surface* and *volume charge densities*

$$\sigma_b = \boldsymbol{P} \cdot \hat{\boldsymbol{n}}, \qquad \rho_b = -\boldsymbol{\nabla} \cdot \boldsymbol{P}, \qquad \text{(7-4), (7-7)}$$

where $\hat{\boldsymbol{n}}$ is a unit vector that is normal to the surface and points *outward.*

In a fluctuating electric field the fluctuating polarization gives a *polarization current density*

$$\boldsymbol{J}_p = \frac{\partial \boldsymbol{P}}{\partial t}. \qquad \text{(7-9)}$$

The law of conservation of charge is

$$\boldsymbol{\nabla} \cdot \boldsymbol{J}_f = -\frac{\partial \rho_f}{\partial t}. \qquad \text{(7-10)}$$

One can calculate $\boldsymbol{E}$ both inside and outside dielectrics by treating both free and bound charges as if they were situated in a vacuum.

In dielectrics, *Gauss's law* states that, for a closed surface of area $\mathcal{A}$,

$$\boxed{\int_{\mathcal{A}} \boldsymbol{E} \cdot d\boldsymbol{\mathcal{A}} = \frac{Q}{\epsilon_0},} \qquad \text{(7-14)}$$

where Q is the total enclosed charge, free plus bound, of densities ρ_f and ρ_b, respectively. The differential form of Gauss's law is

$$\boxed{\boldsymbol{\nabla} \cdot \boldsymbol{E} = \frac{\rho}{\epsilon_0},} \qquad \text{(7-16)}$$

where, again, $\rho = \rho_f + \rho_b$.

As a consequence, Poisson's equation for V is

$$\nabla^2 V = -\frac{\rho}{\epsilon_0}. \qquad \text{(7-17)}$$

By definition, the *electric flux density* is

$$\boldsymbol{D} = \epsilon_0 \boldsymbol{E} + \boldsymbol{P}, \tag{7-21}$$

and thus

$$\nabla \cdot \boldsymbol{D} = \rho_f. \tag{7-22}$$

This is still another form of *Gauss's law.*

In linear and isotropic dielectrics,

$$\boldsymbol{P} = \epsilon_0 \chi_e \boldsymbol{E}, \tag{7-26}$$

where χ_e is a constant, independent of $\boldsymbol{E}$, called the *electric susceptibility.* Then

$$\boldsymbol{D} = \epsilon_0(1 + \chi_e)\boldsymbol{E} = \epsilon_r \epsilon_0 \boldsymbol{E} = \epsilon \boldsymbol{E}, \tag{7-27}$$

where ϵ_r is the *relative permittivity* and ϵ is the *permittivity* of the material. Also,

$$\boldsymbol{P} = \epsilon_0(\epsilon_r - 1)\boldsymbol{E}. \tag{7-29}$$

In a homogeneous, isotropic, and linear dielectric,

$$\rho_b = -\left(1 - \frac{1}{\epsilon_r}\right)\rho_f. \tag{7-32}$$

and Gauss's law is

$$\nabla \cdot \boldsymbol{E} = \frac{\rho_f}{\epsilon_r \epsilon_0}. \tag{7-34}$$

At the interface between a dielectric and a conductor, under static conditions,

$$\sigma_f = D = \epsilon_r \epsilon_0 E, \qquad \sigma_b = -\left(1 - \frac{1}{\epsilon_r}\right)\sigma_f, \qquad \text{(7-35), (7-37)}$$

with the sign conventions of Fig. 7-5.

The *displacement current density* is $\partial \boldsymbol{D}/\partial t$.

PROBLEMS

7-1. (*7.2*) The dipole moment of a carbon atom

A sample of diamond has a density of 3.5×10^3 kilograms/meter^3 and a polarization of 10^{-7} coulomb/meter^2.

(a) Calculate the average dipole moment per atom.

(b) What is the average separation between the centers of positive and negative charge? The carbon nucleus has a charge $+6e$ and is surrounded by six electrons. The diameter of an atom is of the order of 10^{-10} meter.

7-2. (*7.2*) Nonhomogeneous dielectric
Show that, in a nonhomogeneous dielectric, if $\rho_f = 0$, then $\rho_b = -(\epsilon_0/\epsilon_r)\boldsymbol{E} \cdot \boldsymbol{\nabla}\epsilon_r$.

7-3. (*7.3*) The volume and surface bound charge densities ρ_b and σ_b
Consider a block of dielectric with bound charge densities ρ_b and σ_b. Show mathematically that

$$\int_v \rho_b \, dv + \int_{\mathscr{A}} \sigma_b \, d\mathscr{A} = 0,$$

where v is the volume of the dielectric and $\mathscr{A}$ is its surface. In other words, the total net bound charge is zero.

7-4. (*7.4*) The field of a polarized dielectric sphere
The polarization $\boldsymbol{P}$ of a given dielectric sphere is uniform. Here is an elegant way of calculating its field, both inside and outside.

Take the sum of the fields of two spherical charges of densities ρ and $-\rho$, displaced by a distance Δz, one with respect to the other, Δz being the distance between the centers of charge of the individual dipoles. Then $\boldsymbol{P} = \rho \, \Delta z \, \hat{z}$ if the vector $\boldsymbol{P}$ points in the positive direction of the z-axis.

(a) Show that, inside, $\boldsymbol{E} = -\boldsymbol{P}/3\epsilon_0$ and is therefore uniform.
(b) Show that, outside, V and $\boldsymbol{E}$ are those of a dipole of moment $\boldsymbol{P}v$ situated at the center of the sphere of volume v.

7-5. (*7.5*) Coaxial line
Show that $\boldsymbol{\nabla} \cdot \boldsymbol{E} = 0$ in the dielectric of a coaxial line.
Hint: Apply the divergence theorem to a portion of the dielectric.

7-6. (*7.6*) Poisson's equation in nonhomogeneous media
Show that, in a nonhomogeneous medium, $\boldsymbol{\nabla}^2 V + (\boldsymbol{\nabla} V \cdot \boldsymbol{\nabla}\epsilon)/\epsilon = -\rho_f/\epsilon$.

7-7. (*7.7*) The surface charge densities on a dielectric
The surface of a dielectric carries a free charge density σ_f. Inside the dielectric, $\rho_f = 0$.

(a) Find E and D near the surface, on both sides of the interface.
(b) Find the ratio σ_b/σ_f.

7-8. (*7.7*) Electrically charged Lucite block
If high-energy electrons bombard a block of insulating material such as Lucite, the electrons penetrate the material and remain trapped inside. If one then gives the block a sharp rap with a conducting object, say with a center punch, the electrons escape, leaving a beautiful tree-like design where the plastic has broken down.

In one particular instance, a 0.1-microampere beam bombarded an area of 25 centimeter^2 of Lucite ($\epsilon_r = 3.2$) for 1 second, and essentially all the electrons were trapped about 6 millimeters below the surface in a region about 2 millimeters thick. The block was 12 millimeters thick.

In the following calculations, neglect edge effects and assume a uniform density for the trapped electrons. Assume also that both faces of the Lucite are in contact with grounded conducting plates.

(a) What is the bound charge density in the charged region?
(b) What is the bound charge density at the surface of the Lucite?
(c) Sketch graphs of D, E, V as functions of position inside the dielectric.
(d) Show that the potential at the center of the sheet of charge is about 4 kilovolts.
(e) What is the energy stored in the block? Could the block explode?

7-9. (*7.9*) Capacitance with and without a dielectric

A conducting body A, of arbitrary shape, is grounded. Another conducting body B, of arbitrary shape and position, is maintained at a potential V. In air, the capacitance is C_0. Show that the capacitance is $\epsilon_r C_0$ when the bodies are submerged in a large body of dielectric ϵ_r.

7-10. (*7.9*) Variable capacitor utilizing a printed-circuit board

A printed-circuit board is a sheet of plastic, about one millimeter thick, one side of which is covered with a thin coating of copper. Part of the copper can be etched away, leaving conducting paths that serve to interconnect resistors, capacitors, and so forth. The terminals of these components are soldered directly to the copper.

Figure 7-8 shows a variable capacitor in which a sliding conducting plate lies under a printed-circuit board that has been etched to give a prescribed variation of capacitance with position z.

At the position z, the capacitance is

$$C = \epsilon_r \epsilon_0 \frac{1}{t} \int_0^z y \, dz,$$

where t is the thickness of the plastic sheet of relative permittivity ϵ_r.

Set $\epsilon_r = 3$, $t = 1$ millimeter, and find $y(z)$ giving the following values of C, with C expressed in farads and z in meters:

(a) $10^{-9} z$,
(b) $10^{-8} z^2$.

In both cases, draw a curve of $y(z)$ from $z = 0$ to $z = 100$ millimeters, showing the unetched region on the circuit board.

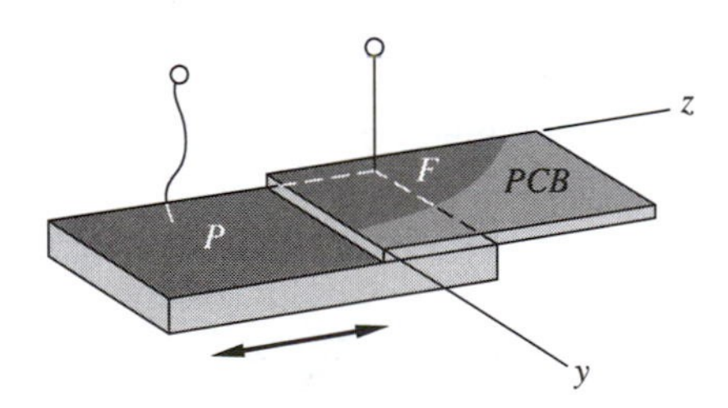

Figure 7-8 Variable capacitor made with a printed-circuit board PCB and a conducting plate P. The capacitance between the terminals depends on the position of the plate P and on the shape of the copper foil F. See Prob. 7-10.

7-11. (*7.9*) Nonhomogeneous dielectrics

Show that a nonhomogeneous dielectric can have a volume density of bound charge in the absence of a free charge density.

7-12. (*7.9*) The coaxial line

Figure 24-2 shows a section of coaxial line. A dielectric ϵ_r fills the space between the two conductors. From Prob. 6-13, the capacitance per unit length C' is $2\pi\epsilon_r\epsilon_0/\ln(R_2/R_1)$. The outer conductor is grounded, and the inner conductor is at the potential V_1.

(a) Calculate the charge per unit length λ' on the inner conductor.

(b) Show that the bound charge per unit length on the inner surface of the dielectric is $-\lambda'(1-1/\epsilon_r)$.

Thus the net charge per unit length at the radius R_1 is λ'/ϵ_r. The bound charge per unit length on the outer surface of the dielectric is similarly $+\lambda'(1-1/\epsilon_r)$, and the net charge per unit length at R_2 is $-\lambda'/\epsilon_r$.

(c) Show that the volume density of bound charge is zero.

(d) Draw graphs of D, E, and V as functions of the radius r from $r = R_1$ to $r = R_2$, for $V_1 = 100$ volts, $R_1 = 1.00$ millimeter, $R_2 = 10.0$ millimeters, $\epsilon_r = 3.00$.

7-13. (*7.9*) A parallel-plate capacitor with a nonuniform dielectric

The dielectric of a parallel-plate capacitor has a permittivity that varies as $\epsilon_{r0} + ax$, where x is the distance from one plate. The area of a plate is $\mathcal{A}$, and their spacing is s.

(a) Find the capacitance.

(b) Show that, if ϵ_r varies from ϵ_{r0} to $2\epsilon_{r0}$, then C is 1.44 times as large as if a were zero.

(c) Find P from the values of D and E for that case.

(d) Deduce the value of ρ_b.

(e) Now calculate ρ_b from the relation given in Prob. 7-2.

(f) Draw curves of E, ρ_b, and P as functions of x for $\epsilon_{r0} = 3.00$, $a = \epsilon_{r0}/s$, $s = 1.00$ millimeter when the applied voltage is 1.00 volt.

7-14. (*7.9*) The resistance and the capacitance between two electrodes

When the space between the plates of a parallel-plate capacitor is filled with a dielectric ϵ, it has a capacitance of C farads. If the dielectric is replaced by a material whose resistivity $\rho = 1/\sigma$ is much larger than that of the electrodes, the resistance between the electrodes is R ohms.

(a) Show that $RC = \rho\epsilon$, neglecting edge effects.

(b) Show that this result also applies to cylindrical and spherical capacitors.

(c) Show that this result applies to any pair of electrodes submerged in a medium whose resistivity ρ is much larger than that of the electrodes.

You should be able to show that the field is unaffected by the conductivity, with the above restriction.

(d) The capacitance per unit length between two parallel wires of diameter d and separated by a distance D is $C' = \pi\epsilon/\cosh^{-1}(D/d)$. Find the conductance $G = 1/R$ between parallel wires 10 millimeters in diameter separated by 100 millimeters and submerged in sea water ($\sigma = 5$).

7-15. (*7.10*) The current between two electrodes in a conducting medium.

See Prob. 7-14. The electrodes of part (c) are initially charged, and discharge through the medium. Show that $J_f + \partial D/\partial t = 0$.

CHAPTER

8

ELECTRIC FIELDS VI

Dielectric Materials B: Real Dielectrics, Continuity Conditions at an Interface. Stored Energy

In this second and last chapter on dielectric materials, we first describe briefly some characteristics of real dielectrics: finite conductivity, frequency and temperature dependence of ϵ_r, anisotropy, ferroelectricity, and hysteresis. Then we study the continuity conditions at the interface between two media. We shall require these conditions on many occasions. Finally, we return to the energy stored in an electric field, this time in the presence of dielectrics.

†Material marked with an asterisk can be omitted without losing continuity.

*8.1 REAL DIELECTRICS

Some dielectrics, such as polyethylene, are close to ideal: they are homogeneous, linear, and isotropic; their conductivity is close to zero; and their relative permittivity is practically independent of frequency. However, most dielectrics exhibit a more complex behavior.

*8.1.1 Lossy Dielectrics

Dielectrics that are somewhat conducting are said to be *lossy.* For example, most natural substances, such as wood or wheat, show a slight conductivity that is associated with the presence of water.

With alternating electric fields, it is convenient to express the conductivity in terms of a complex permittivity as follows. Consider a parallel-plate capacitor containing such a material, as in Fig 8-1. The current is the same as if one had an ideal nonconducting dielectric in the capacitor, with a resistance

$$R = \frac{s}{\sigma \mathcal{A}} \tag{8-1}$$

connected in parallel, σ being the conductivity of the dielectric (Sec. 4.3).

Then, from Sec. 6.6,

$$I = Vj\omega C + \frac{V}{R} = Vj\omega \frac{\epsilon_r \epsilon_0 \mathcal{A}}{s} + V\frac{\sigma \mathcal{A}}{s}. \tag{8-2}$$

It is convenient to define ϵ_r differently and to rewrite this as follows:

$$I = \frac{Vj\omega(\epsilon_r' - j\epsilon_r'')\epsilon_0 \mathcal{A}}{s} \tag{8-3}$$

$$= \frac{Vj\omega \epsilon_r \epsilon_0 \mathcal{A}}{s}. \tag{8-4}$$

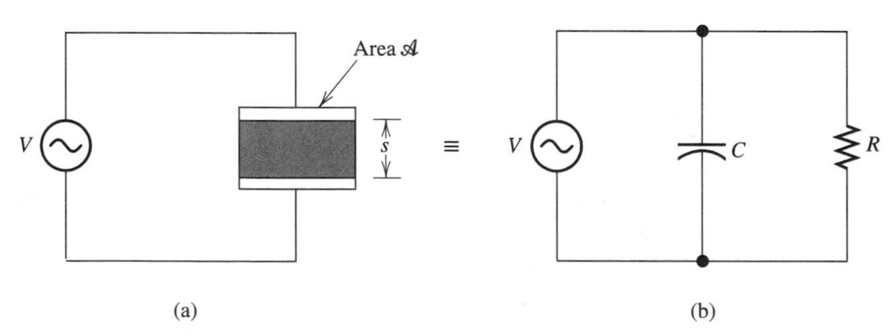

Figure 8-1 A parallel-plate capacitor containing a lossy dielectric as in (a) is equivalent to an ideal lossless capacitor, in parallel with a resistance R, as in (b).

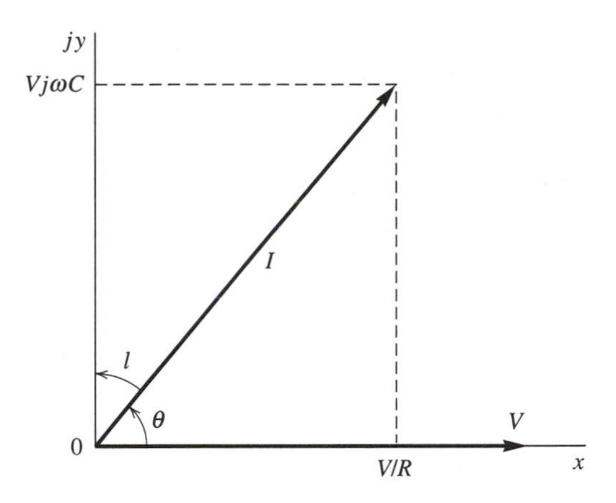

Figure 8-2 The phasor I for the circuit of Fig. 8-1(b).

We have now extended the definition of the relative permittivity to include conduction:

$$\epsilon_r = \epsilon_r' - j\epsilon_r'', \tag{8-5}$$

$$\epsilon_r'' = \frac{\sigma}{\epsilon_0 \omega}. \tag{8-6}$$

This *new* ϵ_r *is the complex relative permittivity* whose real part is the ϵ_r of Eq. 8-2. Note the negative sign before the imaginary part. The imaginary part ϵ_r'' pertains to the dissipation of energy as heat, and the real part ϵ_r' to the storage of energy in the electric field.

Unless specified otherwise, the term *relative permittivity,* as in Table 7-1, refers to the real part of ϵ_r.

Figure 8-2 shows the phasors V and I, the loss angle l, and the complementary angle θ, known as the *power-factor angle* for the circuit of Fig. 8-1(b).

*8.1.2 The Frequency and Temperature Dependence of ϵ_r

Generally speaking, ϵ_r' decreases with frequency, as in Table 7-1, whereas ϵ_r'' increases with frequency. This is only a rough rule of thumb because atoms and molecules exhibit resonances, especially above 1 gigahertz. The frequencies that are used for *dielectric heating* range from a few megahertz to a few gigahertz. Domestic microwave ovens usually operate at 2.45 gigahertz. They act on the water molecules in food.

A given dielectric can polarize through more than one of the processes described in Sec. 7.1, and the relative importance of a given process can vary with frequency. For example, water has a relative permittivity of 81 in an electrostatic field and of 1.8 at optical frequencies. The large static value results from the orientation of the permanent dipole moments, but the rotational inertia of the molecules prevents any significant response at optical frequencies. Similarly, the relative permittivity of sodium chloride is 5.9 in an electrostatic field and 2.3 at optical frequencies. The larger static value comes from ionic motion, which again is impossible at high frequencies.

As a rule, the permittivity of a substance increases by a large factor upon melting. For example, the ϵ_r of nitrobenzene increases from 3 to 35 at the melting point near 279 kelvins. In the solid, the permanent dipoles are frozen in the lattice and cannot rotate under the influence of an applied field.

Observe that since ϵ_r is a function of frequency, it is strictly definable only for a pure sine wave.

*8.1.3 Anisotropy

Crystalline solids commonly possess different dielectric properties in different crystal directions, because the ions can move more easily in some directions than in others. As a result, their susceptibility depends on direction, and $\boldsymbol{P}$ is not necessarily in the same direction as $\boldsymbol{E}$.

Then the x-component of $\boldsymbol{P}$ is of the form

$$P_x = \epsilon_0(\chi_{exx}E_x + \chi_{exy}E_y + \chi_{exz}E_z), \tag{8-7}$$

and similarly for the other two components. All three components of $\boldsymbol{P}$ depend on all three components of $\boldsymbol{E}$. The susceptibility thus has nine components and is a *tensor.* Actually, there are only six independent components, and if one chooses the coordinate axes properly with respect to the crystal, these six components reduce to three.

If the various components of the susceptibility are *not* functions of $\boldsymbol{E}$, then $\boldsymbol{P}$ is a linear function of the components of $\boldsymbol{E}$, and the dielectric is *linear.*

In anisotropic dielectrics, it is still true that

$$\boldsymbol{D} = \epsilon_0\boldsymbol{E} + \boldsymbol{P}, \qquad \boldsymbol{\nabla}\cdot\boldsymbol{D} = \boldsymbol{\nabla}\cdot(\epsilon_0\boldsymbol{E} + \boldsymbol{P}) = \rho_f, \tag{8-8}$$

but the three vectors are usually not parallel.

*8.1.4 Ferroelectricity

Ferroelectric dielectrics have the peculiar property of exhibiting spontaneous polarization over microscopic crystalline regions called *domains.* Figure 8-3 shows one specific case. The word *ferroelectric* originates from the fact that their behavior is in several respects similar to that of *ferromagnetic* substances (Chapter 16). When placed in an electric field, domains that happen to be correctly polarized grow at the expense of neighboring domains and eventually coalesce.

If the temperature of a ferroelectric substance increases slowly, the spontaneous polarization varies, usually in some complex fashion, and eventually disappears above the *Curie temperature.* This temperature is a characteristic of the material.

The main advantage of the ferroelectric dielectrics in common use in *ceramic capacitors* is that they have large relative permittivities, ranging up to 10,000. As a rule, these dielectrics are compounds of titanium.

Some ferroelectric dielectrics possess definite temperature coefficients that are useful in certain applications.

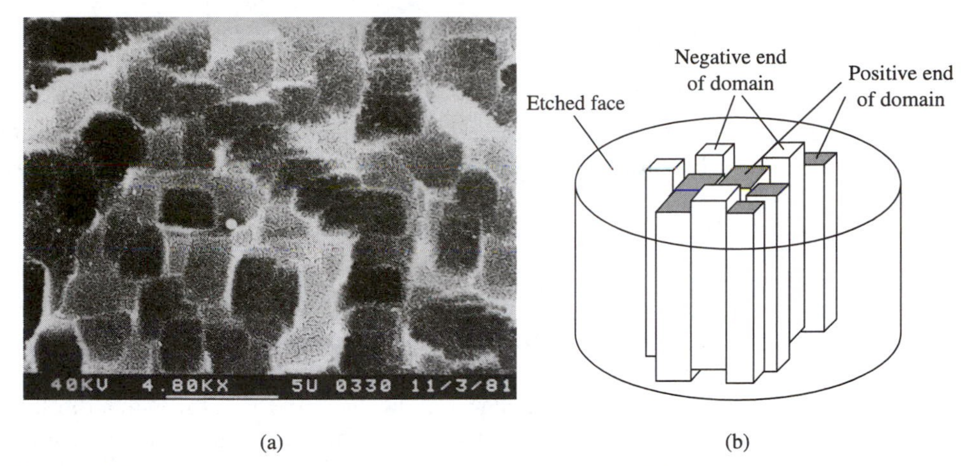

Figure 8-3 (a) Scanning electron microscope micrograph of the ferroelectric domains in Potassium-Sodium-Strontium-Barium niobate. (b) Schematic diagram showing the prismatic domains (Xu, 1991).

*8.1.5 Hysteresis

Ferroelectric materials also exhibit a property called *hysteresis,* illustrated in Fig. 8-4. Then the ratio D/E depends on the previous history of the material. Such materials are thus said to be *nonlinear* because D is not proportional to E.

The relative permittivity $\epsilon_r = D/\epsilon_0 E$ of a given ferroelectric sample thus has no definite value and can even be negative. When one quotes relative permittivities for ferroelectric materials, as we have done above and in Table 7-1, one refers to the order of magnitude of $D/\epsilon_0 E$ some distance from the origin $E = 0$, $D = 0$, in the first and third quadrants.

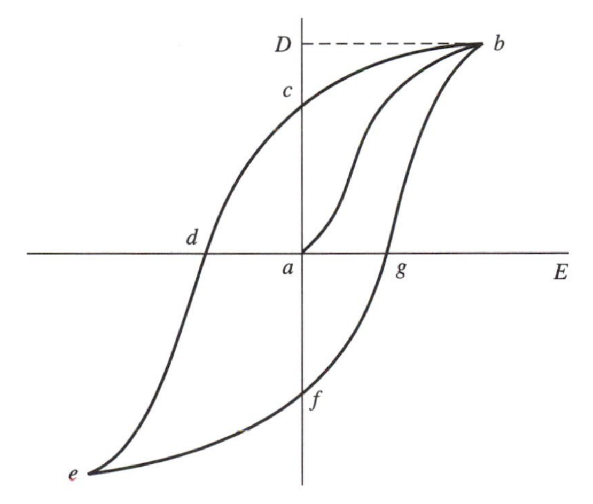

Figure 8-4 Hysteresis curve for a typical ferroelectric material. Starting with an unpolarized sample and $E = 0$ at point a, an increase in E increases D to point b. Then decreasing E to zero leaves a residual D at c, etc. The value of D thus depends not only on the value of E, but also on the previous history of the specimen.

8.2 THE CONTINUITY CONDITIONS AT AN INTERFACE

8.2.1 The Potential V

The potential V is continuous across the boundary between two media. Otherwise, a discontinuity would imply an infinitely large E, which is physically impossible.

8.2.2 The Normal Component of $\boldsymbol{D}$

Consider a short imaginary cylinder spanning the interface, and of cross section $\mathcal{A}$ as in Fig. 8-5. The top and bottom faces of the cylinder are parallel to the boundary and close to it. The interface carries a free surface charge density σ_f.

According to Gauss's law (Sec. 7.7), the net flux of $\boldsymbol{D}$ coming out of the cylinder is equal to the enclosed free charge. Now the only flux of $\boldsymbol{D}$ is that through the top and bottom faces, because the height of the cylinder is small. If now the area $\mathcal{A}$ is not too large, $\boldsymbol{D}$ is approximately uniform over it, and then

$$(D_{2n} - D_{1n})\mathcal{A} = \sigma_f \mathcal{A}, \qquad (\boldsymbol{D}_2 - \boldsymbol{D}_1) \cdot \hat{\boldsymbol{n}} = \sigma_f, \tag{8-9}$$

where $\hat{\boldsymbol{n}}$ is the unit vector normal to the interface and pointing from medium 1 to medium 2.

As a rule, the boundary between two dielectrics does not carry *free* charges, and then the normal component of $\boldsymbol{D}$ is continuous across the interface. Thus the normal component of $\boldsymbol{E}$ is discontinuous.

On the other hand, if one medium is a conductor and the other a dielectric, and if $\boldsymbol{D}$ is not a function of the time, then $\boldsymbol{D} = 0$ in the conductor and $D_n = \sigma_f$ in the dielectric. If $\boldsymbol{D}$ is a function of the time, Eq. 8.9 still applies, but $\boldsymbol{D}$ is not zero in the conductor.

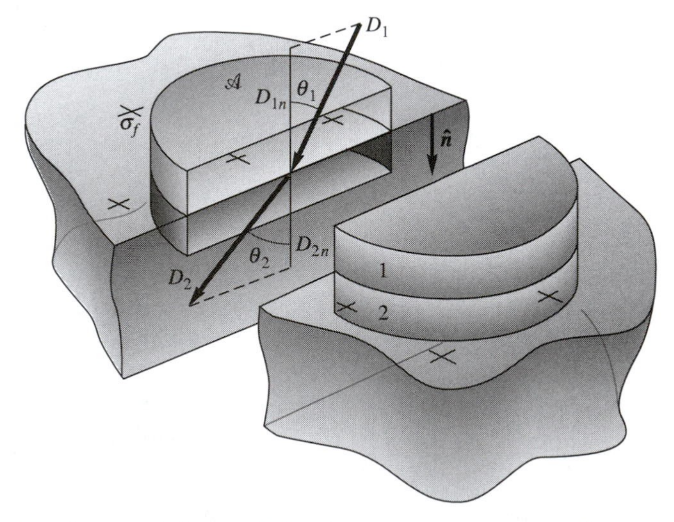

Figure 8-5 Imaginary cylinder straddling the interface between media 1 and 2 and delimiting an area $\mathcal{A}$. The difference $D_{2n} - D_{1n}$ between the *normal* components of $\boldsymbol{D}$ is equal to the free surface charge density σ_f.

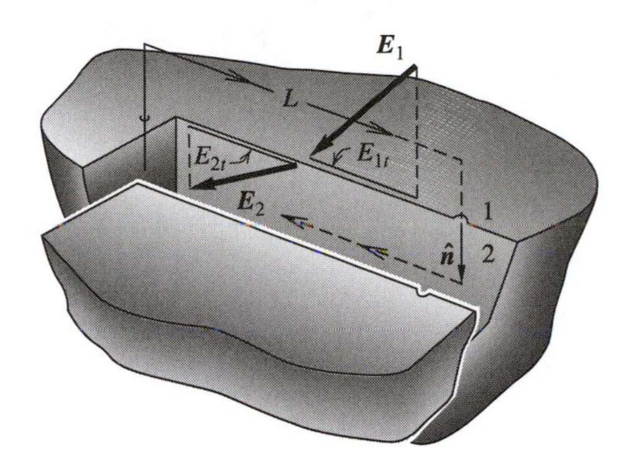

Figure 8-6 Closed path of integration spanning the interface between media 1 and 2. The tangential components of $\boldsymbol{E}$ are equal: $E_{1t} = E_{2t}$.

8.2.3 The Tangential Component of $\boldsymbol{E}$

Consider now the path shown in Fig. 8-6, with two sides of length L parallel to the boundary and close to it. The other two sides are infinitesimal. If L is short, $\boldsymbol{E}$ does not vary significantly over that distance, and integrating over the path yields

$$\oint \boldsymbol{E} \cdot d\boldsymbol{l} = E_{2t}L - E_{1t}L. \tag{8-10}$$

Now, from Sec. 3.4 this line integral is zero, and thus

$$E_{1t} = E_{2t},\dagger \qquad \text{or} \qquad (\boldsymbol{E}_1 - \boldsymbol{E}_2) \times \hat{\boldsymbol{n}} = 0, \tag{8-11}$$

with $\hat{\boldsymbol{n}}$ defined as above. The tangential component of $\boldsymbol{E}$ is continuous across any interface.

8.2.4 Bending of Lines of $\boldsymbol{E}$ at an Interface

The $\boldsymbol{E}$ and $\boldsymbol{D}$ vectors change direction at the interface between two different linear and isotropic dielectrics. In Fig. 8-7, if there is zero free surface charge density, the continuity of the normal component of $\boldsymbol{D}$ requires that

$$D_1 \cos\theta_1 = D_2 \cos\theta_2 \tag{8-12}$$

or that

$$\epsilon_{r1}\epsilon_0 E_1 \cos\theta_1 = \epsilon_{r2}\epsilon_0 E_2 \cos\theta_2 . \tag{8-13}$$

Also, from the continuity of the tangential component of $\boldsymbol{E}$,

$$E_1 \sin\theta_1 = E_2 \sin\theta_2 . \tag{8-14}$$

Dividing the third equation by the second gives

†The line integral of $\boldsymbol{E} \cdot d\boldsymbol{l}$ is zero in electrostatic fields. More generally, it is equal to minus the time derivative of the magnetic flux linking the path of integration, as we shall see in Sec. 18.2. However, since the enclosed area is zero, the flux is zero, and this equation is general.

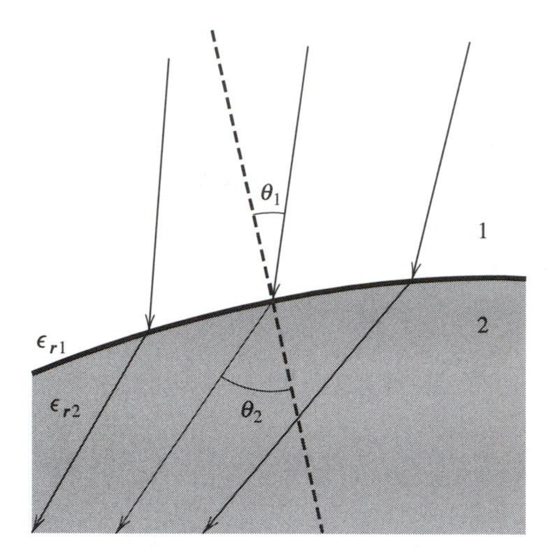

Figure 8-7 Lines of $\boldsymbol{E}$ crossing the interface between media 1 and 2. The permittivity of medium 2 is larger than that of medium 1.

$$\frac{\tan\theta_1}{\epsilon_{r1}} = \frac{\tan\theta_2}{\epsilon_{r2}}. \tag{8-15}$$

The larger angle from the normal is in the medium with the larger permittivity. The lines of $\boldsymbol{D}$ "prefer" the medium with the higher permittivity. This rule does not apply to electrets. See Prob. 8-10.

Since the *normal* component of $\boldsymbol{E}$ is discontinuous, there exists a bound surface charge density. From Gauss's law,

$$\sigma_b = \epsilon_0 E_2 \cos\theta_2 - \epsilon_0 E_1 \cos\theta_1 = \epsilon_0 E_2 \cos\theta_2\left(1 - \frac{\epsilon_{r2}}{\epsilon_{r1}}\right). \tag{8-16}$$

For example, if medium 1 is air and medium 2 is a dielectric, then the bound surface charge density is negative if $\boldsymbol{E}$ points into the surface of the dielectric.

8.3 THE POTENTIAL ENERGY OF A CHARGE DISTRIBUTION IN THE PRESENCE OF DIELECTRICS

In Chap. 6 we calculated the energy stored in an electric field when the charges reside either on conductors or in free space.

The addition of dielectrics to the field constitutes a major complication, but the problem becomes tractable with the following assumptions.

Our first two assumptions are realistic. (1) We assume that both the free and bound charge distributions are of finite extent. This makes the potential V equal to zero at infinity. (2) We assume that the dielectrics are linear ($\boldsymbol{D}$ is proportional to $\boldsymbol{E}$), but not necessarily homogeneous or isotropic.

We also assume that the dielectrics are fixed in position and rigid. In other words, the dielectrics do not distort under the electric forces.

8.3.1 The Potential Energy Expressed in Terms of ρ_f and V, and of $\boldsymbol{E}$

At any point,

$$V = \frac{1}{4\pi\epsilon_0}\int_v \frac{\rho_f + \rho_b}{r}\,dv + \frac{1}{4\pi\epsilon_0}\int_{\mathcal{A}} \frac{\sigma_f + \sigma_b}{r}\,d\mathcal{A}, \tag{8-17}$$

as in Sec. 7.4. Since all the dielectrics in the field are linear, by hypothesis, ρ_b is proportional to ρ_f and σ_b to σ_f (Sec. 7.9, examples).

We start with a zero charge density everywhere and gradually pull in charges from infinity. We disregard surface charges for the moment. Let the final potential and the final volume free charge density at a point be, respectively, V and ρ_f. At a given moment, the free volume charge density is $\alpha\rho_f$, where α increases gradually from zero to unity.

Since

$$\boldsymbol{\nabla}\cdot\boldsymbol{D} = \boldsymbol{\nabla}\cdot(\epsilon_r\epsilon_0\boldsymbol{E}) = -\epsilon_0\boldsymbol{\nabla}\cdot(\epsilon_r\boldsymbol{\nabla}V) = \rho_f, \tag{8-18}$$

V is proportional to ρ_f. Thus the potential is αV when the charge density is $\alpha\rho_f$.

Suppose that α increases to $\alpha + d\alpha$. Then, at a given point,

$$d\rho_f = \rho_f\,d\alpha, \qquad dV = V\,d\alpha. \tag{8-19}$$

The energy required to pull in the extra charge is

$$d\mathscr{E} = \int_v (\rho_f\,d\alpha)V\left(\alpha + \frac{d\alpha}{2}\right)dv, \tag{8-20}$$

where the volume v includes all the charges and where $V(\alpha + d\alpha/2)$ is the average potential during the operation. Disregarding the $(d\alpha)^2$ term, we find that

$$d\mathscr{E} = \int_v (\rho_f V\,dv)\alpha\,d\alpha, \tag{8-21}$$

$$\mathscr{E} = \int_0^1 \alpha\,d\alpha\int_v \rho_f V\,dv = \tfrac{1}{2}\int_v \rho_f V\,dv. \tag{8-22}$$

The potential energy stored in the field is thus equal to one-half of the volume integral of $\rho_f V$, exactly as in Sec. 6.1.2. However, V now depends on the nature, shape, size, and position of the dielectrics situated in the field.

Also, by analogy with Eq. 6-11, we expect that

$$\mathscr{E} = \int_v \frac{\epsilon_r\epsilon_0 E^2}{2}\,dv, \tag{8-23}$$

as long as ϵ_r is both linear and isotropic, and the energy density is

$$\mathscr{E}' = \epsilon_r \epsilon_0 E^2/2. \tag{8-24}$$

If you care to see a general proof, then read Sec. 8.3.2 below.

EXAMPLE The Dielectric-Insulated Parallel-Plate Capacitor

Real capacitors usually comprise two sheets of conducting material separated by a sheet of dielectric. Think of a dielectric-insulated parallel-plate capacitor. One of the plates is connected to ground. What is the energy required to charge the other plate to the potential V? It is $QV/2$, where $Q = VC$. Thus

$$\mathscr{E} = QV/2 = CV^2/2, \tag{8-25}$$

as in the first example in Sec. 6.4.

What is the effect of the dielectric? This is a bit tricky. For a given *charge* Q, the presence of the dielectric *decreases* V by a factor of ϵ_r, and hence *decreases* the stored energy $\mathscr{E}$ by the same factor. But C is proportional to ϵ_r and, for a given *potential* V, $\mathscr{E}$ is proportional to ϵ_r.

Say the parallel-plate capacitor has an area $\mathscr{A}$ and a spacing s. Neglect edge effects. Then

$$\mathscr{E} = \frac{1}{2}CV^2 = \frac{1}{2}\frac{\epsilon_r \epsilon_0 \mathscr{A}}{s}(sE)^2 = \frac{\epsilon_r \epsilon_0 E^2}{2} s\mathscr{A}, \tag{8-26}$$

where $s\mathscr{A}$ is the volume of the dielectric. So the energy density is $\epsilon_r \epsilon_0 E^2/2$, as above.

*8.3.2 The Energy Density Expressed in Terms of *E* and *D*: General Proof

Our thought experiment here will be similar to that of the starred Sec. 6.2.1.

Assume that there is one and only one region where V is minimum, with $V_{min} \le 0$. Similarly, there is one and only one region where $V = V_{max} \ge 0$. The dielectrics are linear.

Now imagine a conducting surface of area $\mathscr{A}$ along the equipotential $V = V_{min}$. There is zero field inside. The conductor expands slowly, from one equipotential to the next, sweeping through the dielectrics and picking up the free charges that stand in its way.

The conductor carries a free surface charge density σ. At the surface, $\boldsymbol{E}$ is normal and its magnitude is equal to σ/ϵ_0 as in Fig. 8-8. If the medium is not isotropic, $\boldsymbol{D}$ is not parallel to $\boldsymbol{E}$. However,

$$\sigma d\mathscr{A} = \boldsymbol{D} \cdot \boldsymbol{d\mathscr{A}}, \tag{8-27}$$

from Gauss's law, $\boldsymbol{D}$ being zero inside the conductor.

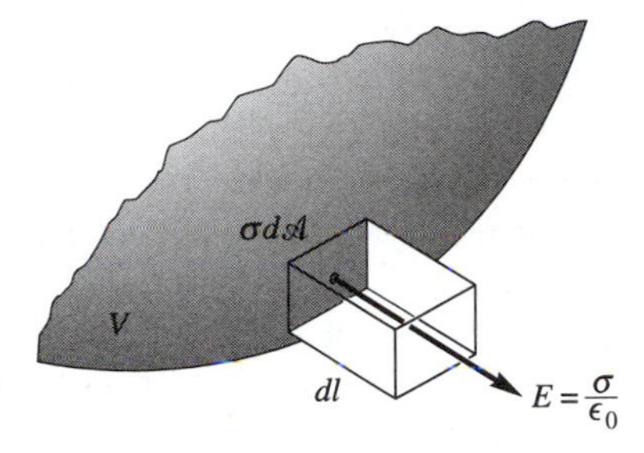

Figure 8-8 Thought experiment for calculating the energy stored in an electric field in the presence of dielectrics. The shaded volume of area $\mathscr{A}$ is a conductor that expands slowly, from one equipotential surface to the next. The surface charge density is σ. The element of area $\boldsymbol{d\mathscr{A}}$ moves out by a distance dl.

Say the element of area $\boldsymbol{d\mathscr{A}}$ moves out by a distance dl as in the figure, $\boldsymbol{dl}$ being normal to the surface. Then the work done *by* the field *on* the charge $\sigma\, d\mathscr{A}$ is

$$d\mathscr{E} = (\sigma\, d\mathscr{A})\left(\frac{E}{2}\right) dl, \tag{8-28}$$

the field acting on $\sigma\, d\mathscr{A}$ being $E/2$, from Sec. 6.7. So

$$d\mathscr{E} = \boldsymbol{D} \cdot \boldsymbol{d\mathscr{A}}\frac{E}{2}\, dl = \frac{\boldsymbol{D} \cdot \boldsymbol{E}}{2}\, d\mathscr{A}\, dl = \frac{\boldsymbol{D} \cdot \boldsymbol{E}}{2}\, dv. \tag{8-29}$$

The conducting surface continues to expand until it reaches the equipotential $V = V_{max}$. In the end, it has swept through all space, the free charges have been either removed to infinity or canceled by the addition of opposite charges on the conducting surface, and $\boldsymbol{E}$ is zero everywhere.

Thus the total work done by the field is

$$\mathscr{E} = \int_v \tfrac{1}{2}\boldsymbol{E} \cdot \boldsymbol{D}\, dv, \tag{8-30}$$

and the energy density is

$$\mathscr{E}' = \tfrac{1}{2}\boldsymbol{E} \cdot \boldsymbol{D}. \tag{8-31}$$

If the dielectrics are not only linear, but also isotropic, then $\boldsymbol{D} = \epsilon_r\epsilon_0\boldsymbol{E}$ and

$$\mathscr{E} = \int_v \frac{\epsilon_r\epsilon_0 E^2}{2}\, dv. \tag{8-32}$$

The energy density is then $\epsilon_r\epsilon_0 E^2/2$. This integral extends over all the volume occupied by the field.

8.4 ELECTRIC FORCES IN THE PRESENCE OF DIELECTRICS

The electric *surface* force density on a conductor is equal to the *volume* energy density in the field, as in Sec. 6.5, even if dielectrics are present. That is a simple rule to remember. Of course, if there is a dielectric, the energy density is $\epsilon_r\epsilon_0 E^2/2$, as in Eq. 8-32, instead of $\epsilon_0 E^2/2$. The method of virtual work for calculating forces applies.

Imagine a parallel-plate capacitor that can be submerged in a liquid dielectric. *For a given E* at the surface of a plate, the presence of the dielectric *increases* the energy density in the field by a factor of ϵ_r, and hence *increases* the electric force density on that plate by the same factor.

For a given free charge density σ_f on the plates, the dielectric has no effect on D: $D_d = D_a$. Then $E_d = E_a/\epsilon_r$, the energy density $\epsilon_r\epsilon_0 E^2/2$ *decreases* by a factor of ϵ_r, and the force density *decreases* by the same factor.

The case of solid dielectrics is best illustrated by Probs. 8-13 to 8-15. Refer also to the third Example of Sec. 7.9.

8.5 SUMMARY

Although some dielectrics are close to ideal, most exhibit a more complex behavior. If the conductivity σ of a dielectric is not negligible, then its *complex relative permittivity* is

$$\epsilon_r = \epsilon_r' - j\epsilon_r'', \qquad \epsilon_r'' = \frac{\sigma}{\epsilon_0 \omega}. \tag{8-5, 8-6}$$

Generally speaking, ϵ_r' *decreases* with frequency and ϵ_r'' *increases*. Temperature effects can be large.

Some dielectrics are *nonisotropic*. Then each component of $\boldsymbol{P}$ depends on the three components of $\boldsymbol{E}$, and the susceptibility is a tensor.

Ferroelectric dielectrics exhibit large permittivities and *hysteresis:* the value of $\boldsymbol{D}$ depends on the previous values of $\boldsymbol{E}$.

At the *interface* between any two media, both V and the tangential component of $\boldsymbol{E}$ are continuous. The normal component of $\boldsymbol{D}$ can be discontinuous, however, the discontinuity being equal to the free surface charge density on the interface.

Lines of $\boldsymbol{E}$ *bend at an interface*, the larger angle from the normal being in the dielectric with the larger permittivity.

The *potential energy* stored in a charge distribution, when dielectrics are present, is given either by

$$\mathscr{E} = \frac{1}{2}\int_v \rho_f V \, dv, \tag{8-22}$$

where v is any volume containing all the free charges of the system, or by

$$\mathscr{E} = \int_v \frac{1}{2}\boldsymbol{E} \cdot \boldsymbol{D} \, dv, \tag{8-30}$$

where v now includes all the regions where the $\boldsymbol{E}$ of the charge distribution exists. For linear and isotropic dielectrics,

$$\mathscr{E} = \int_v \tfrac{1}{2}\epsilon_r\epsilon_0 E^2 dv. \qquad (8\text{-}23, 8\text{-}32)$$

One may therefore ascribe to an electric field an *energy density* $\boldsymbol{E} \cdot \boldsymbol{D}/2$ or, for linear and isotropic dielectrics, $\epsilon_r\epsilon_0 E^2/2$. The electric force density on a conductor is equal to the energy density in the field.

PROBLEMS

8-1. (*8.1.1*) Parallel-plate capacitor with a conducting dielectric

Note: In this problem we call a surface charge density σ_{ch} and a conductivity σ_{co}.

The dielectric of a parallel-plate capacitor has a relative permittivity ϵ_r and a conductivity σ_{co}. The conductivity of the dielectric is much less than that of the plates, which makes $\boldsymbol{E}$ uniform between the plates.

(a) Assume surface charge densities σ_{ch} on the plates, and use Gauss's law to relate $\boldsymbol{E}$ to σ_{ch}. The capacitor is equivalent to a resistor and a capacitor in parallel.

(b) Show that when the capacitor is disconnected, the charge on the capacitor decreases by a factor of e in ϵ/σ_{co} seconds. This is the *relaxation time* of the capacitor.

8-2. (*8.1.1*) Parallel-plate capacitor with two conducting dielectrics

The dielectric of a parallel-plate capacitor is made up of two parts, as in Fig. 8-9. Calculate the surface charge density on the interface.

8-3. (*8.1.1*) Parallel-plate capacitor with a nonuniform dielectric

A parallel-plate capacitor has plates of area $\mathscr{A}$ separated by a distance s. Its dielectric has a conductivity $\sigma = a + bx$, where x is the distance to one plate, and a uniform relative permittivity ϵ_r.

(a) Calculate the resistance R of the capacitor.

(b) Show that with a steady voltage V applied to the electrodes, there is a uniform volume density of free charge.

(c) Sketch lines of $\boldsymbol{E}$ for $b > 0$. The field is not uniform.

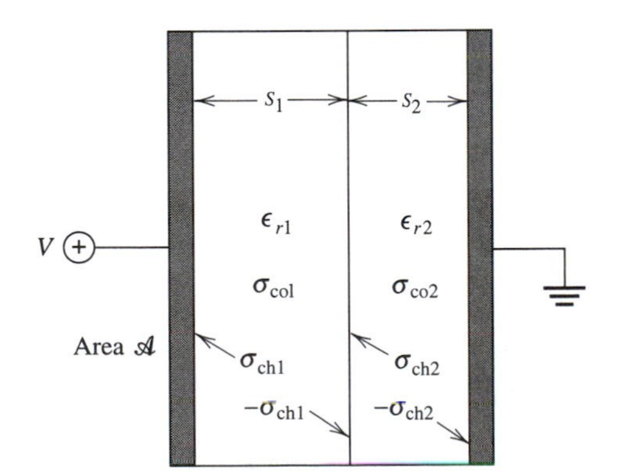

Figure 8-9

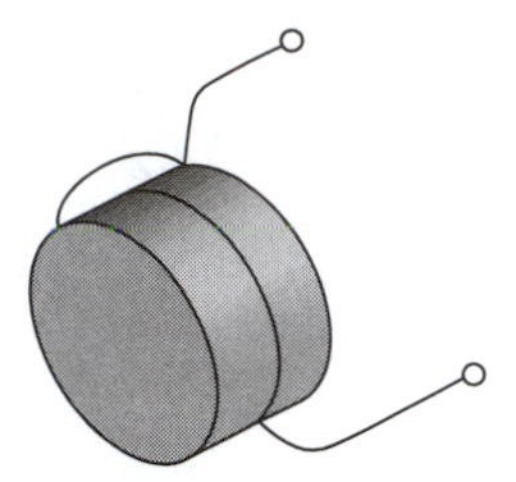

Figure 8-10

(d) With an alternating voltage across the electrodes,

$$I = \mathscr{A}\left(\sigma E + \frac{\partial D}{\partial t}\right) = \mathscr{A} J_t .$$

Show that $\nabla \cdot \boldsymbol{J}_t = 0$. Then J_t is independent of x.

(e) Find ρ_f.

8-4. (*8.1.4*) Ceramic capacitor

One particular ceramic capacitor is cylindrical, with three electrodes as in Fig. 8-10. Each ceramic disk has a diameter of 21 millimeters and a thickness of 0.5 millimeter. The nominal capacitance is 0.05 microfarad within a range of -20% to $+80\%$. What is the approximate value of ϵ_r?

8-5. (*8.1.4*) The hysteresis curve of a ferroelectric material

Figure 8-11 shows a schematic diagram of an instrument that has been used to plot the hysteresis curves of ferroelectric materials. An oscillator applies an alternating voltage V to two capacitors in series, the parallel plate capacitor C_x containing the material and a normal capacitor $C \gg C_x$. The voltage across C goes to the Y input of an XY recorder, while the voltage across $R_2 \ll R_1$ goes to the X input. The ferroelectric sample is a few millimeters thick, and $V \approx 10$ kilovolts, $f \approx 10^{-2}$ hertz.

Explain why the voltage across C is proportional to the D in the sample contained in C_x, while that across R_2 is proportional to E. The recorder draws essentially zero current at its X and Y terminals.

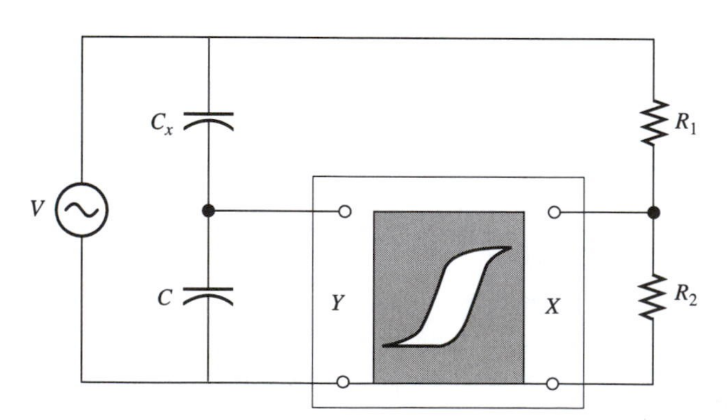

Figure 8-11

8-6. (*8.1.5*) The stored energy density in a nonlinear dielectric

A parallel-plate capacitor whose dielectric is nonlinear is connected to a power supply. The voltage V increases slightly by dV, and an extra charge dQ flows into the capacitor. Show that the density of stored energy increases by $E\, dD$.

8-7. (*8.1.5*) Hysteresis in ferroelectric materials

Show that the area of the hysteresis loop for a ferroelectric material is equal to the energy dissipated per cubic meter and per cycle.

8-8. (*8.2*) The boundary conditions at the interface between two conductors

Show that, at the interface between two conductors,

$$(\boldsymbol{E}_1 - \boldsymbol{E}_2) \times \hat{\boldsymbol{n}} = 0, \qquad |(\boldsymbol{J}_1 - \boldsymbol{J}_2) \times \hat{\boldsymbol{n}}| = \frac{d\sigma_{\text{ch}}}{dt},$$

where $\hat{\boldsymbol{n}}$ is a unit vector that is normal to the interface and points away from conductor 1, and where σ_{ch} is the surface charge density on the interface.

8-9. (*8.3.1*) Capacitive energy storage for a small vehicle

Investigate the possibility of propelling a small vehicle with an electric motor fed by a charged capacitor. Consider only the problem of energy storage.

(a) Show that the maximum energy density in the dielectric of a parallel-plate capacitor is $\epsilon a^2/2$, where a is the *dielectric strength* of the insulator, or the maximum E before breakdown. A good dielectric to use would be Mylar, which has a dielectric strength of 1.6×10^8 volts/meter when in the form of thin sheets, and a relative permittivity of 3.2.

(b) Calculate the energy density and the approximate size and mass of the capacitor that you would need to operate a 1-kilowatt motor for 1 hour.

8-10. (*8.2.4*) The bar electret

Figure 7-4 shows the $\boldsymbol{E}$ and $\boldsymbol{D}$ fields of a bar electret.

(a) Show that the lines of $\boldsymbol{E}$ do not bend at the cylindrical surface, but that the lines of $\boldsymbol{D}$ do bend.

(b) Show that the inverse is true at the end faces.

8-11. (*8.3.1*) Energy storage in capacitors

A one-microfarad capacitor is charged to a potential of one kilovolt.

How high could you lift a one-kilogram mass with the stored energy, if you could achieve 100% efficiency?

8-12. (*8.3.2*) The potential energy of a dipole situated in an electric field

(a) A dipole of *fixed* dipole moment $\boldsymbol{p}$ orients itself in a uniform electric field $\boldsymbol{E}$. Show that its potential energy is $-pE$, assuming that the potential energy is zero when the dipole axis is perpendicular to the field.

(b) A nonpolar molecule acquires a dipole moment in a uniform electric field that gradually increases from zero to E. Show that its potential energy is $+pE/2$.

8-13. (*8.3.2*) An accelerator for neutral molecules

There exist accelerators for *neutral* molecules that operate as follows. Figure 8-12 shows a pair of spheres that carry charges $+Q$ and $-Q$, and a molecule of dipole moment $\boldsymbol{p}$. The molecule accelerates toward the spheres until it reaches their midpoint.

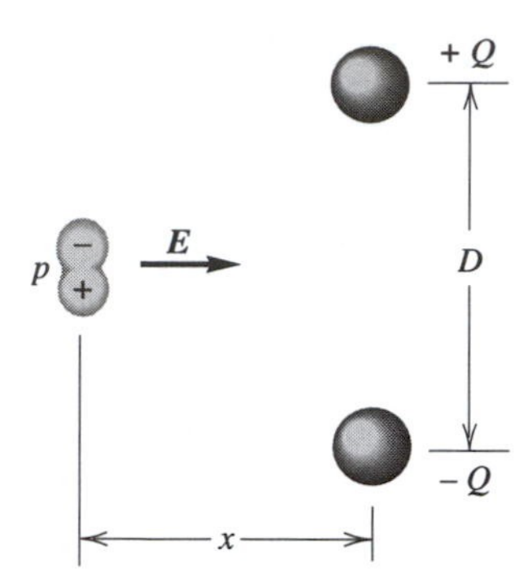

Figure 8-12

At that instant the spheres are discharged, and the molecule continues on its way at a constant velocity. Such accelerators serve to study the processes that occur during molecular collisions.

(a) Find the kinetic energy acquired by the molecule if the distance x in the figure is initially much larger than the distance D between the pair of electrodes. Consider the spheres as point charges, and apply the principle of conservation of energy. Assume that p is constant, and refer to Prob. 8-12. The dipole moment in fact increases as the molecule approaches, so that we have underestimated the energy.

(b) In one particular accelerator the electrode voltages are ±40 kilovolts, their radius is 0.25 millimeter, and D = 1.00 millimeter. Calculate the approximate value of the kinetic energy of a molecule in electronvolts for $p = 2 \times 10^{-29}$ coulomb-meter and for 700 stages. That accelerator has a length of 10 meters.

8-14. (*8.4*) Electrostatic clamps

Electrostatic clamps are used for holding workpieces while they are being machined, for holding silicon wafers during electron beam microfabrication, etc. They comprise an insulated conducting plate maintained at a potential of several thousand volts and covered with a thin insulating sheet. The workpiece or the wafer rests on the sheet and is grounded. It is advisable to apply a film of oil to the sheet to prevent sparking.

One particular type operates at 3000 volts and has a holding power of 2 atmospheres (2×10^5 pascals). If the insulator is Mylar ($\epsilon_r = 3.2$), what is its thickness?

8-15. (*8.4*) Self-clamping capacitor

A certain capacitor consists of two polished circular aluminum plates, 237 millimeters in diameter, separated by a sheet of plastic 0.762 millimeter thick, with a relative permittivity of 3.0. Thin films of air subsist between the electrodes and the plastic. This reduces the capacitance below the rated value, and there is no way of clamping the plates mechanically with sufficient force.

(a) Someone suggests that the electric force alone might be sufficient to clamp the plates, at the operating voltage of 60 kilovolts. What is your opinion?

(b) Show that, if the complete capacitor is submerged in an oil with $\epsilon_r \approx 3$, the force is 3 times less. See the next problem.

8-16. (*8.4*) The clamping force, with and without an air film

In Prob. 8-15 we found that the electric clamping force on a capacitor is larger by a factor of ϵ_r when there are air films between the electrodes and the dielectric, for a given applied voltage, or a given E. This is paradoxical. For a given E, the energy density is ϵ_r times larger in a dielectric than in air. Then the force should be ϵ_r times *larger* when there are no air films.

You can explain this paradox by considering the three capacitors of Fig. 8-13. In (2) the air film is much thinner than the dielectric. The electrode spacing is s in all three capacitors.

8-17. (*8.4*) Electric forces, with and without a dielectric

In the Example in Sec. 8.3.1 we saw that electric forces on conductors immersed in liquid dielectrics are *larger* than in air by the factor ϵ_r if the *voltages* are the same. They are *smaller* than in air by the same factor ϵ_r if the *charges* are the same.

Show that this is not in contradiction with Probs. 8-15 and 8-16, where we had a solid dielectric with thin films of air or oil next to the electrodes.

8-18. (*8.4*) Example of a large electric force

One author claims that he can attain fields of 4×10^7 volts per meter over a 2.5 millimeter gap in purified nitrobenzene ($\epsilon_r = 35$), and that the resulting electric force per unit area on the electrodes is then more than two atmospheres.

Is the force really that large?

One atmosphere equals about 10^5 pascals.

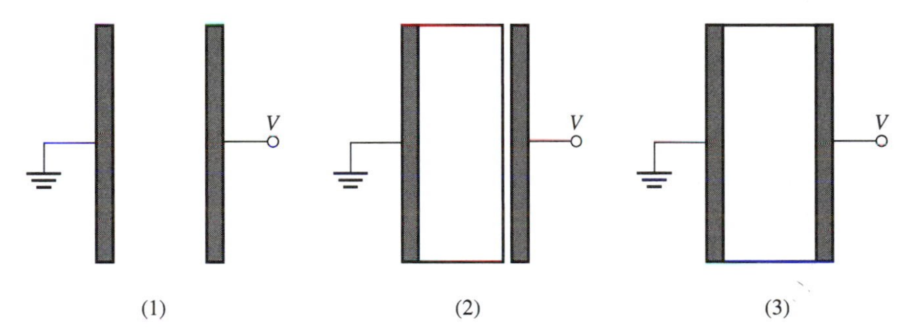

Figure 8-13

CHAPTER

9

*ELECTRIC FIELDS VII†

Images. Laplace's Equation in Rectangular Coordinates

Up to this point we have limited our discussion of electric fields to general considerations and to simple charge distributions. We now calculate more complex fields, both in this chapter and in the next one. This will complete our study of purely electric fields.

*9.1 IMAGES

If an electric charge distribution lies in a uniform dielectric that is in contact with a conducting body, then the method of images often provides the simplest route for calculating the electric field. The method is best explained by examples such as the two given below and is illustrated in Fig. 9-1.

†Material with an asterisk can be skipped without losing continuity.

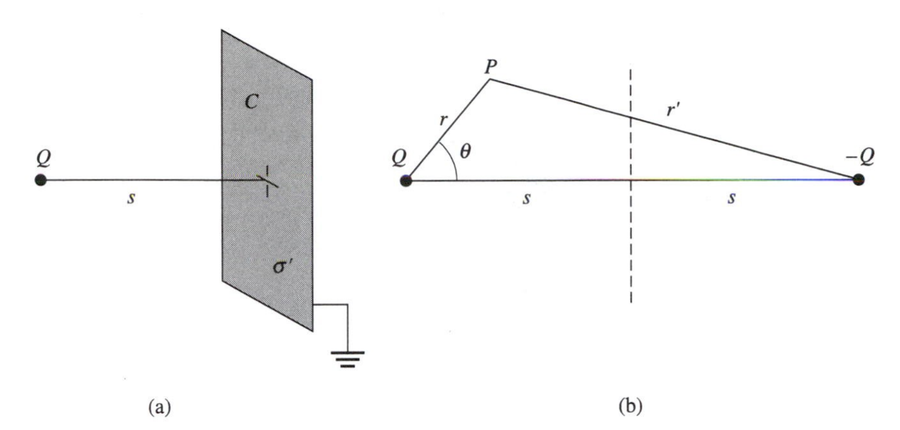

Figure 9-1 (a) Point charge Q near a large grounded, conducting sheet C. (b) We have replaced the conducting sheet by the image charge $-Q$. The field to the *left* of the dashed line is unaffected.

Call the charge distribution Q and the conductor C. One replaces C, on paper, by a second charge distribution such that the original boundary conditions are preserved. Then the field outside C is left undisturbed, according to the uniqueness theorem of Sec. 4.1.1. The second, fictitious, charge distribution is said to be the *image* of Q.

If Q lies in a dielectric that is in contact with a second dielectric, then we can calculate the fields in both dielectrics in a similar fashion, as we shall see in the second Example below.

EXAMPLE Point Charge Near a Grounded Conducting Plate

Figure 9-1(a) shows a point charge Q situated at a distance s from a large grounded conducting plate.†

Clearly, if we remove the plate and add an image charge $-Q$ at a further distance s as in Fig. 9-1(b), then every point in the plane formerly occupied by the plate will be equidistant from the two charges and will thus be at zero potential. So the field in the region to the left of the position formerly occupied by the plate is the same in both figures.

This is remarkable. The image provides a trivial solution to a problem that would, otherwise, be rather difficult.

Figure 9-2 shows lines of $\boldsymbol{E}$ and equipotentials.

†The earth carries a negative charge, and near the surface the electric field strength is about 100 volts/meter. However, the field of Q is independent of the field of the earth, according to the principle of superposition.

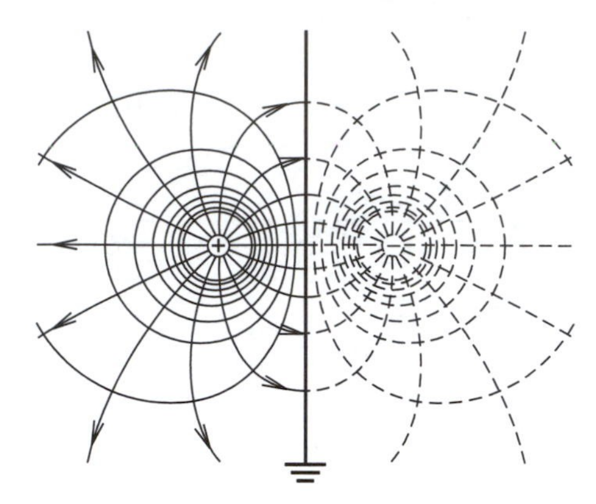

Figure 9-2 Lines of $\boldsymbol{E}$ (arrows) and equipotentials for the field of Fig. 9-1. Equipotentials and field lines near the charge get too close together to be shown at this scale. Rotating the figure about the horizontal axis of symmetry generates equipotential surfaces.

EXAMPLE Point Charge Near a Dielectric

Our point charge Q now lies in air close to a large block of dielectric as in Fig. 9-3. We wish to know $\boldsymbol{E}$ on both sides of the interface.

We could calculate $\boldsymbol{E}$ *without* using images in the following way. The field of Q polarizes the dielectric, and a bound surface charge density σ_b appears on the interface. Then the $\boldsymbol{E}$ at any point on either side of the interface is the same as if one had the point charge Q and the sheet of bound charges of density σ_b situated in a vacuum. As we shall see, it is simple enough to find σ_b. But then calculating $\boldsymbol{E}$ would be rather awkward. The problem is much easier to solve with images.

We first calculate the value of the normal component of $\boldsymbol{E}$ next to the interface and on both sides of it. It is this normal component, E_n, that we shall use as a boundary condition. In the process, we shall find σ_b.

The part of E_n that is ascribable to Q is $[Q/(4\pi\epsilon_0 r^2)](D/r)$. This field is the same on the two sides of the interface, as in Fig. 9-3.

From Gauss's law, σ_b provides equal normal field strengths $\sigma_b/(2\epsilon_0)$ on either side of the interface, oriented as in the figure. We shall find that the sign of σ_b is opposite that of Q.

Therefore, just inside the dielectric, the normal component of $\boldsymbol{E}$ pointing to the right is

$$E_{ni} = \frac{QD}{4\pi\epsilon_0 r^3} + \frac{\sigma_b}{2\epsilon_0}. \tag{9-1}$$

Now, from Eqs. 7-4 and 7-29,

$$\sigma_b = \boldsymbol{P} \cdot \hat{\boldsymbol{n}} = P_n = -\epsilon_0(\epsilon_r - 1)E_{ni}, \tag{9-2}$$

where the unit vector $\hat{\boldsymbol{n}}$ is normal to the interface and points left. Therefore

$$\sigma_b = -\frac{(\epsilon_r - 1)QD}{2\pi(\epsilon_r + 1)r^3}. \tag{9-3}$$

As expected, Q and σ_b have opposite signs.

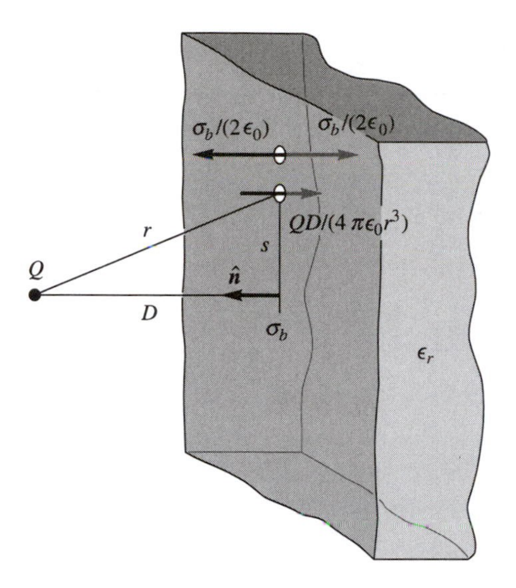

Figure 9-3 Point charge Q near a large block of dielectric. The arrows for the fields are oriented on the assumption that both Q and σ_b are positive. If Q is positive, σ_b is, in fact, negative.

If we take the right-hand direction as positive, the normal components of $\boldsymbol{E}$ outside and inside the dielectric, respectively, are as follows:

$$E_{no} = \left(1 + \frac{\epsilon_r - 1}{\epsilon_r + 1}\right)\frac{QD}{4\pi\epsilon_0 r^3} \tag{9-4}$$

$$= \frac{2\epsilon_r}{\epsilon_r + 1}\frac{QD}{4\pi\epsilon_0 r^3}, \tag{9-5}$$

$$E_{ni} = \left(1 - \frac{\epsilon_r - 1}{\epsilon_r + 1}\right)\frac{QD}{4\pi\epsilon_0 r^3} \tag{9-6}$$

$$= \frac{2}{\epsilon_r + 1}\frac{QD}{4\pi\epsilon_0 r^3}. \tag{9-7}$$

Since $\epsilon_r E_{ni} = E_{no}$, the normal component of $\boldsymbol{D}$ is continuous across the interface. This is to be expected because the free surface charge density is zero.

Thanks to the uniqueness theorem and to images, our problem is nearly solved.

The Field E_{no} on the Left of the Interface

1. The boundary condition is stated in Eqs. 9-4 and 9-5. From Eq. 9-4, we can remove the dielectric and replace it by an image charge Q', as in Fig. 9-4(a), without disturbing E_{no}. Then, from the uniqueness theorem, the field everywhere on the left of the interface is simply the field of Q plus that of Q'.
2. Now, according to Eq. 9-5, E_{no} is also unaffected if one removes the dielectric and replaces Q by a charge

$$Q''' = \frac{2\epsilon_r}{\epsilon_r + 1}Q, \tag{9-8}$$

as in Fig. 9-4(b).

The Field E_{ni} on the Right of the Interface

1. From Eq. 9-6, E_{ni} is the same as if one had, instead of Q and the dielectric, the charge Q and a charge $-Q'$ at a distance D to the right of the interface.
2. From Eq. 9-7, the field on the right of the interface is the same as if there were no dielectric and a charge Q'' were substituted for Q, as in Fig. 9-4(c).

   Figure 9-5 shows lines of $\boldsymbol{D}$ and equipotentials for this field.

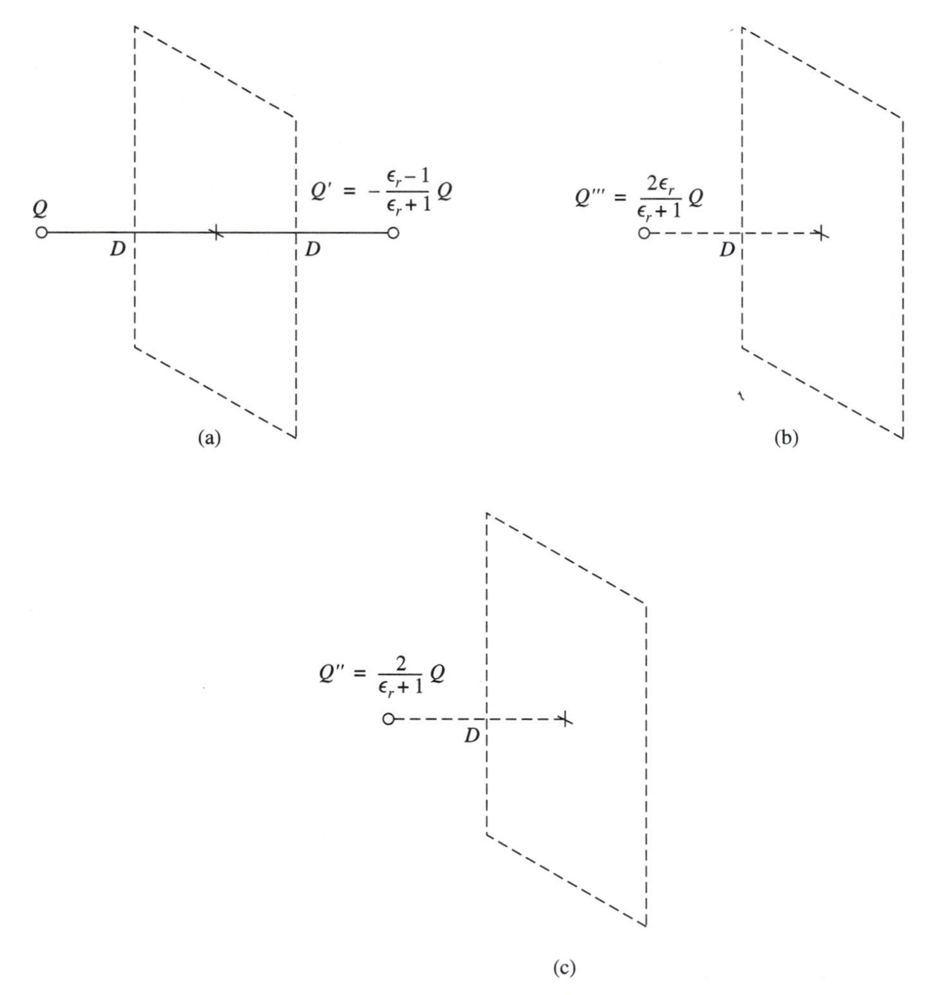

Figure 9-4 (a), (b) The field on the *left* of the interface of Fig. 9-3 is the same as that of these charges. (c) The field on the *right* of the interface is the same as that of Fig. 9-3.

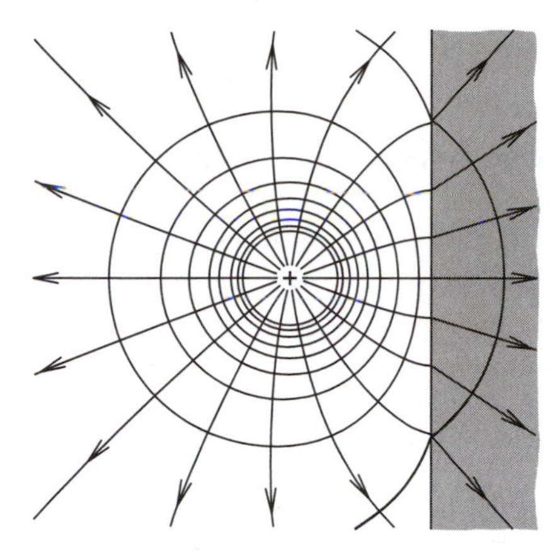

Figure 9-5 Lines of $\mathbf{D}$ (arrows) and equipotentials for a point charge situated near a dielectric. Rotating the figure about the horizontal axis of symmetry generates equipotential surfaces. The figure does not show equipotentials near the point charge.

*9.2 SOLVING LAPLACE'S EQUATION IN RECTANGULAR COORDINATES

Solutions of Laplace's equation $\nabla^2 V = 0$ are known as *harmonic functions*. These functions possess a number of general properties, from among which we shall use the following one. If one can find solutions $V_1, V_2, V_3, \ldots$ of Laplace's equation, then any linear combination $A_1 V_1 + A_2 V_2 + A_3 V_3 + \cdots$, where the A's are constants, is also a solution. This becomes obvious upon substituting the sum into the original equation.

As a rule, one can solve Laplace's equation by *separating the variables*. For example, in Cartesian coordinates, one can seek solutions of the form

$$V = X(x)Y(y)Z(z), \tag{9-9}$$

where X is a function of x only, Y a function of y only, and Z a function of z only. Substituting into Laplace's equation yields

$$YZ\frac{d^2X}{dx^2} + ZX\frac{d^2Y}{dy^2} + XY\frac{d^2Z}{dz^2} = 0. \tag{9-10}$$

Dividing by XYZ gives

$$\frac{1}{X}\frac{d^2X}{dx^2} + \frac{1}{Y}\frac{d^2Y}{dy^2} + \frac{1}{Z}\frac{d^2Z}{dz^2} = 0, \tag{9-11}$$

where the first term depends solely on x, and similarly for the other two terms.

Since these three terms add to zero at any point in the field, each one is equal to a constant and

$$\frac{d^2X}{dx^2} = C_1 X, \qquad \frac{d^2Y}{dy^2} = C_2 Y, \qquad \frac{d^2Z}{dz^2} = C_3 Z, \tag{9-12}$$

with

$$C_1 + C_2 + C_3 = 0. \tag{9-13}$$

We have now separated the variables. Solving the three equations separately yields X, Y, Z, and thus V.

Observe how astute this is: we have transformed a *partial* differential equation in all three variables x, y, z into three simple *ordinary* differential equations in x, and y, and z.

Sometimes, as in the example below, one has to sum an infinite number of such solutions, each multiplied by a suitable coefficient, to fit a given boundary condition.

EXAMPLE The Field Between Two Parallel Conducting Plates at Different Potentials

Here is a really simple example. Figure 9-6 shows two plane parallel electrodes separated by a distance a, with the lower one grounded and the upper one maintained at the potential V_0. The medium between the plates is air. We assume that the field between the plates is independent of both x and z, and we disregard edge effects. Then the voltage V at an arbitrary point between the plates satisfies the equations

$$\partial V/\partial x = 0, \quad \partial V/\partial z = 0, \tag{9-14}$$

and

$$\partial^2 V/\partial x^2 = 0, \quad \partial^2 V/\partial z^2 = 0. \tag{9-15}$$

So Laplace's equation reduces to

$$\partial^2 V/\partial y^2 = 0, \tag{9-16}$$

and

$$\partial V/\partial y = K, \tag{9-17}$$

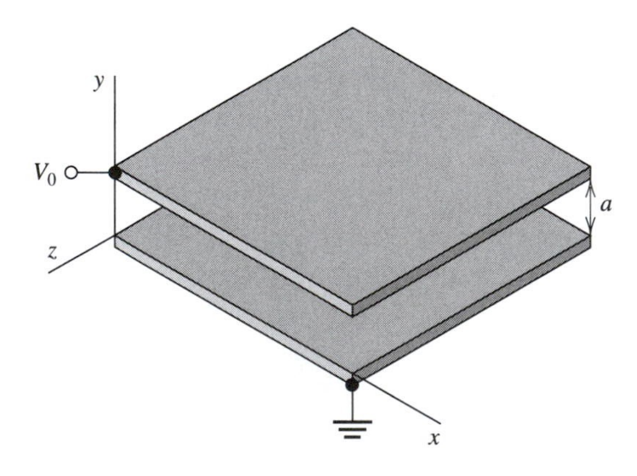

Figure 9-6 Pair of plane parallel electrodes.

where K is a constant, so that

$$V = Ky, \tag{9-18}$$

with $V = 0$ at $y = 0$. But $V = V_0$ at $y = a$. Then $K = V_0/a$ and, at the distance y above the lower plate, with $y \leq a$,

$$V = (V_0/a)y. \tag{9-19}$$

What are the surface charge densities on the electrodes? The positive electrode carries a positive charge of surface density given by Gauss's law of Sec. 3.6: it is $\epsilon_0 E$, or $\epsilon_0 V_0/a$. The negative electrode carries a charge of the same density, but negative.

EXAMPLE **The Field Between Two Grounded Semi-Infinite Parallel Electrodes Terminated by a Plane Electrode Maintained at a Fixed Potential. Fourier Series**

Figure 9-7 shows the electrodes. We wish to find $V(x,y)$ between the plates. By hypothesis, the field is independent of z and $C_3 = 0$. Then

$$\frac{d^2X}{dx^2} = k^2X, \quad \frac{d^2Y}{dy^2} = -k^2Y. \tag{9-20}$$

We have substituted k^2 for C_1 and $-k^2$ for C_2 to avoid square roots in the solution. As we shall see immediately, C_2 must be negative.

We solve the Y equation by setting

$$Y = A \sin ky + B \cos ky, \tag{9-21}$$

where A and B are constants. This solution is easily verified by substitution.

Now V must satisfy the following boundary conditions: (1) $V = 0$ at $y = 0$, $y = b$; (2) $V = V_0$ at $x = 0$; (3) $V \rightarrow 0$ as $x \rightarrow \infty$.

Since $V = 0$ at $y = 0$, then $B = 0$.

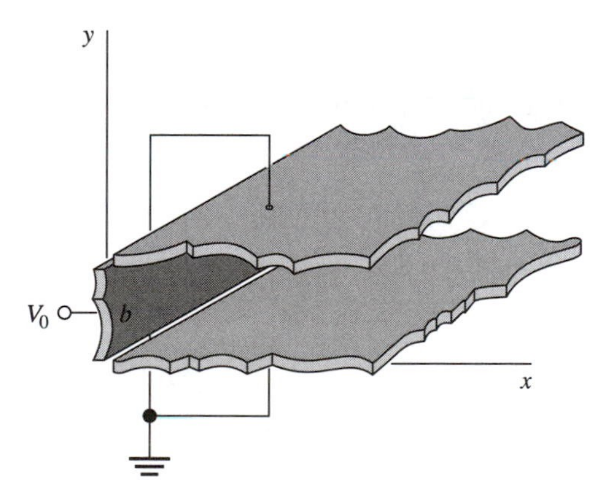

Figure 9-7 Grounded, plane, parallel, and semi-infinite electrodes terminated by a plane electrode at the potential V_0.

Also, the condition $V = 0$ at $y = b$ requires that

$$kb = n\pi, \quad k = \frac{n\pi}{b} \qquad (n = 1, 2, 3, \ldots). \tag{9-22}$$

So

$$Y = A \sin \frac{n\pi y}{b} \qquad (n = 1, 2, 3, \ldots). \tag{9-23}$$

We ignore the value $n = 0$ because it corresponds to zero field.

The differential equation for X is now

$$\frac{d^2X}{dx^2} = \left(\frac{n\pi}{b}\right)^2 X, \tag{9-24}$$

and

$$X = G \exp \frac{n\pi x}{b} + H \exp\left(-\frac{n\pi x}{b}\right), \tag{9-25}$$

where G and H are constants of integration. We can again verify this solution by substitution. Since V tends to zero as x tends to infinity, $G = 0$. Thus

$$V = XY = C \sin \frac{n\pi y}{b} \exp\left(-\frac{n\pi x}{b}\right), \tag{9-26}$$

where C is another constant.

But this is not right! Although this V satisfies conditions (1) and (3), it clearly does not satisfy condition (2). We can also satisfy condition (2) by adding an infinite number of such solutions:

$$V = \sum_{n=1}^{\infty} C_n \sin \frac{n\pi y}{b} \exp\left(-\frac{n\pi x}{b}\right). \tag{9-27}$$

Then we use condition (2) to evaluate the coefficients C_n by setting $x = 0$. So

$$V_0 = \sum_{n=1}^{\infty} C_n \sin \frac{n\pi y}{b} \tag{9-28}$$

for all y between 0 and b.

An infinite series of the form

$$\sum_{n=0}^{\infty} \left(C_n \sin \frac{n\pi y}{b} + D_n \cos \frac{n\pi y}{b}\right),$$

where C_n and D_n are constants, is a *Fourier series* and forms a *complete set:* given a reasonably well-behaved† function $V(y)$ defined in the interval $y = 0$ to $y = b$, there exists a Fourier series that is equal to $V(y)$ in this interval.‡

We can find the values of the C_n coefficients by an ingenious technique devised by Fourier. First, we multiply both sides of Eq. 9-28 by $\sin(p\pi y/b)$, where p is an integer, and then we integrate from $y = 0$ to $y = b$:

$$\int_0^b V_0 \sin\frac{p\pi y}{b}\,dy = \sum_{n=1}^{\infty}\int_0^b C_n \sin\frac{n\pi y}{b}\sin\frac{p\pi y}{b}\,dy. \tag{9-29}$$

On the left-hand side,

$$\int_0^b V_0 \sin\frac{p\pi y}{b}\,dy = \begin{cases} \dfrac{2bV_0}{p\pi} & \text{if } p \text{ is odd} \\ 0 & \text{if } p \text{ is even.} \end{cases} \tag{9-30}$$

The terms of a Fourier series are thus said to be *orthogonal.* On the right,

$$\int_0^b C_n \sin\frac{n\pi y}{b}\sin\frac{p\pi y}{b}\,dy = \begin{cases} 0 & \text{if } p \neq n \\ \dfrac{C_n b}{2} & \text{if } p = n. \end{cases} \tag{9-31}$$

It follows that

$$C_n = \begin{cases} \dfrac{4V_0}{n\pi} & \text{if } n \text{ is odd} \\ 0 & \text{if } n \text{ is even.} \end{cases} \tag{9-32}$$

Finally,

$$V = \frac{4V_0}{\pi}\sum_{1,3,5\ldots}^{\infty}\frac{1}{n}\sin\frac{n\pi y}{b}\exp\left(-\frac{n\pi x}{b}\right). \tag{9-33}$$

The successive terms in the series become progressively less important, because of the factor $1/n$, but mostly because of the exponential function. Figure 9-8 shows the degree of approximation achieved with 1, 3, 10, and 100 terms of the series.

Note how fast the exponential function decreases with x. For $n = 1$ and $x = b$ it is down to $\exp(-\pi)$, or 4%. Roughly speaking, the field of the charged plate does not

†By a "reasonably well-behaved" function, we mean one that is continuous, or at least piecewise continuous, and that does not become infinite at any point. Most boundary conditions encountered in practice possess these characteristics.

‡Except possibly at the endpoints $y = 0$ and $y = b$, but this discrepancy between V and the Fourier series is usually unimportant in practice.

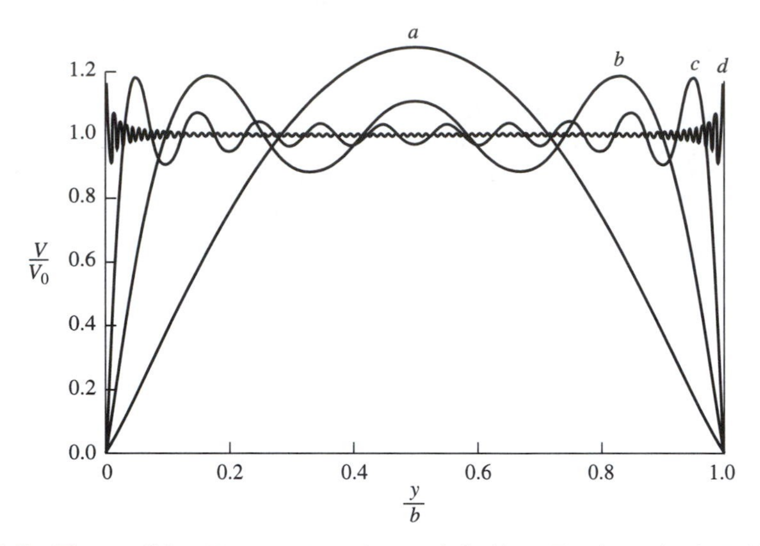

Figure 9-8 The condition $V = V_0$ at $x = 0$, as satisfied by a Fourier series by taking (a) only the first term, (b) the first 3 terms, (c) the first 10 terms, and (d) the first 100 terms. The fit improves as the number of terms increases, but the spikes at $y = 0$ and $y = b$, where V is discontinuous, remain. This is the *Gibbs phenomenon.*

penetrate in the x direction beyond a distance equal to the spacing b between the plates.

Notice also how fast the exponential decreases with n. For $n = 1$ and $x = b$ it is 0.04, but for $n = 2$ and $x = b$ it is 0.002. This means that $V(y/b)$ rapidly becomes a sine curve, with increasing x. Figure 9-9 shows the equipotentials.

*9.3 SOLVING LAPLACE'S EQUATION WITH A SPREADSHEET†

It is only in the simplest situations that Laplace's equation can be solved in closed form; most real physical situations have difficult geometry and require numerical methods of solution.

Linear functions of the form

$$V(x,y) = ax + by + c$$

have zero second-order derivatives and obviously have zero Laplacian, but there are many non-linear functions of two or three variables that have zero Laplacian. In a specific case, the correct function depends on the boundary conditions. One can write the exact solution in the form of an infinite series that can be truncated to achieve the required accuracy, but if a numerical solution is all that is required, the spreadsheet solution is much simpler.

†This section was contributed by Daniel L. Hatten, Department of Physics and Applied Optics, Rose-Hulman Institute of Technology, Terre Haute IN 47803, USA.

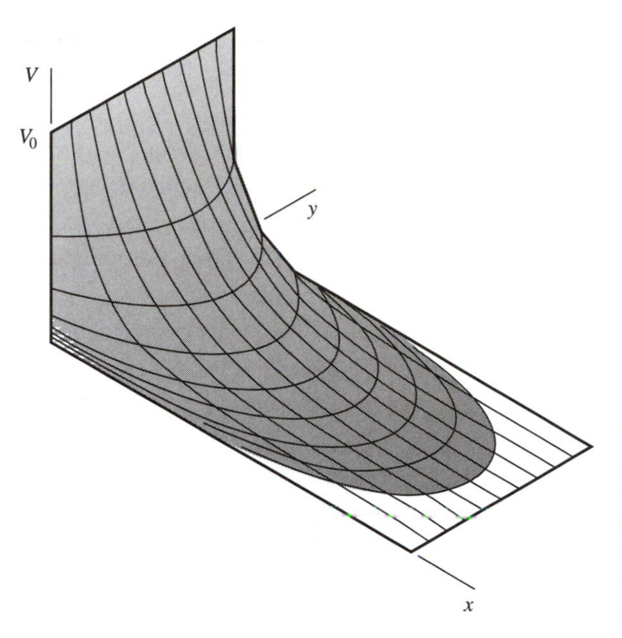

Figure 9-9 Three-dimensional plot of the potential V for the field of Fig. 9-7. The U-shaped curves are equipotentials; the others show the intersections of the potential surface with planes parallel to the xV-plane.

With the widespread availability of computers, numerical solutions have become relatively simple to perform. We describe here a basic algorithm for solving Laplace's equation with a spreadsheet, and we provide two simple Examples.

The algorithm has long been known as the *relaxation method*, and is largely due to Southwell (1940 and 1946). It is applicable to a wide variety of problems, and not only to the solution of Laplace's equation. See, for example, the *second* edition of Lorrain, Corson, and Lorrain (1970), pages 86, 177, and 420, and the two large Southwell books.

A *spreadsheet* is a table with many columns and rows that one can display on a computer screen. The columns and rows need not all be used.

When one uses a spreadsheet in solving Laplace's equation, a cell stands for a point in space, and the content of the cell is the potential at that point. The spreadsheet thus provides a numerical, point-by-point solution of the equation in one, two, or three dimensions. It is possible to plot corresponding curves and surfaces.

Recall from Sec. 4.1 that Laplace's equation is written

$$\nabla^2 V = 0 \tag{9-34}$$

or, in Cartesian coordinates,

$$\frac{\partial^2 V}{\partial x^2} + \frac{\partial^2 V}{\partial y^2} + \frac{\partial^2 V}{\partial z^2} = 0. \tag{9-35}$$

This equation applies whenever there is zero space charge density.

Let us try a one-dimensional case, with the potential V a function of the coordinate x. Call V_i the potential at the point x_i. We wish to find V_i as a function of x_i. Now

$$\frac{\partial V}{\partial x} \approx \frac{\Delta V}{\Delta x} = \frac{V_{i+1} - V_i}{\Delta x}. \tag{9-36}$$

The second derivative can be calculated from the first in the same manner:

$$\frac{\partial^2 V}{\partial x^2} \approx \frac{1}{\Delta x}\Delta\left(\frac{\Delta V}{\Delta x}\right) \tag{9-37}$$

$$= \frac{1}{\Delta x}\left[\left(\frac{V_{i+1} - V_i}{\Delta x}\right) - \left(\frac{V_i - V_{i-1}}{\Delta x}\right)\right] \tag{9-38}$$

$$= \frac{V_{i+1} - 2V_i + V_{i-1}}{(\Delta x)^2}. \tag{9-39}$$

Laplace's equation will be satisfied if

$$V_i = \frac{V_{i+1} + V_{i-1}}{2}. \tag{9-40}$$

Thus, if x_i designates the position at the center of cell i, then the potential V_i is equal to the average of the potentials at the centers of the cells on either side.

This simple relation applies whatever the boundary conditions, as long as there is zero space charge density.

Clearly, if the potential V is a function of only one coordinate, say x, then the equation

$$\partial^2 V/\partial x^2 = 0$$

implies that

$$V(x) = ax + b,$$

and V is a linear function of x. In that case, for any given Δx, Eqs. 9-39 and 9-40 are satisfied exactly, not only approximately. The two- and three-dimensional cases, however, are not as simple, as we shall see.

The spreadsheet is an ideal tool for carrying out this type of calculation. As a first approximation, you set the given potentials at the boundaries of the field, and $V_i = 0$ (or any other value or values) everywhere else. Then you apply the above formula to all cells that are not on the boundaries, and repeat the operation many times. The value of V in the cells converge to values that satisfy Eq. 9-40 to a better and better degree of approximation. *Do not attempt to do this immediately on your own*, unless you have done similar calculations before; we explain below exactly how to do this with a spreadsheet.

The method can also be applied to two-dimensional fields. Suppose that V is a function of both x and y. Setting $\Delta x = \Delta y = h$, and proceeding as we did to find Eq. 9-39,

$$\frac{\partial^2 V}{\partial x^2} + \frac{\partial^2 V}{\partial y^2} \approx \frac{V(x+h, y) - 2V(x, y) + V(x-h, y)}{h^2} + \frac{V(x, y+h) - 2V(x, y) + V(x, y-h)}{h^2} \tag{9-41}$$

$$= \frac{V(x+h, y) + V(x-h, y) + V(x, y+h) + V(x, y-h) - 4V(x, y)}{h^2}. \tag{9-42}$$

If the Laplacian is equal to zero, then

$$V(x, y) = \frac{V(x+h, y) + V(x-h, y) + V(x, y+h) + V(x, y-h)}{4}. \tag{9-43}$$

To represent this potential field on a spreadsheet, use a rectangular block of cells. As a first approximation, set the given potentials on the boundary, and set all the other cells of the block to zero, or to any other value. Then, by applying Eq. 9-43 repeatedly, you obtain in all the non-boundary cells values of V that are approximately equal to the average of the *four* adjoining V's.

For three-dimensional fields, you use several blocks of cells, either on a single sheet or on several sheets. In that case, the value of V in a non-boundary cell is approximately equal to the average of the values of V in the *six* adjoining cells.

The method can run into a problem when one or more of the boundary conditions is that the potential is zero *at infinity*. Since the region of space modeled is necessarily finite, the cells at the border of the simulation look to empty cells one space over to participate in the averaging during the iteration process. Most spreadsheets assume that an empty cell has a value zero in it, which amounts to implicitly establishing a boundary condition of $V = 0$ everywhere at the border of the simulation. This does not come up in the Example below because the exterior cells have constants in them.

There are two ways of solving the problem of zero potential at infinity. The first way is to revise the formula in the boundary cells so that they do not reference the empty cells outside the simulation area. Then the average used to find the potential in these border cells is carried out over fewer cells. This goes a long way toward fixing the problem in most cases. The second approach requires knowing something about the expected behavior of the potential at the border of the simulation. If something is known, then suitable constants can be inserted to form the border.

EXAMPLE A One-Dimensional Field: The Potential Inside a Parallel-Plate Capacitor

A parallel-plate capacitor has one plate at a potential of 5 volts, and the other plate is grounded, as in Fig. 9-6. This field can be calculated as an infinite series that one can truncate to achieve a given accuracy, but the numerical method presented here makes the problem quite easy.

Table 9-1

5.00	0.00	0.00	0.00	0.00	0.00
5.00	2.50	0.00	0.00	0.00	0.00
5.00	2.50	1.25	0.00	0.00	0.00
5.00	3.13	1.25	0.63	0.00	0.00
5.00	3.13	1.88	0.63	0.31	0.00
5.00	3.44	1.88	1.09	0.31	0.00
5.00	3.44	2.27	1.09	0.55	0.00
5.00	3.63	2.27	1.41	0.55	0.00
5.00	3.63	2.52	1.41	0.70	0.00
5.00	3.76	2.52	1.61	0.70	0.00
5.00	3.76	2.69	1.61	0.81	0.00
5.00	3.84	2.69	1.75	0.81	0.00
5.00	3.84	2.79	1.75	0.87	0.00
5.00	3.90	2.79	1.83	0.87	0.00
5.00	3.90	2.87	1.83	0.92	0.00
5.00	3.93	2.87	1.89	0.92	0.00
5.00	3.93	2.91	1.89	0.95	0.00
5.00	3.96	2.91	1.93	0.95	0.00

Set the number of cells equal to 6. Column A will stand for the 5-volt electrode, and column F for the grounded electrode. In row 1, put the value 5 in A1, and 0 in cells B1 to F1.

The next step is to enter the formula for the potential into one of the cells in the box. In row 2, type = A1 in cell A2, and the formula = (A1 + C1)/2 in cell B2, which is the average of the two cells that border B1. Do not forget the equal signs!

Then select B2 and replicate its formula in cells C2 to E2. Finally, type = F1 in cell F2.

The spreadsheet fills in the other four cells automatically. For example, it sets D2 = (C1 + E1)/2.

Now select cells A2 to F2, and replicate these cells downward fifty lines or more. Table 9-1 shows the numbers obtained with 17 iterations. There is convergence. A personal computer can do hundreds of iterations in seconds, leading to $V = 5,4,3,2,1,0$, as expected. This solution is better than that obtained by keeping the first three terms of the analytic solution, and is very much simpler to program.

With 101 cells, and one plate at 100 volts and the other at 0 volts, the potentials converge to 100, 99, 98, 97, . . . 0, again as expected.

EXAMPLE A Two-Dimensional Field: The Potential Inside a Long Conducting Box of Square Cross-Section

Two-dimensional fields are more difficult. To work out the present example, you will need the help of an instructor, or at least the help of a good manual on the use of spreadsheets for scientific calculations.

A long conducting box of square cross-section has one side at 5 volts, and the other three sides are grounded. Then you start out by choosing a square area on the

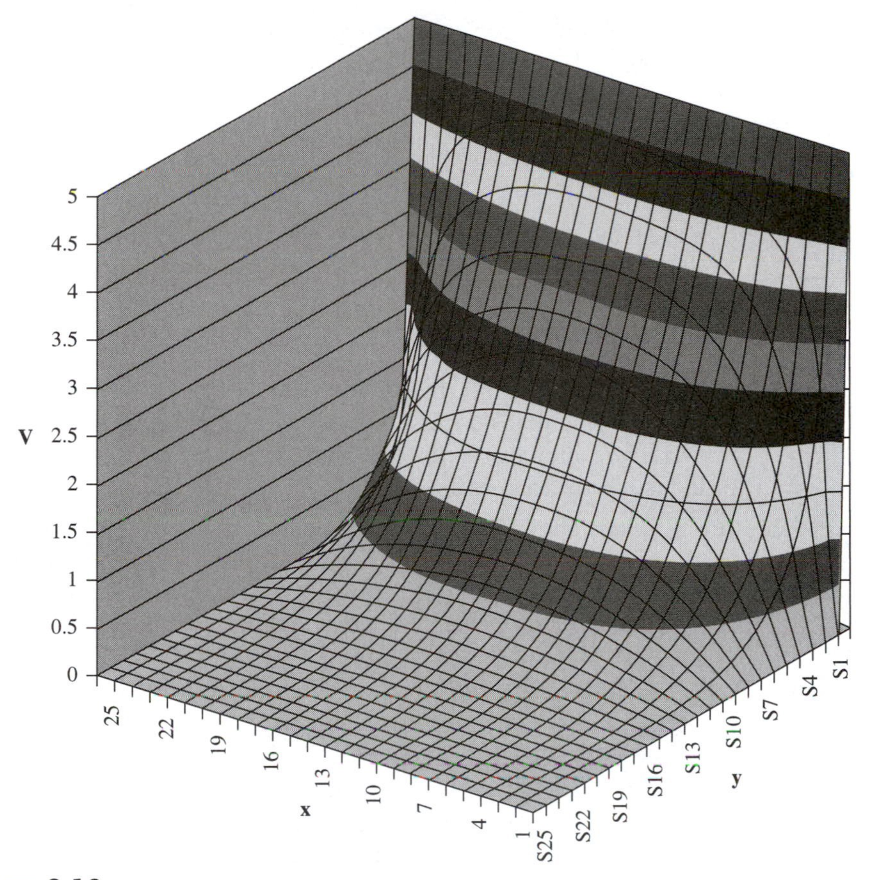

Figure 9-10

spreadsheet with, say, A1 to Z1 equal to 5, A2 to A26 equal to 0, B26 to Z26 equal to 0, and Z2 to Z26 also equal to 0.

Figure 9-10 shows a three-dimensional plot of V as a function of x and y.

*9.4 SUMMARY

The *method of images* can often simplify the calculation of electric fields that involve interfaces between different media. It consists in setting up, on paper, a different field on the other side of the interface, with fictitious, or image, charges.

Solutions of Laplace's equation $\nabla^2 V = 0$ are termed *harmonic functions*. It is often possible to reduce Laplace's equation to a set of three independent, ordinary differential equations, one for each coordinate. Then V is of the form

$$V(x, y, z) = X(x)Y(y)Z(z), \tag{9-9}$$

where X is a function of x alone, etc. This operation is called *separating the variables*.

For certain fields one can fit the boundary conditions only by summing an infinite number of harmonic functions. We used a *Fourier series*

$$\sum_{n=1}^{\infty}\left(C_n \sin\frac{n\pi y}{b} + D_n \cos\frac{n\pi y}{b}\right),$$

where C_n and D_n are constants.

*PROBLEMS

9-1. The uniqueness theorem of Sec. 4.1.1

According to this theorem, the Poisson equation $\nabla^2 V = -\rho_t/\epsilon_0$ can have only one solution if the potential V is defined at the boundaries of the field.

Show that two solutions can differ at most by a constant if the normal component of ∇V is defined at the boundaries.

9-2. (*9.2*) Point charge above a conducting plate

A point charge Q lies at a distance D above a large grounded conducting plate.

(a) Calculate the surface charge density induced on the plate as a function of the radius r from the foot of the perpendicular drawn from the charge.

(b) Show that the total induced charge is $-Q$.

9-3. (*9.1*) Line charge near a conducting plate

A copper wire, parallel to a large conducting plate and at a distance a, carries a charge of λ coulombs/meter.

(a) Find **E** in a plane perpendicular to the wire.

(b) Find the force of attraction per meter of wire between the wire and the plate by integrating $(1/2)\epsilon_0 E^2$ over the surface of the plate.

9-4. (*9.1*) The field of a charge inside a hollow conducting sphere

A hollow grounded conducting sphere of radius a contains a point charge Q at the radius b as in Fig. 9-11.

(a) Show that the field inside the sphere is the same as if there were no sphere and, instead, a charge $Q' = -(a/b)Q$ at $D = (a/b)a$. You can prove this by showing that the V of Q plus Q' is uniform over the surface of the sphere.

(b) Calculate the force of attraction.

(c) Calculate the surface charge density on the inside surface of the conducting sphere.

9-5. (*9.1*) The field of a point charge near a block of dielectric

Show that the potential at the surface of the dielectric, opposite Q in Fig. 9-3, is the same whether one calculates the field in air, as in Fig. 9-4(a), or in the dielectric, as in Fig. 9-4(c). Set $\epsilon_r = 3$.

9-6. (*9.2*) Solutions of Laplace's equation can be of the form $X(x) + Y(y) + Z(z)$

Show that there exist solutions of Laplace's equation that are of the form $X(x) + Y(y) + Z(z)$.

9-7. (*9.2*) The function $1/r$ and its derivatives are solutions of Laplace's equation

(a) Show that $\nabla^2(1/r) = 0$.

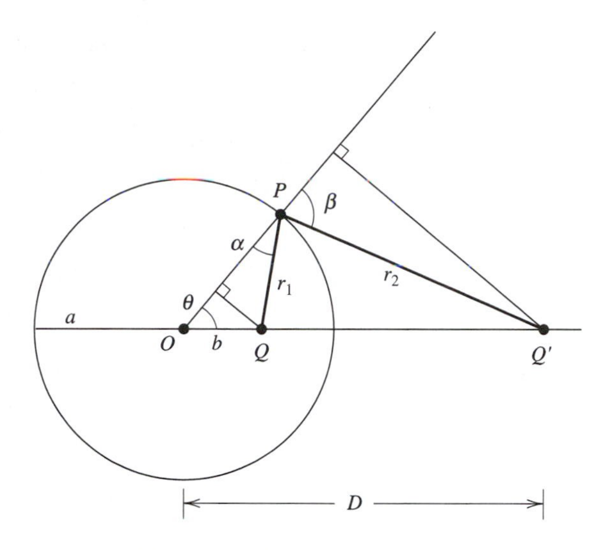

Figure 9-11

(b) Use this fact to show that

$$\frac{\partial}{\partial x}\frac{1}{r}, \qquad \frac{\partial^2}{\partial x^2}\frac{1}{r}, \qquad \frac{\partial^2}{\partial x \partial y}\frac{1}{r}$$

are also solutions of Laplace's equation.

9-8. *(9.2)* The field of Fig. 9-7

Plot $\exp(-n\pi x/b)$ as a function of x/b from $x/b = 0$ to 1 and for $n = 1, 2$, and 3.

9-9. (*9.2*) The field of Fig. 9-7

Plot V/V_0 as a function of y/b for the field of Fig. 9-9 for $x = 0$, $x = 0.5b$, and $x = b$, up to $n = 100$. You can truncate the series when $\exp(-n\pi x/b)$ is less than 0.01.

Note: You might need the help of an instructor to work on spreadsheets. Also, different software vendors have different types of spreadsheets.

9-10. (*9.3*) Solving a two-dimensional problem with a spreadsheet: box of square cross-section with its four sides at different potentials.

The box has sides at 1, 2, 3, and 4 volts, respectively, clockwise. Use a spreadsheet to plot equipotentials. Then sketch electric field lines.

Hints. (a) Use a block of 25 × 25 cells. That is a good compromise: a larger number of cells would require more iterations, and, with a smaller number of cells, you would lose accuracy. (b) Set the initial value of V_i in the inner cells equal to 2.5 volts, instead of 0 volts. (The average potential is $(1 + 2 + 3 + 4)/4 = 10/4$ volts.) (c) Do 400 iterations. (d) Set the potential difference between two neighboring equipotentials equal to 0.25 volt.

9-11. (*9.3*) The field between two plates at different potentials

Two plates intersect at right angles, with one plate grounded and the other at 5 volts. Use a spreadsheet to plot equipotentials. Then sketch electric field lines.

Hints. (a) Use a block of 25 × 25 cells, with $V_i = 0$ on one side, and $V_i = 5$ on a neighboring side. (b) Set the initial value of V_i in the inner cells equal to 2.5 volts, which is the average potential. (c) For the cells that are on the periphery of the block of cells and not on the plates, use the average of the three neighboring cells of the preceding block. For the single corner cell that is not on the plates, use the average of the two neighboring cells of the preceding block. (d) Iterate 400 times. (e) In your last block of cells, use only the 10×10 block at the corner where the plates meet, to plot the equipotentials.

9-12. (*9.3*) The field near a lightning rod.

Use a spreadsheet to plot equipotentials. Then sketch electric field lines.

Hints. (a) This is a 3D problem, but you can find an approximate solution by making it into a 2D problem with the rod replaced by a plate, say at 5 volts. (b) Use a block of 25×25 cells, say A1 to Y25, with the potential equal to 5 volts in cells K13 to Y13, and zero elsewhere. (c) Refer to the third hint for Prob. 9-11. (d) Iterate 400 times. In your last block of cells, use only the 9×9 block centered on the tip of the plate, to plot the equipotentials. (e) Repeat the calculation, setting the initial voltage at points outside the plate equal to 1 volt, and then to 2 volts. (f) Compare the three fields.

9-13. (*9.3*) A point inside a grounded box is held at 5 volts.

Plot the equipotentials, then sketch electric field lines.

Hints. (a) This is a 3D field that can be approximated by a 2D field if the point is replaced by a rod, perpendicular to the paper. (b) Use a block of 25×25 cells with the central cell at 5 volts. (c) Set the other cells at 1 volt, which is closer to the final voltage than 0 volt, and set the boundary cells at 0 volt. (d) In this case, 100 iterations provide a good approximation to the field, but you will have more accuracy with 400 iterations. (e) Set $\Delta V = 0.25$ or 0.333 volt between equipotentials so as to better see the shape of the field.

CHAPTER

10

*ELECTRIC FIELDS VIII†

*Laplace's Equation in Spherical Coordinates. Poisson's Equation for **E***

In this last chapter on electric fields we shall see how to solve Laplace's equation in spherical coordinates. This will require the Legendre polynomials that we discussed in the starred Sec. 5.4.1. Then we shall deduce and discuss briefly Poisson's equation for $\mathbf{E}$, which leads to an integral for $\mathbf{E}$ that is strangely different from the one that follows from Coulomb's law.

*10.1 SOLVING LAPLACE'S EQUATION IN SPHERICAL COORDINATES. LEGENDRE'S EQUATION

Some electric fields are best treated in spherical polar coordinates (Sec. 1-10). Solutions of Laplace's equation expressed in these coordinates are known as *spherical harmonic functions.*

We limit ourselves to fields possessing axial symmetry and therefore to fields that are independent of the azimuthal angle ϕ. Then Laplace's equation takes the following form:

†Material marked with an asterisk can be skipped without losing continuity.

$$\frac{\partial}{\partial r}\left(r^2\frac{\partial V}{\partial r}\right) + \frac{1}{\sin\theta}\frac{\partial}{\partial\theta}\left(\sin\theta\frac{\partial V}{\partial\theta}\right) = 0. \tag{10-1}$$

To separate the variables as in Chap. 9, we set

$$V = R(r)\Theta(\theta), \tag{10-2}$$

where R is a function of r only and Θ a function of θ only. Then, by substitution,

$$\Theta\frac{\partial}{\partial r}\left(r^2\frac{\partial R}{\partial r}\right) + \frac{R}{\sin\theta}\frac{\partial}{\partial\theta}\left(\sin\theta\frac{\partial\Theta}{\partial\theta}\right) = 0. \tag{10-3}$$

Dividing by $R\Theta$, we get

$$\frac{1}{R}\frac{d}{dr}\left(r^2\frac{dR}{dr}\right) + \frac{1}{\Theta\sin\theta}\frac{d}{d\theta}\left(\sin\theta\frac{d\Theta}{d\theta}\right) = 0. \tag{10-4}$$

We have now written total instead of partial derivatives because R and Θ are both functions of a single variable.

The second term is independent of r. Then the first term is also independent of r and is equal to a constant:

$$\frac{1}{R}\frac{d}{dr}\left(r^2\frac{dR}{dr}\right) = k. \tag{10-5}$$

Then

$$\frac{1}{\Theta\sin\theta}\frac{d}{d\theta}\left(\sin\theta\frac{d\Theta}{d\theta}\right) = -k, \tag{10-6}$$

since the sum of the two constants must equal zero.

Let us examine the R equation first. Multiplying both sides by R and differentiating the term enclosed in parentheses, we have that

$$r^2\frac{d^2R}{dr^2} + 2r\frac{dR}{dr} - kR = 0. \tag{10-7}$$

The solution of this equation is of the form

$$R = Ar^n + Br^{-(n+1)}, \tag{10-8}$$

with

$$n(n+1) = k, \tag{10-9}$$

as you can check by substitution.

Rewriting the Θ equation yields

$$\frac{d}{d\theta}\left(\sin\theta\frac{d\Theta}{d\theta}\right) + n(n+1)\Theta\sin\theta = 0. \tag{10-10}$$

Now set

$$\mu = \cos\theta, \tag{10-11}$$

remembering that, for any function $f(\mu)$,

$$\frac{df}{d\theta} = \frac{df}{d\mu}\frac{d\mu}{d\theta} = -\sin\theta\frac{df}{d\mu} = -(1-\mu^2)^{1/2}\frac{df}{d\mu}. \tag{10-12}$$

Then the Θ equation becomes *Legendre's equation:*

$$\frac{d}{d\mu}\left[(1-\mu^2)\frac{d\Theta}{d\mu}\right] + n(n+1)\Theta = 0. \tag{10-13}$$

When n is an integer, its solutions are the Legendre polynomials of Sec. 5.4.1:

$$\Theta = P_n(\cos\theta). \tag{10-14}$$

Table 5-1 shows the first five Legendre polynomials.

We shall use the following property of the Legendre polynomials. Since

$$n(n+1) = n'(n'+1), \qquad \text{if } n' = -(n+1), \tag{10-15}$$

then

$$P_{-(n+1)}(\cos\theta) = P_n(\cos\theta). \tag{10-16}$$

It follows that, for every solution of Laplace's equation of the form

$$V = Ar^n P_n(\cos\theta), \tag{10-17}$$

there exists another one of the form

$$V = Br^{-(n+1)}P_{-(n+1)}(\cos\theta) = Br^{-(n+1)}P_n(\cos\theta). \tag{10-18}$$

Observe that this result is in agreement with Eq. 10-8.

So the general solution of Laplace's equation for fields possessing axial symmetry is

$$V = \sum_{n=0}^{\infty}\left[A_n r^n + B_n r^{-(n+1)}\right]P_n(\cos\theta). \tag{10-19}$$

This series is analogous to a Fourier series (Example, Sec. 9.2) in the following ways. First, the expressions under the summation sign form a complete set: the series can satisfy any reasonably well-behaved boundary condition exhibiting axial symmetry. Second,

$$\int_{-1}^{+1} P_m(\cos\theta)P_n(\cos\theta)\,d(\cos\theta) = \begin{cases} 0 & \text{if } m \neq n \\ \dfrac{2}{2n+1} & \text{if } m = n. \end{cases} \tag{10-20}$$

Legendre polynomials are thus orthogonal. Third, we can use this orthogonality to calculate the values of the A_n and B_n coefficients.

*EXAMPLE Uncharged Conducting Sphere in a Previously Uniform Electric Field

An insulated and uncharged conducting sphere is situated in a previously uniform electric field $\boldsymbol{E}_0$. This applied field originates in remotely situated charge distributions that are unaffected by the presence of the sphere.

At any point in space, either inside or outside the sphere, $\boldsymbol{E} = \boldsymbol{E}_0 + \boldsymbol{E}_i$, where $\boldsymbol{E}_i$ is the field of the charges induced on the sphere. The induced charges arrange themselves so as to render the net field inside equal to zero.

We calculate the field outside the sphere by solving Laplace's equation in two different ways.

1. We first use spherical polar coordinates, with the origin at the center of the sphere and the polar axis along $\boldsymbol{E}_0$. The boundary conditions are then as follows:

$$\text{(a) } V_{r=a} = 0, \qquad \text{(b) } V_{r\to\infty} = -E_0 z = -E_0 r\cos\theta. \tag{10-21}$$

From boundary condition (a) and from Eq. 10-19,

$$0 = \sum_{n=0}^{\infty} A_n a^n P_n(\cos\theta) + \sum_{n=0}^{\infty} B_n a^{-(n+1)} P_n(\cos\theta). \tag{10-22}$$

We now evaluate the A and B coefficients in a manner analogous to that of the example in Sec. 9.2. We multiply both sides of the equation by $P_m(\cos\theta)$ and integrate from $\cos\theta = -1$ to $\cos\theta = +1$:

$$0 = \sum_{n=0}^{\infty} \int_{-1}^{+1} A_n a^n P_n(\cos\theta) P_m(\cos\theta)\, d(\cos\theta)$$

$$+ \sum_{n=0}^{\infty} \int_{-1}^{+1} B_n a^{-(n+1)} P_n(\cos\theta) P_m(\cos\theta)\, d(\cos\theta). \tag{10-23}$$

According to Eq. 10-20, the only nonvanishing terms are those for which $m = n$. Then each summation reduces to a single term:

$$0 = A_n a^n \int_{-1}^{+1} P_n^2(\cos\theta)\, d(\cos\theta) + B_n a^{-(n+1)} \int_{-1}^{+1} P_n^2(\cos\theta)\, d(\cos\theta) \tag{10-24}$$

$$= \frac{A_n a^n + B_n a^{-(n+1)}}{n + \frac{1}{2}}. \tag{10-25}$$

So

$$B_n = -A_n a^{2n+1}. \tag{10-26}$$

Substituting into Eq. 10-19 yields

$$V = \sum_{n=0}^{\infty} A_n \left[r^n - a^{2n+1} r^{-(n+1)} \right] P_n(\cos\theta). \tag{10-27}$$

Now boundary condition (b) concerns the value of V at infinity where all inverse powers of r tend to zero. Thus, at $r \to \infty$,

$$V = -E_0 r \cos\theta = -E_0 r P_1(\cos\theta) = \sum_{n=0}^{\infty} A_n r^n P_n(\cos\theta) \tag{10-28}$$

for all θ. By inspection, the only nonzero term on the right is that for which $n = 1$. See Table 10-1. Then

$$A_1 = -E_0 \tag{10-29}$$

and all the other As are zero. Also, from Eq. 10-26, all the Bs are zero except B_1:

$$B_1 = -A_1 a^3 = E_0 a^3. \tag{10-30}$$

Finally, at any point outside the sphere,

$$V = -E_0 r \cos\theta + \frac{E_0 a^3 \cos\theta}{r^2} = -\left(1 - \frac{a^3}{r^3}\right) E_0 r \cos\theta, \tag{10-31}$$

Table 10-1 Solutions of Laplace's equation in spherical polar coordinates for fields possessing axial symmetry

n	$r^n P_n(\cos\theta)$	$r^{-(n+1)} P_n \cos\theta$
0	1	r^{-1}
1	$r\cos\theta$	$r^{-2}\cos\theta$
2	$r^2 \dfrac{(3\cos^2\theta - 1)}{2}$	$r^{-3} \dfrac{(3\cos^2\theta - 1)}{2}$
3	$r^3 \dfrac{(5\cos^3\theta - 3\cos\theta)}{2}$	$r^{-4} \dfrac{(5\cos^3\theta - 3\cos\theta)}{2}$
4	$r^4 \dfrac{(35\cos^4\theta - 30\cos^2\theta + 3)}{8}$	$r^{-5} \dfrac{(35\cos^4\theta - 30\cos^2\theta + 3)}{8}$
5	$r^5 \dfrac{(63\cos^5\theta - 70\cos^3\theta + 15\cos\theta)}{8}$	$r^{-6} \dfrac{(63\cos^5\theta - 70\cos^3\theta + 15\cos\theta)}{8}$

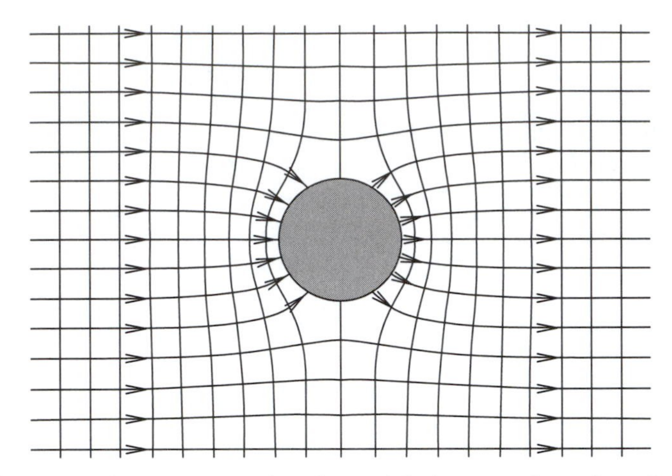

Figure 10-1 Lines of $\boldsymbol{E}$ (arrows) and equipotentials for an uncharged *conducting* sphere situated in a previously uniform electric field. The lines of $\boldsymbol{E}$ are normal to the surface, and there is zero field inside. Observe that the field is hardly disturbed at distances larger than one radius from the surface of the sphere. The origin of the spherical polar coordinates used for the calculation is at the center of the sphere.

$$E_r = -\frac{\partial V}{\partial r} = \left(1 + \frac{2a^3}{r^3}\right)E_0 \cos\theta, \tag{10-32}$$

$$E_\theta = -\frac{1}{r}\frac{\partial V}{\partial \theta} = -\left(1 - \frac{a^3}{r^3}\right)E_0 \sin\theta. \tag{10-33}$$

The surface charge density on the conducting sphere is

$$\sigma = \epsilon_0 E_{r(r=a)} = 3\epsilon_0 E_0 \cos\theta. \tag{10-34}$$

In the above expression for V, observe that the first term comes from E_0, while the second is the V of a point dipole (Sec. 5.1) situated at the origin and oriented along the z-axis, of moment $4\pi\epsilon_0 a^3 E_0$.

2. We can also find V by a much less formal method, as follows.

We require the term $-E_0 r\cos\theta$ so as to fit the condition at infinity. No other function with a positive power of r is permissible. This one term, however, is inadequate to fit the condition at $r = a$, where V must be independent of θ. We must therefore add another function that also includes the cos θ factor so that the coefficient of cos θ will be zero at $r = a$. Then, from Eq. 10-19,

$$V = -E_0 r\cos\theta + \frac{B\cos\theta}{r^2}. \tag{10-35}$$

We finally set $B = E_0 a^3$ to make $V = 0$ at $r = a$. Our solution satisfies both Laplace's equation and the boundary conditions; it is therefore the correct solution, according to the uniqueness theorem of Sec. 4.1.1. See Fig. 10-1.

*EXAMPLE Dielectric Sphere in a Previously Uniform Electric Field

We now have a previously uniform field $\boldsymbol{E}_0$, as in the preceding section, and the following boundary conditions, where a is the radius of the sphere:

1. V is continuous at $r = a$.
2. The normal component of $\boldsymbol{D}$ is continuous at $r = a$.
3. $V_{r \to \infty} = -E_0 r \cos\theta$.

Instead of going through a formal solution, as we did in part 1 of the previous example, we proceed as in part 2 and devise a combination of spherical harmonics that satisfies all three conditions.

There now exists a field inside the sphere. So we require two solutions, one that is valid inside and one that is valid outside. The field *outside* must satisfy boundary condition 3. So we require a term $-E_0 r \cos\theta$ and no other positive power of r. All negative powers of r qualify. *Inside* the sphere, there must be no negative powers of r, because V cannot become infinite at $r = 0$.

Let V_o and V_i be the potentials outside and inside the sphere, respectively. Then

$$V_o = -E_0 r \cos\theta + \sum_{n=0}^{\infty} B_n r^{-(n+1)} P_n(\cos\theta), \tag{10-36}$$

$$V_i = \sum_{n=0}^{\infty} C_n r^n P_n(\cos\theta). \tag{10-37}$$

Boundary conditions require that, at $r = a$,

$$V_o = V_i, \qquad -\frac{\partial V_o}{\partial r} = -\epsilon_r \frac{\partial V_i}{\partial r}. \tag{10-38}$$

Then we have the following two equations:

$$-E_0 a \cos\theta + \frac{B_0}{a} + \frac{B_1 \cos\theta}{a^2} + \frac{B_2 P_2(\cos\theta)}{a^3} + \cdots$$
$$= C_0 + C_1 a \cos\theta + C_2 a^2 P_2(\cos\theta) + \cdots \tag{10-39}$$

and

$$E_0 \cos\theta + \frac{B_0}{a^2} + \frac{2B_1 \cos\theta}{a^3} + \frac{3B_2 P_2(\cos\theta)}{a^4} + \cdots$$
$$= -\epsilon_r [C_1 \cos\theta + 2C_2 a P_2(\cos\theta) + \cdots]. \quad (10\text{-}40)$$

These two equations apply for all values of θ. It follows that the coefficient of a given function of θ on the left must be equal to the coefficient of the same function on the right. From the first equation,

$$\frac{B_0}{a} = C_0, \qquad -E_0 a + \frac{B_1}{a^2} = C_1 a, \qquad \frac{B_2}{a^3} = C_2 a^2, \qquad \ldots. \quad (10\text{-}41)$$

From the second,

$$\frac{B_0}{a^2} = 0, \qquad E_0 + \frac{2B_1}{a^3} = -\epsilon_r C_1, \qquad \frac{3B_2}{a^4} = -2\epsilon_r C_2 a, \qquad \ldots. \quad (10\text{-}42)$$

Combining these equations gives

$$B_0 = C_0 = 0, \qquad B_1 = \frac{\epsilon_r - 1}{\epsilon_r + 2} E_0 a^3, \quad (10\text{-}43)$$

$$C_1 = -\frac{3}{\epsilon_r + 2} E_0, \qquad B_n = C_n = 0 \qquad (n > 1). \quad (10\text{-}44)$$

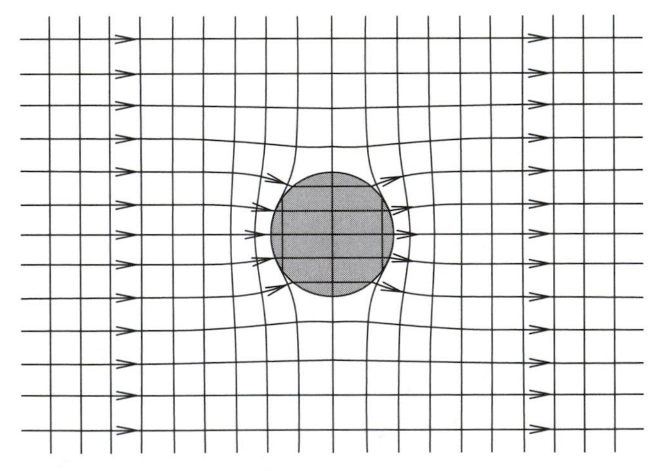

Figure 10-2 Lines of $\boldsymbol{D}$ (arrows) and equipotentials for a *dielectric* sphere ($\epsilon_r = 3$) situated in a previously uniform electric field. The lines of $\boldsymbol{D}$ crowd into the sphere, and $\boldsymbol{D}$ is *larger* inside than outside. The equipotentials spread out inside, so $\boldsymbol{E}$ is smaller inside than outside. There exists a bound surface charge density, and $\boldsymbol{E}$ is discontinuous at the surface. As for the conducting sphere (Fig. 10-1), the field is hardly disturbed at distances larger than one radius from the surface. The field inside is uniform.

Finally,

$$V_o = -\left[1 - \frac{\epsilon_r - 1}{\epsilon_r + 2}\frac{a^3}{r^3}\right]E_o r\cos\theta, \tag{10-45}$$

$$V_i = -\frac{3}{\epsilon_r + 2}E_0 r\cos\theta. \tag{10-46}$$

Of course, $r\cos\theta = z$.

Notice that, if $\epsilon_r \gg 1$, then V_o is approximately the same as the V of the conducting sphere of the previous Example.

Also, the $\boldsymbol{E}$ inside the sphere is uniform:

$$\boldsymbol{E}_i = -\boldsymbol{\nabla}V_i = \frac{3}{\epsilon_r + 2}E_0\hat{\mathbf{z}}, \qquad \boldsymbol{D}_i = \frac{3\epsilon_r}{\epsilon_r + 2}E_0\hat{\mathbf{z}}. \tag{10-47}$$

See Fig. 10-2.

*10.2 POISSON'S EQUATION FOR *E* IN ELECTROSTATIC FIELDS

We discussed Poisson's equation for V in Sec. 4.1. There also exists a Poisson equation for $\boldsymbol{E}$. From Sec. 1.10,

$$\boldsymbol{\nabla}\times(\boldsymbol{\nabla}\times\boldsymbol{E}) = -\boldsymbol{\nabla}^2\boldsymbol{E} + \boldsymbol{\nabla}(\boldsymbol{\nabla}\cdot\boldsymbol{E}). \tag{10-48}$$

Now, in electrostatic fields, the curl of $\boldsymbol{E}$ is zero, so that

$$\boldsymbol{\nabla}^2\boldsymbol{E} = \boldsymbol{\nabla}(\boldsymbol{\nabla}\cdot\boldsymbol{E}) = \frac{\boldsymbol{\nabla}\rho}{\epsilon_0}, \tag{10-49}$$

from Sec. 7.5, where ρ is the total charge density, free plus bound.

We can solve this equation by first separating it into its three components, the x-component being

$$\boldsymbol{\nabla}^2 E_x = \frac{\partial}{\partial x}\frac{\rho}{\epsilon_0}. \tag{10-50}$$

This scalar equation is similar to the Poisson equation for V (Eq. 4-1), whose solution is the V of Eq. 3-34. Thus

$$E_x = -\frac{1}{4\pi\epsilon_0}\int_{v'}\frac{\partial\rho/\partial x'}{r}dv', \tag{10-51}$$

and similarly for the other two components. As usual, the primes refer to the charge distribution, and r is the distance between the field point $P(x, y, z)$ and the source point $P'(x', y', z')$. The solution of Poisson's equation for $\boldsymbol{E}$ is thus

$$\boldsymbol{E} = -\frac{1}{4\pi\epsilon_0}\int_{v'}\frac{\boldsymbol{\nabla}'\rho}{r}dv'. \tag{10-52}$$

This equation relates $\boldsymbol{E}$ to the *gradient* of the total charge density. Note the *negative sign* and the *first* power of r in the denominator. This equation is valid whatever the nature of the media that are present in the field, as long as the gradient is definable.

The more usual expression for the $\boldsymbol{E}$ of a volume distribution of charge is a consequence of Coulomb's law, and it is the one that we found in Sec. 3.3:

$$\boldsymbol{E} = \frac{1}{4\pi\epsilon_0}\int_{v'} \frac{\rho\hat{\boldsymbol{r}}}{r^2}\, dv'. \tag{10-53}$$

Although the two integrals for $\boldsymbol{E}$ are equal, the integrands are obviously unequal. One integral applies only where $\boldsymbol{\nabla}\rho$ is not zero, while the other extends over the complete charge distribution.

***EXAMPLE** The Field of a Sphere of Charge Whose Density is a Function of the Radius

Imagine a sphere of charge of uniform density, except near the periphery, where the density gradually falls to zero. With Eq. 10-52, it is only the region near the surface that contributes to the integral, since the gradient of the charge density is zero everywhere else. Nevertheless, this equation leads to the same result as Eq. 10-53. See Prob. 10-6.

*10.3 SUMMARY

In spherical coordinates, if V is independent of the ϕ coordinate, we set

$$V(r, \theta) = R(r)\Theta(\theta). \tag{10-2}$$

Then Laplace's equation becomes a pair of ordinary differential equations

$$r^2\frac{d^2R}{dr^2} + 2r\frac{dR}{dr} - n(n+1)R = 0, \qquad \text{(10-7), (10-9)}$$

and Legendre's equation

$$\frac{d}{d\mu}\left[(1-\mu^2)\frac{d\Theta}{d\mu}\right] + n(n+1)\Theta = 0, \tag{10-13}$$

where $\mu = \cos\theta$. The first equation provides a solution of the type

$$R = Ar^n + \frac{B}{r^{n+1}}, \tag{10-8}$$

while the Legendre polynomials of Sec. 5.4.1 are solutions of Legendre's equation.

The *general solution of Laplace's equation in spherical coordinates,* for *axial symmetry* is thus

$$V = \sum_{n=0}^{\infty}\left[A_n r^n + B_n r^{-(n+1)}\right]P_n(\cos\theta). \tag{10-19}$$

Poisson's equation for $\boldsymbol{E}$ is

$$\nabla^2 \boldsymbol{E} = \frac{\nabla \rho}{\epsilon_0}, \tag{10-49}$$

and its solution is

$$\boldsymbol{E} = -\frac{1}{4\pi\epsilon_0}\int_{v'} \frac{\nabla' \rho}{r}\, dv'. \tag{10-52}$$

For electrostatic fields, we can calculate $\boldsymbol{E}$ either as above or through Coulomb's law:

$$\boldsymbol{E} = \frac{1}{4\pi\epsilon_0}\int_{v'} \frac{\rho \hat{\boldsymbol{r}}}{r^2}\, dv'. \tag{10-53}$$

*PROBLEMS

10-1. (*10.1*) Grounded cylindrical conductor in a uniform $\boldsymbol{E}$

A grounded, infinite, circular cylindrical conductor of radius a lies in a previously uniform electric field, with its axis perpendicular to $\boldsymbol{E}_0$ as in Fig 10-3.

(a) Show that $V = -E_0(1 - a^2/\rho^2)\rho\cos\phi$ satisfies Laplace's equation.

(b) Find $\boldsymbol{E}$ at $\rho^2 \gg a^2$ and at $\rho = 0$. Do those values make sense?

10-2. (*10.2*) The volume integral of $\nabla' \times (\boldsymbol{E}/r)$ is zero

Show that $\int_{v'} \nabla' \times (\boldsymbol{E}/r) dv' = 0$ for any finite charge distribution. Use identity 19 from the page facing the front cover.

10-3. (*10.2*) Equality of the two integrals for $\boldsymbol{E}$

Show that the two integrals for $\boldsymbol{E}$,

$$\boldsymbol{E} = \frac{1}{4\pi\epsilon_0}\int_{v'} \frac{\rho}{r^2}\hat{\boldsymbol{r}}\, dv' \quad \text{and} \quad \boldsymbol{E} = -\frac{1}{4\pi\epsilon_0}\int_{v'} \frac{\nabla' \rho}{r}\, dv'$$

are equal, at least if v' is finite. Use identities 15 and 18 from the page facing the front cover.

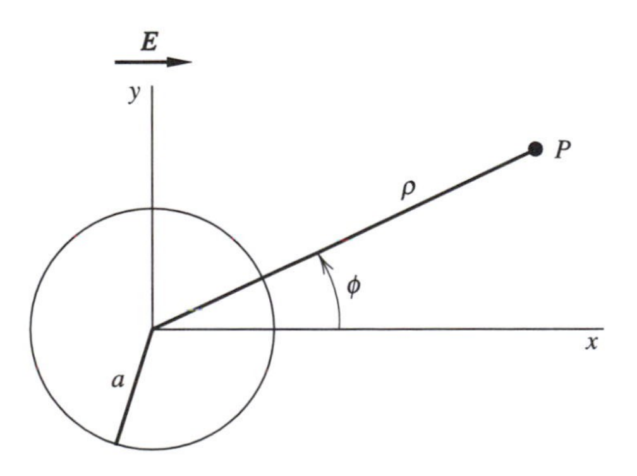

Figure 10-3

10-4. (*10.2*) The volume integral of $\nabla'(\rho/r)$ is zero

We have shown that, for static fields

$$\mathbf{E} = \frac{1}{4\pi\epsilon_0}\int_{v'} \frac{\rho}{r^2}\hat{\mathbf{r}}\, dv' = -\frac{1}{4\pi\epsilon_0}\int_{v'} \frac{\nabla'\rho}{r}\, dv'.$$

Show that, as a consequence,

$$\int_{v'} \nabla'\left(\frac{\rho}{r}\right) dv' = 0.$$

10-5. (*10.2*) The Dirac positron

Dirac proposed at one time that a positron could be considered as a hole in an infinite sea of negative electrons. Assume that positrons and electrons have finite dimensions.

Deduce the field of a positron from Coulomb's law on these assumptions. You can find this field without having to integrate!

10-6. (*10.2*) The electric field of an atomic nucleus

The density of a sphere of charge is ρ_0 up to the radius a and then decreases linearly to zero at the radius b. This corresponds roughly to the charge distribution inside an atomic nucleus.

Let r' be the distance between the center O of the nucleus and a point P' inside, r the distance between O and a point P outside, r'' the distance PP', and θ' the angle POP'.

(a) Use Gauss's law to find E at a distance $r > b$ from the center of the nucleus.

(b) Show that the expression for $\mathbf{E}$ given in Eq. 10-52 leads to the same result. Both calculations remain valid when the "skin thickness" $(b - a)$ tends to zero.

CHAPTER

11

*RELATIVITY I†

The Lorentz Transformation

We have now studied electric fields at quite some length. At this point you can either skip to Chapter 14, or you can study Chapters 11 through 13, which deal with Relativity.

There are two reasons for this digression. First, Relativity reveals the relation between electric and magnetic fields. Second, there are many phenomena, mostly related to high speeds, that are baffling without a grasp of Relativity.

†This whole chapter can be skipped without losing continuity.

The longer path is more interesting, as always, but it may not be the better one. The three chapters on Relativity do not replace the more conventional approach that comes afterward. Selecting one path or the other is a matter of time and of personal taste.

Chapters 11 and 12 set forth the fundamentals of Relativity, with little reference to electromagnetic phenomena. Their function is to set the stage for Chapter 13. In that last chapter on Relativity we show that, if a *given phenomenon* is investigated by two observers that move, one with respect to the other, then their observations can differ radically: if a field is purely electric for one observer, then it is both an electric and a magnetic field for the other.†

This first chapter on Relativity concerns the Lorentz transformation and some of its weird consequences: the *Lorentz contraction* makes an object shortened in the direction of its motion, while the *Lorentz time dilation* makes a time interval on a moving object appear longer for a stationary observer. We end this chapter with the transformation of a velocity and with electric charge invariance.

*11.1 REFERENCE FRAMES AND OBSERVERS

Consider the two rigid reference frames S and S' of Fig. 11-1, where S' moves at some constant velocity $\mathcal{V}$ with respect to S.

Both frames are inertial, by assumption. We define an *inertial frame* as one that does not accelerate and that does not rotate. In other words, an inertial frame is one with respect to which there are no inertial forces.‡

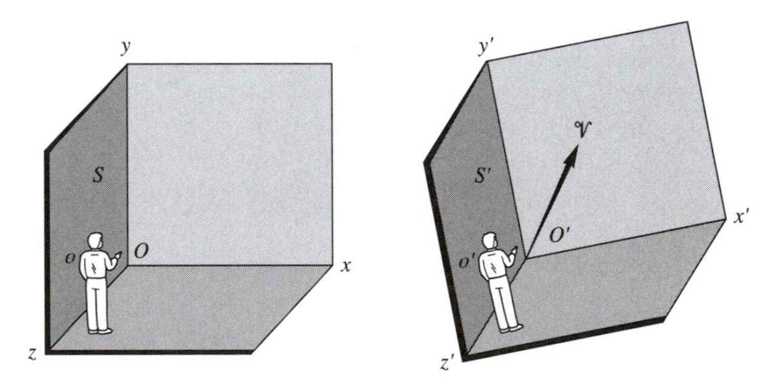

Figure 11-1 Reference frames S and S', with S' moving at some arbitrary constant velocity $\mathcal{V}$ with respect to S, *without* rotation. Observers o and o' perform various experiments, each one in his or her own frame.

†For a more detailed introduction to Relativity, see the delightful book by Taylor and Wheeler (1992). See also van Bladel (1984).

‡The forces that one feels while stationary inside an accelerating vehicle are *inertial forces*. The inertial force on a person of mass m is $-ma$, where a is the acceleration of the vehicle.

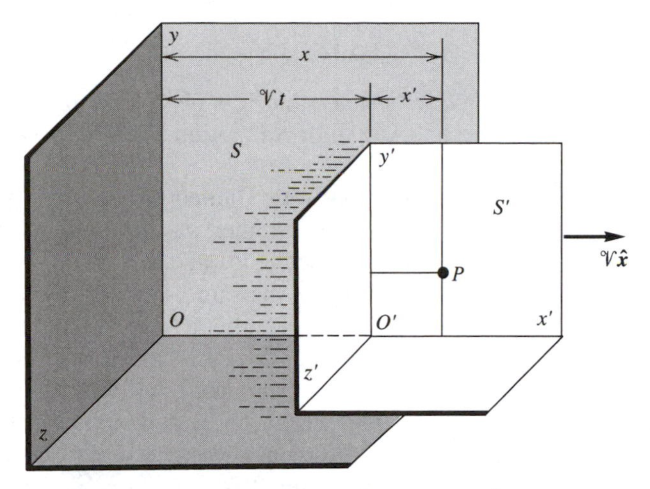

Figure 11-2 The reference frame S' moves at a velocity $\mathcal{V}\hat{x}$ with respect to S.

Each reference frame carries an *observer,* which is either a human being equipped with instruments, or some device that can take readings or transmit images, either automatically or under remote control. Observer o is situated in S, and o' in S'.

The *special theory of relativity* concerns measurements made by these two observers. As a rule, we refer to the two specific frames of Fig. 11-2.

We are *not* concerned with the *general theory of relativity,* which deals with accelerated frames and gravitation.

*11.2 THE GALILEAN TRANSFORMATION

According to *classical* physics, observers in reference frames S and S' of Fig. 11-2 may select the same time scale. If they set $t = 0$ at the instant when their frames overlap, then the coordinates of a given point in space, with respect to the two frames, satisfy the equations

$$x = x' + \mathcal{V}t, \qquad y = y', \qquad z = z', \qquad t = t'. \tag{11-1}$$

This is the *Galilean transformation.*

In particular, suppose observer o' finds that the speed of a certain light pulse relative to S' is c, in the positive direction of the common x-axis. Then, according to the Galilean transformation and according to elementary mechanics, observer o in reference frame S should find that the speed of that particular light pulse is $c + \mathcal{V}$.

The Galilean transformation proves to be in error for high speeds. Experiments show that the speed of the light pulse is the *same* for the two observers. This fact goes against common sense, of course. The Galilean transformation also proves to be incompatible with electromagnetism.

*11.3 THE PRINCIPLE OF RELATIVITY

The *principle of relativity* states that it is physically impossible to detect whether an inertial frame is at rest or in uniform motion from observations made entirely within that frame.

This principle was formulated by Galileo in 1638. It is quite clear in itself, but it is so fundamental that we state it in another way, to emphasize its meaning. The principle means that *any* experiment leads to precisely the same result, whether it is performed in S or in S'. The result is the same, whether the experiment takes place in a stationary or in a moving vehicle, as long as the vehicle moves in a straight line at a constant velocity. The principle applies even at high speeds.

For example, observer o' in reference frame S' can measure the period of a pendulum suspended at some point in S', or the collision of billiard balls on a table at rest in S'. In all instances, the phenomenon observed by o' is precisely the same as if o' performed the experiment in reference frame S.

This principle is firmly established. In particular, it is an experimental fact that the speed of light is the same for all observers traveling at a constant velocity.

To illustrate, consider the law

$$F = \frac{d}{dt}(mv), \tag{11-2}$$

as observed in reference frame S. According to the principle of relativity, there exists an identical law

$$F' = \frac{d}{dt'}(m'v') \tag{11-3}$$

that applies in S'. Physical *laws* are thus said to be *invariant*.

It follows that there exist mathematical relations that *transform* unprimed quantities into primed quantities, and inversely. We shall discover several such *transformations* in these three chapters on relativity.

*11.4 THE LORENTZ TRANSFORMATION

The *Lorentz transformation* relates the space and time coordinates of reference frame S to those of S', and inversely. For the frames of Fig. 11-2,

$$x = \frac{x' + \mathcal{V}t'}{(1 - \mathcal{V}^2/c^2)^{1/2}}, \qquad x' = \frac{x - \mathcal{V}t}{(1 - \mathcal{V}^2/c^2)^{1/2}}, \tag{11-4}$$

$$y = y', \qquad y' = y, \tag{11-5}$$

$$z = z', \qquad z' = z, \tag{11-6}$$

$$t = \frac{t' + (\mathcal{V}/c^2)x'}{(1 - \mathcal{V}^2/c^2)^{1/2}}, \qquad t' = \frac{t - (\mathcal{V}/c^2)x}{(1 - \mathcal{V}^2/c^2)^{1/2}}, \tag{11-7}$$

where c is the speed of light in a vacuum.

With the *Galilean* transformation, $F = F'$, $m = m'$, $t = t'$, $v = v' + \mathcal{V}$, and the principle of relativity applies. However, nature is not that simple; these are, in fact, approximations. Relativistic transformations show that $F \neq F'$, $m \neq m'$, $t \neq t'$, and $v \neq v' + \mathcal{V}$!

The above set of eight equations forms the basis of special relativity.† We spend the rest of this chapter, and the next two, discussing some of its strange consequences. For the moment, we note six fairly obvious features.

1. The right-hand column is identical to the left-hand column, except that the primed and unprimed quantities are interchanged, and that $-\mathcal{V}$ replaces $\mathcal{V}$. The reason is simply that it is immaterial whether S' moves at the velocity $\mathcal{V}\hat{\mathbf{x}}$ with respect to S, or whether S moves at the velocity $-\mathcal{V}\hat{\mathbf{x}}$ with respect to S'.
2. There are only four independent equations, because the right-hand column follows from the left-hand one, and inversely. You can easily check this.
3. The transformation assumes that the origins O and O' coincide at $t = t' = 0$ because, if $x = y = z = t = 0$, then $x' = y' = z' = t' = 0$, and inversely.
4. The Lorentz transformation reduces to the Galilean transformation (Eq. 11-1) if we set the speed of light c equal to infinity. However, the Galilean transformation is perfectly adequate in everyday life, where $\mathcal{V}^2 \ll c^2$.
5. The relative velocity $\mathcal{V}$ of the two frames cannot exceed c, for otherwise either x and t or x' and t' become imaginary.
6. The fact that $t \neq t'$ means that, if observer o measures a time interval T between two events, then o' will, in general, measure a *different* time interval T' between the *same* pair of events. In particular, if two events are simultaneous for o, then they are not necessarily simultaneous for o'.

It is the custom to set

$$\beta = \frac{\mathcal{V}}{c} \qquad \text{and} \qquad \gamma = \frac{1}{(1-\beta^2)^{1/2}}. \tag{11-8}$$

Note that $\beta \leq 1$ and $\gamma \geq 1$. Then

$$x = \gamma(x' + \mathcal{V}t'), \qquad x' = \gamma(x - \mathcal{V}t), \tag{11-9}$$

$$y = y', \qquad y' = y, \tag{11-10}$$

$$z = z', \qquad z' = z, \tag{11-11}$$

$$t = \gamma\left(t' + \frac{\mathcal{V}x'}{c^2}\right), \qquad t' = \gamma\left(t - \frac{\mathcal{V}x}{c^2}\right). \tag{11-12}$$

†The Lorentz transformation follows from the principle of relativity applied to the velocity of light. We omit the proof, for lack of space. See, for example, Taylor and Wheeler (1992).

Table 11-1 Lorentz transformation

$\mathbf{r} = \gamma(\mathbf{r}'_{\parallel} + \mathcal{V}t') + \mathbf{r}'_{\perp}$	$\mathbf{r}' = \gamma(\mathbf{r}_{\parallel} - \mathcal{V}t) + \mathbf{r}_{\perp}$
$t = \gamma\left(t' + \frac{\mathcal{V}r'_{\parallel}}{c^2}\right)$	$t' = \gamma\left(t - \frac{\mathcal{V}r_{\parallel}}{c^2}\right)$

Points 1 and 2 above illustrate two general rules.

1. The relation between quantities in one inertial reference frame and the corresponding quantities in another inertial frame can always be expressed by either one of two equations, or sets of equations, that are equivalent.
2. One obtains the inverse relation by adding primes to unprimed quantities, deleting primes from primed quantities, and changing the sign of $\mathcal{V}$.

It is often more convenient to express the transformation in vector form, as in Table 11-1, where the subscripts $\parallel$ and $\perp$ refer, respectively, to the components that are either parallel or perpendicular to $\mathcal{V}$.

*EXAMPLE The Transformation of an Element of Area $d\mathcal{A}$

Say the element of area is a rigid parallelogram of sides $\mathbf{dl}_{10}$ and $\mathbf{dl}_{20}$ of arbitrary magnitudes and orientations in its own reference frame. Then

$$d\mathcal{A}_0 = \mathbf{dl}_{10} \times \mathbf{dl}_{20}. \tag{11-13}$$

This vector is normal to the element of area.

Now if S is an inertial reference frame with respect to which the element of area moves at velocity $\mathcal{V}$, the element of area suffers a Lorentz contraction (Sec. 11.5) in the direction of $\mathcal{V}$:

$$d\mathcal{A} = \mathbf{dl}_1 \times \mathbf{dl}_2 = d\mathcal{A}_{0\parallel} + \frac{d\mathcal{A}_{0\perp}}{\gamma}. \tag{11-14}$$

*EXAMPLE The Invariance of $x^2 + y^2 + z^2 - c^2t^2$ Under a Lorentz Transformation

A *quantity* is said to be *invariant* if its numerical value is the same in all inertial frames. For example, under a *Galilean* transformation, the distance r_{ab} between two points a and b is the same whether one performs the measurement in S or in S'.

Note that the word *invariant* is *not* synonymous with the word *constant.* Indeed, if point a is fixed in S, and b is fixed in S', then r_{ab} is a function of the time but is still invariant under a Galilean transformation.

With the *Lorentz* transformation, r_{ab} is *not* invariant, but $r^2 - c^2t^2$ is. You can easily check that

$$x^2 + y^2 + z^2 - c^2t^2 = x'^2 + y'^2 + z'^2 - c^2t'^2,$$

$$\text{or} \quad r^2 - c^2t^2 = r'^2 - c^2t'^2. \tag{11-15}$$

This quantity is the *interval* between two *events*, squared.

***EXAMPLE** The Photon

Imagine a flash of light emitted at O or O' at the instant when the two origins coincide. According to the principle of relativity, the light propagates in all directions at the *same* velocity c in both reference frames. Therefore the position of a photon satisfies the equation

$$\frac{x^2 + y^2 + z^2}{t^2} = \frac{x'^2 + y'^2 + z'^2}{t'^2} = c^2, \tag{11-16}$$

in agreement with the previous equations.

***EXAMPLE** Causality and Maximum Signal Speed

The *order* in which two events F and G occur can be different in different frames because

$$t'_G - t'_F = \gamma\left[(t_G - t_F) - \left(\frac{\mathcal{V}}{c^2}\right)(x_G - x_F)\right], \tag{11-17}$$

and the signs of $t'_G - t'_F$ and $t_G - t_F$ can be different. This is disturbing because, according to the *principle of causality,* a cause necessarily precedes its effect.

For example, imagine that observer o throws a ball in the positive direction of the x-axis. After a flight of a few seconds, the ball breaks a windowpane. The Lorentz transformation surely cannot mean that these events could occur backward, for certain observers.

Say event F occurs at origins O and O' at the instant when they coincide. Event F is the cause of event G, which occurs at x_G at a later time t_G in S and t'_G in S'. Event G cannot occur before event F in S'. Then t'_G must not be negative. Now, if v is the speed at which a "signal" propagates from F to G,

$$t'_G = \gamma\left[t_G - \left(\frac{\mathcal{V}}{c^2}\right)x_G\right] = \gamma\left[t_G - \left(\frac{\mathcal{V}}{c^2}\right)vt_G\right] \tag{11-18}$$

$$= \gamma t_G\left(1 - \frac{\mathcal{V}v}{c^2}\right), \tag{11-19}$$

For the above example, v is the horizontal velocity of the ball. Because t'_G is positive, $\mathcal{V}v \leq c^2$. Also, since $\mathcal{V} \leq c$, a "signal" cannot propagate at a speed v that is larger than c.

***EXAMPLE** Simultaneity

Refer to the preceding example and set $T_F = T_G$: the events F and G are simultaneous in the unprimed frame. Are they also simultaneous in the primed frame? In other words, is $T'_F = T'_G$? According to Eq. 11-17, the answer is No, unless two conditions are satisfied: a) the two events occur at the same place, or $x_F = x_G$, and $\gamma \approx 1$. So, even if the two events occur at the same time and place in one reference frame, they are not, in general, simultaneous in another frame!

*11.5 TRANSFORMATION OF A LENGTH. THE LORENTZ CONTRACTION

Imagine that observer o' in reference frame S' fixes a ruler of length l_0 on the x'-axis of S' so that its extremities are at $x' = 0$ and $x' = l_0$. The quantity l_0 is the length of the ruler, as measured in its own reference frame, and is called its *proper length.*

What is the length of the *same* ruler for observer o on S? That observer performs his measurement by noting the positions of the two extremities of the ruler at the same time, say $t = 0$. At the right-hand end, $x = l$, and $t = 0$. From the Lorentz transformation,

$$l_0 = x' = \gamma(x - \mathcal{V}t) = \gamma l, \qquad l = \frac{l_0}{\gamma}. \tag{11-20}$$

Thus a ruler moving in the direction of its length at a velocity $\mathcal{V}$ relative to an observer appears to be shortened by the factor $1/\gamma$. Remember, that $\gamma \geq 1$. The *Lorentz contraction* is independent of the sign of $\mathcal{V}$. Of course, the Lorentz contraction applies if the ruler is anywhere else along the x'-axis.

If the ruler lies on the x-axis in S, then o' finds it shortened by the same factor $1/\gamma$.

Thus o says that the meters of o' are too short, and o' maintains that those of o are too short. This is not absurd because the comparisons are quite complex; they involve eight separate measurements.

If the ruler moves relative to o in a direction perpendicular to its length, then you can show that there is no Lorentz contraction: its length, measured by o, is equal to its proper length l_0. A length l therefore transforms as follows:

$$\mathbf{l}' = \frac{\mathbf{l}_{0\parallel}}{\gamma} + \mathbf{l}_{0\perp}, \tag{11-21}$$

where $l_{0\parallel}$ is the component of l_0 that is parallel to the motion and $l_{0\perp}$ is the orthogonal component.

*11.6 TRANSFORMATION OF A TIME INTERVAL. TIME DILATION

Observer o' measures the duration of a certain phenomenon that occurs at $x' = 0$ in S'. The phenomenon starts at $t' = 0$, and it ends at $t' = T_0$. The time T_0 measured in the reference frame of the phenomenon is the *proper time.* Observer o' has a single clock located at $x' = 0$.

Observer o measures the same time interval with two identical and synchronized clocks, one at $x = 0$, where the phenomenon starts, and the other at the position of O' at the end of the time interval.

According to the Lorentz transformation, the event for which $x' = 0$ and $t' = T_0$ occurs at $t = \gamma T_0$ for o in S. So

$$T = \gamma T_0. \tag{11-22}$$

For o, the time interval is longer than T_0. This is the phenomenon called *time dilation*. In other words, a moving clock appears to run slow by the factor γ if one measures its rate as above.†

*EXAMPLE The Time Read on a Rapidly Moving Clock

We have seen that, if o uses two identical clocks at two different distances x on S to measure a given time interval that o' measures at a fixed point on S' with a single clock, then o finds a time interval that is longer than that of o' by the factor γ.

What if o uses a single clock and *looks* at the moving clock of o' as in Fig. 11-3? Let the primed clock be at O', and let both $\mathscr{V}$ and t be positive. Then O' is to the right of O and moves *away.*

Imagine that o has a set of identical and synchronized clocks all along his x-axis. As the primed clock goes by each one, the relation $t = \gamma t'$ holds. But o stands at the origin O of S, and the light from the primed clock at O' takes some time to reach O. Suppose observer o reads a time t' on the moving clock. What time is it on his own

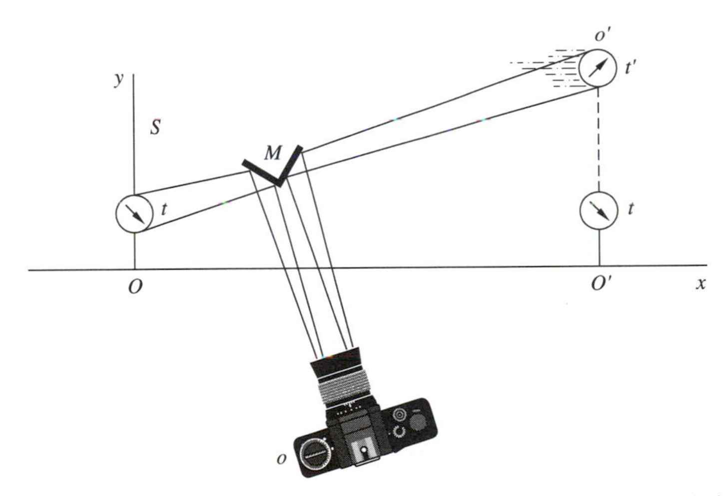

Figure 11-3 Observer o uses a double mirror M to photograph simultaneously both his own clock and the moving clock. The relation $t = \gamma t'$ holds, but the light from the moving clock does not reach the camera until a later time $t'' > t$.

†See the film entitled "Time Dilation: An Experiment on Mu-Mesons" by F. Friedman, D. Frisch, and J. Smith, produced by the Educational Development Center, Newton, Mass.

clock? Call this time t''. Then t'' is the above t, plus the time required for light to travel the distance $\mathcal{V}t$:

$$t'' = t + \frac{\mathcal{V}t}{c} = \left(1 + \frac{\mathcal{V}}{c}\right)t = (1 + \beta)t = (1 + \beta)\gamma t', \tag{11-23}$$

with $\beta = \mathcal{V}/c$, as usual. Thus, when o reads t' on a clock that is moving away, her own time is

$$t'' = \frac{1 + \beta}{(1 - \beta^2)^{1/2}} t' = \left(\frac{1 + \beta}{1 - \beta}\right)^{1/2} t'. \tag{11-24}$$

With $\mathcal{V}$ and β both positive, $t'' > t'$.

Therefore, if one *looks* at a clock that is moving *away*, the moving clock appears to run even slower.

What if the moving clock is *approaching* with $\mathcal{V}$ positive? Then the origin O' is to the left of O. With $\mathcal{V}$ positive, both t and t' are negative and

$$t'' = t + \frac{|\mathcal{V}t|}{c} = t - \frac{\mathcal{V}t}{c} = (1 - \beta)t = (1 - \beta)\gamma t' \tag{11-25}$$

is also negative, since $\beta \leq 1$. Thus, when o reads a time t' on an approaching clock, his own time is

$$t'' = \left(\frac{1 - \beta}{1 + \beta}\right)^{1/2} t'. \tag{11-26}$$

Does the approaching clock appear to run fast, or slow, with respect to the stationary clock? Say observer o reads a time $t' = -1$ second on the approaching clock, and the above square root equals 0.9. Then $t'' = -0.9$ second. When the moving clock reaches 0, both t' and t'' will be zero. So, in the interval, the moving clock will have advanced by 1 second, and the fixed clock by 0.9 second. The approaching clock appears to run *fast*.

***EXAMPLE** The Relativistic Doppler Effect for Electromagnetic Waves

The *Doppler effect* is the frequency shift observed when a source of waves moves with respect to a detector. This phenomenon is well known in the field of acoustics.

Imagine a source of electromagnetic waves of proper frequency f_0 situated at O' and a detector at O. What is the apparent frequency at O?

This problem is identical to the clock problem that we just solved, because the source beats periods $1/f_0$, instead of seconds. Therefore, with a receding source, the period measured by a fixed observer at O is

$$T = \left(\frac{1 + \beta}{1 - \beta}\right)^{1/2} T_0, \tag{11-27}$$

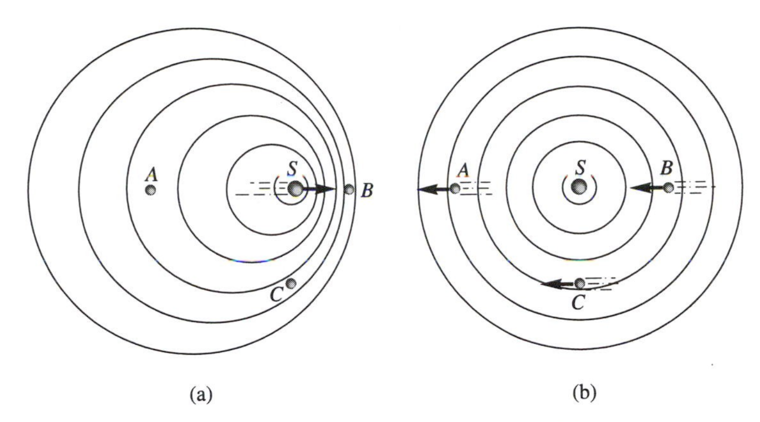

Figure 11-4 The Doppler effect. (a) Source S moves to the right with respect to stationary receivers A, B, C. (b) The source is stationary, and the receivers all move to the left. The frequency shifts are the same in the two figures. The measured frequency is lower than the proper frequency at A, and higher at B. At C, the measured frequency is either lower or higher than the proper frequency, as in Eq. 11-30.

and

$$f = \frac{1}{T} = \left(\frac{1-\beta}{1+\beta}\right)^{1/2} \frac{1}{T_0} = \left(\frac{1-\beta}{1+\beta}\right)^{1/2} f_0 < f_0. \tag{11-28}$$

For an approaching source,

$$f = \left(\frac{1+\beta}{1-\beta}\right)^{1/2} f_0 > f_0. \tag{11-29}$$

Note that it is only the *relative* velocity that counts. See Fig. 11-4.

If, at a given instant, the relative velocity forms a right angle with the line joining the source to the detector, there is still a Doppler effect because of time dilation. This is a purely relativistic effect. The frequency measured at the receiver is either larger or smaller than f_0, depending on the reference frame in which the angle is 90°:

$$f = \gamma f_0 \text{ (source)}, \qquad \text{or} \qquad f = \frac{f_0}{\gamma} \text{ (detector)}. \tag{11-30}$$

*11.7 TRANSFORMATION OF A VELOCITY

Observer o' on reference frame S' notes that an object moves at some velocity $\boldsymbol{v}'$. The velocity need not be constant. What is the velocity of this same object for observer o on reference frame S?

From the vector form of the Lorentz transformation (Table 11-1),

$$\boldsymbol{v} = \frac{d\boldsymbol{r}}{dt} = \frac{\gamma(d\boldsymbol{r}'_{\parallel} + \boldsymbol{\mathcal{V}}\,dt') + d\boldsymbol{r}'_{\perp}}{\gamma(dt' + \mathcal{V}\,dr'_{\parallel}/c^2)}. \tag{11-31}$$

Dividing numerator and denominator by $\gamma\, dt'$ gives

$$\boldsymbol{v} = \frac{\boldsymbol{v}'_{\parallel} + \boldsymbol{\mathcal{V}} + \boldsymbol{v}'_{\perp}/\gamma}{1 + v'_{\parallel}\mathcal{V}/c^2}. \qquad (11\text{-}32)$$

The Galilean relation $\boldsymbol{v} = \boldsymbol{v}' + \boldsymbol{\mathcal{V}}$ is therefore valid only if $\gamma \approx 1$ and if $v'_{\parallel}\mathcal{V} \ll c^2$. It follows from Sec. 11.4 that

$$\boldsymbol{v}' = \frac{\boldsymbol{v}_{\parallel} - \boldsymbol{\mathcal{V}} + \boldsymbol{v}_{\perp}/\gamma}{1 - v_{\parallel}\mathcal{V}/c^2}. \qquad (11\text{-}33)$$

EXAMPLE Transforming the Speed of Light

A photon travels at the speed c in some arbitrary direction in reference frame S'. Then

$$v'^2_{\parallel} + v'^2_{\perp} = c^2 \qquad (11\text{-}34)$$

and

$$v^2_{\parallel} + v^2_{\perp} = \left(\frac{v'_{\parallel} + \mathcal{V}}{1 + v'_{\parallel}\mathcal{V}/c^2}\right)^2 + \left(\frac{v'_{\perp}/\gamma}{1 + v'_{\parallel}\mathcal{V}/c^2}\right)^2. \qquad (11\text{-}35)$$

Expanding and using Eq. 11-34 twice, we find that the speed of the photon in S is also c, whatever the value of $\mathcal{V}$, which is consistent with the principle of relativity of Sec. 11.3.

This is also in agreement with the second example in Sec. 11.4, where we discussed the invariance of $x^2 + y^2 + z^2 - c^2t^2$.

*11.8 INVARIANCE OF ELECTRIC CHARGE

Electric charge is invariant: a body carries the same electric charge for all observers. For example, an electron carries the same charge of 1.602177×10^{-19} coulomb, whatever its speed: an electron that is part of a copper atom at room temperature carries the same electric charge as an electron that has an energy of several gigaelectronvolts in a cosmic-ray shower.

One proof of the invariance of electric charge is the experimental fact that a metal object does not acquire an electric charge when it is either heated or cooled (excluding thermionic emission). One might argue that the speeds of the electrons and protons are affected equally by a change in temperature, but that is incorrect: the speeds of the conduction electrons are much *less* affected than the speeds of the atoms (Kittel 1976). It is because the enormous positive and negative charges in matter cancel perfectly at all temperatures that ordinary matter remains neutral when the temperature changes.

***EXAMPLE**

Ten kilograms of copper contain about 10^{26} atoms and $10^{26} \times 1.6 \times 10^{-19}$ coulombs of conduction electrons (1.6×10^{7} coulombs!). If the positive charge increased or decreased by only one part in 10^{15} upon heating, the copper would acquire a net charge of 1.6×10^{-8} coulomb, and the potential of a 10-kilogram copper sphere, with a radius of about 65 millimeters, would either rise or fall by about 2 kilovolts. Such an effect has never been observed.

*11.9 SUMMARY

An *inertial reference frame* does not accelerate and does not rotate.

The *special theory of relativity* concerns observations and measurements on a given phenomenon made by observers situated in two inertial frames, one of which moves at a constant velocity with respect to the other. As a rule, the two frames referred to are those of Fig. 11-2.

The *principle of relativity* states that it is physically impossible to detect whether an *inertial* reference frame is at rest or in uniform motion from observations made entirely within that frame.

Thus *physical laws are invariant:* the law describing a given phenomenon is mathematically the same, in whatever inertial frame the phenomenon occurs.

The *Lorentz transformation* relates x, y, z, t in S to x', y', z', t' in S':

$$x = \gamma(x' + \mathcal{V}t'), \qquad x' = \gamma(x - \mathcal{V}t), \tag{11-9}$$

$$y = y', \qquad y' = y, \tag{11-10}$$

$$z = z', \qquad z' = z, \tag{11-11}$$

$$t = \gamma\left(t' + \frac{\mathcal{V}x'}{c^2}\right), \qquad t' = \gamma\left(t - \frac{\mathcal{V}x}{c^2}\right), \tag{11-12}$$

where $\gamma = (1 - \mathcal{V}^2/c^2)^{-1/2}$.

Table 11-1 shows these equations in vector form. The subscripts $\|$ and $\perp$ refer, respectively, to components that are parallel or perpendicular to the velocity $\mathcal{V}$ of S' with respect to S.

A *quantity* is said to be *invariant* if its numerical value is the same in all inertial reference frames.

A signal cannot propagate at a speed greater than c.

An object has a proper length l_0, oriented in some arbitrary direction. Then, in another reference frame moving at some constant velocity $\mathcal{V}$ with respect to the object,

$$l' = \frac{l_{0\|}}{\gamma} + l_{0\perp}. \tag{11-21}$$

Lengths parallel to the motion are shorter by the factor γ. This is the *Lorentz contraction.* Lengths orthogonal to the motion are unaffected.

A process lasts a proper time T_0 in its own reference frame. In another frame as above, the time interval is γ times larger:

$$T = \gamma T_0. \tag{11-22}$$

This is *time dilation.*

If a source of electromagnetic waves of proper frequency f_0 moves *away* at a velocity $\mathcal{V} = \beta c$ from an observer, the apparent frequency is

$$f = \left(\frac{1-\beta}{1+\beta}\right)^{1/2} f_0 < f_0. \tag{11-28}$$

For an *approaching* source, the sign before β changes,

$$f = \left(\frac{1+\beta}{1-\beta}\right)^{1/2} f_0 > f_0. \tag{11-29}$$

This is the *Doppler effect.* Only the *relative* velocity between source and detector matters.

A *velocity* transforms as follows:

$$\mathbf{v} = \frac{\mathbf{v}'_{\parallel} + \boldsymbol{\mathcal{V}} + \mathbf{v}'_{\perp}/\gamma}{1 + v'_{\parallel}\mathcal{V}/c^2}, \tag{11-32}$$

$$\mathbf{v}' = \frac{\mathbf{v}_{\parallel} - \boldsymbol{\mathcal{V}} + \mathbf{v}_{\perp}/\gamma}{1 - v_{\parallel}\mathcal{V}/c^2}. \tag{11-33}$$

Electric charge is invariant: a body carries the same electric charge for all observers.

*PROBLEMS

11-1. †(*11.4*) The Lorentz transformation

(a) For what value of β is the value of γ equal to 1.01?

(b) Calculate γ, β, and v for a conduction electron whose energy is 10 electronvolts. The rest energy of the electron is 5.11×10^5 electronvolts.

(c) Show that

$$x' + ct' = \left(\frac{1-\beta}{1+\beta}\right)^{1/2}(x + ct).$$

11-2. (*11.4*) Signaling problems with a fast train

Three persons A, O', and B, ride on a train moving at a velocity $\mathcal{V}$, with A in front, O' in the middle, and B in the rear. A fourth person, O, stands beside the rails. At the moment O' passes O, light signals from A and B reach both O and O'. Persons O and O' are asked who emitted her light signal first. What do they answer?

---

†Several of the problems on relativity are adapted, with permission, from Taylor and Wheeler (1992).

11-3. (*11.4*) Transformation of an angle

A straight line passing through the origin O' of S' forms an angle α' with the x'-axis.

(a) Find a relation between α and α'.

(b) What is the value of α when $\mathcal{V}$ tends to c?

11-4. (*11.4*) Things that move faster than a photon?

The Lorentz transformation implies that the relative velocity $\mathcal{V}$ of two frames of reference cannot exceed the speed of light c. We have also shown that a signal cannot exceed the speed of light. Discuss the following cases.

(a) A long, straight rod forms a small angle θ with another rod, which is horizontal and stationary. The first rod moves downward at a velocity v.

What is the speed of the point of intersection of the lower edge of the moving rod with the fixed rod? Can this speed be greater than c? Can the point of intersection be used to transmit a signal?

(b) The upper rod is initially at rest with the point of intersection at the origin. The rod is struck a downward blow at the origin with a hammer.

Can the motion of the point of intersection be used to transmit a signal at a speed greater than the speed of light?

(c) A powerful laser rotates rapidly about an axis perpendicular to its length.

Can the azimuthal speed of the beam exceed the speed of light? Can the beam transmit a signal between two points at a speed greater than c?

(d) The manufacturers of some oscilloscopes claim writing speeds in excess of the speed of light. Is this possible?

11-5. (*11.4*) c is the ultimate speed

Imagine a series of reference frames S, S', S'', S''', etc., with S' moving at a velocity $\mathcal{V}\hat{\boldsymbol{x}}$ with respect to S, S'' moving at the same velocity with respect to S', etc. According to the Galilean transformation, a particle at rest in the last frame moves at a velocity $n\mathcal{V}\hat{\boldsymbol{x}}$ with respect to S, where $n\mathcal{V}$ is arbitrarily large.

Show that, according to special relativity, the velocity of that particle with respect to S is always less than c.

11-6. (*11.4*) The speed of light in a moving medium

Light moves more slowly through a material medium than through a vacuum, its phase velocity v being c/n, where n is the index of refraction of the medium.

If now the medium itself moves at a velocity $\mathcal{V} \ll c$ with respect to the laboratory, show that the phase velocity of the light with respect to the laboratory is approximately $c/n + \mathcal{V}(1 - 1/n^2)$.

11-7. (*11.4*) The headlight effect

A source of light moves at a velocity $\mathcal{V}\hat{\boldsymbol{x}}$. Consider a ray that forms an angle θ' with respect to the x'-axis. See Fig. 11-5.

(a) Show that in the reference frame S, $\tan\theta = \sin\theta'/[\gamma(\cos\theta' + \beta)]$.

Then $\tan\theta' = \sin\theta/[\gamma(\cos\theta - \beta)]$, from Sec. 11.4.

(b) Plot θ as a function of θ' for $\beta = 0, 0.5, 0.9, 0.9999$.

Observe that, for large values of β, θ is much smaller than θ', except for values of θ' near π. If the source radiates isotropically in its own reference frame, then, for a stationary observer, the source radiates mostly in the forward direction. This is the *headlight effect*.

(c) An isotropic source of light moves at a speed $c/3$ with respect to an observer. Calculate the solid angle defined by a cone that points forward and that contains 25% of the total light flux.

11-8. (*11.4*) The collimator paradox

Figure 11-5 shows a source of light and a collimator C fixed in a reference frame S' that moves at a velocity $\mathcal{V}\hat{x}$ with respect to a fixed frame S. The detector D measures the light that goes through the collimator.

According to Prob. 11-7, the angle θ formed by the beam of light is *less* than θ' because of the headlight effect. However, the Lorentz contraction on the collimator makes its angle θ *larger* than θ'. But this is absurd! If light reaches D in S', then it does so in S.

You can solve this paradox by using the Lorentz transformation.

11-9. (*11.4*) Three reference frames

We have three reference frames A, B, C. Frames B and C move, respectively, at velocities $\mathcal{V}\hat{x}/2$ and $\mathcal{V}\hat{x}$ with respect to A. Use subscripts to identify the velocities: $v_{BA} = \mathcal{V}/2$, $v_{CA} = \mathcal{V}$.

Show that v_{CB} (velocity of C with respect to B) is larger than $\mathcal{V}/2$. Thus, with respect to B, C moves away faster than A.

11-10. (*11.5*) Lorentz contraction for a length

A one-meter ruler moves at a speed $c/2$. In its own reference frame it forms an angle of 45° with its velocity.

What is its length, as measured by a fixed observer?

11-11. (*11.6*) Two successive events at a given point

Two events occur at the same place in the laboratory at an interval of 3 seconds.

What is the spatial distance between these two events in a moving frame with respect to which the events occur 5 seconds apart, and what is the relative speed of the moving and laboratory frames?

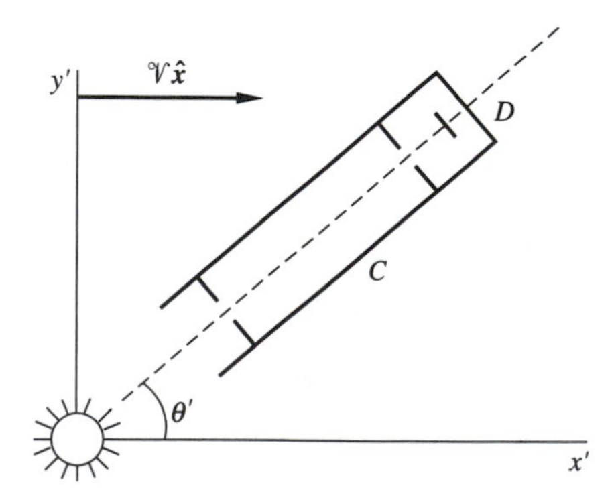

Figure 11-5

11-12. (*11.6*) The red shift

The radio galaxy 3C295 has a red shift of 46%. Astronomers mean by this that the observed wavelength is 1.46 times the wavelength of the same radiation originating in the laboratory.

(a) Calculate the radial velocity of the galaxy.

(b) Some quasars have red shifts of 200%. What is their radial velocity?

11-13. (*11.6*) The case of the speeding physicist

A physicist is arrested for going through a red traffic light. In court she pleads that she approached at such a speed that the red light appeared green. The judge, a graduate of a physics class, changes the charge to speeding and fines the defendant $1 for every kilometer per hour she exceeded the speed limit of 50 kilometers/hour.

What is the fine ($\lambda_{green} \approx 5.3 \times 10^{-7}$ meter, $\lambda_{red} \approx 6.5 \times 10^{-7}$ meter)?

11-14. (*11.6*) The twin paradox

On his twenty-first birthday, Peter leaves his twin Paul behind on the earth and goes off in a straight line for 7 years of his time at a speed of $0.96c$. Peter then reverses direction and returns at the same speed.

What are the ages of Peter and Paul at the moment of reunion?

11-15. (*11.8*) The invariance of electric charge

Imagine that electric charge is not invariant and that $Q = Q_0[1 - (v^2/c^2)]^{1/2}$. (Remember that charge is, in fact, invariant, according to all experiments performed to date). The charge Q_0 is that measured by an observer moving with the charge, and Q is the charge for an observer moving at a velocity $\mathcal{V}$ with respect to it.

If the electrons in a given sample have an average kinetic energy of 100 electron-volts, what percentage increase in their charge must we expect if their velocity increases by 1%?

CHAPTER

12

*RELATIVITY II†

Mass, Momentum, Force, and Energy

This second chapter on relativity concerns mechanics, and its main object is to derive the transformation of a force. The reason for this will become clear in the next chapter: an electric force becomes both an electric force and a magnetic force when seen in a different reference frame. So electric and magnetic forces are closely related.

As we shall see in this chapter, all the basic concepts of mechanics, such as mass, momentum, force, and energy, crumble, so that classical mechanics sinks to the rank of an approximate theory. It is a curious fact that classical electromagnetism, by contrast, is completely compatible with relativity, and is thus spared.

*12.1 RELATIVISTIC MASS

In its own reference frame, an object has a rest mass m_0. If the object moves at a speed v with respect to an observer, then, for *that* observer,

†This whole chapter can be skipped without losing continuity.

$$m = \frac{m_0}{[1 - (v^2/c^2)]^{1/2}} = \gamma m_0 > m_0, \tag{12-1}$$

where m is the *relativistic mass*. We take this fairly well-known result for granted (Taylor and Wheeler, 1992). All mass measurements are in agreement with it. For example, the mass of a high-energy electron in an accelerator is γ times that of a slow electron.

The relativistic mass tends to infinity as v approaches c. Then, what about the photon? It has a relativistic mass m, and a (presumably) zero rest mass. See Sec. 12.6.

*12.1.1 Transformation of a Mass

An object has a mass m in reference frame S. What is its mass in frame S' that moves at the velocity $\mathcal{V}$ with respect to S? If the object is at rest in S, its mass in that frame is m_0, and its mass in S' is given by Eq. 12-1, with $v = \mathcal{V}$.

But what if the mass has a velocity $\boldsymbol{v} = \boldsymbol{v}_\parallel + \boldsymbol{v}_\perp$ in S? Then, in S,

$$m = \frac{m_0}{\{1 - [(v_\parallel^2 + v_\perp^2)/c^2]\}^{1/2}}, \tag{12-2}$$

and m' is given by a similar expression with primes on m and on the v's. Now how do you transform m into m', and inversely?

Using the equation for the transformation of a velocity of Sec. 11.7,

$$v_\parallel^2 + v_\perp^2 = \frac{(v'_\parallel + \mathcal{V})^2 + v'^2_\perp/\gamma^2}{[1 + (v'_\parallel \mathcal{V}/c^2)]^2}. \tag{12-3}$$

Recall that $1/\gamma^2 = 1 - (\mathcal{V}^2/c^2)$. After simplification, we find that

$$m = \gamma m'\left(1 + \frac{v'_\parallel \mathcal{V}}{c^2}\right), \tag{12-4}$$

and, from the second rule of Sec. 11.4,

$$m' = \gamma m\left(1 - \frac{v_\parallel \mathcal{V}}{c^2}\right). \tag{12-5}$$

*12.2 RELATIVISTIC MOMENTUM

The *relativistic momentum* $\boldsymbol{p}$ of a mass m moving at a velocity $\boldsymbol{v}$ is defined as in classical mechanics:

$$\boldsymbol{p} = m\boldsymbol{v} = \frac{m_0 \boldsymbol{v}}{[1 - (v^2/c^2)]^{1/2}} = \gamma m_0 \boldsymbol{v}. \tag{12-6}$$

It is this quantity that is conserved in collisions, and not the momentum $m_0\boldsymbol{v}$ of classical mechanics.

*12.2.1 Transformation of a Momentum

Let us first transform $p_\parallel$:

$$p_\parallel = mv_\parallel = \gamma m'\left(1 + \frac{v'_\parallel \mathcal{V}}{c^2}\right)\left[\frac{v'_\parallel + \mathcal{V}}{1 + (v'_\parallel \mathcal{V}/c^2)}\right] \tag{12-7}$$

$$= \gamma m'(v'_\parallel + \mathcal{V}) = \gamma(p'_\parallel + m'\mathcal{V}). \tag{12-8}$$

Now transform $p_\perp$:

$$p_\perp = mv_\perp = \gamma m'\left(1 + \frac{v'_\parallel \mathcal{V}}{c^2}\right)\left\{\frac{v'_\perp}{\gamma[1 + (v'_\parallel \mathcal{V}/c^2)]}\right\} = m'v'_\perp = p'_\perp. \tag{12-9}$$

The perpendicular components of $\boldsymbol{p}$ and $\boldsymbol{p}'$ are equal.

So

$$\boldsymbol{p} = \gamma(\boldsymbol{p}'_\parallel + m'\boldsymbol{\mathcal{V}}) + \boldsymbol{p}'_\perp, \tag{12-10}$$

$$\boldsymbol{p}' = \gamma(\boldsymbol{p}_\parallel - m\boldsymbol{\mathcal{V}}) + \boldsymbol{p}_\perp. \tag{12-11}$$

Note the similarity between these equations and those for $\boldsymbol{r}$ and $\boldsymbol{r}'$ shown in Table 11-1.

*12.3 RELATIVISTIC FORCE

The *relativistic force* is also defined as in classical mechanics:

$$\boldsymbol{F} = \frac{d\boldsymbol{p}}{dt}, \tag{12-12}$$

but $\boldsymbol{p}$ is now the relativistic momentum.

*EXAMPLE A High-Energy Particle

Recall from Sec. 11.8 that electric charge is invariant. A high-energy particle of rest mass m_0, velocity $\boldsymbol{v}$, and charge Q crosses a region where the electric field strength is $\boldsymbol{E}$. The electric force is

$$\boldsymbol{F} = \frac{d\boldsymbol{p}}{dt} = \frac{d}{dt}(m\boldsymbol{v}) = \frac{d}{dt}\left\{\frac{m_0\boldsymbol{v}}{[1 - (v^2/c^2)]^{1/2}}\right\} = Q\boldsymbol{E}. \tag{12-13}$$

We defer the transformation of a force until Sec. 12.5.

*12.4 RELATIVISTIC ENERGY $\mathscr{E} = mc^2$

If a mass m moving at a velocity $\boldsymbol{v}$ is subjected to a force $\boldsymbol{F}$, then the mechanical energy expended by $\boldsymbol{F}$ on m, per second, is $\boldsymbol{F} \cdot \boldsymbol{v}$. The mass acquires an increment of relativistic energy $d\mathscr{E}$ and

$$\frac{d\mathscr{E}}{dt} = \boldsymbol{F} \cdot \boldsymbol{v} = \frac{d\boldsymbol{p}}{dt} \cdot \boldsymbol{v} \tag{12-14}$$

$$= \frac{d(m\boldsymbol{v})}{dt} \cdot \boldsymbol{v} = \left(m\frac{d\boldsymbol{v}}{dt} + \boldsymbol{v}\frac{dm}{dt}\right) \cdot \boldsymbol{v}. \tag{12-15}$$

To prove that $\mathscr{E} = mc^2$ we must now show that

$$\frac{d\mathscr{E}}{dt} = \frac{d}{dt}(mc^2) = c^2\frac{dm}{dt}, \tag{12-16}$$

or that

$$\left(m\frac{d\boldsymbol{v}}{dt} + \boldsymbol{v}\frac{dm}{dt}\right) \cdot \boldsymbol{v} = c^2\frac{dm}{dt}. \tag{12-17}$$

Grouping the dm/dt terms,

$$m\frac{d\boldsymbol{v}}{dt} \cdot \boldsymbol{v} = (c^2 - v^2)\frac{dm}{dt}, \tag{12-18}$$

$$\frac{m}{c^2}\frac{d\boldsymbol{v}}{dt} \cdot \boldsymbol{v} = \left(1 - \frac{v^2}{c^2}\right)\frac{d}{dt}\left\{\frac{m_0}{[1 - (v^2/c^2)]^{1/2}}\right\}. \tag{12-19}$$

It is easy to show that this is an identity except that, to find the time derivative of v^2, one must set $v^2 = \boldsymbol{v} \cdot \boldsymbol{v}$, so that

$$\frac{dv^2}{dt} = 2\boldsymbol{v} \cdot \frac{d\boldsymbol{v}}{dt}. \tag{12-20}$$

This proves Eq. 12-16.

More generally, in *any* physical process, an increase of energy $d\mathscr{E}$ results in an increase of mass dm such that

$$d\mathscr{E} = c^2 dm, \tag{12-21}$$

and

$$\mathscr{E} = mc^2 \tag{12-22}$$

is the *relativistic energy*, or *mass-energy*, of an object of mass m. A particle at rest in a given reference frame has a *rest energy* m_0c^2 in that frame.

***EXAMPLES**

(a) Say a mixture of hydrogen and oxygen explodes. What happens to the mass? Before the reaction, the molecules may be assumed to be at rest. Let their total mass be m_{0b}. Just after the reaction, the relativistic mass is unaltered because no *external* energy has been fed into the gas. Then

$$m_a = m_{0b} = m_{0a} + KE, \qquad (12\text{-}23)$$

where KE is the kinetic energy of the high-temperature steam. The new *rest* mass m_{0a} is smaller than the initial *rest* mass m_{0b}.

After a while, the steam cools and becomes water at room temperature, KE tends to zero, and m_a decreases to m_{0a}. In the process, the kinetic energy has spread out to neighboring bodies, thereby increasing their masses. The *rest* mass of the water is less than that of the original mixture, but the relativistic mass of the Universe is unchanged.

Fission and fusion reactions are qualitatively similar to exothermic chemical reactions: there is a loss of *rest* mass and a release of thermal energy.

(b) What happens when you drop a brick? The *rest* mass of the brick depends on its position, because of its potential energy: the higher it is, the larger is its rest mass. As the brick falls, its rest mass decreases and its kinetic energy increases. When the brick hits the ground, its kinetic energy becomes thermal energy. It has lost rest mass but, again, the relativistic mass of the Universe is unchanged.

Strictly, one should think of the potential energy of the brick-earth system and take into account the upward motion of the earth as the brick falls.

(c) When a positron meets an electron, both particles disappear to form a pair of photons, as in Fig. 12-1. Charge, mass-energy, and momentum are all conserved.

Inversely, a photon can create a positron-electron pair. There is conservation of charge, but the conservation of mass-energy and momentum requires the presence of another particle.

*12.4.1 Kinetic Energy

By definition, the *kinetic energy* of a point mass m moving at a speed v with respect to a given reference frame is equal to the energy expended in increasing its speed from zero to v in that frame. This is

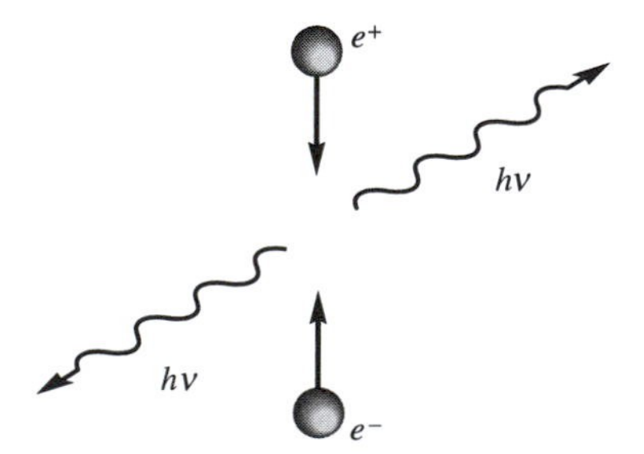

Figure 12-1 A positive and a negative electron annihilate to form two photons.

$$mc^2 - m_0c^2 = m_0c^2\left[\frac{1}{(1 - v^2/c^2)^{1/2}} - 1\right] \quad (12\text{-}24)$$

$$= m_0c^2\left(\frac{v^2}{2c^2} + \frac{3v^4}{8c^4} + \cdots\right) \quad (12\text{-}25)$$

$$= \frac{m_0v^2}{2} + \frac{3m_0v^4}{8c^2} + \cdots. \quad (12\text{-}26)$$

The first term on the right is the kinetic energy of classical mechanics. It is equal to the relativistic kinetic energy if $v^2 \ll c^2$.

*12.4.2 Transformation of An Energy

Because of the above relations, a relativistic energy transforms like a relativistic mass:

$$\mathscr{E} = \gamma\mathscr{E}'\left(1 + \frac{v'_{\parallel}\mathscr{V}}{c^2}\right), \quad (12\text{-}27)$$

$$\mathscr{E}' = \gamma\mathscr{E}\left(1 - \frac{v_{\parallel}\mathscr{V}}{c^2}\right). \quad (12\text{-}28)$$

*12.5 TRANSFORMATION OF A FORCE

The transformation equations for a force will serve when we discuss magnetic fields in Sec. 13.2.

First, set $m = \mathscr{E}/c^2$ and $m' = \mathscr{E}'/c^2$ in Eqs. 12-10 and 12-11. Then, from Eqs. 12-10 to 12-12,

$$\boldsymbol{F} = \frac{d\boldsymbol{p}}{dt} = \frac{\gamma(d\boldsymbol{p}'_{\parallel} + \boldsymbol{\mathscr{V}}\,d\mathscr{E}'/c^2) + d\boldsymbol{p}'_{\perp}}{\gamma(dt' + \mathscr{V}\,dr'_{\parallel}/c^2)}. \quad (12\text{-}29)$$

Dividing above and below by $\gamma dt'$,

$$\boldsymbol{F} = \frac{\boldsymbol{F}'_{\parallel} + (\boldsymbol{\mathscr{V}}/c^2)d\mathscr{E}'/dt' + \boldsymbol{F}'_{\perp}/\gamma}{1 + v'_{\parallel}\mathscr{V}/c^2}, \quad (12\text{-}30)$$

where $\boldsymbol{v}'$ is the velocity of the point of application of $\boldsymbol{F}'$ in reference frame S'.

Now $d\mathscr{E}'/dt'$ is the rate at which the relativistic energy builds up in S' because of $\boldsymbol{F}'$. Then

$$\frac{d\mathscr{E}'}{dt'} = \boldsymbol{F}' \cdot \boldsymbol{v}' = F'_{\parallel}v'_{\parallel} + \boldsymbol{F}'_{\perp} \cdot \boldsymbol{v}'_{\perp}. \quad (12\text{-}31)$$

Substituting into Eq. 12-30 and simplifying leads to the transformation equation

$$\boldsymbol{F} = \boldsymbol{F}'_{\parallel} + \frac{\gamma(\mathcal{V}/c^2)\boldsymbol{F}'_{\perp}\cdot\boldsymbol{v}'_{\perp} + \boldsymbol{F}'_{\perp}}{\gamma(1 + v'_{\parallel}\mathcal{V}/c^2)}. \tag{12-32}$$

The inverse transformation is

$$\boldsymbol{F}' = \boldsymbol{F}_{\parallel} + \frac{-\gamma(\mathcal{V}/c^2)\boldsymbol{F}_{\perp}\cdot\boldsymbol{v}_{\perp} + \boldsymbol{F}_{\perp}}{\gamma(1 - v_{\parallel}\mathcal{V}/c^2)}. \tag{12-33}$$

Note that the transformation equations do *not* involve the point of application of the force.

***EXAMPLES**

If $\boldsymbol{F}$ is parallel to $\mathcal{V}$, then $\boldsymbol{F}_{\perp} = 0$ and $\boldsymbol{F} = \boldsymbol{F}'$.

Two forces that are equal in one frame are not necessarily so in another frame. They are equal only if their points of application have equal velocities.

*12.6 THE PHOTON

The photon has presumably a zero rest mass. Its speed is c† and its energy is

$$\mathcal{E} = h\nu, \tag{12-34}$$

where h is Planck's constant, 6.626×10^{-34} joule-second, and ν is the frequency of the associated wave. Its relativistic mass is

$$m = \frac{\mathcal{E}}{c^2}, \tag{12-35}$$

and its momentum is

$$p = mc = \frac{\mathcal{E}}{c} = \frac{h\nu}{c} = \frac{h}{\lambda}, \tag{12-36}$$

where λ is the wavelength of the associated wave. The relation between p and $\mathcal{E}$ for a photon has been checked by measuring both the radiation pressure and the energy flux in a light beam.

The kinetic energy of a photon is $(m - m_0)c^2 = mc^2$, and *not* $mc^2/2$, as one would expect from classical mechanics.

Although a given photon has a speed c for all observers, its energy $\mathcal{E}$ and its momentum p are not invariant because of the Doppler effect (Sec. 11.6). For example, if

†Photons always travel at the speed c, even inside matter. See Sec. 22.5. For the moment, we can think of photons traveling in a vacuum.

the reference frame S' moves in the same direction as the photon, then $\nu' < \nu$, and thus

$$h\nu' < h\nu,\ \mathscr{E}' < \mathscr{E},\ p' < p. \tag{12-37}$$

*12.7 SUMMARY

The *relativistic mass* of an object depends on its speed with respect to the observer:

$$m = \frac{m_0}{[1 - (v^2/c^2)]^{1/2}} = \gamma m_0 > m_0. \tag{12-1}$$

The transformation equations are the following:

$$m = \gamma m'\left(1 + \frac{v'_{\parallel}\mathscr{V}}{c^2}\right), \tag{12-4}$$

$$m' = \gamma m\left(1 - \frac{v_{\parallel}\mathscr{V}}{c^2}\right). \tag{12-5}$$

The *relativistic momentum* of a point mass is

$$\boldsymbol{p} = m\boldsymbol{v} = \gamma m_0 \boldsymbol{v}, \tag{12-6}$$

and

$$\boldsymbol{p} = \gamma(\boldsymbol{p}'_{\parallel} + m'\boldsymbol{\mathscr{V}}) + \boldsymbol{p}'_{\perp}, \tag{12-10}$$

$$\boldsymbol{p}' = \gamma(\boldsymbol{p}_{\parallel} - m\boldsymbol{\mathscr{V}}) + \boldsymbol{p}_{\perp}. \tag{12-11}$$

The *relativistic force* is

$$\boldsymbol{F} = \frac{d\boldsymbol{p}}{dt}, \tag{12-12}$$

and

$$\boldsymbol{F} = \boldsymbol{F}'_{\parallel} + \frac{\gamma(\boldsymbol{\mathscr{V}}/c^2)\boldsymbol{F}'_{\perp}\cdot\boldsymbol{v}'_{\perp} + \boldsymbol{F}'_{\perp}}{\gamma(1 + v'_{\parallel}\mathscr{V}/c^2)}, \tag{12-32}$$

$$\boldsymbol{F}' = \boldsymbol{F}_{\parallel} + \frac{-\gamma(\boldsymbol{\mathscr{V}}/c^2)\boldsymbol{F}_{\perp}\cdot\boldsymbol{v}_{\perp} + \boldsymbol{F}_{\perp}}{\gamma(1 - v_{\parallel}\mathscr{V}/c^2)}. \tag{12-33}$$

The *relativistic energy* $\mathscr{E}$ is equal to mc^2. In any physical process, an increase of energy $d\mathscr{E}$ results in an increase of mass dm and

$$\mathscr{E} = \gamma\mathscr{E}'\left(1 + \frac{v'_{\parallel}\mathscr{V}}{c^2}\right), \tag{12-27}$$

$$\mathscr{E}' = \gamma\mathscr{E}\left(1 - \frac{v_{\|}\mathscr{V}}{c^2}\right). \qquad (12\text{-}28)$$

The *rest energy* is m_0c^2.

In *any* interaction there is conservation of mass-energy and momentum.

Kinetic energy is $mc^2 - m_0c^2$. The kinetic energy of classical mechanics, $(1/2)mv^2$ applies on the condition that $v^2 \ll c^2$.

The *photon* has a presumably zero rest mass and

$$\mathscr{E} = h\nu, \qquad (12\text{-}34)$$

where h is Planck's constant and ν is the frequency. Also,

$$m = \mathscr{E}/c^2. \qquad (12\text{-}35)$$

Because of the Doppler effect, the frequency, and hence the mass-energy and the momentum of a photon are not the same for all observers.

*PROBLEMS

12-1. (*12.1*) Transformation of a mass density

We use the symbol τ for volume in this problem.

A small element in an object has a proper mass dm_0, a proper volume $d\tau_0$, and a proper mass density $\rho_0 = dm_0/d\tau_0$. With respect to S and S' the mass densities are $\rho = dm/d\tau$ and $\rho' = dm'/d\tau'$, respectively, and the parallel velocities of the element are $v_{\|}$ and $v'_{\|}$.

Show that, if $\mathscr{V}$ is the speed of S' with respect to S, then

(a) $dm = \gamma\left(1 + v'_{\|}\frac{\mathscr{V}}{c^2}\right)dm'$, (b) $d\tau = \dfrac{d\tau'}{\gamma(1 + v'_{\|}\mathscr{V}/c^2)}$,

(c) $\rho = \gamma^2\left(1 + \dfrac{v'_{\|}\mathscr{V}}{c^2}\right)^2\rho'$.

12-2. (*12.3*) The relativistic force

In classical mechanics, $\boldsymbol{F} = m\boldsymbol{a}$, if the mass is constant.

Show that with relativity,

$$\boldsymbol{F} = \frac{m_0}{(1 - v^2/c^2)^{3/2}}\boldsymbol{a}_{\|} + \frac{m_0}{(1 - v^2/c^2)^{1/2}}\boldsymbol{a}_{\perp} = \gamma^2 m\boldsymbol{a}_{\|} + m\boldsymbol{a}_{\perp},$$

where $\boldsymbol{a}_{\|}$ and $\boldsymbol{a}_{\perp}$ are the components of the acceleration that are, respectively, parallel and perpendicular to the velocity $\boldsymbol{v}$ of the point of application of the force.

A force that is perpendicular to $\boldsymbol{v}$ changes the direction of $\boldsymbol{v}$, but not the mass, and hence not the speed, as we could expect because the force does no work. Then, if $\boldsymbol{F}$ is perpendicular to $\boldsymbol{v}$, $\boldsymbol{F} = m\boldsymbol{a}$ applies!

However, a force parallel to $\boldsymbol{v}$ changes the magnitude of $\boldsymbol{v}$ and hence the mass also. The resistance to acceleration is larger because of the γ^2 term.

The quantity m is sometimes called the *transverse inertial mass*, and $\gamma^2 m$ the *longitudinal inertial mass*.

12-3. (*12.4*) Burning hydrogen

Ignite a mixture of hydrogen and oxygen inside a closed vessel, and then allow the water vapor to cool.

Sketch graphs of mc^2 and of m_0c^2 as functions of time.

12-4. (*12.4*) The mass of a high-energy proton

A proton has a kinetic energy of 500 million electronvolts. Find its mass and speed.

12-5. (*12.4*) The conservation laws for colliding particles

In the course of a collision between two particles there is conservation of mass-energy and conservation of momentum.

(a) Show that, if these conservation laws apply in one inertial reference frame, then they apply in any other inertial frame.

(b) Given that, in a given reaction, relativistic mass-energy is conserved in all inertial frames, show that $\boldsymbol{p}$ is also conserved, and inversely.

12-6. (*12.4*) Relativistic effects with 40-GeV electrons

A linear accelerator accelerates electrons up to energies of 40 gigaelectronvolts (40×10^9 electronvolts).

(a) Calculate the mass of an electron that has the full energy. How does this mass compare with that of a proton at rest? See the page facing the back cover.

(b) What is the length of the accelerator in the reference frame of an electron that has the full energy? The length of the accelerator, as measured on the ground, is 3000 meters.

(c) How much time would such an electron take to go from one end of the accelerator to the other (i) in the laboratory frame and (ii) in the electron's frame of reference?

12-7. (*12.4*) The gravitational red shift

A photon of energy $h\nu_0$ leaves the surface of a star of radius R and mass M.

(a) Show that, after the photon has escaped to infinity, $\Delta\nu/\nu_0$ is equal to $GM/(Rc^2)$, where G is the gravitational constant. This change is so small that you can set $\nu = \nu_0$ in your calculation of the change in potential energy. What is the sign of $\Delta\nu$? This change of frequency is the *gravitational red shift*.

(b) Calculate $\Delta\nu/\nu$ for the sun and for the earth. See the page facing the back cover.

(c) Calculate $\Delta\nu/\nu$ for a photon that travels from the surface of the sun to the surface of the earth, taking into account both gravitational fields.

(d) Sirius and a smaller star revolve around each other. The mass of the smaller star is about equal to that of the sun, but its light has a $\Delta\nu/\nu$ of 7×10^{-4}. What is its average density?

(e) The period of rotation of the sun is 24.7 days. What is the Doppler shift for 500-nanometer light emitted from the edge of the sun's disk, at its equator? Compare this Doppler shift with the gravitational red shift.

(f) The sun ejects ionized hydrogen. How does the mass of a proton vary as it flies away from the sun?

12-8. (*12.4*) The ultimate spaceship

The thrust of a spaceship engine is equal to the product $m'v$, where m' is the mass of propellant ejected per second and v is the exhaust velocity with respect to the ship. The ultimate spaceship would transform all its propellant into radiation and eject photons backward at the speed of light. The mass of the propellant would then be minimum.

(a) Show that the power-to-thrust ratio P/F for a photon engine is c.

Since P/F and dM/dt, where m is the mass of the spaceship, are independent of the frequency, the source of radiation need not be monochromatic.

(b) Then a photon ship burning 1 gram of matter per second would have a thrust of 3×10^5 newtons. The difficulty is to transform an appreciable fraction of the propellant mass into radiation, as the following example will show.

An ordinary flashlight has a capacity of about 2 ampere-hours at about 2 volts. Show that its terminal velocity is of the order of 10^{-4} meter/second

12-9. (*12.4*) Is interstellar travel possible?

(a) First, time should be dilated by, say, a factor of 10. Then $\gamma = 10$. Calculate v/c.

(b) Imagine a spaceship equipped with a photon motor. See Prob. 12-8. You can find the fraction f of the initial mass that remains, after the ship has attained the proper β, from the conservation of energy and the conservation of momentum. Take into account the energy and momentum of the radiation. You should find that $f = 0.05$.

The spaceship must then brake to a stop. This requires 95% of the remaining mass. At the end of the return trip we are left with $(0.05)^4 = 6.25 \times 10^{-6}$ of the initial mass. If the mass of the ship and its payload is 1 ton, then the propellant has a mass of about 200,000 tons.

(c) In principle, the spaceship could collect and annihilate interstellar matter. There is about one atom of hydrogen per cubic centimeter.

Calculate the mass of hydrogen collected during 1 year if the ship sweeps out a volume 1000 square meters in cross-section at the speed of light.

(d) Now this hydrogen must first be brought up to speed. This slows the ship. Show that the net gain is positive up to $\beta = 0.707$ and negative afterward.

12-10. (*12.5*) Transformation of a force

Show that

$$\mathbf{F} = \mathbf{F}'_{\parallel} + \gamma\left[\frac{\boldsymbol{\mathcal{V}} \times (\mathbf{F}' \times \boldsymbol{\mathcal{V}})}{\mathcal{V}^2} - \frac{\boldsymbol{v} \times (\mathbf{F}' \times \boldsymbol{\mathcal{V}})}{c^2}\right].$$

CHAPTER

13

*RELATIVITY III†

The E and B Fields of a Moving Electric Charge

The simplest field of all is that of a stationary point charge. The next simplest one is the field of a charge Q that moves at a constant velocity. That field is the subject of this third and last chapter on relativity.

We consider two point charges as in Fig. 13-1. The charge Q is fixed at the origin of reference frame S', so that it moves at the constant velocity $\mathcal{V}\hat{\boldsymbol{x}}$ with respect to the reference frame S. Charge q moves at some unspecified velocity $\boldsymbol{v}$ with respect to S, and at the velocity $\boldsymbol{v}'$ with respect to S'. The velocity of q need not be constant.

There are no other electric charges or currents in the vicinity.

The force exerted *by Q on q* is $\boldsymbol{F}_{Qq}$. That force will give us the field of Q at the position of q. We calculate this force, first in S', and then in S.

†This whole chapter can be skipped without losing continuity.

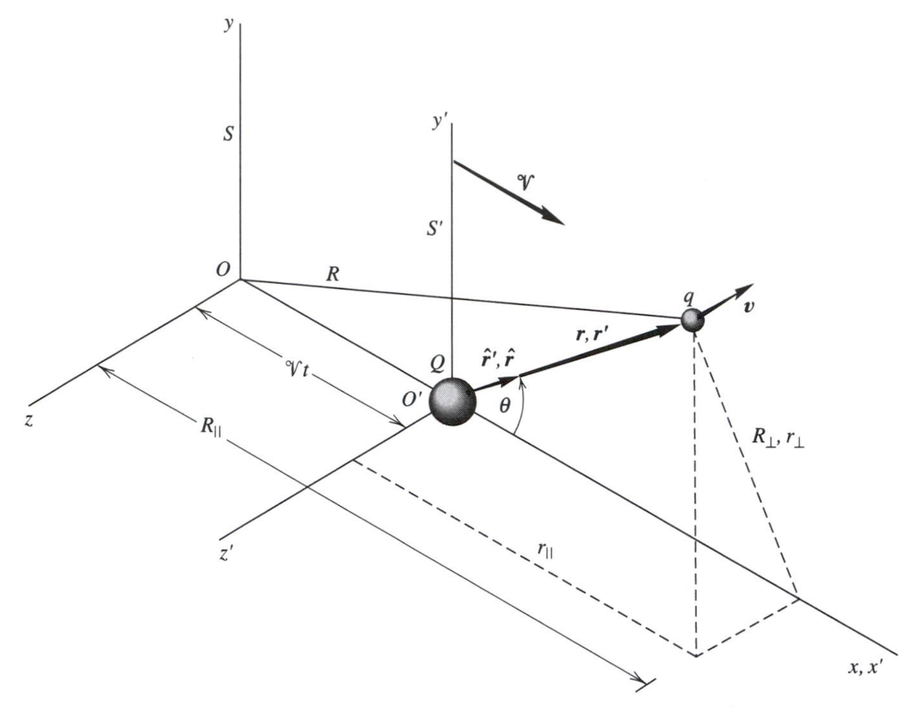

Figure 13-1 The velocity of the charge Q at the origin O' of S' is $\mathcal{V}\hat{\boldsymbol{x}}$ with respect to S. The velocity $\boldsymbol{v}$ of the charge q with respect to S is arbitrary. All the unprimed variables shown are measured with respect to S.

*13.1 THE FORCE ON q

*13.1.1 The Force $\boldsymbol{F}'_{Qq}$

In S', charge Q is stationary. Now it is a well-established experimental fact that the force exerted *by a stationary* charge *on a moving* charge is independent of the latter's velocity (Sec. 3.1). We may therefore calculate $\boldsymbol{F}'_{Qq}$ from Coulomb's law:

$$\boldsymbol{F}'_{Qq} = \frac{Qq\hat{\boldsymbol{r}}'}{4\pi\epsilon_0 r'^2} = \frac{Qq\boldsymbol{r}'}{4\pi\epsilon_0 r'^3}, \tag{13-1}$$

where $\boldsymbol{r}'$ and $\hat{\boldsymbol{r}}'$ are defined as in Fig. 13-1.

*13.1.2 The Force $\boldsymbol{F}_{Qq}$

Referring now to Eq. 12-32, we can deduce $\boldsymbol{F}_{Qq}$ from $\boldsymbol{F}'_{Qq}$:

$$\boldsymbol{F}_{Qq} = \frac{Qq}{4\pi\epsilon_0 r'^3}\left[\boldsymbol{r}'_{\parallel} + \frac{\gamma(\mathcal{V}/c^2)\boldsymbol{r}'_{\perp}\cdot\boldsymbol{v}'_{\perp} + \boldsymbol{r}'_{\perp}}{\gamma(1 + v'_{\parallel}\mathcal{V}/c^2)}\right]. \tag{13-2}$$

We still have to transform the primed quantities on the right. From Table 11-1 and from Fig. 13-1,

$$\boldsymbol{r}'_{\parallel} = \gamma(\boldsymbol{R}_{\parallel} - \boldsymbol{\mathcal{V}}t) = (\gamma)\boldsymbol{r}_{\parallel}, \qquad \boldsymbol{r}'_{\perp} = \boldsymbol{R}_{\perp} = \boldsymbol{r}_{\perp}, \tag{13-3}$$

$$r'^2 = \gamma^2(R_{\parallel} - \mathcal{V}t)^2 + R_{\perp}^2 = \gamma^2[(R_{\parallel} - \mathcal{V}t)^2 + R_{\perp}^2(1 - \beta^2)]. \tag{13-4}$$

Now

$$r^2 = (R_{\parallel} - \mathcal{V}t)^2 + R_{\perp}^2. \tag{13-5}$$

Thus

$$r'^2 = \gamma^2(r^2 - \beta^2 R_{\perp}^2) = \gamma^2 r^2(1 - \beta^2\sin^2\theta), \tag{13-6}$$

where θ is measured in reference frame S. Also, from Sec. 11-7,

$$\boldsymbol{v}'_{\parallel} = \frac{\boldsymbol{v}_{\parallel} - \boldsymbol{\mathcal{V}}}{1 - v_{\parallel}\mathcal{V}/c^2}, \qquad \boldsymbol{v}'_{\perp} = \frac{\boldsymbol{v}_{\perp}}{\gamma(1 - v_{\parallel}\mathcal{V}/c^2)}. \tag{13-7}$$

Substituting and simplifying, we find that

$$\boldsymbol{F}_{Qq} = \frac{Qq}{4\pi\epsilon_0}\frac{\boldsymbol{r} + (\boldsymbol{\mathcal{V}}/c^2)\boldsymbol{r}_{\perp}\cdot\boldsymbol{v}_{\perp} - (\mathcal{V}/c^2)v_{\parallel}\boldsymbol{r}_{\perp}}{\gamma^2 r^3(1 - \beta^2\sin^2\theta)^{3/2}}, \tag{13-8}$$

where, from Identity 2 on the page inside the front cover,

$$\boldsymbol{\mathcal{V}}(\boldsymbol{r}_{\perp}\cdot\boldsymbol{v}_{\perp}) - \mathcal{V}v_{\parallel}\boldsymbol{r}_{\perp} = \boldsymbol{\mathcal{V}}(\boldsymbol{v}\cdot\boldsymbol{r}_{\perp}) - \boldsymbol{r}_{\perp}(\boldsymbol{v}\cdot\boldsymbol{\mathcal{V}}) \tag{13-9}$$

$$= \boldsymbol{v}\times(\boldsymbol{\mathcal{V}}\times\boldsymbol{r}_{\perp}) = \boldsymbol{v}\times(\boldsymbol{\mathcal{V}}\times\boldsymbol{r}). \tag{13-10}$$

Finally,

$$\boldsymbol{F}_{Qq} = \frac{Qq}{4\pi\epsilon_0}\frac{\boldsymbol{r} + \boldsymbol{v}\times(\boldsymbol{\mathcal{V}}\times\boldsymbol{r})/c^2}{\gamma^2 r^3(1 - \beta^2\sin^2\theta)^{3/2}}. \tag{13-11}$$

*13.1.3 The Lorentz Force

The force $\boldsymbol{F}_{Qq}$ comprises two terms. The first term is independent of $\boldsymbol{v}$ and is the *electric force.* The second term does depend on $\boldsymbol{v}$ and is the *magnetic force.* So we have the *Lorentz force*

$$\boldsymbol{F}_{Qq} = q(\boldsymbol{E}_Q + \boldsymbol{v}\times\boldsymbol{B}_Q), \tag{13-12}$$

where $\boldsymbol{v}$ is the velocity of q in S, and where

$$\boldsymbol{E}_Q = \frac{Q}{4\pi\epsilon_0\gamma^2 r^2(1 - \beta^2\sin^2\theta)^{3/2}}\hat{\boldsymbol{r}} \tag{13-13}$$

is the *electric field strength* in the field of Q at the position of q, $\beta = \mathcal{V}/c$, and

$$\boldsymbol{B}_Q = \frac{Q(\boldsymbol{\mathcal{V}} \times \hat{\boldsymbol{r}})/c^2}{4\pi\epsilon_0\gamma^2 r^2(1 - \beta^2\sin^2\theta)^{3/2}} \tag{13-14}$$

is the *magnetic flux density* in the field of Q at the position of q. The unit of magnetic flux is the *tesla* (weber/meter^2).

In these three equations, all quantities concern the *same* reference frame S.

More generally, if a charge q moves at a velocity $\boldsymbol{v}$ in a field $\boldsymbol{E}$, $\boldsymbol{B}$, the Lorentz force is

$$\boldsymbol{F} = q(\boldsymbol{E} + \boldsymbol{v} \times \boldsymbol{B}). \tag{13-15}$$

The velocity $\boldsymbol{v}$ need *not* be constant.

Note that, for an observer in S, the charge Q exerts on q both an electric force $q\boldsymbol{E}$ and a magnetic force $q\boldsymbol{v} \times \boldsymbol{B}$. However, for an observer in the reference frame S' of Q, the charge Q is stationary and the force exerted *by Q on q* is purely electric, as in Eq. 13-1.

The fact that the magnetic force exerted on an electric charge is proportional to the vector product of its velocity $\boldsymbol{v}$ by the local magnetic flux density $\boldsymbol{B}$ has four obvious consequences. In a given reference frame, the magnetic force on a point charge (1) exists only if the charge moves with respect to that frame, (2) is independent of the component of $\boldsymbol{v}$ that is parallel to $\boldsymbol{B}$, (3) is perpendicular to both $\boldsymbol{v}$ and $\boldsymbol{B}$, (4) does not affect the kinetic energy of q.

Then, if $\mathcal{E}$ is the relativistic energy mc^2 of a particle of mass m and charge Q moving at a velocity $\boldsymbol{v}$ in a field $\boldsymbol{E}$, $\boldsymbol{B}$,

$$\frac{d\mathcal{E}}{dt} = \boldsymbol{F} \cdot \boldsymbol{v} = q(\boldsymbol{E} + \boldsymbol{v} \times \boldsymbol{B}) \cdot \boldsymbol{v} = q\boldsymbol{E} \cdot \boldsymbol{v}. \tag{13-16}$$

Only an electric field can affect the kinetic energy of a charged particle; a magnetic field can deflect such a particle, but cannot change its speed.

*13.2 THE ELECTRIC AND MAGNETIC FIELDS

It is convenient to set

$$\frac{1}{\epsilon_0 c^2} = \mu_0 \equiv 4\pi \times 10^{-7} \text{ weber/ampere-meter}. \tag{13-17}$$

This is the *permeability of free space*.

We now rewrite the fields $\boldsymbol{E}$ and $\boldsymbol{B}$ at a point P as in Fig. 13-2 near a charge Q moving at a constant velocity $\boldsymbol{\mathcal{V}}$:

$$\boldsymbol{E} = \frac{Q\hat{\boldsymbol{r}}}{4\pi\epsilon_0\gamma^2 r^2(1 - \beta^2\sin^2\theta)^{3/2}}, \tag{13-18}$$

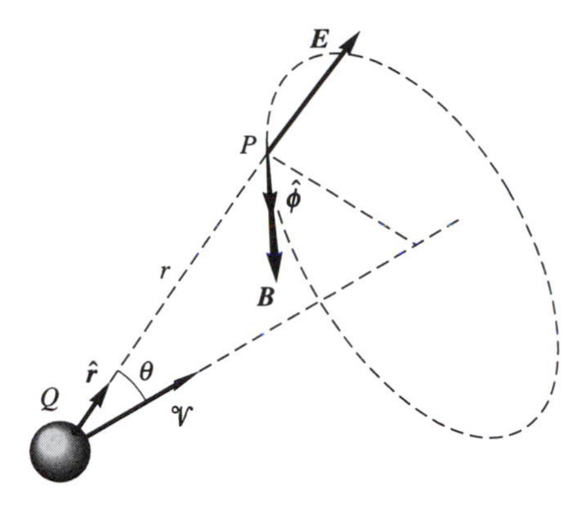

Figure 13-2 The $\boldsymbol{E}$ and $\boldsymbol{B}$ vectors at a point P in the field of a charge Q moving at a constant velocity $\boldsymbol{\mathcal{V}}$ with respect to a fixed reference frame S. The distance r and the angle θ are both measured in S.

$$\boldsymbol{B} = \frac{\mu_0 Q \mathcal{V} \sin\theta \hat{\boldsymbol{\phi}}}{4\pi\gamma^2 r^2 (1 - \beta^2 \sin^2\theta)^{3/2}} = \frac{\mu_0 Q \boldsymbol{\mathcal{V}} \times \hat{\boldsymbol{r}}}{4\pi\gamma^2 r^2 (1 - \beta^2 \sin^2\theta)^{3/2}}. \tag{13-19}$$

See Figs. 13-3 to 13-6 and Prob. 13-1. See also Sec. 21.1.

Observe that $\boldsymbol{E}$ is *radial*, as if the information concerning the position of the charge traveled at an infinite velocity! Actually, it is only when the velocity of the charge is constant that the electric field is radial. If the charge accelerates, the lines are not radial.

The lines of $\boldsymbol{B}$ are circles centered on the trajectory of Q.

At a given distance r, both E and B are maximum at $\theta = \pi/2$ and minimum at $\theta = 0$, $\theta = \pi$.

Write out $\boldsymbol{E}$ and $\boldsymbol{B}$ for low speeds where $\beta \approx 0$, $\gamma \approx 1$.

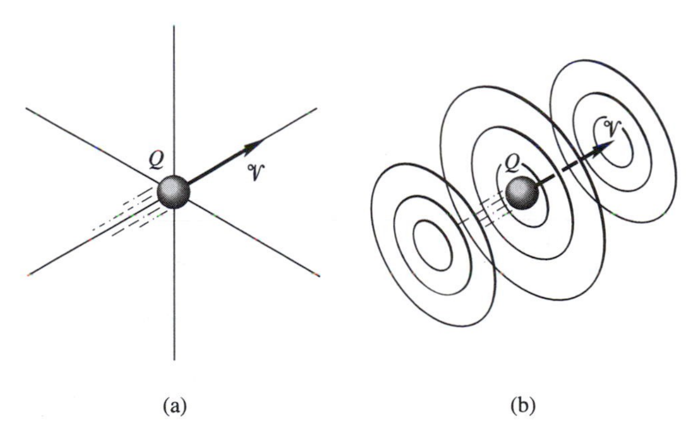

Figure 13-3 (a) Lines of $\boldsymbol{E}$ and (b) lines of $\boldsymbol{B}$ for a point charge Q moving at a constant velocity $\boldsymbol{\mathcal{V}}$, as seen by a stationary observer.

*EXAMPLE The Field of a 10-gigaelectronvolt electron

What are the maximum values of E and of B 10 millimeters from the path of a single 10-gigaelectronvolt electron? One gigaelectronvolt is 10^9 electron volts.

Both E and B are maximum at $\theta = 90°$, or when the line joining the particle to the point of observation is perpendicular to the trajectory in the reference frame S of the laboratory. Then $\sin\theta = 1$ and

$$\mathbf{E} = \frac{Q\hat{\mathbf{r}}}{4\pi\epsilon_0 r^2(1/\gamma)}, \tag{13-20}$$

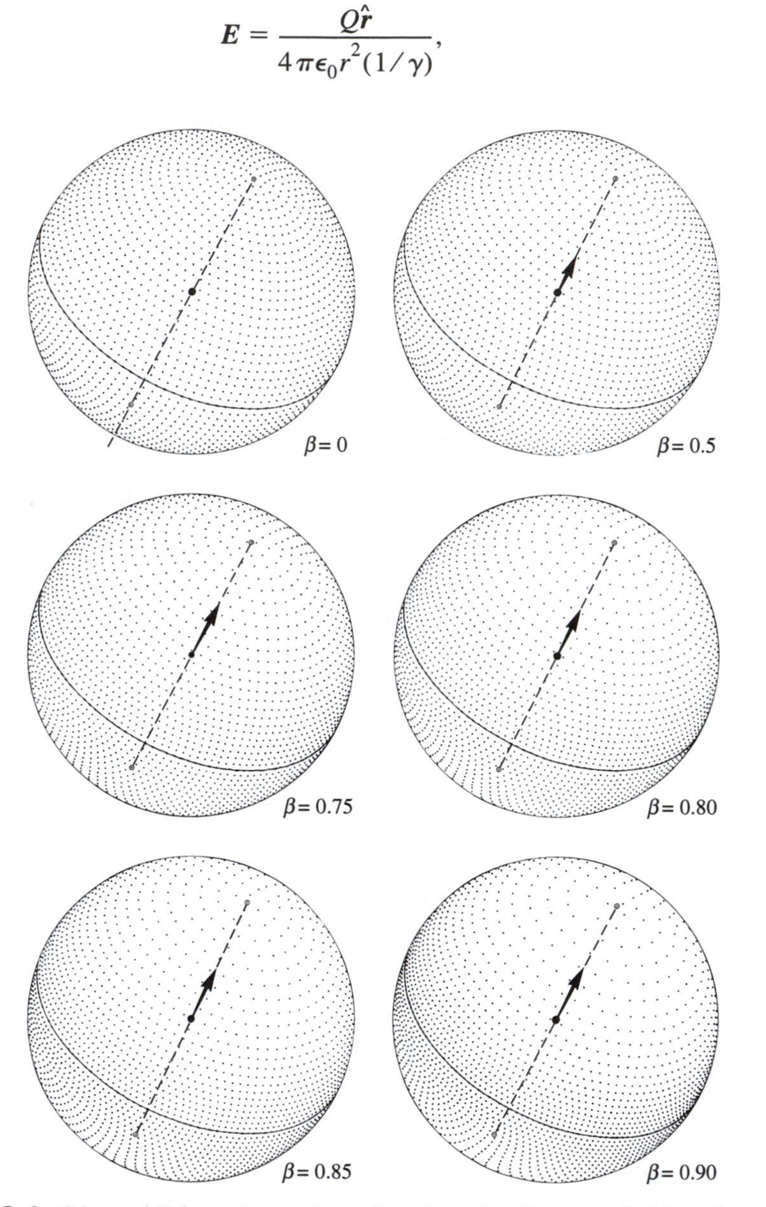

Figure 13-4 Lines of $\mathbf{E}$ for a charge Q moving along the diameter of an imaginary stationary sphere. The dots show where the lines emerge from the sphere at the instant when the charge goes through its center. The density of the dots is a measure of the magnitude of $\mathbf{E}$. The total number of dots is the same in all six figures, to satisfy Gauss's law (Sec. 3.6). Note how the field shifts away from $\theta = 0$ as the velocity increases. For $\beta = 1$ the field is all concentrated at $\theta = 90°$.

or γ times larger than if the electron were stationary,

$$E_{\max} = \frac{\gamma Q}{4\pi\epsilon_0 r^2} \tag{13-21}$$

in the radial direction, and

$$B_{\max} = \frac{\mu_0 \gamma Q \mathcal{V}}{4\pi r^2} \tag{13-22}$$

in the azimuthal direction.

Since the relativistic kinetic energy is 10 gigaelectronvolts,

$$(m - m_0)c^2 = m_0(\gamma - 1)c^2 = 10^{10} \text{ electronvolts} \tag{13-23}$$

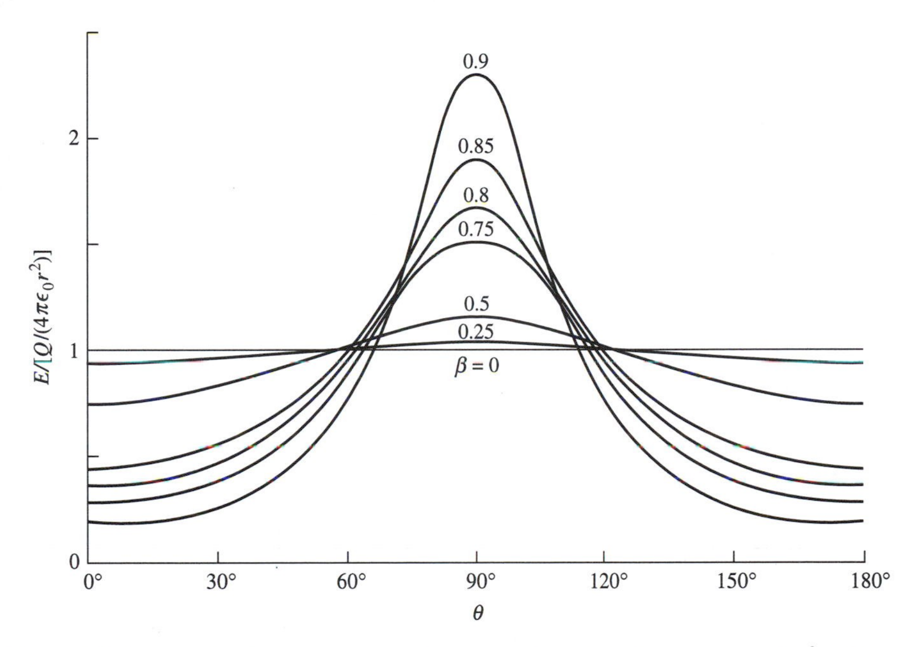

Figure 13-5 The E of a moving point charge as a function of the polar angle θ of Fig. 13-2 for seven values of $\beta = \mathcal{V}/c$. The observer is stationary and sees the charge moving at the uniform speed $\mathcal{V}$. For $\beta = 0$ the field is isotropic. It is hardly disturbed at $\beta = 0.25$. As the speed increases, the field increases near $\theta = 90°$ and decreases both ahead of the charge (near $\theta = 0°$) and behind it (near $\theta = 180°$). At extremely high speeds, most of the electric field concentrates near $\theta = 90°$. These curves explain qualitatively the validity of Gauss's law for moving charges: as the speed increases, the flux of E shifts from the region where $\theta \approx 0$ and $\theta \approx 180°$ to $\theta \approx 90°$, and the total flux of E remains constant. (Then why are the areas under the curves not equal?) Note that the electric field is symmetric about 90°. Thus there is no way of telling from the shape of the field whether the charge is moving to the right or to the left. The maximum ordinate on any curve is γ.

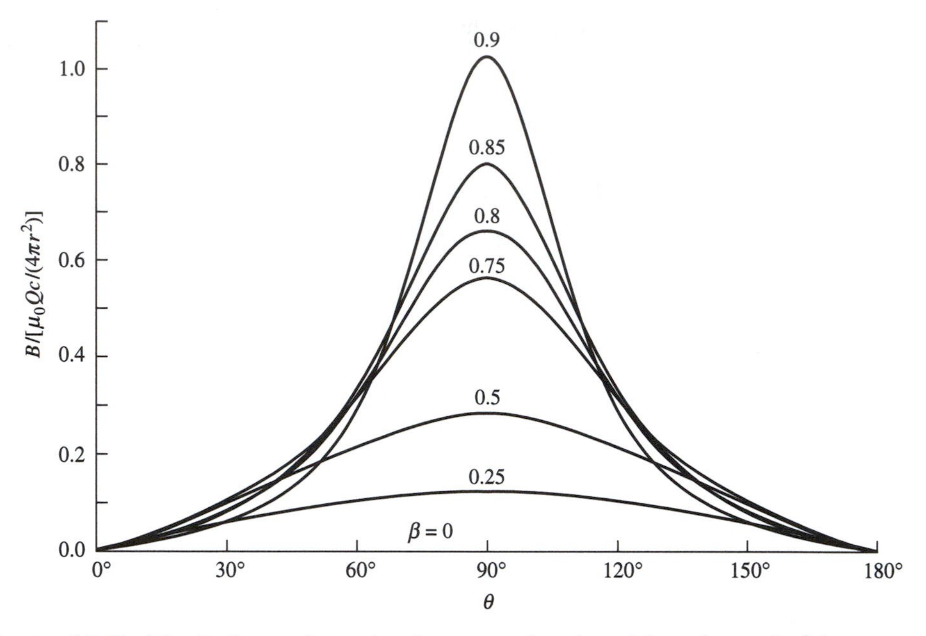

Figure 13-6 The B of a moving point charge as a function of the polar angle θ for seven values of $\beta = \mathcal{V}/c$. For $\beta = 0$ there is no magnetic field. As β increases, B first increases at all angles. Then B continues to increase near $\theta = 90°$, while decreasing both ahead of the charge and behind it. At extremely high speeds, most of the magnetic field concentrates near the plane $\theta = 90°$. The magnetic field, like the electric field, is symmetric about 90°.

with $m_0c^2 = 5.11 \times 10^5$ electronvolts. Thus

$$\gamma = \frac{10^{10}}{5.11 \times 10^5} + 1 \approx 2.0 \times 10^4, \tag{13-24}$$

and $\mathcal{V} \approx c$. Then

$$E_{max} = \frac{(9 \times 10^9)(2 \times 10^4)(1.6 \times 10^{-19})}{10^{-4}} = 0.29 \text{ volt/meter}, \tag{13-25}$$

$$B_{max} = \frac{(10^{-7})(2 \times 10^4)(1.6 \times 10^{-19})(3 \times 10^8)}{10^{-4}} = 9.6 \times 10^{-10} \text{ tesla}. \tag{13-26}$$

■

*EXAMPLE The Magnetic Field Near a Straight Wire Carrying a Steady Current

Figure 13-7 shows a positive charge Q moving at a velocity $\boldsymbol{v}$ parallel to a stationary wire carrying a current I at a distance y. Let us calculate the force exerted by the wire on Q.

In the reference frame S of the wire, the net linear charge density in the wire is zero:

$$\lambda_p + \lambda_n = 0, \tag{13-27}$$

where p refers to the lattice of positive charges, and n to the conduction electrons. There are surface charges, but we can disregard them because they just superpose another electric field over the one that we are interested in here. This extra field depends on the resistivity and geometry of the wire, as well as on the current flowing through it.

The wire is on the x-axis, and current flows in the negative direction as in Fig. 13-7. Then the conduction electrons flow at the velocity $v_n\hat{\boldsymbol{x}}$ in the positive direction of the x-axis.

The linear charge density of the conduction electrons, in their own reference frame S', is λ'_n. What is their linear charge density λ_n in S? Recall from Sec. 11.8 that electric charge is invariant. Then, because of the Lorentz contraction (Sec. 11.5), their linear charge density in S is larger by the factor γ_n:

$$\lambda_n = \gamma_n \lambda'_n, \qquad \gamma_n = \left(1 - \frac{v_n^2}{c^2}\right)^{-1/2}. \tag{13-28}$$

In S, the positive charges are stationary, and they exert on a positive charge Q a repulsive force in the y direction

$$F_p = QE = Q\left(\frac{\lambda_p}{2\pi\epsilon_0 y}\right). \tag{13-29}$$

Similarly, the conduction electrons exert in their own frame an attractive force

$$F'_n = \frac{Q\lambda'_n}{2\pi\epsilon_0 y} = \frac{Q\lambda_n}{\gamma_n 2\pi\epsilon_0 y}, \tag{13-30}$$

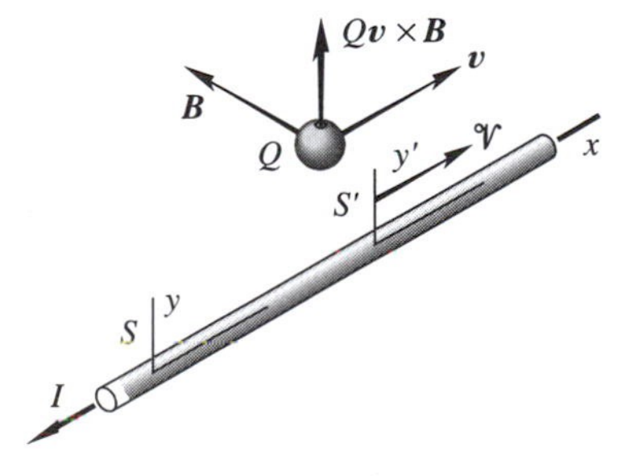

Figure 13-7 A positive charge Q moves at a velocity $\boldsymbol{v}$ parallel to a straight wire carrying a current I. The magnetic force $Q\boldsymbol{v} \times \boldsymbol{B}$ is in the direction shown because, for the charge Q, the wire is positively charged. With Q positive, the force is repulsive.

where λ'_n is negative and $y' = y$. This force also points in the y direction. From Eq. 12-32,

$$F_n = F'_n \gamma_n \left(1 - \frac{v_n v}{c^2}\right) = \frac{Q\lambda_n}{2\pi\epsilon_0 y}\left(1 - \frac{v_n v}{c^2}\right). \tag{13-31}$$

This force does not quite cancel F_p.

The net force exerted *by* the wire *on* the charge Q, in the reference frame S of the wire, is

$$F = F_p + F_n = \frac{Q\lambda_p}{2\pi\epsilon_0 y} + \frac{Q\lambda_n}{2\pi\epsilon_0 y} - \frac{Q(\lambda_n v_n)v}{2\pi\epsilon_0 y c^2}. \tag{13-32}$$

Now $\lambda_n v_n$ is equal to the current that flows to the *right,* and $-\lambda_n v_n = I$, as in the figure. So, substituting μ_0 for $1/\epsilon_0 c^2$ and using Eq. 13-27,

$$F = Qv\frac{\mu_0 I}{2\pi y}. \tag{13-33}$$

Expressing this as a magnetic force $Q\boldsymbol{v} \times \boldsymbol{B}$, we find that

$$\boldsymbol{B} = \frac{\mu_0 I}{2\pi y}\hat{\boldsymbol{\phi}} \tag{13-34}$$

in the direction shown in the figure. We shall rediscover this relation in Sec. 17.1.

This magnetic force is infinitesimal compared to the force between Q and the conduction electrons in the wire or between Q and the lattice of positive ions.

Say we have a copper wire with a cross-sectional area of 1 millimeter^2 and carrying a current of 1 ampere. Then suppose that Q is an electron traveling at the drift velocity of conduction electrons in the wire. We can calculate these forces as follows.

We found in Sec. 4.5.1 that the drift velocity is about 10^{-4} meter/second or about 40 centimeters/hour. Then γ is equal to unity within 1 part in 10^{25}.

Copper contains 10^{29} atoms per cubic meter and one conduction electron per atom. Then 1 meter of the wire contains 1.6×10^4 coulombs of conduction electrons. If y is 1 centimeter, then the force of attraction between the electron and the positive lattice is

$$F = \frac{\lambda_p}{2\pi\epsilon_0 y}Q = \frac{1.6 \times 10^4}{2\pi \times 8.85 \times 10^{-12} \times 10^{-2}} 1.6 \times 10^{-19} \tag{13-35}$$

$$= 4.6 \times 10^{-3} \text{ newton}. \tag{13-36}$$

This force, if acting alone, would impart to the electron an acceleration of 5×10^{27} meters/second^2! The force of repulsion between Q and the conduction electrons in the wire is slightly larger.

The *net* force on an electron whose speed is 10^{-4} meter/second is the magnetic force of Eq. 13-33:

$$F = \frac{1.6 \times 10^{-19} \times 10^{-4}}{4\pi \times 10^{-7} \times 1/(2\pi \times 10^{-2})} = 3.2 \times 10^{-28} \text{ newton.} \tag{13-37}$$

So the magnetic force, in this particular instance, is smaller than the electrostatic forces by 25 orders of magnitude! The magnetic force is a purely relativistic effect that takes place for $v/c = 10^{-4}/(3 \times 10^8) \approx 3 \times 10^{-13}$.

Of course, the force exerted on a second current-carrying *wire* is appreciable, for the simple reason that it also contains an enormous number of conduction electrons.

*13-3 THE FORCE ON Q

To calculate the force exerted *by q on Q* in Fig. 13-1, we interchange the roles of q and Q in Eq. 13-11. This means: (1) replacing $\hat{\boldsymbol{r}}$ by $-\hat{\boldsymbol{r}}$, as in Fig. 13-8; (2) interchanging $\boldsymbol{\mathcal{V}}$ and $\boldsymbol{v}$; (3) replacing θ by θ'; and (4) replacing $\beta = \mathcal{V}/c$ by $\beta' = v/c$ and γ by γ'.

Thus

$$\boldsymbol{F}_{qQ} = -\frac{qQ}{4\pi\epsilon_0} \frac{\boldsymbol{r} + \boldsymbol{\mathcal{V}} \times (\boldsymbol{v} \times \boldsymbol{r})/c^2}{\gamma'^2 r^3 (1 - \beta'^2 \sin^2 \theta')^{3/2}}. \tag{13-38}$$

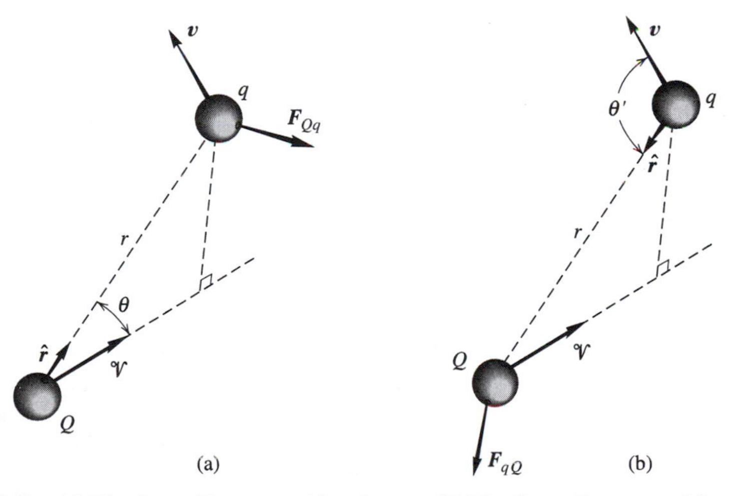

Figure 13-8 (a) The force $\boldsymbol{F}_{Qq}$ exerted by Q on q. (b) The force $\boldsymbol{F}_{qQ}$ exerted by q on Q. These two forces are *not* equal and opposite, as one would expect from classical mechanics.

Clearly, $\boldsymbol{F}_{qQ} \neq -\boldsymbol{F}_{Qq}$. The electric forces are in opposite directions, but their magnitudes are not the same because of the γ', β', θ' terms. The magnetic forces are totally different. The forces $\boldsymbol{F}_{Qq}$ and $\boldsymbol{F}_{qQ}$ are equal if $\mathcal{V}^2 \ll c^2$ and if $\mathcal{V} v \ll c^2$.

There is a difference between $\boldsymbol{F}_{qQ}$ and $\boldsymbol{F}_{Qq}$ that is noteworthy. The expression for $\boldsymbol{F}_{Qq}$ is valid if the velocity $\boldsymbol{\mathcal{V}}$ of Q is constant, while that for $\boldsymbol{F}_{qQ}$ applies if $\boldsymbol{v}$ is constant.

The fact is that $\boldsymbol{F}_{qQ} \neq \boldsymbol{F}_{Qq}$ is not peculiar to electrical phenomena. It is a relativistic effect that occurs with any type of force.

*13.4 TRANSFORMATION OF Q, $\boldsymbol{E}$, $\boldsymbol{B}$, $\boldsymbol{A}$, AND V

We have just seen that a purely electric field in one reference frame becomes both an electric and a magnetic field, when observed in another frame. We now state the general rules for transforming electric and magnetic fields (Lorrain, Corson, and Lorrain, 1988).

As usual, we have two reference frames, a fixed frame S and a frame S' that moves at the velocity $\boldsymbol{\mathcal{V}}$ with respect to S. Both frames are inertial, or non-accelerated. Observers in S and in S' compare their measurements for the *same* phenomenon.

In the general case where the speed $\mathcal{V}$ of S' with respect to S can assume any value up to the speed of light c, the transformation equations are those of Table 13-1 where, as usual, $\gamma = [1 - (v^2/c^2)]^{-1/2}$ and $\gamma \geq 1$.

If $v^2 \ll c^2$, or for $v \leq 0.1c = 3 \times 10^7$ meters/second, then $\gamma \approx 1$.

Observe that, in the equations for the $\boldsymbol{E}$'s and for the $\boldsymbol{B}$'s, the parentheses concern the components that are *perpendicular* to the velocity $\boldsymbol{\mathcal{V}}$. But, in the equations for the $\boldsymbol{A}$'s and for the V's, the parentheses concern the *parallel* components.

Now compare the transformation of $\boldsymbol{A}$ and of V/c with the Lorentz transformation of Table 11-1: the vector $\boldsymbol{A}$ plays the same role as $\mathbf{r}$, and the scalar V/c the same role as t. Quantities such as $(\boldsymbol{A}, V/c)$, or $(A_x, A_y, A_z, V/c)$, which transform like $(\mathbf{r}, t)$, or (x, y, z, t), are called *four-vectors*.

*EXAMPLE The Parallel-Plate Capacitor

In its own reference frame, a charged capacitor has an electric field, and it has no magnetic field. If the capacitor moves at a velocity $\boldsymbol{\mathcal{V}}$ parallel to the plates, what field

Table 13-1 Transformation of electric and magnetic quantities

$Q = Q'$,	$Q' = Q$
$\boldsymbol{E} = \boldsymbol{E}'_{\parallel} + \gamma(\boldsymbol{E}'_{\perp} - \boldsymbol{\mathcal{V}} \times \boldsymbol{B}')$,	$\boldsymbol{E}' = \boldsymbol{E}_{\parallel} + \gamma(\boldsymbol{E}_{\perp} + \boldsymbol{\mathcal{V}} \times \boldsymbol{B})$,
$\boldsymbol{B} = \boldsymbol{B}'_{\parallel} + \gamma\left(\boldsymbol{B}'_{\perp} + \dfrac{\boldsymbol{\mathcal{V}} \times \boldsymbol{E}'}{c^2}\right)$,	$\boldsymbol{B}' = \boldsymbol{B}_{\parallel} + \gamma\left(\boldsymbol{B}_{\perp} - \dfrac{\boldsymbol{\mathcal{V}} \times \boldsymbol{E}}{c^2}\right)$,
$\boldsymbol{A} = \gamma\left(\boldsymbol{A}'_{\parallel} + \dfrac{\boldsymbol{\mathcal{V}}}{c^2}V'\right) + \boldsymbol{A}'_{\perp}$,	$\boldsymbol{A}' = \gamma\left(\boldsymbol{A}_{\parallel} - \dfrac{\boldsymbol{\mathcal{V}}}{c^2}V\right) + \boldsymbol{A}_{\perp}$,
$\dfrac{V}{c} = \gamma\left(\dfrac{V'}{c} + \dfrac{\mathcal{V}A'_{\parallel}}{c}\right)$,	$\dfrac{V'}{c} = \gamma\left(\dfrac{V}{c} - \dfrac{\mathcal{V}A_{\parallel}}{c}\right)$.

does a fixed observer measure? Assume that the plates are large, and neglect edge effects. Also, disregard the electric field outside the capacitor. The fixed observer measures the voltage difference between the plates through a pair of sliding contacts. Both observers are equipped with B-meters.

Choose axes as in Fig. 13-9.

(a) The $\mathbf{E}$'s. In S', $\mathbf{E}$ has only a y-component:

$$E'_x = 0, \quad E'_y = E', \quad E'_z = 0. \tag{13-39}$$

Then, from Table 13-1,

$$E_x = 0, \quad E_y = \gamma E', \quad E_z = 0. \tag{13-40}$$

Now why is that? In the reference frame S' of the capacitor, the electric charge densities are $\pm\sigma' = \pm\epsilon_0 E'$, from Gauss's law (Sec. 3.6). In the fixed reference frame S, *the charges are the same*, because electric charges are invariant, but *the plates are shorter* in the direction of the motion by a factor $1/\gamma$ (Sec. 11.5), and therefore

$$\sigma = \gamma\sigma'. \tag{13-41}$$

So a fixed observer, in S, measures an electric field that is larger than the one measured by an observer who moves with the capacitor in S', by the factor γ:

$$E = \gamma E'. \tag{13-42}$$

(b) The V's. In S' the potential difference between the plates is

$$V'_0 = E's. \tag{13-43}$$

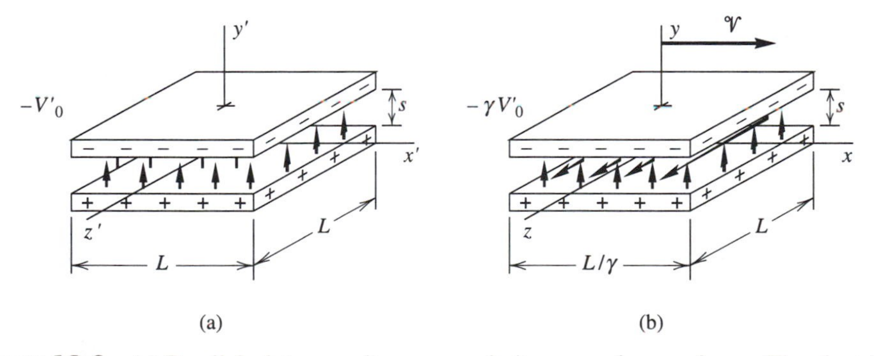

Figure 13-9 (a) Parallel-plate capacitor, as seen in its own reference frame. The electric field strength is E', and $B' = 0$. (b) The same capacitor moves to the right at the velocity $\mathcal{V}$. Here $E = \gamma E'$, and $B = \gamma(\mathcal{V}/c^2)E'$ in the positive direction of the z-axis. The origins of S and S' are on the lower plate. We have disregarded edge effects.

Now the spacing s between the plates is the same in both frames, because that length is perpendicular to $\mathcal{V}$. But we have just seen that $\sigma = \gamma\sigma'$. Then the potential difference V_0 between the plates, in the fixed frame S, is given by

$$V_0 = Es = \gamma E's = \gamma V'_0. \tag{13-44}$$

The potential differences are not the same: $V_0 > V'_0$! Also,

$$V = \gamma V'. \tag{13-45}$$

(c) The capacitances. The capacitance is the charge on one plate divided by the potential between the plates. So

$$C = \frac{Q}{V_0}, \quad C' = \frac{Q}{V'_0} = \frac{Q}{V_0/\gamma} = \gamma C. \tag{13-46}$$

The capacitance in the reference frame of the capacitor is larger than in the fixed frame by a factor of γ. That is in agreement with the fact that, even though the charges are the same, V_0 is larger than V'_0 by the factor γ. In both cases, the capacitance is proportional to the area and inversely proportional to the spacing, as in Sec. 6.4, and the area is γ times larger in S' than in S.

(d) The $\boldsymbol{B}$'s. In the reference frame S' of the capacitor, there is no magnetic field:

$$B'_x = B'_y = B'_z = 0. \tag{13-47}$$

Then, from the equation for $\boldsymbol{B}$ in Table 13-1,

$$B_x = 0, \quad B_y = 0, \quad B_z = \gamma\frac{\mathcal{V}}{c^2}E'. \tag{13-48}$$

The fixed observer, in frame S, measures a magnetic field because he sees a current flowing to the right in the lower plate, and an equal current flowing to the left in the upper plate. The field B_z points in the $+\mathcal{V} \times \boldsymbol{E}$ direction, and thus in the $+z$ direction.

(e) The $\boldsymbol{A}$'s. In S', $\boldsymbol{B}' = 0$ and $\boldsymbol{A}' = 0$. Now, in the equation for $\boldsymbol{A}$ in Table 13-1, γ times the parenthesis gives the parallel component of $\boldsymbol{A}$, and

$$A_{\parallel} = \gamma\frac{\mathcal{V}}{c^2}V'. \tag{13-49}$$

Similarly, the equation for $\boldsymbol{A}'$ yields two results:

$$A_{\parallel} = \frac{\mathcal{V}}{c^2}V = \gamma\frac{\mathcal{V}}{c^2}V', \tag{13-50}$$

as above, and

$$A_{\perp} = 0, \tag{13-51}$$

which is in agreement with the fact that the currents are parallel to $\mathcal{V}$.

From the equation for V, since $\boldsymbol{A}' = 0$, $V = \gamma V'$, as above. Finally, from the equation for V',

$$V' = \gamma(V - \mathcal{V}A_{\parallel}) = \gamma\left(V - \mathcal{V}\frac{\mathcal{V}}{c^2}V\right) = \gamma V\left(1 - \frac{\mathcal{V}^2}{c^2}\right) = \frac{V}{\gamma}, \qquad (13\text{-}52)$$

again as above.

*13-5 SUMMARY

A charge Q moves at a constant velocity $\mathcal{V}$, as in Figs. 13-1 to 13-6. The Lorentz force exerted *by Q on* a charge q that moves at an arbitrary velocity $\boldsymbol{v}$ is

$$\boldsymbol{F}_{Qq} = q(\boldsymbol{E}_Q + \boldsymbol{v} \times \boldsymbol{B}_Q), \qquad (13\text{-}12)$$

where $\boldsymbol{E}_Q$ and $\boldsymbol{B}_Q$ are, respectively, the electric and magnetic fields of Q at the position of q, with

$$\boldsymbol{E} = \frac{Q\hat{\boldsymbol{r}}}{4\pi\epsilon_0\gamma^2 r^2(1 - \beta^2\sin^2\theta)^{3/2}}, \qquad (13\text{-}18)$$

$$\boldsymbol{B} = \frac{\mu_0 Q\mathcal{V} \times \hat{\boldsymbol{r}}}{4\pi\gamma^2 r^2(1 - \beta^2\sin^2\theta)^{3/2}}, \qquad (13\text{-}19)$$

and $\beta = \mathcal{V}/c$.

The force exerted by Q on q is *not* equal to the force exerted by q on Q, unless $\mathcal{V}^2 \ll c^2$.

The magnetic field of a current-carrying wire is a relativistic effect that occurs, typically, for $v/c \approx 10^{-13}$, where v is the drift speed of the conduction electrons. Then $\gamma \approx 1$ within one part in 10^{25}.

Electric charges are invariant:

$$Q = Q'. \qquad (13\text{-}39)$$

The transformation equations for $\boldsymbol{E}$ and $\boldsymbol{B}$ are as follows:

$$\boldsymbol{E} = \boldsymbol{E}'_{\parallel} + \gamma(\boldsymbol{E}'_{\perp} - \mathcal{V} \times \boldsymbol{B}'), \qquad (13\text{-}42)$$

$$\boldsymbol{E}' = \boldsymbol{E}_{\parallel} + \gamma(\boldsymbol{E}_{\perp} + \mathcal{V} \times \boldsymbol{B}), \qquad (13\text{-}43)$$

$$\boldsymbol{B} = \boldsymbol{B}'_{\parallel} + \gamma\left(\boldsymbol{B}'_{\perp} + \frac{\mathcal{V} \times \boldsymbol{E}'}{c^2}\right), \qquad (13\text{-}44)$$

$$\boldsymbol{B}' = \boldsymbol{B}_{\parallel} + \gamma\left(\boldsymbol{B}_{\perp} - \frac{\mathcal{V} \times \boldsymbol{E}}{c^2}\right). \qquad (13\text{-}45)$$

*PROBLEMS

13-1. (*13.2*) Alternate expressions for $\boldsymbol{E}$ and $\boldsymbol{B}$

Show that

$$\boldsymbol{E} = \frac{\gamma Q \boldsymbol{r}}{4\pi\epsilon_0(\gamma^2 x^2 + y^2 + z^2)^{3/2}}, \qquad \boldsymbol{B} = \frac{\mu_0 \gamma Q \boldsymbol{\mathcal{V}} \times \boldsymbol{r}}{4\pi(\gamma^2 x^2 + y^2 + z^2)^{3/2}}.$$

13-2. (*13.2*) The field of a 10-megaelectronvolt proton

Plot E and B as functions of the time at a point P one centimeter away from the path of a 10-megaelectronvolt proton. Set P at (0, 0.01, 0), with the charge at $(\mathcal{V}t, 0, 0)$.

13-3. (*13.2*) The force between electrons moving side by side

Calculate the force, as observed in the laboratory, between two electrons moving side by side along parallel paths 1 millimeter apart if each has a kinetic energy of (a) 1 electron volt and (b) 1 megaelectronvolt.

13-4. (*13.2*) $\boldsymbol{E} \cdot \boldsymbol{B}$ is invariant

Show that $\boldsymbol{E} \cdot \boldsymbol{B}$ is invariant under a Lorentz transformation.

13-5. (*13.2*) $B^2 - E^2/c^2$ is invariant

Show that $B^2 - E^2/c^2$ is invariant under a Lorentz transformation.

13-6. (*13.2*) The angle between $\boldsymbol{E}$ and $\boldsymbol{B}$ is *not* invariant

Show that the angle between $\boldsymbol{E}$ and $\boldsymbol{B}$ is not invariant.

13-7. (*13.2*) Time-independent magnetic field.

Assume that, in S, there is a constant magnetic field and no electric field.

Show that $\boldsymbol{E}' = \boldsymbol{\mathcal{V}} \times \boldsymbol{B}'$ in S'. Note the prime on the right hand side. So $\boldsymbol{E}'$ is perpendicular to both $\boldsymbol{\mathcal{V}}$ and $\boldsymbol{B}'$.

CHAPTER

14

MAGNETIC FIELDS I

*The Magnetic Flux Density **B** and the Vector Potential **A***

Imagine a set of charges moving around in space.‡ At any point $\boldsymbol{r}$ in space and at any time t there exists an electric field strength $\boldsymbol{E}(\boldsymbol{r}, t)$ and a magnetic flux density $\boldsymbol{B}(\boldsymbol{r}, t)$ that are defined as follows. If a charge Q moves at velocity $\boldsymbol{v}$ at $(\boldsymbol{r}, t)$ in this field, then it suffers a Lorentz force

$$\boldsymbol{F} = Q(\boldsymbol{E} + \boldsymbol{v} \times \boldsymbol{B}). \tag{14-1}$$

The *electric force* $Q\boldsymbol{E}$ is proportional to Q but independent of $\boldsymbol{v}$, while the *magnetic force* $Q\boldsymbol{v} \times \boldsymbol{B}$ is orthogonal to both $\boldsymbol{v}$ and $\boldsymbol{B}$.

In Chaps. 3 to 10 we studied the $\boldsymbol{E}$ fields of charges that are fixed in position or that move slowly. Fixed charges have no magnetic field; a magnetic field exists only if charges move with respect to the observer.

†Material marked with an asterisk can be skipped without losing continuity.

‡If you have not studied Chaps. 11 to 13 on Relativity, simply disregard references to them from here on.

In this chapter we study the magnetic fields of constant electric currents. We assume that the electric charge density ρ is also constant. Thus $\partial\rho/\partial t = 0$, and hence, from Sec. 4.2, $\boldsymbol{\nabla} \cdot \boldsymbol{J} = 0$. We also assume that there are no magnetic materials, and no moving materials, in the field.

*14.1 MAGNETIC MONOPOLES

We assume here that magnetic fields arise solely from the motion of electric charges.

However, Dirac postulated in 1931 that magnetic fields can also arise from magnetic "charges," called *magnetic monopoles*. Such particles have not been observed to date (2000). The theoretical value of the elementary magnetic charge is

$$\frac{h}{e} = 4.1356692 \times 10^{-15} \text{ weber,}\dagger \tag{14-2}$$

where h is Planck's constant and e is the charge of the electron. See the table inside the back cover.

At a distance r from a stationary magnetic monopole of "charge" Q^*, we would have that

$$\boldsymbol{B} = \frac{Q^*}{4\pi r^2}\hat{\boldsymbol{r}}. \tag{14-3}$$

Also, the force of attraction or repulsion between two monopoles Q_a^* and Q_b^* would be

$$\boldsymbol{F} = \frac{Q_a^* Q_b^*}{4\pi\mu_0 r^2}\hat{\boldsymbol{r}}. \tag{14-4}$$

A magnetic field would exert a force $Q^*\boldsymbol{B}/\mu_0$ on a monopole in free space.

14.2 THE MAGNETIC FLUX DENSITY *B*. THE BIOT-SAVART LAW

In the neighborhood of an electric circuit C carrying a steady current I there exists a magnetic field and, at a point P in space, as in Fig. 14-1,

$$\boldsymbol{B} = \frac{\mu_0 I}{4\pi}\oint_C \frac{d\boldsymbol{l}' \times \hat{\boldsymbol{r}}}{r^2}. \tag{14-5}$$

†This is the Dirac charge; the Schwinger charge is twice as large.

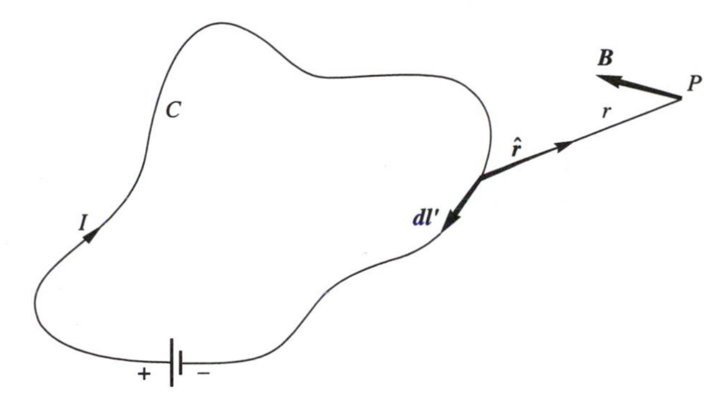

Figure 14-1 Circuit C carrying a current I, and a point P in its field. At P the magnetic flux density is $\boldsymbol{B}$.

As usual, the unit vector $\hat{\boldsymbol{r}}$ points *from* the source *to* the point of observation P. This is the *Biot-Savart law.* The integration can be carried out analytically only for the simplest geometries.

By *definition,*

$$\mu_0 \equiv 4\pi \times 10^{-7} \text{weber/ampere-meter.} \tag{14-6}$$

This is the *permeability of free space.*

Equation 14-5 also applies to the fields of alternating currents, as long as the time r/c, where c is the speed of light, is a small fraction of one period.

The unit of magnetic flux density is the *tesla.* We can find the dimensions of the tesla as follows. As we saw in the introduction to this chapter, vB has the dimensions of E. Then

$$\text{tesla} = \frac{\text{volt}}{\text{meter}} \frac{\text{second}}{\text{meter}} = \frac{\text{weber}}{\text{meter}^2}. \tag{14-7}$$

One volt-second is defined as 1 *weber.*

We have assumed a current I flowing through a thin wire. If the current flows over a finite volume, we substitute $\boldsymbol{J} \cdot d\boldsymbol{\mathcal{A}}'$ for I, $\boldsymbol{J}$ being the current density in amperes per square meter at a point and $d\boldsymbol{\mathcal{A}}'$ an oriented element of area, as in Fig. 14-2. Then $\boldsymbol{J} d\boldsymbol{\mathcal{A}}' \cdot \boldsymbol{dl}'$ is $\boldsymbol{J} dv'$ and, at a point P,

$$\boldsymbol{B} = \frac{\mu_0}{4\pi} \int_{v'} \frac{\boldsymbol{J} \times \hat{\boldsymbol{r}}}{r^2} \, dv', \tag{14-8}$$

in which v' is any volume enclosing all the currents and r is the distance between the element of volume dv' and the point P.

The current density $\boldsymbol{J}$ encompasses moving free charges, polarization currents in dielectrics (Sec. 7.3), and equivalent currents in magnetic materials (Sec. 16.3).

Can this integral serve to calculate $\boldsymbol{B}$ at a point *inside* a current distribution? The integral appears to diverge because r goes to zero when dv' is at P. The integral does

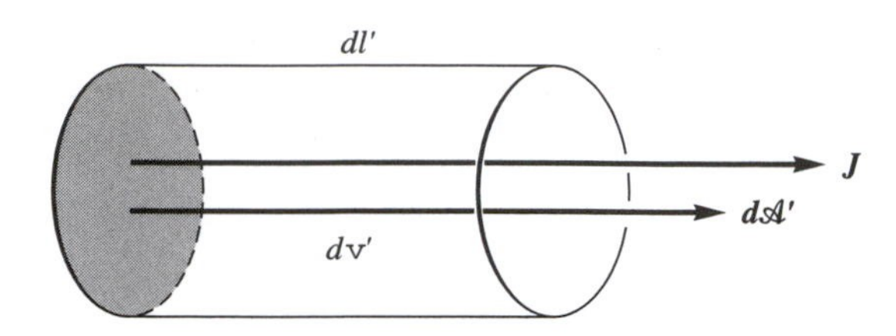

Figure 14-2 At a given point in a volume distribution of current, the current density is $\boldsymbol{J}$. The vector $d\boldsymbol{\mathcal{A}}'$ specifies the magnitude and orientation of the shaded area. Shifting this element of area to the right by the distance dl' along $\boldsymbol{J}$ sweeps out a volume $d\mathcal{A}'dl' = dv'$.

not, in fact, diverge: it does apply even if the point P lies inside the conducting body. We encountered the same problem when we calculated the value of $\boldsymbol{E}$ inside a charge distribution in Sec. 3.5.

Lines of $\boldsymbol{B}$ point everywhere in the direction of $\boldsymbol{B}$. They are useful "thinking crutches," like lines of $\boldsymbol{E}$, but no more. Many authors call them "lines of force" which is a misnomer because the magnetic force (Sec. 19.1) is perpendicular to, and not parallel to, $\boldsymbol{B}$.

It is also a common error to discuss the geometry of imaginary magnetic field lines without ever considering the required currents. There is of course no point in postulating a magnetic field configuration without first making sure that the required electric currents make sense.

As with electric fields again, a great deal of convenience attends the use of the concept of flux. The *magnetic flux* through a surface of area $\mathcal{A}$ is

$$\Phi = \int_{\mathcal{A}} \boldsymbol{B} \cdot d\boldsymbol{\mathcal{A}} \text{ webers.} \tag{14-9}$$

The surface is usually open; if it is closed, then $\Phi = 0$, as we shall see below.

The above integrals for $\boldsymbol{B}$ imply that the net $\boldsymbol{B}$ is the sum of the $\boldsymbol{B}$'s of the elements of current $I\,\boldsymbol{dl}'$, or $\boldsymbol{J}\,dv'$. The *principle of superposition* applies to magnetic fields as well as to electric fields (Sec. 3.3): if there exist several current distributions, then the net $\boldsymbol{B}$ is the vector sum of the individual $\boldsymbol{B}$'s.

EXAMPLE A Long Straight Wire

An element dl' of a long, straight wire carrying a current I, as in Fig. 14-3, gives, at the point $P(r, \theta, \phi)$, a magnetic flux density

$$\boldsymbol{dB} = \frac{\mu_0}{4\pi}\frac{I\,dl' \sin\theta}{r^2}\hat{\boldsymbol{\phi}} = \frac{\mu_0}{4\pi}\frac{I\,dl' \cos\alpha}{r^2}\hat{\boldsymbol{\phi}}. \tag{14-10}$$

The relative orientations of $\boldsymbol{Idl}'$ and $\boldsymbol{B}$ satisfy the right-hand screw rule. Here

$$l' = \rho \tan\alpha, \qquad dl' = \frac{r\,d\alpha}{\cos\alpha} = \frac{r^2 d\alpha}{\rho}. \tag{14-11}$$

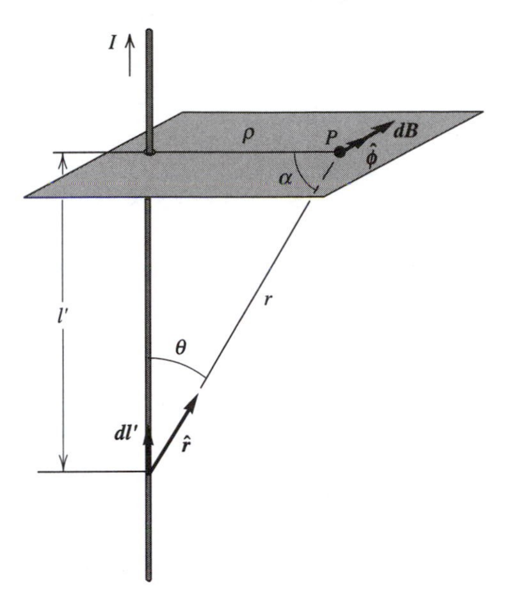

Figure 14-3 Long, straight wire carrying a current I. At the point P the element $I\,\boldsymbol{dl'}$ contributes a $\boldsymbol{dB}$ in the direction shown. A line of $\boldsymbol{B}$ is a circle centered on the wire.

Thus

$$\boldsymbol{B} = \frac{\mu_0 I}{4\pi\rho}\int_{-\pi/2}^{+\pi/2} \cos\alpha\, d\alpha\ \hat{\boldsymbol{\phi}} = \frac{\mu_0 I}{2\pi\rho}\ \hat{\boldsymbol{\phi}}. \tag{14-12}$$

Lines of $\boldsymbol{B}$ are circles lying in a plane perpendicular to the wire and centered on it. The magnitude of $\boldsymbol{B}$ falls off as $1/\rho$.

EXAMPLE The Circular Loop

To calculate the value of $\boldsymbol{B}$ on the axis of a circular loop of radius a, refer to Fig 14-4. The figure shows the $\boldsymbol{dB}$ of an element of current $I\,\boldsymbol{dl'}$. By symmetry, the total $\boldsymbol{B}$ points along the axis and

$$dB_z = \frac{\mu_0}{4\pi}\frac{I\,\boldsymbol{dl'}}{r^2}\cos\theta, \tag{14-13}$$

$$B_z = \frac{\mu_0}{4\pi}\frac{2\pi a I}{r^2}\cos\theta = \frac{\mu_0 I a^2}{2(a^2+z^2)^{3/2}}. \tag{14-14}$$

Along the axis, $B = \mu_0 I/(2a)$ at $z = 0$ and falls off as $1/z^3$ for $z^2 \gg a^2$.

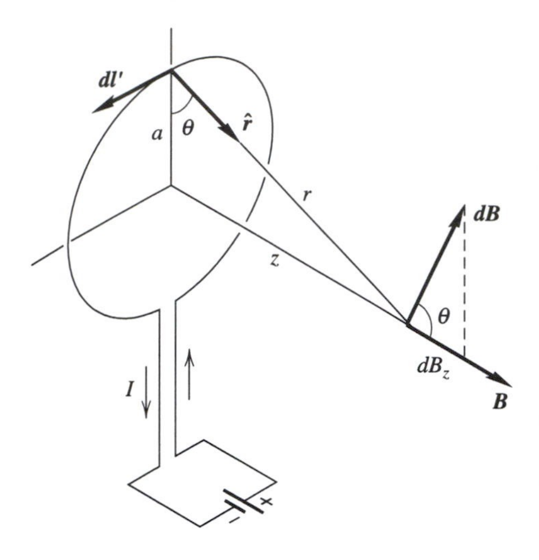

Figure 14-4 Coil of wire of radius a carrying a current *I*, the field ***dB*** on the *z-axis* that originates in the element *I* ***dl′***, and the total field ***B***.

EXAMPLE The Solenoid

The above result can serve to calculate ***B*** on the axis of the solenoid of Fig. 14-5 by summing the ***dB***'s of the individual turns. The solenoid is close-wound, of length *L*, with N' turns per meter, and its radius is *R*. At the center,

$$B = \frac{\mu_0}{2}\int_{-L/2}^{+L/2} \frac{R^2 N' I dz}{(R^2 + z^2)^{3/2}} \tag{14-15}$$

$$= \frac{\mu_0}{2} N' I \frac{L}{(R^2 + L^2/4)^{1/2}} = \mu_0 N' I \sin\theta_m. \tag{14-16}$$

See Fig. 14-5 for the definition of θ_m and θ_e. At one end, again on the axis,

$$B = \frac{\mu_0 N' I \sin\theta_e}{2}. \tag{14-17}$$

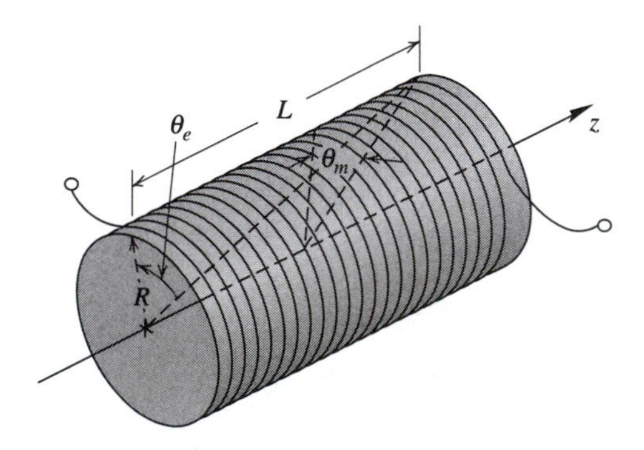

Figure 14-5 Thin solenoid.

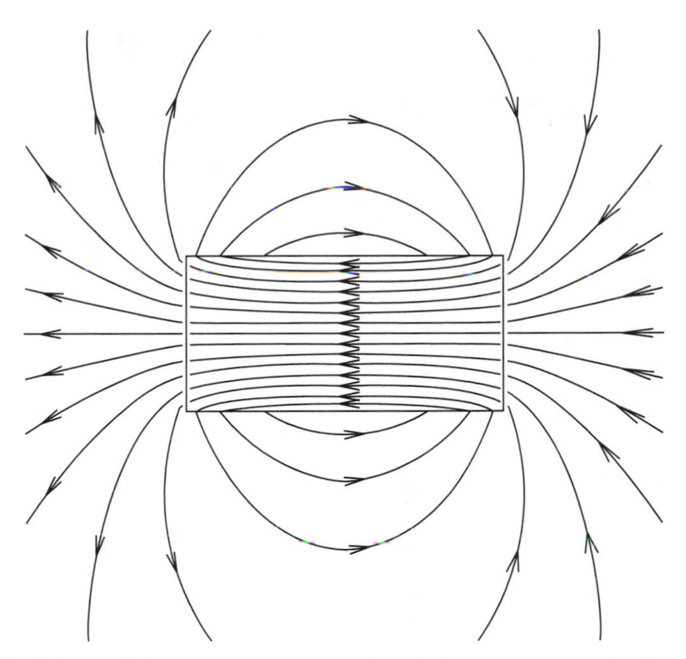

Figure 14-6 Lines of $\boldsymbol{B}$ for a solenoid whose length is equal to twice its diameter.

The magnetic flux density is larger at the center than at the ends because the lines of $\boldsymbol{B}$ flare out at the ends, as in Fig. 14-6. Inside a *long* solenoid, at points remote from the ends, $B = \mu_0 N' I$, even off the axis (Sec. 15.4).

Calculating $\boldsymbol{B}$ off the axis of a short solenoid would be much more difficult.

EXAMPLE Solar Coronal Loops. The Magnetic Field on the Axis of a Solenoid, Far Away

Let us find an approximate expression for B_z at a point P on the axis of the thin solenoid of Fig. 14-5, at a distance D that is much larger than either its length L or its radius a.

This example concerns *solar coronal loops*, which are luminous arcs that extend up to 250 megameters above the surface of the sun. (The diameter of the earth is about 13 megameters). See Fig. 14-7. See also Zirin (1988), and Golub and Pasachoff (1997). Loops are definitely not ballistic, first because they are often shaped like horseshoes, and second because some loops are nearly horizontal. Coronal loops more or less follow magnetic field lines, because they connect regions of opposite magnetic polarity at the surface of the sun, for example sunspots. (Sunspots exhibit magnetic fields as if they were solenoids that stand more or less vertically at the surface. See the Example in Sec. 17.5.) There are innumerable coronal loops at the surface of the sun. The same type of phenomenon undoubtedly also occurs on other stars.

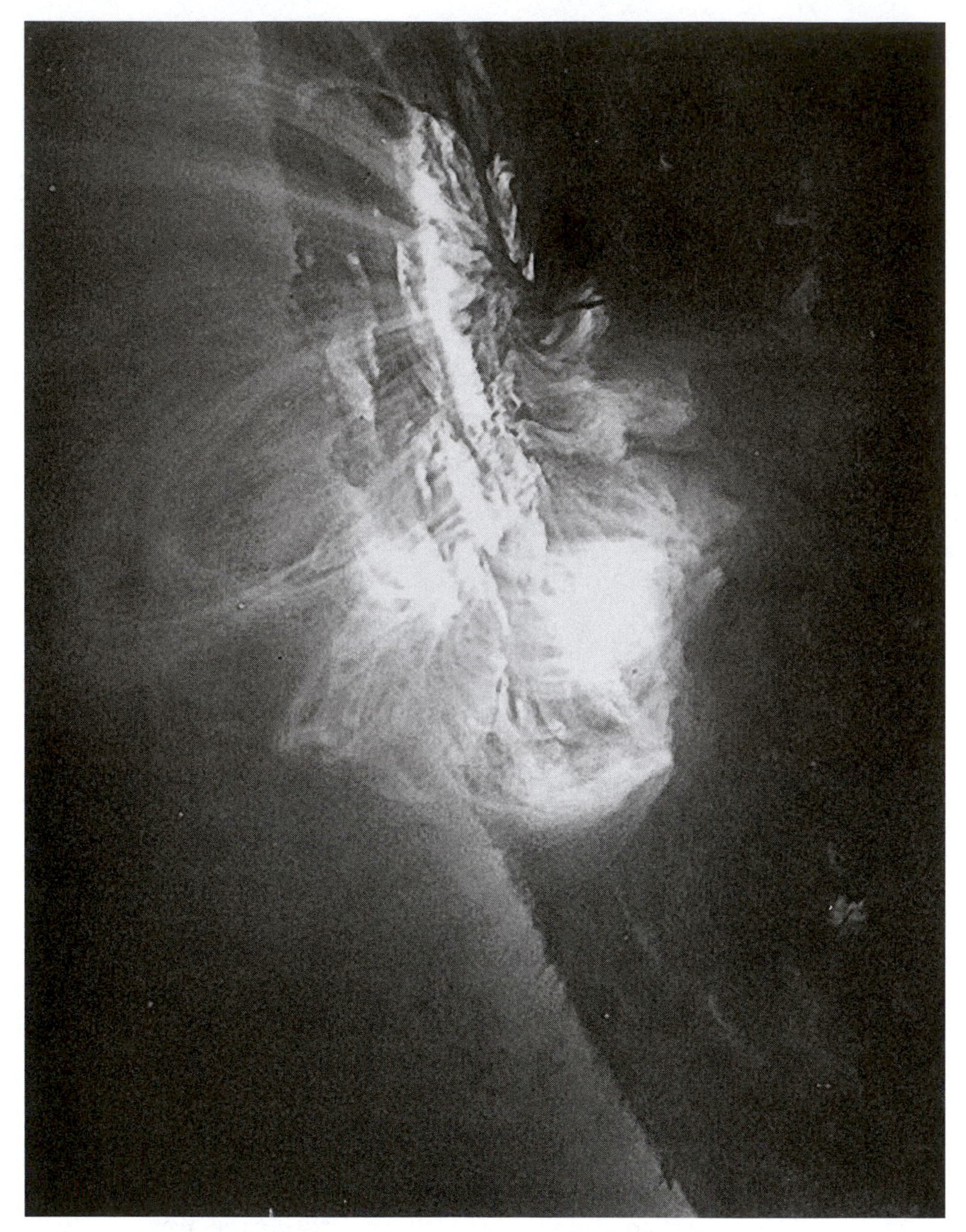

Figure 14-7 Solar coronal loops that you can see, in full color (in 2000), on the Web: type http://vestige.lmsal.com/TRACE/Public/AGU, then select image trace.425.jpg. See also the front cover. One sees the *limb* (edge) of the sun. The loops emerge from an active region, where there are intense magnetic fields. The sun has a radius of 7×10^8 meters. Up to what altitude do these loops extend?

There do not exist satisfactory models for coronal loops. Explaining this spectacular phenomenon is particularly difficult. Loops are possibly charged-particle beams that follow magnetic field lines, but then, how can they have a uniform width throughout their lengths, while magnetic field lines diverge rapidly above sunspots, as at the end of a solenoid? Recall Fig. 14-6. Also, neighboring loops do not seem to interact! If loops are charged particle beams, then each one has its own magnetic field, and they should either attract or repel. A further baffling fact is that, the guiding magnetic field decreases by many orders of magnitude with increasing altitude, as we shall see here.

We really need to calculate the *horizontal* field a long distance above, and half way between, a pair of parallel solenoids of opposite polarity, but it will suffice, here, to calculate the *axial* field far above a single solenoid.

See Fig. 14-8. Set the origin at the center O of a thin solenoid of radius R, D the distance to the point P, and z the position of the element of winding $N'dz$, where N' is the number of turns per meter. Then, from Eq. 14-14, at P, with $a = R$,

$$B_{z,D} = \int_{-L/2}^{L/2} \frac{\mu_0 I R^2}{2[R^2 + (D-z)^2]^{3/2}} N'dz \tag{14-18}$$

$$= \frac{\mu_0 I N'}{2}\left[\frac{-(2D-L)}{(4R^2 + 4D^2 + L^2 - 4DL)^{1/2}}\right]$$

$$+ \frac{\mu_0 I N'}{2}\left[\frac{2D+L}{(4R^2 + 4D^2 + L^2 + 4DL)^{1/2}}\right]. \tag{14-19}$$

As a first approximation, if $D \gg L$ and $D \gg R$, then $B_z \approx 0$. That is correct, but of no use.

Let us find an approximate expression for the first bracket in the above equation; the second bracket can be treated similarly. Note that the denominator in the first bracket is equal to

$$[4R^2 + (2D-L)^2]^{1/2}.$$

We can therefore rewrite the first bracket of Eq. 14-19 as

$$-\left[\frac{1}{\{1 + [4R^2/(2D-L)^2]\}^{1/2}}\right].$$

Since the second term in the denominator of this fraction is much smaller than unity, we expand as a power series, and the bracket becomes approximately equal to

$$-\left[1 - \frac{1}{2}\frac{4R^2}{(2D-L)^2}\right] = -\left[1 - \frac{1}{2}\frac{4R^2}{4D^2}\left(\frac{1}{1 - L/(2D)}\right)^2\right]. \tag{14-20}$$

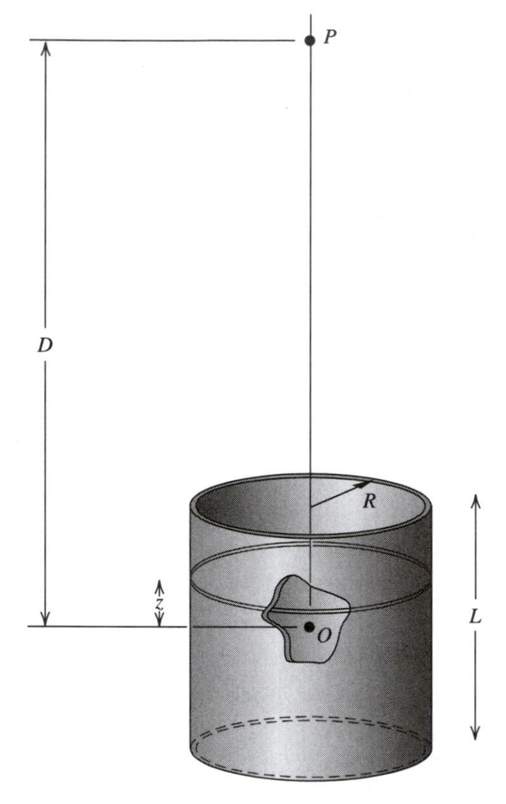

Figure 14-8 Thin solenoid of radius R and length L, with N' turns per meter. The point P is at a distance D above the center O of the solenoid.

But since $L/(2D) \ll 1$, we can again expand as a power series. The first bracket of Eq. 14-19 then becomes approximately equal to

$$-\left[1 - \frac{R^2}{2D^2}\left(1 + 2\frac{L}{2D}\right)\right].$$

Similarly, the second bracket in Eq. 14-19 is approximately equal to

$$+\left[1 - \frac{R^2}{2D^2}\left(1 - 2\frac{L}{2D}\right)\right].$$

Finally,

$$B_{z,D} = \frac{\mu_0 IN'}{2}\frac{R^2 L}{D^3}. \tag{14-21}$$

Now, at one end of the solenoid, from Eq. 14-17,

$$B_{z,L/2} = \frac{\mu_0 N'I}{2}\frac{L}{(R^2 + L^2)^{1/2}}, \tag{14-22}$$

so that

$$\frac{B_{z,D}}{B_{z,L/2}} = \frac{R^2(L^2 + R^2)^{1/2}}{D^3}. \tag{14-23}$$

Set $L = KR$, with K of the order of a few. Then this ratio is equal to

$$(K^2 + 1)^{1/2} \; (R/D)^3 \approx (R/D)^3 \ll 1.$$

Near the top of a typical coronal loop, $(R/D)^3 \approx (1/30)^3 \approx 4 \times 10^{-5}$, and the guiding magnetic field is weaker than at the surface of the sun by that factor!

14.3 THE DIVERGENCE OF *B*

Assuming that magnetic monopoles do not exist (Sec. 14.1), or at least that the net magnetic charge density is everywhere zero, all magnetic fields result from electric currents, and the lines of $\boldsymbol{B}$ for each element of current are circles, as in Fig 14-3. Thus the net outward flux of $\boldsymbol{B}$ through any closed surface is zero:

$$\oint_{\mathcal{A}} \boldsymbol{B} \cdot d\boldsymbol{\mathcal{A}} = 0. \tag{14-24}$$

Applying the divergence theorem, it follows that

$$\boxed{\boldsymbol{\nabla} \cdot \boldsymbol{B} = 0.} \tag{14-25}$$

These are alternate forms of one of Maxwell's equations. Observe that Eq. 14-25 establishes a relation between the space derivatives of $\boldsymbol{B}$ at a given *point*. Equation 14-24, on the contrary, concerns the magnetic flux over a *closed surface*.

14.4 THE VECTOR POTENTIAL *A*

Contrary to what is often stated or implied in textbooks, the vector potential $\boldsymbol{A}$ is *not* just a mathematical artifice; it is just as important as the magnetic flux density $\boldsymbol{B}$, and just as important as the electric potential V. As we shall see in Chapter 20, the only correct way of expressing the Faraday induction law in stationary systems is to write that $\boldsymbol{E} = -\partial \boldsymbol{A}/\partial t$ for the electric field of magnetic origin.

We have just seen that $\boldsymbol{\nabla} \cdot \boldsymbol{B} = 0$. Then

$$\boldsymbol{B} = \boldsymbol{\nabla} \times \boldsymbol{A}, \tag{14-26}$$

where $\boldsymbol{A}$ is the *vector potential,* as opposed to V, which is the *scalar potential.* The divergence of $\boldsymbol{B}$ is then automatically equal to zero because the divergence of a curl is zero.

It is immediately apparent that, for a given $\boldsymbol{B}$, there exist an infinite number of possible $\boldsymbol{A}$'s. Indeed, one can add to $\boldsymbol{A}$ any quantity whose curl is zero, for example

$25\hat{\boldsymbol{x}}$, without affecting $\boldsymbol{B}$. The magnetic flux density is a measurable quantity, but $\boldsymbol{A}$ is known only within an additive term.†

Note the analogy with the relation $\boldsymbol{E} = -\boldsymbol{\nabla} V$ of electrostatics.

Notice also that $\boldsymbol{B}$ is a function of the space derivatives of $\boldsymbol{A}$, just as $\boldsymbol{E}$ is a function of the space derivatives of V. Thus, to deduce the value of $\boldsymbol{B}$ from $\boldsymbol{A}$ at a given point P, one must know the value of $\boldsymbol{A}$ in the *region* around P.

We now deduce the integral for $\boldsymbol{A}$, starting from the Biot-Savart law of Sec. 14.2:

$$\boldsymbol{B} = \frac{\mu_0}{4\pi}\int_{v'} \frac{\boldsymbol{J} \times \hat{\boldsymbol{r}}}{r^2}\, dv' = \frac{\mu_0}{4\pi}\int_{v'} \left(\boldsymbol{\nabla}\frac{1}{r}\right) \times \boldsymbol{J}\, dv', \tag{14-27}$$

from Identity 16 inside the front cover. Applying now Identity 11, we find that

$$\left(\boldsymbol{\nabla}\frac{1}{r}\right) \times \boldsymbol{J} = \boldsymbol{\nabla} \times \frac{\boldsymbol{J}}{r} - \frac{\boldsymbol{\nabla} \times \boldsymbol{J}}{r}, \tag{14-28}$$

where the second term on the right is zero because $\boldsymbol{J}$ is a function of x', y', z', while $\boldsymbol{\nabla}$ involves derivatives with respect to x, y, z. Thus

$$\boldsymbol{B} = \frac{\mu_0}{4\pi}\int_{v'} \left(\boldsymbol{\nabla} \times \frac{\boldsymbol{J}}{r}\right) dv' = \boldsymbol{\nabla} \times \left(\frac{\mu_0}{4\pi}\int_{v'} \frac{\boldsymbol{J}}{r}\, dv'\right), \tag{14-29}$$

and

$$\boldsymbol{A} = \frac{\mu_0}{4\pi}\int_{v'} \frac{\boldsymbol{J}}{r}\, dv'. \tag{14-30}$$

This expression for $\boldsymbol{A}$ has a definite value for a given current distribution.

This integral, like that for $\boldsymbol{B}$, appears to diverge inside a current carrying conductor, because of the r in the denominator. Actually, it is well-behaved, like the integral for V inside a charge distribution.

If a current I flows in a circuit C that is not necessarily closed, then, at a point $P(x, y, z)$ in space,

$$\boldsymbol{A} = \frac{\mu_0 I}{4\pi}\int_{C} \frac{\boldsymbol{dl'}}{r}, \tag{14-31}$$

where the element $\boldsymbol{dl'}$ of circuit C is at $P'(x', y', z')$, and r is the distance between P and P'.

These two integrals also apply to the fields of alternating currents if the time delay r/c is a small fraction of one period.

†See Feynman, Leighton, and Sands (1964) for a discussion of some quantum mechanical aspects of $\boldsymbol{A}$.

EXAMPLE $\boldsymbol{A}$ and $\boldsymbol{B}$ Near a Long, Straight Wire

We first calculate $\boldsymbol{A}$ and then deduce $\boldsymbol{B}$ in the field of the long, straight, current-carrying wire of Fig. 14-9. We should find the same value of $\boldsymbol{B}$ as in the first example in Sec. 14.2.

At a distance ρ

$$dA = \frac{\mu_0 I}{4\pi}\frac{dl'}{r}. \tag{14-32}$$

The vector $\boldsymbol{A}$ is *parallel to the wire* and points in the direction of the current.

For an infinitely long conductor, $\boldsymbol{dA}$ is proportional to dl'/l' for large values of r where $r \approx l'$. Then $\boldsymbol{A}$ tends to infinity logarithmically. However, the fact that a function is infinite does *not* mean that its derivatives are infinite; that is, $\boldsymbol{B}$ can be finite even though $\boldsymbol{A}$ is infinite.

We can circumvent this infinite value of $\boldsymbol{A}$ by first calculating $\boldsymbol{A}$ and $\boldsymbol{B}$ for a wire of finite length L and then setting $L \gg \rho$ at the end of the calculation. Referring to Fig. 14-9,

$$A = \frac{\mu_0 I}{4\pi}\int_{-L/2}^{L/2}\frac{dl'}{(\rho^2 + l'^2)^{1/2}} = \frac{\mu_0 I}{2\pi}\left(\ln\frac{l' + (\rho^2 + l'^2)^{1/2}}{\rho}\right)_0^{L/2} \tag{14-33}$$

$$= \frac{\mu_0 I}{2\pi}\ln\frac{(L/2)[1 + (1 + 4\rho^2/L^2)^{1/2}]}{\rho} \approx \frac{\mu_0 I}{2\pi}\ln\frac{L}{\rho} \qquad (4\rho^2 \ll L^2) \tag{14-34}$$

$$\approx \frac{\mu_0 I}{2\pi}\ln\frac{\mathcal{R}}{\rho} \qquad (4\rho^2 \ll L^2). \tag{14-35}$$

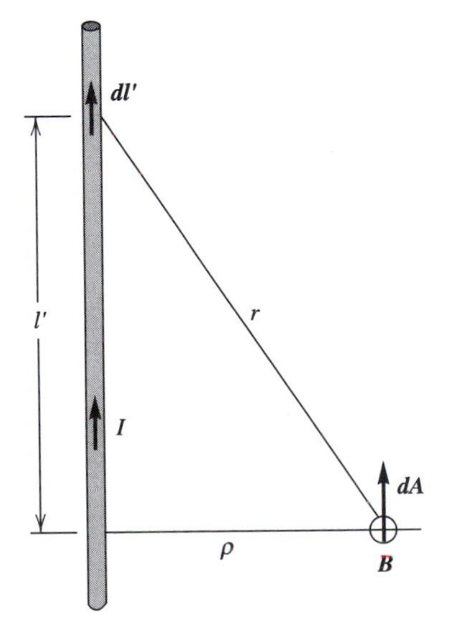

Figure 14-9 The element of vector potential $\boldsymbol{dA}$ due to the element $I\boldsymbol{dl}'$. The vector $\boldsymbol{B}$ is azimuthal, as in Fig. 14-3.

In this last expression we have neglected a term in $\ln(L/\mathcal{R})$ and we have arbitrarily set $\boldsymbol{A} = 0$ at $\rho = \mathcal{R}$.

To calculate $\boldsymbol{B} = \nabla \times \boldsymbol{A}$, we use cylindrical coordinates, keeping in mind that $\boldsymbol{A}$ is parallel to the z-axis and independent of both ϕ and z. From the expression for $\nabla \times \boldsymbol{A}$ on the back of the front cover,

$$\boldsymbol{B} = \nabla \times \boldsymbol{A} = \frac{\mu_0 I}{2\pi\rho}\hat{\boldsymbol{\phi}} \qquad (4\rho^2 \ll L^2) \tag{14-36}$$

as in the first example in Sec. 14.2.

EXAMPLE Pair of Long Parallel Currents

Figure 14-10 shows two long parallel wires separated by a distance D and carrying equal currents I in opposite directions. To calculate $\boldsymbol{A}$, we use the above result for the $\boldsymbol{A}$ of a single wire and add the two vector potentials:

$$A = \frac{\mu_0 I}{2\pi}\left(\ln\frac{L}{\rho_a} - \ln\frac{L}{\rho_b}\right) = \frac{\mu_0 I}{2\pi}\ln\frac{\rho_b}{\rho_a} = \frac{\mu_0 I}{4\pi}\ln\frac{x^2 + (D-y)^2}{x^2 + y^2}. \tag{14-37}$$

The vector $\boldsymbol{A}$ points in the direction of the current that is closer to P; it is zero in the plane $\rho_a = \rho_b$. Then

$$B_x = \frac{\partial A}{\partial y} = -\frac{\mu_0 I}{2\pi}\left(\frac{D-y}{\rho_b^2} + \frac{y}{\rho_a^2}\right), \tag{14-38}$$

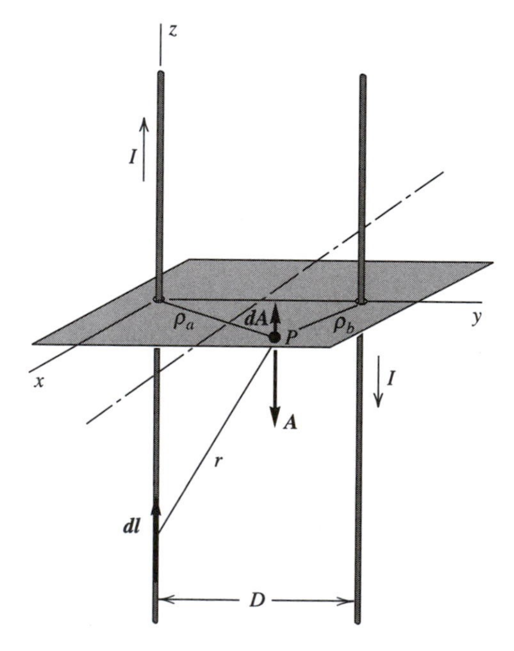

Figure 14-10 Pair of long parallel wires carrying currents of the same magnitude in opposite directions. The vector $\boldsymbol{A}$ is zero in the vertical plane that passes through the dashed line, and it points upward on the left and downward on the right.

$$B_y = -\frac{\partial A}{\partial x} = \frac{\mu_0 I}{2\pi}\left(\frac{1}{\rho_a^2} - \frac{1}{\rho_b^2}\right)x, \tag{14-39}$$

$$B_z = 0. \tag{14-40}$$

Along the line midway between the two wires,

$$B_x = -\frac{2\mu_0 I}{\pi D}, \qquad B_y = 0, \qquad B_z = 0. \tag{14-41}$$

Figure 24-1 shows magnetic field lines for the parallel-wire line.

EXAMPLE $\boldsymbol{A}$ **and** $\boldsymbol{B}$ **in the Field of a Magnetic Dipole**

A magnetic dipole is a loop of wire carrying a current I, as in Fig. 14-11. This will lead us to an interesting relationship between the $\boldsymbol{B}$-field of a magnetic dipole and the $\boldsymbol{E}$-field of an electric dipole.

We calculated the field on the axis of a circular loop in the second example in Sec. 14.2. We now calculate $\boldsymbol{A}$ and $\boldsymbol{B}$ at any remote point in space, at distances r that are much larger than the radius a of the loop. Figure 14-12(b) shows the field close to and inside the loop. The field in that region, away from the axis, is difficult to calculate.

At the point P in Fig. 14-11,

$$\boldsymbol{A} = \frac{\mu_0 I}{4\pi}\oint_C \frac{\boldsymbol{dl'}}{r}. \tag{14-42}$$

By symmetry, $\boldsymbol{A}$ is azimuthal: for any value of r' we have two symmetric $\boldsymbol{dl'}$'s whose y-components add and whose x-components cancel. Then we need calculate only the y-component of the $\boldsymbol{A}$ in the figure, and

$$A = \frac{\mu_0 I}{4\pi}\int_0^{2\pi} \frac{a\, d\phi \cos\phi}{r'}. \tag{14-43}$$

We can express r' as a function of r and ϕ in the following way. Refer to the figure. First,

$$r'^2 = r^2 + a^2 - 2ar\cos\psi. \tag{14-44}$$

Now

$$x\cos\phi = r\cos\psi \tag{14-45}$$

and

$$r'^2 = r^2 + a^2 - 2ax\cos\phi, \tag{14-46}$$

$$r' = r\left\{1 + \left[\frac{a^2}{r^2} - 2\frac{a}{r}\left(\frac{x}{r}\cos\phi\right)\right]\right\}^{1/2} = r\{1 + [\ \ ]\}^{1/2}. \tag{14-47}$$

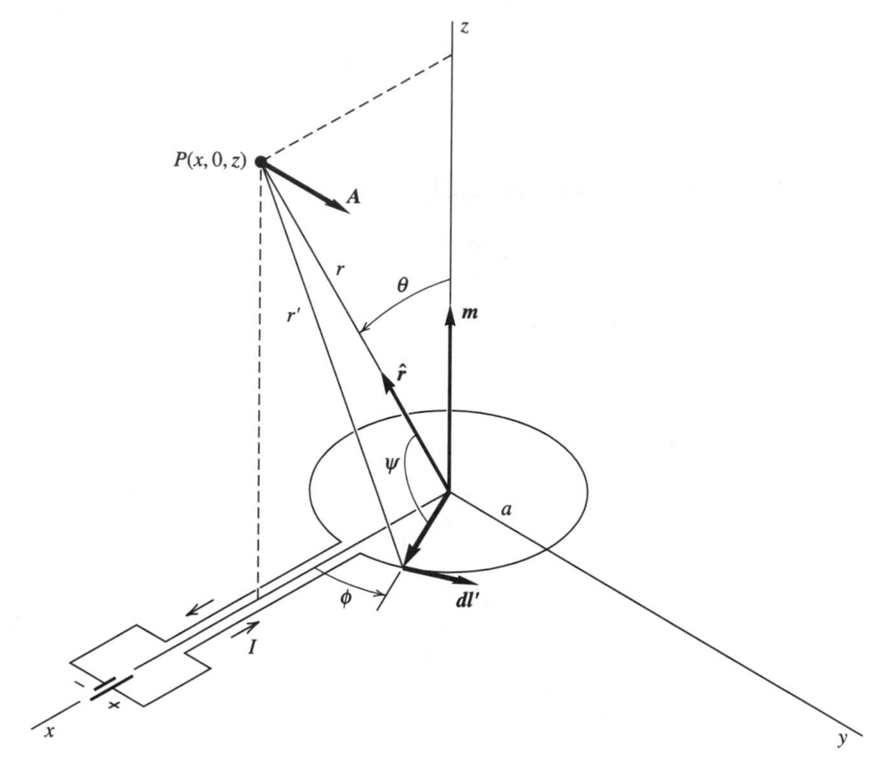

Figure 14-11 Magnetic dipole. The vector $\boldsymbol{A}$ is azimuthal.

Observe that $(x/r)\cos\phi \leq 1$.

Since we are interested in the field only at points where $r \gg a$, we expand $1/r'$ as an infinite series and disregard terms involving higher powers of a/r. Thus

$$\frac{1}{r'} = \frac{1}{r}\left\{1 - \frac{1}{2}[\quad] + \frac{3}{8}[\quad]^2 - \cdots\right\}. \tag{14-48}$$

Setting

$$\left(\frac{x}{r}\cos\phi\right) = (\quad), \tag{14-49}$$

$$\frac{1}{r'} = \frac{1}{r}\left\{1 - \frac{1}{2}\left[\frac{a^2}{r^2} - 2\frac{a}{r}(\quad)\right] + \frac{3}{8}\left[\frac{a^4}{r^4} - 4\frac{a^3}{r^3}(\quad) + 4\frac{a^2}{r^2}(\quad)^2\right] - \cdots\right\}. \tag{14-50}$$

Discarding now all terms containing the third and higher powers of a/r,

$$\frac{1}{r'} = \frac{1}{r}\left\{1 + \frac{a}{r}(\quad) - \left[\frac{1}{2} - \frac{3}{2}(\quad)^2\right]\frac{a^2}{r^2}\right\}. \tag{14-51}$$

Finally, substituting into Eq. 14-43 yields

$$A = \frac{\mu_0 I a}{4\pi r}\int_0^{2\pi}\left[1 + \frac{a}{r}\left(\frac{x}{r}\cos\phi\right) - \left(\frac{1}{2} - \frac{3}{2}\frac{x^2}{r^2}\cos^2\phi\right)\frac{a^2}{r^2}\right]\cos\phi\, d\phi. \tag{14-52}$$

Only the second term between the brackets survives, and

$$A = \frac{\mu_0 I a^2 x}{4r^3} = \frac{\mu_0 I a^2 \sin\theta}{4r^2} \qquad (r^3 \gg a^3). \tag{14-53}$$

By definition,

$$\boldsymbol{m} = \pi a^2 I\, \hat{\boldsymbol{z}} \tag{14-54}$$

is the *magnetic dipole moment* of the loop. If there are N turns, then $\boldsymbol{m}$ is N times larger.

Since $\boldsymbol{A}$ is azimuthal,

$$\boldsymbol{A} = \frac{\mu_0}{4\pi}\frac{\boldsymbol{m}\times\hat{\boldsymbol{r}}}{r^2} \quad (r^3 \gg a^3). \tag{14-55}$$

The condition $r^3 \gg a^3$ is easy to satisfy: at $r = 5a$, $a^3/r^3 = \frac{1}{125}$.

The value of $\boldsymbol{B} = \boldsymbol{\nabla}\times\boldsymbol{A}$ follows immediately:

$$\boldsymbol{B} = \frac{\mu_0 m}{4\pi r^3}(2\cos\theta\,\hat{\boldsymbol{r}} + \sin\theta\,\hat{\boldsymbol{\theta}}). \tag{14-56}$$

The analogy with the field of the electric dipole of Sec. 5.1 is obvious. The analogy, however, applies solely at distances r that are large compared to the sizes of the dipoles. Figure 14-12 shows the near fields.

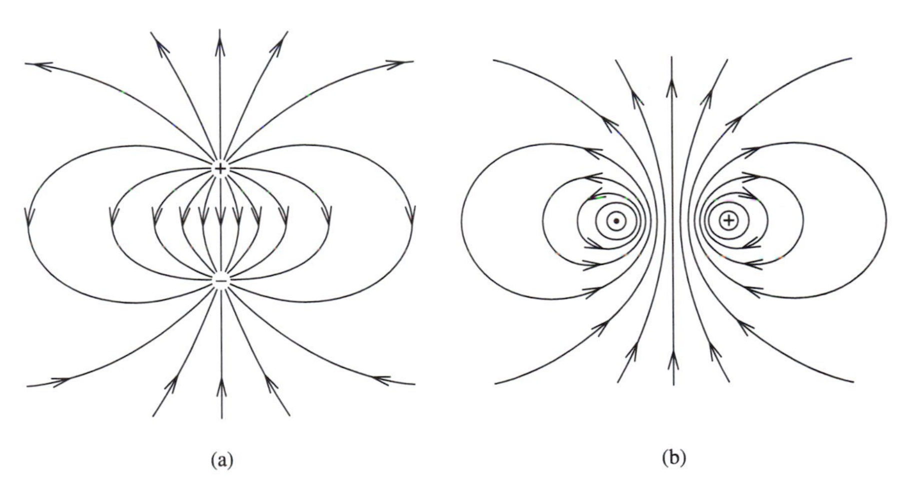

Figure 14-12 The fields (a) of an electric dipole and (b) of a magnetic dipole, in the immediate vicinity of the dipoles.

14.5 SUMMARY

A charge Q moving at a velocity $\boldsymbol{v}$ in the field of an arbitrary distribution of charges and currents is subjected to a *Lorentz force*

$$\boldsymbol{F} = Q(\boldsymbol{E} + \boldsymbol{v} \times \boldsymbol{B}), \tag{14-1}$$

where $Q\boldsymbol{E}$ is the *electric force*, $Q\boldsymbol{v} \times \boldsymbol{B}$ is the *magnetic force*, and $\boldsymbol{B}$ is the *magnetic flux density*, expressed in *teslas*.

In the field of a current I flowing through a circuit C,

$$\boldsymbol{B} = \frac{\mu_0 I}{4\pi} \oint_C \frac{d\boldsymbol{l}' \times \hat{\boldsymbol{r}}}{r^2}. \tag{14-5}$$

In the field of a volume current distribution,

$$\boldsymbol{B} = \frac{\mu_0}{4\pi} \int_{v'} \frac{\boldsymbol{J} \times \hat{\boldsymbol{r}}}{r^2} dv'. \tag{14-8}$$

Lines of $\boldsymbol{B}$ point everywhere in the direction of $\boldsymbol{B}$.

The *magnetic flux* through an open surface of area $\mathcal{A}$ is

$$\Phi = \int_{\mathcal{A}} \boldsymbol{B} \cdot d\boldsymbol{\mathcal{A}} \qquad \text{webers.} \tag{14-9}$$

The principle of superposition applies to magnetic fields.

The net magnetic flux through a closed surface is zero:

$$\oint_{\mathcal{A}} \boldsymbol{B} \cdot d\boldsymbol{\mathcal{A}} = 0. \tag{14-24}$$

Hence

$$\boxed{\boldsymbol{\nabla} \cdot \boldsymbol{B} = 0.} \tag{14-25}$$

These are, respectively, the integral and the differential forms of one of Maxwell's equations.

The equation

$$\boldsymbol{B} = \boldsymbol{\nabla} \times \boldsymbol{A} \tag{14-26}$$

defines the *vector potential* $\boldsymbol{A}$. For a volume current distribution,

$$A = \frac{\mu_0}{4\pi}\int_{v'} \frac{J}{r}\, dv' \tag{14-30}$$

while, for a current I flowing through a circuit C,

$$A = \frac{\mu_0 I}{4\pi}\int_C \frac{dl'}{r}. \tag{14-31}$$

In the field of a *magnetic dipole,*

$$A = \frac{\mu_0}{4\pi}\frac{m \times \hat{r}}{r^2}, \tag{14-55}$$

where the magnitude of the *magnetic dipole moment* m is the area of the loop, times the number of turns, times the current. The vector m is normal to the plane of the loop, in the direction defined by the right-hand screw rule. Refer to Fig. 14.11. Also,

$$B = \frac{\mu_0 m}{4\pi r^3}(2\cos\theta\,\hat{r} + \sin\theta\,\hat{\theta}). \tag{14-56}$$

PROBLEMS

***14-1.** (*14.1*) The force on a magnetic monopole situated in a magnetic field

Show that the equation $F = Q_a^* Q_b^* /(4\pi\mu_0 r^2)$ is dimensionally correct. This means that $F = Q^* B/\mu_0$, and not $Q^* B$, as stated by some authors.

14-2. (*14.1*) The field of two parallel wires

Two parallel wires of radius R and separated by a distance $2D$ carry a current I in opposite directions.

(a) Find B along a perpendicular line passing through the wires.

(b) Plot B for $R = 1.00$ millimeter, $D = 10.0$ millimeters, and $I = 1$ ampere.

14-3. (*14.2*) Saddle coils

Figure 14-13 shows two views of a pair of saddle coils. In (*a*) we have shown just one turn in each coil, and (*b*) shows a cross section at C. More generally, we could have the current distribution of Fig. 14-13(*c*), where the two parts carry equal current *densities*. There is zero current in the central region. We could also have the current distribution of Fig. 14-13(*d*). As we shall see, the magnetic fields in the cavities are uniform.

(a) Show that $B = \mu_0 J \times r/2$ inside a conductor of circular cross section. The origin of r is at the center of the cross section.

(b) In Fig. 14-13(a), B is the same as if each conductor occupied a full circle, with opposite currents in the central region.

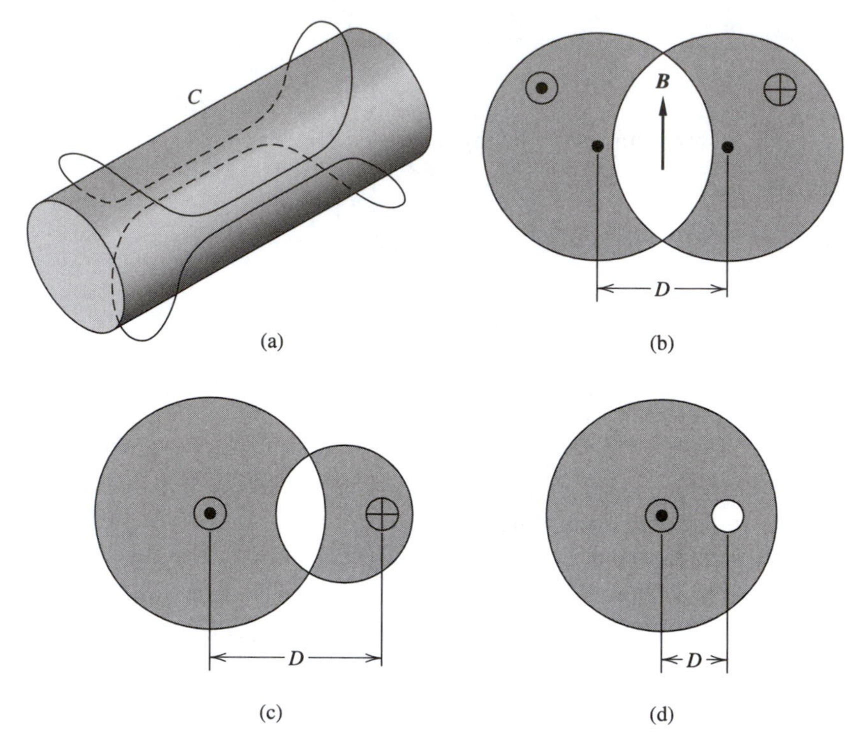

Figure 14-13

Find $\boldsymbol{B}$ in the central region.

(c) Find $\boldsymbol{B}$ in the cavities of Figs. 14-13(c) and (d).

14-4. (*14.2*) Magnetic field on the axis of a circular loop

Plot a curve of B as a function of z on the axis of a circular loop of 100 turns having a mean radius of 100 millimeters and carrying a current of 1 ampere.

14-5. (*14.2*) The Bohr magneton

According to the old Bohr model of the atom, electrons describe orbits around the nucleus. Atomic and molecular magnetic moments are expressed in Bohr magnetons.

(a) Find the magnetic moment of an electron on a circular orbit of radius r.

(b) According to the Bohr postulate, the angular momentum is quantized: $mvr = n\hbar = nh/(2\pi) = n \times 1.0546 \times 10^{-34}$, where n is an integer and a quantum number.

Calculate the value of the *Bohr magneton* μ_B, which is the magnetic moment of an electron orbit for which $n = 1$. The number of Bohr magnetons per atom or per molecule is of the order of a few and is, in fact, not an even number.

14-6. (*14.2*) Rotating magnetic field

Three identical coils, oriented as in Fig. 14-14(a), carry three-phase alternating current. Their magnetic fields are

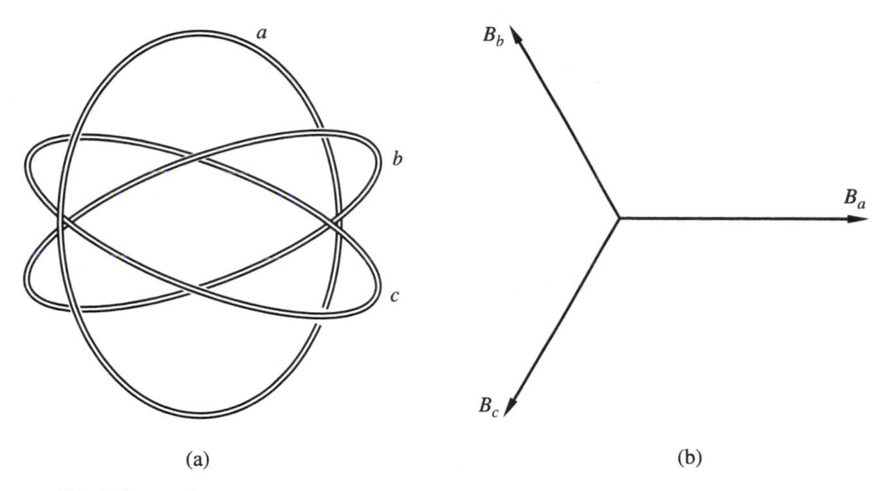

Figure 14-14

$$B_a = B_m \cos \omega t, \qquad B_b = B_m \cos\left(\omega t + \frac{2\pi}{3}\right), \qquad B_c = B_m \cos\left(\omega t + \frac{4\pi}{3}\right),$$

and point as in Fig. 14-14(b).

(a) Show that the resulting field has a magnitude of $1.5B_m$ and rotates at an angular velocity ω. This is the method used to generate rotating magnetic fields in large electric motors.

(b) Does the field rotate clockwise or anticlockwise?

14-7. (*14.2*) The Fabry equation for solenoids

A solenoid has an inner radius R_1, an outer radius R_2, and a length $2L$. The current is I.

(a) Show that at the center

$$B = \mu_0 nIL \ln \frac{\alpha + (\alpha^2 + \beta^2)^{1/2}}{1 + (1 + \beta^2)^{1/2}},$$

where n is the number of turns per square meter ($\approx$ 1/cross section of the wire), $\alpha = R_2/R_1$, and $\beta = L/R_1$.

(b) Show that the length of the wire is

$$l = nV = 2\pi n(\alpha^2 - 1)\beta R_1^3,$$

where V is the volume of the winding.

(c) Check the *Fabry equation*, which states that at the center of any solenoid

$$B = G\left(\frac{P\lambda\sigma}{R_1}\right)^{1/2}.$$

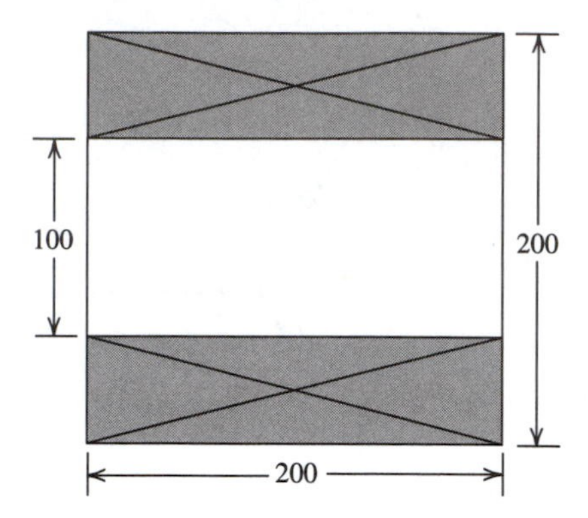

Figure 14-15

Here G depends on the geometry, P is the dissipated power, $\lambda = n\pi r^2$ is the *filling factor*, or the fraction of the coil cross section occupied by the conductor, r is the radius of the wire, and σ the resistivity.

14-8. (*14.2*) A short, thick solenoid

Figure 14-15 shows the cross section of a coil. The dimensions shown are in millimeters. The wire has a square cross section of 2 millimeters^2 and a resistance of 8.93 ohms/kilometer. The current is 1 ampere. See the preceding problem.

(a) Calculate B at the center. Use the formulas given in Prob. 14-7.

(b) Calculate the power and the applied voltage.

(c) Plot B as a function of z along the axis, from $z = -0.3$ to $z = 0.3$ meter.

14-9. (*14.2*) Helmholtz coils provide a uniform field

The *Helmholtz coils* of Fig. 14-16(a) provide a simple means of obtaining a uniform magnetic field over a given volume. Roughly speaking, B_z, is uniform within 10% inside a sphere of radius 0.la.

(a) Find B as a function of z along the axis.

Expanding this expression about $z = 0$,

$$B = B_0\left(1 - \frac{144z^4}{125a^4} + \cdots\right).$$

This means that the first, second, and third derivatives of B with respect to z are zero at $z = 0$. So the curve of $B(z)$ is exceptionally flat near the center.

(b) Plot $B/(\mu_0 NI/a)$ as a function of z/a from $z/a = -0.5$ to $z/a = 0.5$. Figure 14-16(b) shows $B_z(z)$ for values of r ranging up to 0.16a.

14-10. (*14.2*) A Maxwell pair provides a uniform *gradient* of B

Plot $B/(\mu_0 NI/2a)$ as a function of z/a for a pair of coils like those of Fig. 14-16(a) but with a spacing of 2a, instead of a, and with currents flowing in opposite directions. This is a *Maxwell pair.* You will find that dB/dz is surprisingly linear between about $z = -0.7a$ and $z = 0.7a$.

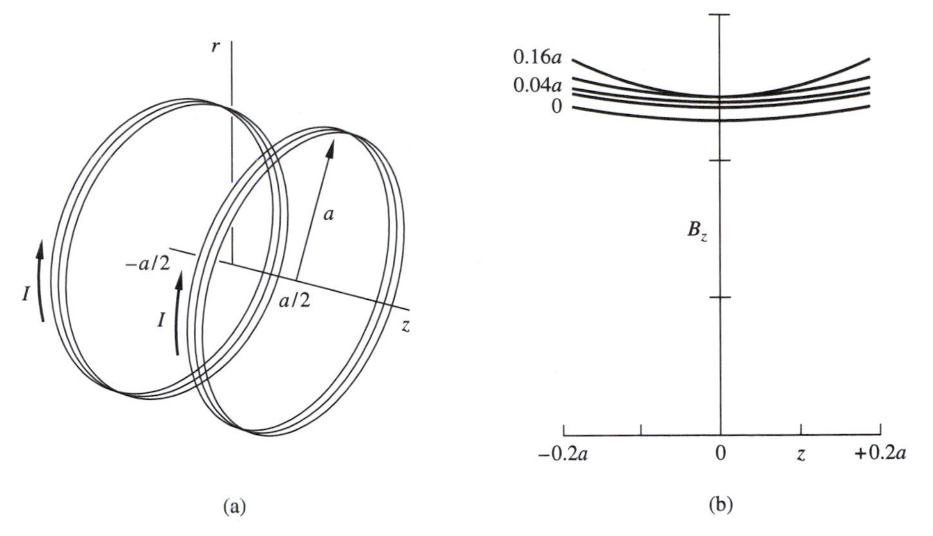

Figure 14-16

14-11. (*14.2*) The space derivatives of $\boldsymbol{B}$ in a static field

Consider the field of a circular loop, as in Fig. 14-12. Set the z-axis vertical and pointing up along the axis of symmetry, with the y-axis pointing right, in the plane of the loop.

Above the plane of the loop, the lines of $\boldsymbol{B}$ flare out, so B_z decreases with z and $\partial B_z/\partial z$ is negative. At the same point, B_y increases with y, so $\partial B_y/\partial y$ is positive. Similarly, $\partial B_x/\partial x$ is positive.

(a) Why is this, mathematically? How are these derivatives related?
(b) Compare these derivative for other points on Fig. 14-12.

14-12. (*14.4*) The vector potential $\boldsymbol{A}$

In a given region, $\boldsymbol{B} = B\hat{\boldsymbol{z}}$. Suggest possible $\boldsymbol{A}$'s and a characteristic of the corresponding current distribution.

14-13. (*14.4*) In two-dimensional magnetic fields a line of constant A is a line of $\boldsymbol{B}$

A certain magnetic field has a zero z-component.

(a) Show that $\boldsymbol{A} = A\hat{\boldsymbol{z}}$ is one possible value of $\boldsymbol{A}$.
(b) Show that a line of constant A is a line of $\boldsymbol{B}$.
(c) Show that this applies to the field of a straight current-carrying wire.

14-14. (*14.5*) The magnetic field of a spinning electrically charged sphere

A conducting sphere of radius R is charged to a potential V and spun about a diameter at an angular velocity ω.

(a) Show that the surface current density is $\alpha = \epsilon_0 \omega V \sin\theta = M \sin\theta$, where M is $\epsilon_0 \omega V$.
(b) Find the magnetic flux density B_0 at the center.

(c) What is the numerical value of B_0 for a sphere 100 millimeters in radius, charged to 10.0 kilovolts, and spinning at 10,000 turns per minute?
(d) Show that the dipole moment is $\frac{4}{3}\pi R^3 M\hat{z}$, where $\hat{z}$ is a unit vector along the axis, related to the direction of rotation by the right-hand screw rule.
(e) What current flowing through a loop 100 millimeters in diameter would have the same dipole moment?

CHAPTER

15

MAGNETIC FIELDS II

*The Vector Potential **A***
Ampère's Circuital Law

In this second chapter on magnetic fields, we first derive a direct consequence of the definition of the vector potential $\boldsymbol{A}$: the line integral of $\boldsymbol{A} \cdot d\boldsymbol{l}$ over a closed curve is equal to the encircled magnetic flux. This result is general. However, the rest of the chapter applies only to static fields. The expressions that we shall find here for $\nabla^2 \boldsymbol{A}$, $\nabla \cdot \boldsymbol{A}$, and $\nabla \times \boldsymbol{B}$, are all truncated: they all lack time-dependent terms. It is only in Chap. 23 that we find the full-fledged expressions.

15.1 THE LINE INTEGRAL OF $\boldsymbol{A} \cdot d\boldsymbol{l}$ AROUND A CLOSED CURVE

Consider first a simple closed curve, C as in Fig. 15-1. The line integral of $\boldsymbol{A} \cdot d\boldsymbol{l}$ around C is equal to the magnetic flux linking C:

$$\oint_C \boldsymbol{A} \cdot d\boldsymbol{l} = \int_{\mathcal{A}} (\nabla \times \boldsymbol{A}) \cdot d\boldsymbol{\mathcal{A}} = \int_{\mathcal{A}} \boldsymbol{B} \cdot d\boldsymbol{\mathcal{A}} = \Phi, \tag{15-1}$$

where $\mathcal{A}$ is the area of any surface bounded by C. We have used Stokes's theorem.

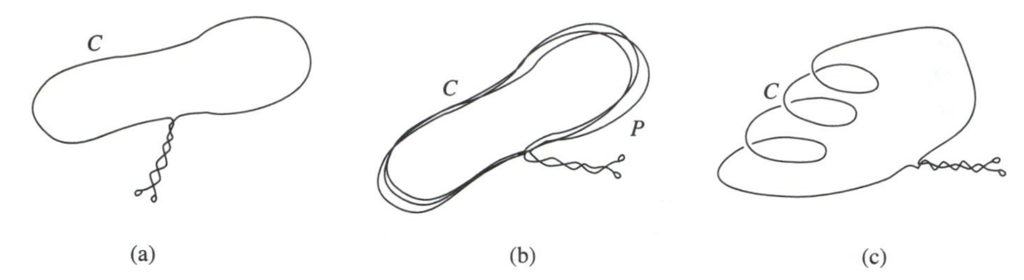

Figure 15-1 (a) A simple closed circuit C. (b) An N-turn coil. The turns are all close together. (c) A more complex closed circuit.

Now suppose the coil has N turns wound close together, as in Fig. 15-1(b). Over any cross section of the coil, say at P, the various turns are all exposed to approximately the same $\boldsymbol{A}$. Then

$$\oint_C \boldsymbol{A} \cdot d\boldsymbol{l} = N \int_{\mathcal{A}} \boldsymbol{B} \cdot d\boldsymbol{\mathcal{A}} = N\Phi = \Lambda, \tag{15-2}$$

where Λ is the *flux linkage* and $\mathcal{A}$ is the area of any surface bounded by the coil.

The unit of flux linkage is the *weber-turn*.

What if one has a circuit such as that of Fig. 15-1(c)? Then

$$\oint_C \boldsymbol{A} \cdot d\boldsymbol{l} = \int_{\mathcal{A}} \boldsymbol{B} \cdot d\boldsymbol{\mathcal{A}} = \Lambda, \tag{15-3}$$

except that now it is difficult to devise a surface bounded by C. Luckily enough, this surface is of no interest because the flux linkage Λ is easily measurable (Sec. 19.2).

EXAMPLE The Vector Potential A in the Field of a Long Solenoid

1. Let us first see, qualitatively, how $\boldsymbol{A}$ varies with position, both inside and outside a long solenoid. Remember the integral for $\boldsymbol{A}$ that we found in Sec. 14.4:

$$\boldsymbol{A} = \frac{\mu_0 I}{4\pi} \oint_C \frac{d\boldsymbol{l}'}{r}. \tag{15-4}$$

At a point on the axis of the solenoid, the $d\boldsymbol{A}$ of an element $I d\boldsymbol{l}'$ situated somewhere on the solenoid cancels the $d\boldsymbol{A}$ of the element $I d\boldsymbol{l}'$ situated diametrically opposite the first one. On the axis of a long solenoid, $\boldsymbol{A}$ is therefore zero.

At a point P inside the solenoid, but off the axis, the elements $I d\boldsymbol{l}'$ closest to P will contribute most. So $\boldsymbol{A}$ is azimuthal, as in Fig. 15-2, and it increases with the radius r.

At a point P' outside the solenoid, $\boldsymbol{B}$ is zero, as we shall see in the second example in Sec. 15.4. But $\boldsymbol{A}$ is clearly not zero because the elements $I d\boldsymbol{l}'$ closest to P' contribute most.

A vector potential $\boldsymbol{A}$ *can therefore exist in a region where* $\boldsymbol{B}$ *is zero.* This simply means that $\boldsymbol{\nabla} \times \boldsymbol{A} = 0$ for $\boldsymbol{A} \neq 0$, which is perfectly sensible. For example, if $\boldsymbol{A} = k\hat{\boldsymbol{x}}$,

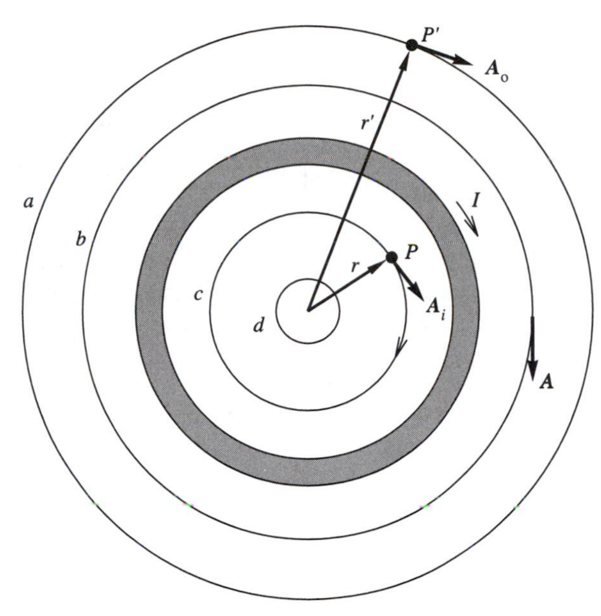

Figure 15-2 Long solenoid seen endwise and lines of $\boldsymbol{A}$ inside and outside. The magnitude of $\boldsymbol{A}$ is proportional to r inside and inversely proportional to r' outside.

where k is independent of the coordinates, then $\nabla \times \boldsymbol{A} = 0$. We are already familiar with a similar situation in electrostatics: V can take any uniform value in a region where $\boldsymbol{E} = -\nabla V = 0$.

Outside a long solenoid, the vector $\boldsymbol{A}$ is azimuthal, and it decreases with r.

2. Now let us calculate $\boldsymbol{A}$.

First, consider the field *inside* the solenoid. At a point remote from the ends, $\boldsymbol{B}$ is parallel to the axis, uniform, and equal to $\mu_0 N' I$, where N' is the number of turns per meter, as in the example on the solenoid in Sec. 14.2. Then, at a radius r as in the figure,

$$2\pi r A_i = \pi r^2 \mu_0 N' I, \qquad A_i = \frac{\mu_0 N' I r}{2}. \tag{15-5}$$

Outside a solenoid of radius R, at the radius r',

$$2\pi r' A_o = \pi R^2 \mu_0 N' I, \qquad A_o = \frac{\mu_0 N' I R^2}{2r'}. \tag{15-6}$$

Think how laborious it would be to calculate $\boldsymbol{A}$ by integrating $I\,\boldsymbol{dl}/r$ over the winding!

The curve of A as a function of the radius is qualitatively the same as that of B. (See Fig. 15-5.)

So, outside a long solenoid, $\boldsymbol{A} \neq 0$, despite the fact that $\boldsymbol{B} = 0$. We return to this fact in the first example of Sec. 18.3. We discuss the $\boldsymbol{B}$ of a long solenoid in Sec. 15.4 below. ∎

15.2 THE LAPLACIAN OF A

You will recall from Secs. 3.4.1 and 4.1 that

$$V = \frac{1}{4\pi\epsilon_0}\int_{v'} \frac{\rho}{r} dv', \qquad \nabla^2 V = -\frac{\rho}{\epsilon_0}. \tag{15-7}$$

The first equation relates the potential V at the point $P(x, y, z)$ to the complete charge distribution, ρ being the total volume charge density $\rho_f + \rho_b$ at $P'(x', y', z')$ and r the distance PP'. The second equation expresses the relation between the *space derivatives* of V at any point to the volume charge density ρ *at that point.*

There exists an analogous pair of equations for the vector potential $\boldsymbol{A}$. We have already found the integral for $\boldsymbol{A}$ in Sec. 14.4:

$$\boldsymbol{A} = \frac{\mu_0}{4\pi}\int_{v'} \frac{\boldsymbol{J}}{r} dv', \tag{15-8}$$

where v' is any volume enclosing all the currents. The x-component of this equation is

$$A_x = \frac{\mu_0}{4\pi}\int_{v'} \frac{J_x}{r} dv'. \tag{15-9}$$

Then, by analogy with Eq. 15-7,

$$\nabla^2 A_x = -\mu_0 J_x. \tag{15-10}$$

Of course, similar equations apply to the y- and z-components, and

$$\nabla^2 \boldsymbol{A} = -\mu_0 \boldsymbol{J}. \tag{15-11}$$

This equation applies only to *static* fields.

15.3 THE DIVERGENCE OF A

We can prove that, for *static* fields and for currents of finite extent, the divergence of $\boldsymbol{A}$ is zero. First,

$$\nabla \cdot \boldsymbol{A} = \nabla \cdot \left(\frac{\mu_0}{4\pi}\int_{v'} \frac{\boldsymbol{J}}{r} dv'\right) = \frac{\mu_0}{4\pi}\int_{v'} \nabla \cdot \left(\frac{\boldsymbol{J}}{r}\right) dv', \tag{15-12}$$

where the del operator acts on the unprimed coordinates (x, y, z) of the field point, while $\boldsymbol{J}$ is a function of the source point (x', y', z'). The integral operates on the primed coordinates. As usual, r is the distance between these two points, and the integration covers any volume enclosing all the currents.

We now use successively Identities 15, 16, and 6 from the back of the front cover:

$$\boldsymbol{\nabla}\cdot\boldsymbol{A} = \frac{\mu_0}{4\pi}\int_{v'}\left(\boldsymbol{\nabla}\frac{1}{r}\right)\cdot\boldsymbol{J}dv' = -\frac{\mu_0}{4\pi}\int_{v'}\left(\boldsymbol{\nabla}'\frac{1}{r}\right)\cdot\boldsymbol{J}\,dv' \tag{15-13}$$

$$= \frac{\mu_0}{4\pi}\int_{v'}\left(-\boldsymbol{\nabla}'\cdot\frac{\boldsymbol{J}}{r} + \frac{\boldsymbol{\nabla}'\cdot\boldsymbol{J}}{r}\right)dv'. \tag{15-14}$$

In a time-independent field, $\partial\rho/\partial t = 0$ and, from the conservation of charge (Sec. 4.2), $\boldsymbol{\nabla}'\cdot\boldsymbol{J} = 0$. Therefore

$$\boldsymbol{\nabla}\cdot\boldsymbol{A} = -\frac{\mu_0}{4\pi}\int_{v'}\boldsymbol{\nabla}'\cdot\frac{\boldsymbol{J}}{r}dv' = -\frac{\mu_0}{4\pi}\int_{\mathcal{A}'}\frac{\boldsymbol{J}}{r}\cdot d\boldsymbol{\mathcal{A}}' = 0, \tag{15-15}$$

where $\mathcal{A}'$ is the area of the surface enclosing the volume v'. We have used the divergence theorem to transform the first integral into the second. The second integral is zero because, over $\mathcal{A}'$, $\boldsymbol{J}$ is either zero or tangential.

15.4 THE CURL OF $\boldsymbol{B}$. AMPÈRE'S CIRCUITAL LAW†

From Identity 13 on the page facing the front cover,

$$\boldsymbol{\nabla}\times\boldsymbol{B} = \boldsymbol{\nabla}\times(\boldsymbol{\nabla}\times\boldsymbol{A}) = \boldsymbol{\nabla}(\boldsymbol{\nabla}\cdot\boldsymbol{A}) - \boldsymbol{\nabla}^2\boldsymbol{A}. \tag{15-16}$$

Thus, from Secs. 15.2 and 15.3,

$$\boldsymbol{\nabla}\times\boldsymbol{B} = \mu_0\boldsymbol{J}. \tag{15-17}$$

This equation is valid only for *static* fields. The more general form of this equation is given in Sec. 21.1.

The line integral of $\boldsymbol{B}\cdot d\boldsymbol{l}$ around a closed curve C is important:

$$\oint_C \boldsymbol{B}\cdot d\boldsymbol{l} = \int_{\mathcal{A}}(\boldsymbol{\nabla}\times\boldsymbol{B})\cdot d\boldsymbol{\mathcal{A}} = \mu_0\int_{\mathcal{A}}\boldsymbol{J}\cdot d\boldsymbol{\mathcal{A}} = \mu_0 I. \tag{15-18}$$

In this set of equations we first used Stokes's theorem, $\mathcal{A}$ being the area of any surface bounded by C. Then we used the relation $\boldsymbol{\nabla}\times\boldsymbol{B} = \mu_0\boldsymbol{J}$ that we found above. Finally, I is the net current that crosses any surface bounded by the closed curve C. The right-hand screw rule applies to the direction of I and to the direction of integration around C, as in Fig. 15-3(a).

This is *Ampère's circuital law:* the line integral of $\boldsymbol{B}\cdot d\boldsymbol{l}$ around a closed curve C is equal to μ_0 times the net current linking C. This result is again valid only for constant fields.

†It seems that this law should not be attributed to Ampère. See Erlichson (1999).

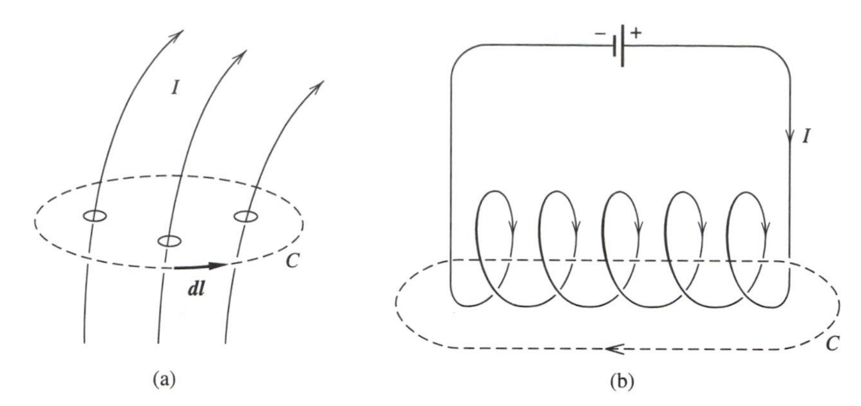

Figure 15-3 (a) Closed path of integration C linked by a current I. Ampère's circuital law states that the line integral of $\boldsymbol{B} \cdot \boldsymbol{dl}$ over C is equal to $\mu_0 I$. (b) Here the line integral of $\boldsymbol{B} \cdot \boldsymbol{dl}$ over the dashed curve is equal to $6\mu_0 I$.

Sometimes the same current crosses the surface bounded by C several times. For example, with a solenoid, the closed curve C could follow the axis and return outside the solenoid, as in Fig. 15-3(b). The total current linking C is then the current in one turn, multiplied by the number of turns, or the number of *ampere-turns*.

The circuital law can be used to calculate B, when B is uniform along the path of integration. This law is analogous to Gauss's law, which we used to calculate an E that is uniform over a surface.

EXAMPLE Long Cylindrical Conductor

Let us apply Ampère's circuital law to calculate $\boldsymbol{B}$ inside and outside the long, straight cylindrical conductor of Fig. 15-4 carrying a current I uniformly distributed over its cross section. We use cylindrical coordinates with the z-axis along the conductor.

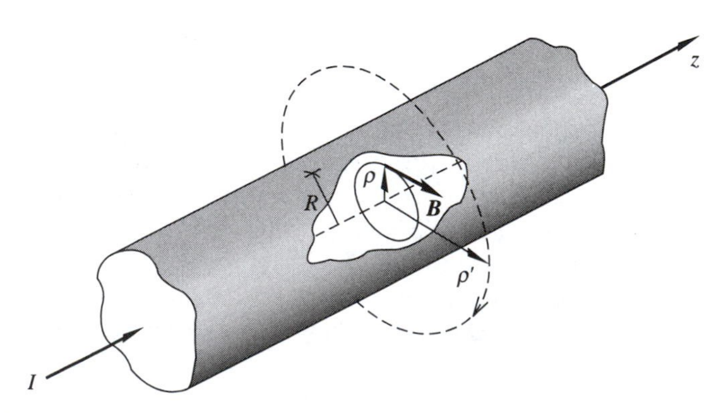

Figure 15-4 Long cylindrical conductor of circular cross section carrying a current I. The circles are paths of integration for calculating $\boldsymbol{B}$.

Outside the conductor, $\boldsymbol{B}$ is azimuthal and independent of ϕ. Then

$$B = \frac{\mu_0 I}{2\pi\rho'}. \tag{15-19}$$

Inside the conductor, for a circuital path of radius ρ,

$$B = \mu_0 \frac{[I/(\pi R^2)]\pi\rho^2}{2\pi\rho} = \frac{\mu_0 I \rho}{2\pi R^2}. \tag{15-20}$$

See Fig. 15-5.

EXAMPLE The Long Solenoid

We return to the long solenoid and recalculate $\boldsymbol{B}$ inside, in a region remote from the ends. This renders end effects negligible. See Fig. 15-6. We assume that the pitch of the winding is small. We again use cylindrical coordinates. The figure shows a solenoid of circular cross section. However, our main conclusions will be valid for any cross section.

First note that $\boldsymbol{B}$ possesses the following general characteristics:

1. By symmetry, $\boldsymbol{B}$ is everywhere independent of z and of ϕ.
2. Imagine a coaxial cylinder d as in the figure. Its radius is either smaller or larger than that of the solenoid. The integrals of $\boldsymbol{B} \cdot d\mathcal{A}$ over the end faces cancel. Then the integral of $\boldsymbol{B} \cdot d\mathcal{A}$ over the cylindrical surface is zero, from Gauss's law for $\boldsymbol{B}$ (Sec. 14.3). Then, both inside and outside, $B_\rho = 0$.

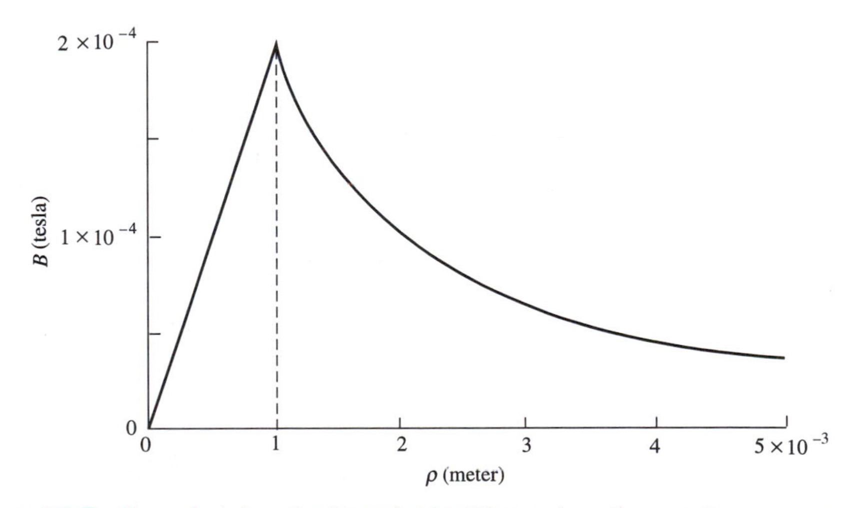

Figure 15-5 B as a function of ρ for a wire 1 millimeter in radius carrying a current of 1 ampere.

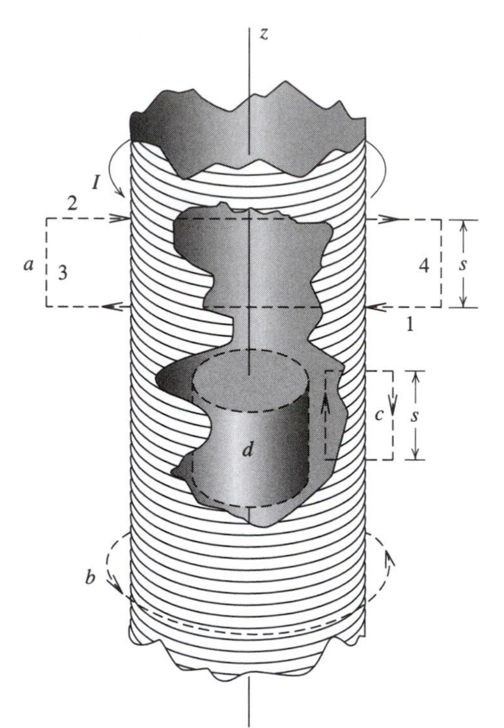

Figure 15-6 Long solenoid carrying a current *I*, with paths of integration *a*, *b*, *c*.

3. The curl of $\boldsymbol{B}$ is zero everywhere except inside the wire, where $\boldsymbol{J} \neq 0$. Then, from the expression for the curl in cylindrical coordinates, outside the wire, $\partial B_z/\partial \rho = 0$. By symmetry, $\partial B_z/\partial \phi$ is also zero. Then B_z *is uniform inside the solenoid and it is also uniform outside, neglecting end effects.*

Now consider the field *outside* the solenoid.

1. We can show that $B_z = 0$ by considering path a in the figure. The net current linking this path is zero, and the line integral of $\boldsymbol{B} \cdot \boldsymbol{dl}$ around it is therefore also zero. Now, since the line integrals along sides 1 and 2 are zero ($B_\rho = 0$), the line integrals along sides 3 and 4 cancel. But sides 3 and 4 can each be situated at any distance from the solenoid, so B_z is either zero, or nonzero and independent of ρ. Now the flux outside is equal to the finite flux inside. Therefore, *outside,* B_z *tends to zero.*
2. A path such as b is linked once by the current. Thus, *outside the solenoid,* $B_\phi = \mu_0 I/(2\pi\rho)$. This flux is usually negligible.

Now let us look *inside* the solenoid.

1. There is zero flux in the ϕ-direction inside because the line integral of $B_\phi dl$ over a circle of radius ρ, say the top edge of the small cylinder shown in the figure, is $2\pi\rho B_\phi$; and this is zero, because the path encloses zero current.

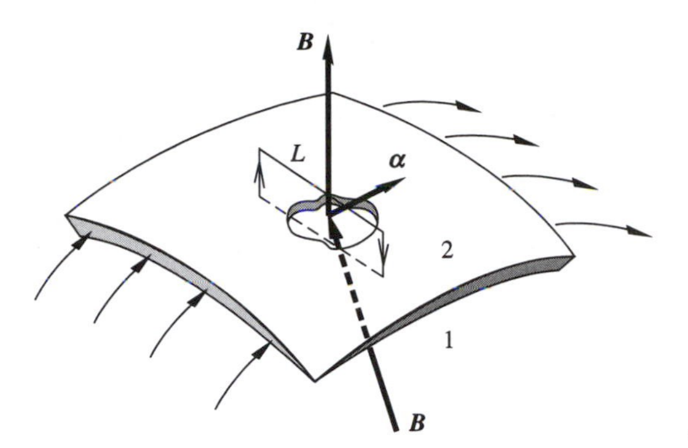

Figure 15-7 Refraction of a line of $\boldsymbol{B}$ crossing a current sheet.

2. Consider now path c in the figure. Remembering that $B_\rho = 0$ both inside and outside, and that $B_z = 0$ outside, we see that, *if there are N' turns/meter, $B_z s = \mu_0 N' I s$ and $B_z = \mu_0 N' I$.* The axial field inside, at points remote from the ends, is uniform.

EXAMPLE The Refraction of Lines of $\boldsymbol{B}$ at a Current Sheet

A thin conducting sheet carries a surface current density of $\boldsymbol{\alpha}$ amperes/meter, as in Fig. 15-7. In passing through the sheet, the lines of $\boldsymbol{B}$ bend as follows.

Since $\boldsymbol{\nabla} \cdot \boldsymbol{B} = 0$, the normal component of $\boldsymbol{B}$ is the same on the two sides: $B_{1n} = B_{2n}$.

Applying Ampère's circuital law to the path of length L shown in the figure,

$$B_{1t}L - B_{2t}L = \mu_0 \alpha L, \qquad B_{2t} = B_{1t} - \mu_0 \alpha. \tag{15-21}$$

A line of $\boldsymbol{B}$ therefore deflects in the clockwise direction for an observer looking in the direction of $\boldsymbol{\alpha}$.

We can also arrive at this result in another way. The magnetic flux density $\boldsymbol{B}$ results from the existence of a current in the sheet *and* to currents flowing elsewhere. According to Ampère's circuital law, the magnetic field of the sheet, just below the sheet, is $\mu_0 \alpha / 2$ and points left. Just above, the field is again $\mu_0 \alpha / 2$, but it points right. Adding this field to that of the other currents leads to tangential components that differ as above.

15.5 SUMMARY

The line integral of $\boldsymbol{A} \cdot d\boldsymbol{l}$ around a closed curve C is equal to the magnetic flux Λ linking C:

$$\oint_C \boldsymbol{A} \cdot d\boldsymbol{l} = \int_{\mathcal{A}} \boldsymbol{B} \cdot d\boldsymbol{\mathcal{A}} = \Lambda, \tag{15-3}$$

where $\mathcal{A}$ is the area of a surface bounded by C.

For *static* fields,

$$\nabla^2 \boldsymbol{A} = -\mu_0 \boldsymbol{J}, \tag{15-11}$$

$$\nabla \cdot \boldsymbol{A} = 0, \tag{15-15}$$

$$\nabla \times \boldsymbol{B} = \mu_0 \boldsymbol{J}. \tag{15-17}$$

Ampère's circuital law states that

$$\oint_C \boldsymbol{B} \cdot \boldsymbol{dl} = \mu_0 I, \tag{15-18}$$

where I is the net current that crosses any open surface bounded by the curve C, in the direction given by the right-hand screw rule.

PROBLEMS

15-1. (*15.1*) The vector potential inside a current-carrying conductor
Show that, inside a straight current-carrying conductor of radius R,

$$A = \frac{\mu_0 I}{4\pi}\left(1 - \frac{\rho^2}{R^2}\right),$$

if A is set equal to zero at $\rho = R$.

15-2. (*15.4*) Van de Graaff high-voltage generator
In a Van de Graaff generator, a charged insulating belt transports electric charge to the high-voltage electrode.
(a) Calculate the current carried by a 500-millimeter-wide belt driven by a 100-millimeter-diameter pulley that rotates at 60 revolutions/second, if $E = 2 \times 10^6$ volts/meter at the surface of the belt.
(b) Calculate the B close to the belt.

15-3. (*15.4*) The toroidal coil
Figure 15-8 shows a toroidal coil of square cross section. There are N turns, and the current is I.
Find (a) the azimuthal field along paths a and c, (b) B inside the toroid, and (c) the line integral of $\boldsymbol{B}$ along path d.

15-4. (*15.4*) $\boldsymbol{B}$ near a conducting sheet
A conducting sheet carries a surface current density of $\boldsymbol{\alpha}$ amperes/meter. There are no other currents in the vicinity.
(a) What is the value of $\boldsymbol{B}$, close to the sheet?
(b) How is $\boldsymbol{B}$ oriented with respect to $\boldsymbol{\alpha}$?
(c) A conducting body carries a high-frequency current that is confined near the surface. The surface current density is $\boldsymbol{\alpha}$ amperes/meter. Show that, in the air near

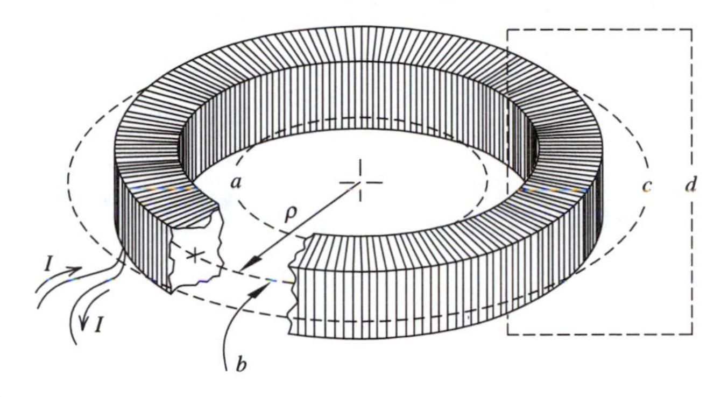

Figure 15-8

the conductor, $\boldsymbol{B} = \mu_0 \boldsymbol{\alpha} \times \hat{\boldsymbol{n}}$, where $\hat{\boldsymbol{n}}$ is a unit vector normal to the surface and pointing outward.

15-5. (*15.4*) The magnetic field near the axis of a circular loop

A circular loop carries a current I. Choose the axis of symmetry as the z-axis, and calculate B_ρ and B_z near the axis.

15-6. (*15.4*) Definition of μ_0

Show that the permeability of free space μ_0 can be defined as follows: If an infinitely long solenoid carries a current density $N'I$ of one ampere per meter, then the magnetic flux density in teslas inside the solenoid is numerically equal to μ_0.

15-7. (*15.4*) The average $\boldsymbol{B}$ over a sphere is equal to $\boldsymbol{B}$ at the center

Refer to Prob. 3-23 concerning the average $\boldsymbol{E}$ over a spherical volume.

Show that, in a region where there are no currents, the average $\boldsymbol{B}$ over a spherical volume is equal to the $\boldsymbol{B}$ at the center.

15-8. (*15.4*) The field of a short thick solenoid, compared to that of a long solenoid

The value of B at the center of a short, thick solenoid given in Prob. 14-7 can be written as $B = \mu_0 N' I g$, where $\mu_0 N' I$ is the field of a long solenoid. Find g.

CHAPTER

16

*MAGNETIC FIELDS III†

*Magnetic Materials

All atoms contain spinning electrons that give rise to magnetic fields through a strictly quantum-mechanical effect. It is our purpose to express these fields in macroscopic terms.

In dielectric materials, individual atoms or molecules can possess electric dipole moments that, when properly oriented, confer a net electric moment to a macroscopic body. Magnetic materials are analogous in that their atoms can act as magnetic dipoles that can also be oriented. The body is then said to be *magnetized.* Magnetic effects are weak in all but ferromagnetic substances.

The *magnetization* ***M*** is the magnetic moment per unit volume of magnetized material at a point. If there are N atoms per unit volume, each possessing a magnetic dipole moment ***m*** oriented in a given direction, then

$$\boldsymbol{M} = N\boldsymbol{m}. \tag{16-1}$$

†This chapter can be omitted without losing continuity.

The magnetization $\boldsymbol{M}$ in magnetic media corresponds to the polarization $\boldsymbol{P}$ in dielectrics. The unit of magnetization is the ampere per meter.

*16.1 TYPES OF MAGNETIC MATERIAL

There exist three main types of magnetic material.

1. All materials are *diamagnetic*. This magnetism originates from the fact that the application of an external magnetic field induces moments according to the Faraday induction law (Chapter 18). This effect is usually imperceptible, and it disappears upon removal of the external field.
2. In most atoms the magnetic moments resulting from the orbital and spinning motions of the electrons cancel. If the cancellation is not complete, the material is *paramagnetic*. Thermal agitation causes the individual moments to be randomly oriented, but the application of a magnetic field brings about a partial orientation.
3. In *ferromagnetic* materials such as iron, the magnetization can be orders of magnitude larger than in either diamagnetic or paramagnetic substances. This effect comes from electron spin.

These materials are partitioned into microscopic *domains*, as in Fig. 16-1, within which the spins are all spontaneously aligned in the same direction, even in the absence of an external field. In the unmagnetized state the spins of the various domains are randomly oriented, and the net macroscopic field is zero.

Upon application of an external magnetic field, those domains that happen to be oriented in the right direction grow at the expense of their neighbors by the migration

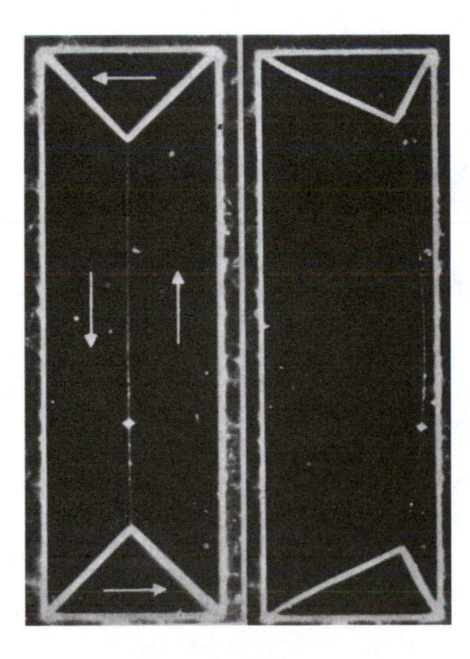

Figure 16-1 Domains in a single-crystal platelet of nickel-cobalt, 500 nanometers thick, 112 micrometers $\times$ 35 micrometers, before and after application of a vertical magnetic field pointing down. The sample is first polished and then covered with an aqueous suspension of Fe_3O_4. The iron oxide particles collect at the domain boundaries, where the field *gradient* is maximum (Tebble, 1969).

of the domain walls. Eventually, near saturation, all the domains are oriented in the imposed direction.

Ferromagnetic substances are grossly non-linear, lossy, and usually anisotropic. Magnetic fields involving such materials do not therefore lend themselves to rigorous mathematical analyses, but computer codes are available for performing numerical calculations and drawing field lines.

Recall Sec. 8.1.4 on ferro*electric* materials.

See the Example on eddy currents in Sec. 18.3.

*16.2 THE EQUIVALENT CURRENTS. THE BIOT-SAVART LAW REVISITED. THE DIVERGENCE OF *B*

Here is a simple-minded calculation of the magnetic field outside a magnetized body, say the cylinder of Fig. 16-2.

From the third Example in Sec. 14.4, the vector potential at a point P located at a distance r from a current loop of magnetic moment $\boldsymbol{m}$ is

$$\boldsymbol{A} = \frac{\mu_0}{4\pi}\frac{\boldsymbol{m} \times \hat{\boldsymbol{r}}}{r^2}. \tag{16-2}$$

The unit vector $\hat{\boldsymbol{r}}$ points in the direction of P, from the center of the loop, and r is large compared to the largest dimension of the loop. Then, for a volume v' of magnetized material,

$$\boldsymbol{A} = \frac{\mu_0}{4\pi}\int_{v'} \frac{\boldsymbol{M} \times \hat{\boldsymbol{r}}}{r^2}\, dv' = \frac{\mu_0}{4\pi}\int_{v'} \boldsymbol{M} \times \boldsymbol{\nabla}'\left(\frac{1}{r}\right) dv'. \tag{16-3}$$

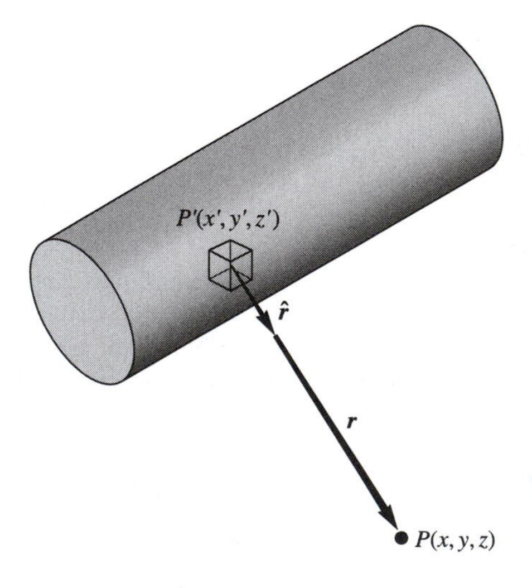

Figure 16-2 Element of volume inside a magnetized body and an external point P.

We have used Identity 15 from the back of the front cover. The volume of magnetized material is v', and its surface has an area $\mathcal{A}'$.

Then, from Identities 11 and 19,

$$\boldsymbol{A} = -\frac{\mu_0}{4\pi}\int_{v'}\left(\boldsymbol{\nabla}' \times \frac{\boldsymbol{M}}{r}\right)dv' + \frac{\mu_0}{4\pi}\int_{v'}\frac{\boldsymbol{\nabla}' \times \boldsymbol{M}}{r}\,dv' \tag{16-4}$$

$$= \frac{\mu_0}{4\pi}\int_{\mathcal{A}'}\frac{\boldsymbol{M} \times \hat{\boldsymbol{n}}}{r}\,d\mathcal{A}' + \frac{\mu_0}{4\pi}\int_{v'}\frac{\boldsymbol{\nabla} \times \boldsymbol{M}}{r}\,dv', \tag{16-5}$$

$\hat{\boldsymbol{n}}$ being the unit vector normal to the surface of area $\mathcal{A}$ of the magnetized material and pointing *outward.* We may omit the prime on the del that operates on $\boldsymbol{M}$, since that del clearly operates on the coordinates x', y', z' of the point where the magnetization is $\boldsymbol{M}$.

These expressions for $\boldsymbol{A}$ are all equivalent, but the last one lends itself to a simple physical interpretation. It is clear that the vector potential in the neighborhood of a piece of magnetized material is the same as if one had, instead, equivalent volume and surface current densities

$$\boldsymbol{J}_e = \boldsymbol{\nabla} \times \boldsymbol{M} \quad \text{and} \quad \boldsymbol{\alpha}_e = \boldsymbol{M} \times \hat{\boldsymbol{n}}. \tag{16-6}$$

Inside a magnetized material none of the above integrals diverge.

More generally,

$$\boldsymbol{A} = \frac{\mu_0}{4\pi}\int_{v'}\frac{1}{r}\left(\boldsymbol{J}_f + \frac{\partial \boldsymbol{P}}{\partial t} + \boldsymbol{\nabla} \times \boldsymbol{M}\right)dv', \tag{16-7}$$

where $\boldsymbol{J}_f$ is the current density of free charges and $\partial \boldsymbol{P}/\partial t$ is the polarization current density of Sec. 7.3.†

Thus a more general form of the Biot-Savart law (Sec. 14.2) is

$$\boldsymbol{B} = \frac{\mu_0}{4\pi}\int_{v'}\frac{(\boldsymbol{J}_f + \partial \boldsymbol{P}/\partial t + \boldsymbol{\nabla} \times \boldsymbol{M}) \times \hat{\boldsymbol{r}}}{r^2}\,dv'. \tag{16-8}$$

†The question is sometimes raised as to whether the term $\epsilon_0 \partial E/\partial t$ should be included under the integral sign. Then one would replace the polarization current density by the displacement current density $\partial D/\partial t$ (Sec 7.10). The term $\epsilon_0 \partial E/\partial t$ does *not* belong here because magnetic fields arise solely from the motion of charge. This is confirmed by the fact that one can calculate the field of a transmitting antenna in free space from the currents flowing in it, disregarding displacement currents in free space.

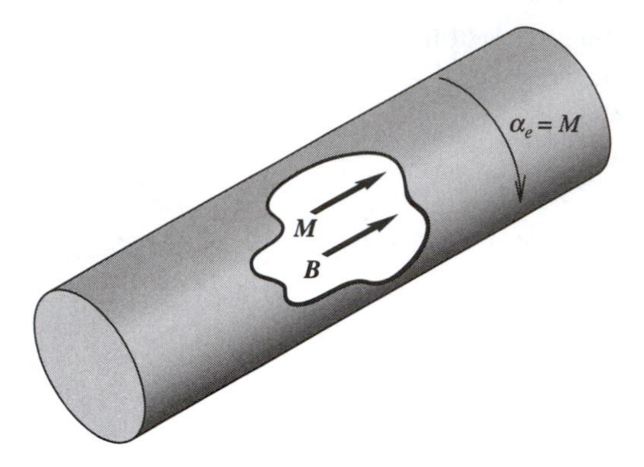

Figure 16-3 A uniformly magnetized rod acts as a solenoid carrying a surface current density $\alpha_e = M$.

*EXAMPLE A Uniformly Magnetized Rod

Suppose the magnetization $\boldsymbol{M}$ is uniform and parallel to the axis. (This is an idealized situation because the elementary dipoles orient themselves along $\boldsymbol{B}$, which is only approximately axial.) Since $\boldsymbol{M}$ is uniform, $\nabla \times \boldsymbol{M} = 0$, and there are no equivalent volume currents. Also, since $\boldsymbol{M}$ is parallel to the axis, the current density on the cylindrical surface is M, in the direction shown in Fig. 16-3, and there are no currents on the end faces. The rod therefore acts as a solenoid with $N'I = M$, where N' is the number of turns per meter. Observe that, inside, $\boldsymbol{B}$ and $\boldsymbol{M}$ point in the same general direction.

Since magnetic fields originate either in the macroscopic motion of charge or in equivalent currents, the relation

$$\boxed{\nabla \cdot \boldsymbol{B} = 0} \tag{16-9}$$

that we found in Sec. 14.3 applies even in the presence of magnetic materials. This is one of Maxwell's equations.

*16.3 THE MAGNETIC FIELD STRENGTH $\boldsymbol{H}$. THE CURL OF $\boldsymbol{H}$. AMPÈRE'S CIRCUITAL LAW REVISITED

In Sec. 15.4 we found that, for static fields in the absence of magnetic materials,

$$\nabla \times \boldsymbol{B} = \mu_0 \boldsymbol{J}_f. \tag{16-10}$$

Henceforth we shall use $\boldsymbol{J}_f$, instead of the unadorned $\boldsymbol{J}$, for the current density related to the motion of free charges.

In the presence of magnetized materials,

$$\nabla \times \boldsymbol{B} = \mu_0 (\boldsymbol{J}_f + \boldsymbol{J}_e). \tag{16-11}$$

This equation, of course, applies only in regions where the space derivatives exist, that is, inside magnetized materials, but not at their surfaces. Then

$$\nabla \times \boldsymbol{B} = \mu_0(\boldsymbol{J}_f + \nabla \times \boldsymbol{M}), \tag{16-12}$$

$$\nabla \times \left(\frac{\boldsymbol{B}}{\mu_0} - \boldsymbol{M}\right) = \boldsymbol{J}_f. \tag{16-13}$$

The vector within the parentheses is the *magnetic field strength*:

$$\boldsymbol{H} = \frac{\boldsymbol{B}}{\mu_0} - \boldsymbol{M}. \tag{16-14}$$

Both $\boldsymbol{H}$ and $\boldsymbol{M}$ are expressed in amperes/meter. Thus

$$\boldsymbol{B} = \mu_0(\boldsymbol{H} + \boldsymbol{M}) \tag{16-15}$$

and, even inside magnetized materials,

$$\nabla \times \boldsymbol{H} = \boldsymbol{J}_f \tag{16-16}$$

for *static* fields.

Let us integrate Eq. 16-16 over an open surface of area $\mathcal{A}$ bounded by a closed curve C:

$$\int_{\mathcal{A}} (\nabla \times \boldsymbol{H}) \cdot d\boldsymbol{\mathcal{A}} = \int_{\mathcal{A}} \boldsymbol{J}_f \cdot d\boldsymbol{\mathcal{A}}, \tag{16-17}$$

or, using Stokes's theorem on the left-hand side,

$$\oint_C \boldsymbol{H} \cdot d\boldsymbol{l} = I_f, \tag{16-18}$$

where I_f is the current of free charges linking C. The right-hand screw rule applies to the direction of integration and to the direction of I. Note that I_f does *not* include the equivalent currents. The term on the left is the *magnetomotance*.

This is a more general form of *Ampère's circuital law* of Sec. 15.4, in that it can serve to calculate $\boldsymbol{H}$ even in the presence of magnetic materials. It is rigorously valid, however, only for steady currents.

*16.3.1 Dielectric and Magnetic Materials Compared

Compare the above equation for $\boldsymbol{B}$ with the corresponding Eq. 7-21 for dielectrics,

$$\boldsymbol{E} = \frac{1}{\epsilon_0}(\boldsymbol{D} - \boldsymbol{P}). \tag{16-19}$$

Note the difference in sign: minus $\boldsymbol{P}$ instead of plus $\boldsymbol{M}$.

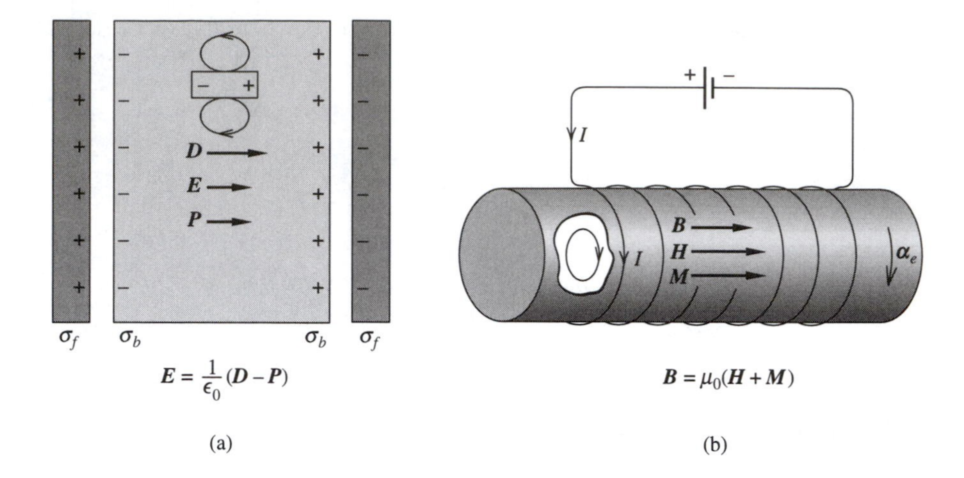

Figure 16-4 (a) Plane-parallel capacitor. The plates are insulated and carry fixed free surface charge densities σ_f. Introducing the dielectric *reduces* E by the factor of ϵ_r, neglecting edge effects. Note the orientation of the small electric dipole and its field. (b) Solenoid carrying a fixed current I. Introducing the magnetic core *increases* B by a factor of μ_r, neglecting end effects. Note the orientation of the small magnetic dipole and of its field.

Figure 16-4 illustrates the difference. In Fig. 16-4(a) the capacitor plates are insulated and carry free charges that are not affected by the presence of the dielectric, neglecting edge effects. So $D = \sigma_f$ is fixed. All three vectors $\boldsymbol{D}$, $\boldsymbol{E}$, $\boldsymbol{P}$ point to the right. Without the dielectric, E would be equal to D/ϵ_0. With the dielectric, E is *smaller* because the field of the bound charges *opposes* that of the free charges.

In Fig. 16-4(b), the coil applies a given $\boldsymbol{H}$. All three vectors $\boldsymbol{B}$, $\boldsymbol{H}$, $\boldsymbol{M}$ point to the right again. Without the magnetic core, $\boldsymbol{B} = \mu_0\boldsymbol{H}$. With the core, $\boldsymbol{B}$ is *larger* because the field of the equivalent currents *adds to* that of the free currents.

Remember that we are concerned here solely with the space- and time-averaged fields inside matter. Remember also that the similarity between the fields of electric and magnetic dipoles exists only at points remote from the dipoles. Closer in, the fields are totally different, as shown in Fig. 14-12.

*16.4 THE MAGNETIC SUSCEPTIBILITY χ_m AND THE RELATIVE PERMEABILITY μ_r

It is convenient to define a *magnetic susceptibility* χ_m such that†

$$\boldsymbol{M} = \chi_m \boldsymbol{H}. \tag{16-20}$$

†Compare with $\boldsymbol{P} = \chi_e \epsilon_0 \boldsymbol{E}$ (Sec. 7.8).

Then

$$\boldsymbol{B} = \mu_0(\boldsymbol{H} + \boldsymbol{M}) = \mu_0(1 + \chi_m)\boldsymbol{H} = \mu_0\mu_r \boldsymbol{H} = \mu\boldsymbol{H}, \tag{16-21}$$

where

$$\mu_r = 1 + \chi_m \tag{16-22}$$

is the *relative permeability* and

$$\mu = \mu_r\mu_0 \tag{16-23}$$

is the *permeability* of a material. Both χ_m and μ_r are pure numbers.

Thus

$$\boldsymbol{M} = \chi_m\frac{\boldsymbol{B}}{\mu} = \frac{\mu_r - 1}{\mu_r\mu_0}\boldsymbol{B}. \tag{16-24}$$

The magnetic susceptibility of purely diamagnetic materials is negative and of the order of 10^{-5}. In paramagnetic materials, χ_m varies from about 10^{-5} to 10^{-3}. The susceptibility of ferromagnetic substances can be as large as 10^6.

The equation $\boldsymbol{B} = \mu_0(\boldsymbol{H} + \boldsymbol{M})$ is general, but equations involving either μ_r or χ_m assume that the material is both isotropic and linear. In other words, they assume that $\boldsymbol{M}$ is proportional to $\boldsymbol{H}$ and in the same direction.

In ferromagnetic materials, $\boldsymbol{B}$ and $\boldsymbol{H}$ do not always point in the same direction, and when they do, μ_r can vary by orders of magnitude, depending on the value of $\boldsymbol{H}$ and on the previous history of the material (Sec. 16.6). In a permanent magnet, $\boldsymbol{B}$ and $\boldsymbol{H}$ point in roughly *opposite* directions.

*16.5 BOUNDARY CONDITIONS FOR *B* AND *H*

Both $\boldsymbol{B}$ and $\boldsymbol{H}$ obey boundary conditions at the interface between two media. We proceed as in Sec. 8.2. There are no surface currents on the interface.

Figure 16-5(a) shows a short, cylindrical Gaussian volume at an interface. From Gauss's law, the flux leaving through the top equals that entering the bottom and

$$B_{1n} = B_{2n}. \tag{16-25}$$

The normal component of $\boldsymbol{B}$ is therefore continuous across an interface.

Consider now Fig. 16-5(b). The small rectangular path pierces the interface. From the circuital law of Sec. 16.3, the line integral of $\boldsymbol{H} \cdot \boldsymbol{dl}$ around the path is equal to the current I linking the path. With the two long sides of the path infinitely close to the interface, I is zero and the tangential component of $\boldsymbol{H}$ is continuous across the interface:

$$H_{1t} = H_{2t}. \tag{16-26}$$

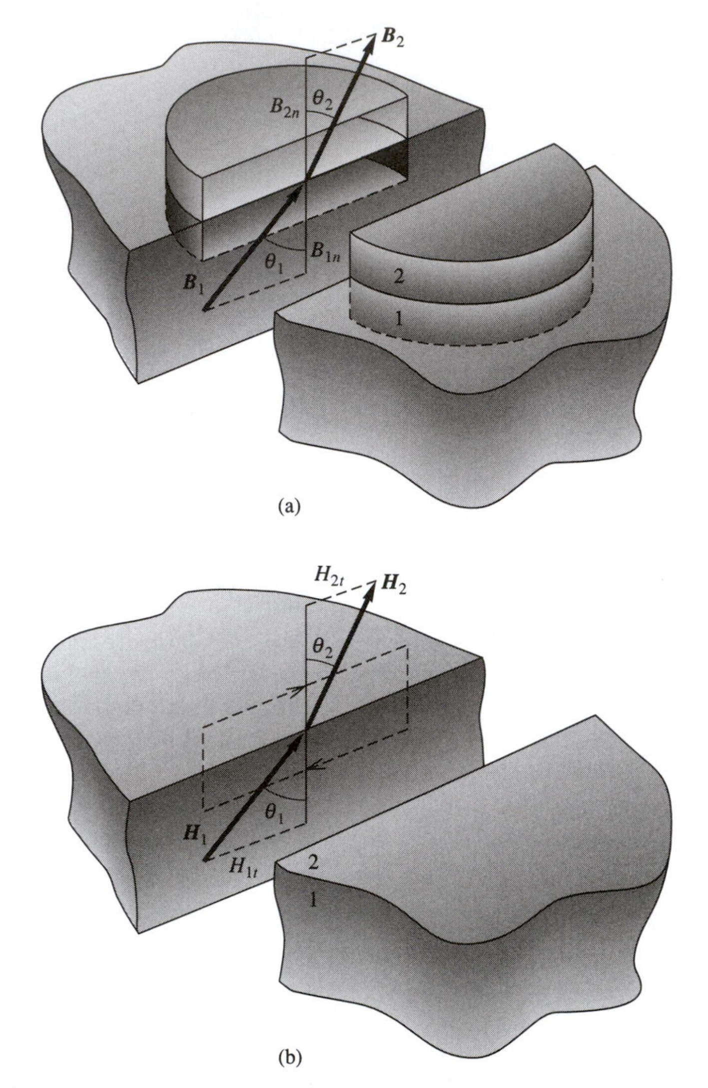

Figure 16-5 (a) Gaussian surface straddling the interface between media 1 and 2. The normal components of the $\boldsymbol{B}$'s are equal. (b) Closed path piercing the interface. The tangential components of the $\boldsymbol{H}$'s are equal, assuming that there are no surface currents.

Setting $\boldsymbol{B} = \mu\boldsymbol{H}$ for both media, the permeabilities being those that correspond to the actual fields, and assuming that the materials are isotropic, then the above two equations imply that

$$\frac{\tan\theta_1}{\tan\theta_2} = \frac{\mu_{r1}}{\mu_{r2}}. \tag{16-27}$$

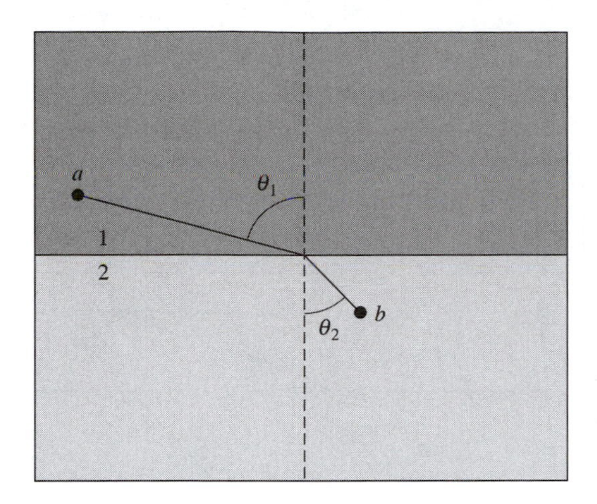

Figure 16-6 Line of ***B*** crossing the interface between linear and isotropic media 1 and 2. The permeability of medium 1 is larger than that of medium 2. Points a and b are at equal distances from the interface. The path in higher-permeability medium 1 is longer than the path in 2: the line of ***B*** "prefers" to be in the higher-permeability material.

We therefore have the following rule for linear and isotropic media: lines of ***B*** lie farther away from the normal in the medium possessing the larger permeability. In other words, the lines "prefer" to pass through the more permeable medium, as in Fig. 16-6. You will recall from Sec. 8.2.4 that we had a similar situation with dielectrics.

*16.6 B VS H IN FERROMAGNETIC MATERIALS

How can you observe the relation between H and B for a given sample of magnetic material? A simple-minded method would be to use a solenoid as in Fig. 16-4(b). Then H would be equal to the number of ampere-turns per meter on the solenoid, and B could be measured with a tesla-meter based on the Hall effect (Sec. 17.1), at an end face. The method is instructive, but it is impractical because of the large end-effects. Fortunately, sophisticated instruments are available commercially.

If we start with an unmagnetized sample, increasing H increases B but, after a while, saturation sets in and B increases no further. Figure 16-7 shows *magnetization curves* for several magnetic materials.

Refer now to Fig. 16-8. Start with an unmagnetized sample at O. Increasing H increases B along the curve ab. Reducing now the current I in the solenoid to zero, B decreases along bc. Then, reversing the sign of I, B decreases to zero at d. On further increasing the current in that direction, a point e, symmetric to point b, is reached. If I now decreases, then reverses and increases, point b is again reached. The closed curve $bcdefgb$ is a *hysteresis loop.*

Hysteresis loops are always of the same general shape, but they take many forms. They can be narrow as in the figure, or broad, or even rectangular, with nearly horizontal and vertical sides.

Soft magnetic materials possess a high permeability and a narrow hysteresis loop. They serve in electromagnets, transformers, motors, etc. *Hard* materials have broad loops and are mainly used for permanent magnets.

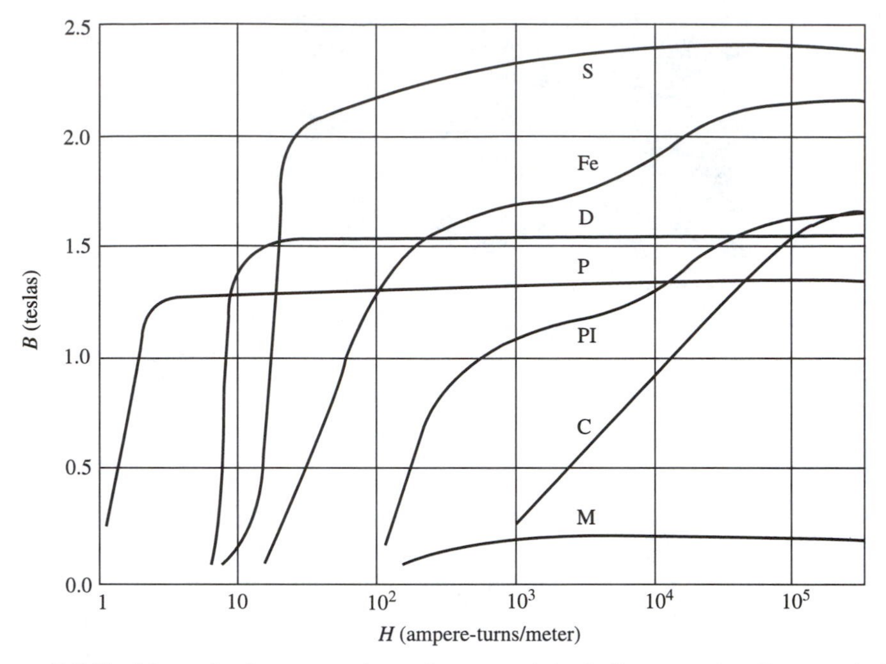

Figure 16-7 Magnetization curves for various materials: S, Supermendur; Fe, annealed pure iron; D, Deltamax; P, Permalloy; PI, powdered iron; C, gray cast iron; M, Monel.

It turns out that the energy dissipated in the material in going around a hysteresis loop is equal to the area of the loop expressed in tesla-ampere-turns per meter, or in weber-ampere-turns per cubic meter, or in joules per cubic meter.

One can demagnetize an object, for example a magnetic tape, by exposing it to a coil carrying an alternating current, and then gradually removing the object from the field, so as to describe smaller and smaller hysteresis loops.

*16.7 THE BAR MAGNET

As we saw in the example in Sec. 16.2, the $\boldsymbol{B}$ field of an idealized bar magnet that is magnetized uniformly parallel to its length is the same as that of a solenoid of identical dimensions and bearing a current density $N'I$ equal to the magnetization M. Figure 16-9 shows the general features of this field. Observe how the lines of $\boldsymbol{B}$ break at the cylindrical surface. However, the lines of $\boldsymbol{H}$ pierce the cylindrical surface undisturbed. The normal component of $\boldsymbol{B}$ and the tangential component of $\boldsymbol{H}$ are continuous at the cylindrical surface and at the ends, as in Sec. 16.5.

Outside, $\boldsymbol{H} = \boldsymbol{B}/\mu_0$, while inside $\boldsymbol{H} = \boldsymbol{B}/\mu_0 - \boldsymbol{M}$.

Observe that inside the magnet $\boldsymbol{B}$ and $\boldsymbol{H}$ point in approximately *opposite* directions.

Where is the point (H, B) on the hysteresis loop? Suppose you wind a coil all around an unmagnetized *ring* of magnetic material. You vary the current in the coil so as to go from point a on the hysteresis loop to point b, and then to c. At c, the current in the coil is zero, $\boldsymbol{H} = 0$, and $\boldsymbol{B} = \mu_0\boldsymbol{M}$.

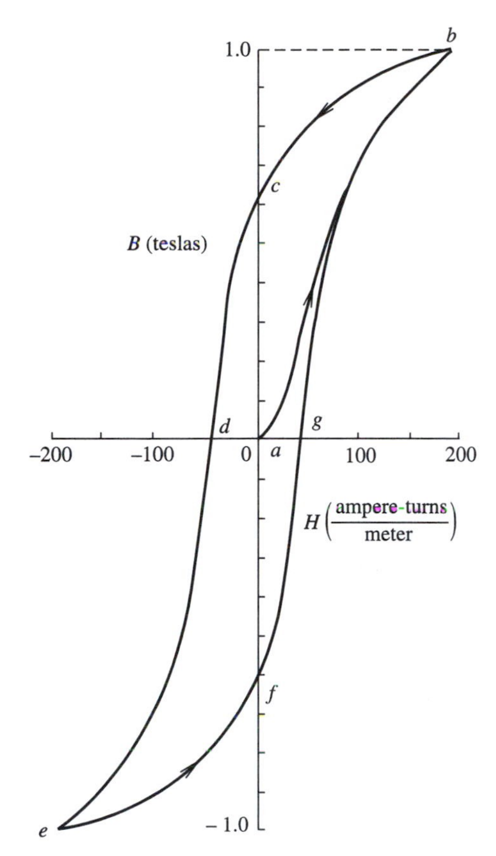

Figure 16-8 Magnetization curve *ab* and hysteresis loop *bcdefgb* for one type of transformer iron.

Now cut out a section of the ring as in Fig. 16-10 to form a bar magnet. From Ampère's circuital law, the line integral of $\boldsymbol{H} \cdot d\boldsymbol{l}$ around path C is zero, because there are no free currents linking C. Outside, $\boldsymbol{H}$ points in the same direction as $\boldsymbol{B}$, since $\boldsymbol{B} = \mu_0\boldsymbol{H}$. Upon removing a section of the ring as in Fig. 16-10, there appears inside

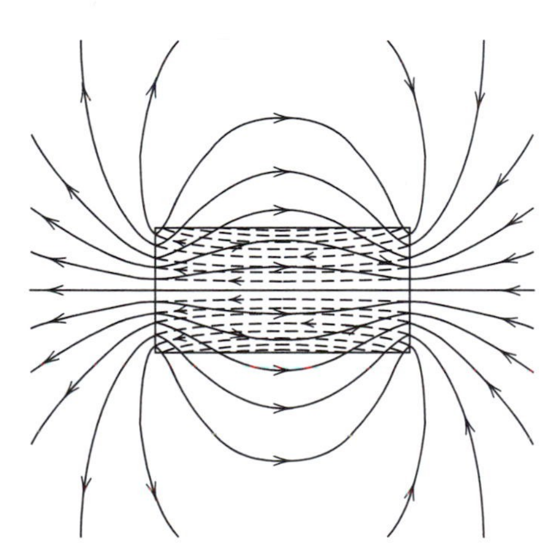

Figure 16-9 Lines of $\boldsymbol{H}$ (solid) for a uniformly magnetized cylinder, with $\boldsymbol{M}$ uniform, parallel to the axis, and pointing to the left. Lines of $\boldsymbol{B}$ are shown broken inside the magnet; outside they follow the lines of $\boldsymbol{H}$. This figure is identical to Fig. 7-4.

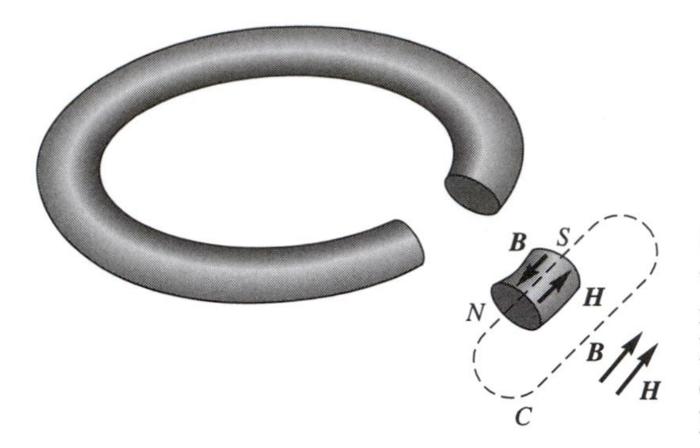

Figure 16-10 Thought experiment in which one cuts out a portion of a magnetized ring to obtain a bar magnet. Inside the bar, ***B*** and ***H*** point in opposite directions. Lines of ***B*** emerge from the North pole *N* and end at the South pole *S*.

an ***H*** that points in the direction opposite to ***B***. The operating point of a permanent magnet therefore lies in the second quadrant of the hysteresis loop.

The operating point of a long slender magnet is close to *c*, and that of a stubby magnet approaches *d*, on the hysteresis loop.

The field of a real bar magnet is not that simple. Since the magnetic moments of the individual atoms tend to align themselves with the ***B*** field, the magnetization ***M***, and hence the equivalent current density on the cylindrical surface, are weaker near the ends. Moreover, since ***M*** is not uniform, there are equivalent currents inside the magnet. The end faces also carry equivalent currents since $\boldsymbol{M} \times \hat{\boldsymbol{n}}$ at the faces is zero only on the axis. The net result is that there are "poles" near the ends of the magnet from which lines of ***B***, outside the magnet, appear to radiate in all directions. The poles are most conspicuous if the bar magnet is long and thin.

*16.7.1 The Bar Magnet and the Bar Electret Compared

It is instructive to compare the field of our uniformly magnetized cylinder of magnetic material, Fig. 16-11(a), with that of its electrical equivalent, the uniformly polarized cylinder of dielectric of Fig. 16-11(b). We discussed the field of the bar electret in the second example of Sec. 7.7.

The ***B*** field of the magnet is that of an equivalent solenoid, Fig. 16-11(c), that carries a surface current density $N'I = M$. Then, to find the ***H*** field, we use the relation

$$\boldsymbol{B} = \mu_0(\boldsymbol{H} + \boldsymbol{M}). \tag{16-28}$$

The ***H*** field of the magnet is the same as if one had magnetic pole densities $+M$ and $-M$ on the end faces.

The ***E*** field of the electret is that of the bound charges on the end faces, as in Fig. 16-11(d), or of a pair of parallel and oppositely charged disks carrying uniform surface charge densities $+P$ and $-P$. The ***D*** field then follows from

$$\boldsymbol{E} = \frac{1}{\epsilon_0}(\boldsymbol{D} - \boldsymbol{P}). \tag{16-29}$$

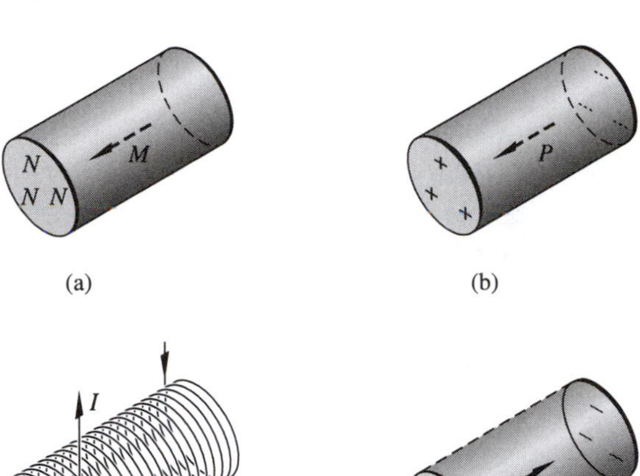

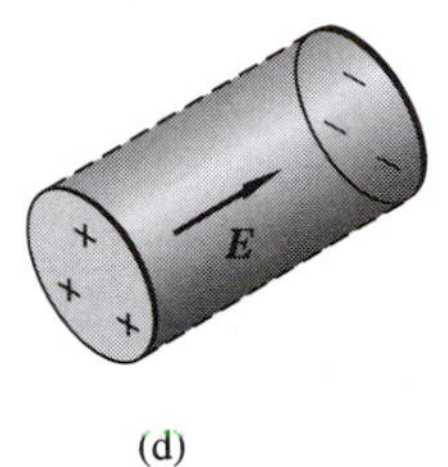

(c)

Figure 16-11 (a) Bar magnet. (b) Bar electret. (c) Solenoid whose $\boldsymbol{B}$ field is the same as that of the bar magnet. (d) Pair of electrically charged plates whose $\boldsymbol{E}$ field is the same as that of the bar electret.

Mathematically, the fields obey similar equations:

Electret		Magnet	
$\boldsymbol{E} = \dfrac{1}{\epsilon_0}(\boldsymbol{D} - \boldsymbol{P})$	(16-30)	$\boldsymbol{B} = \mu_0(\boldsymbol{H} + \boldsymbol{M})$	(16-31)
$\nabla \cdot \boldsymbol{D} = 0$	(16-32)	$\nabla \cdot \boldsymbol{B} = 0$	(16-33)
$\nabla \times \boldsymbol{E} = 0$	(16-34)	$\nabla \times \boldsymbol{H} = 0$	(16-35)

*16.8 SUMMARY

The *magnetization* $\boldsymbol{M}$ is the magnetic dipole moment per unit volume in magnetized material. The magnetic field, both outside and inside, is the same as if the material were replaced by its *equivalent volume* and *surface current densities*:

$$\boldsymbol{J}_e = \nabla \times \boldsymbol{M} \qquad \text{and} \qquad \boldsymbol{\alpha}_e = \boldsymbol{M} \times \hat{\boldsymbol{n}}. \tag{16-6}$$

The *divergence of* $\boldsymbol{B}$ is zero even in the presence of magnetic materials:

$$\boxed{\nabla \cdot \boldsymbol{B} = 0.} \tag{16-9}$$

The *magnetic field strength* $\boldsymbol{H}$ has the same dimensions as $\boldsymbol{M}$, and

$$\boldsymbol{H} = \frac{\boldsymbol{B}}{\mu_0} - \boldsymbol{M}, \qquad \boldsymbol{B} = \mu_0(\boldsymbol{H} + \boldsymbol{M}), \qquad (16\text{-}14), (16\text{-}15)$$

$$\nabla \times \boldsymbol{H} = \boldsymbol{J}_f, \tag{16-16}$$

where $\boldsymbol{J}_f$ is the current density attributable to free charges. This last equation applies only to *static* fields. It follows that, for *static* fields,

$$\oint_C \boldsymbol{H} \cdot \boldsymbol{dl} = I_f, \tag{16-18}$$

where C is a closed curve linked by the current I_f. This is *Ampère's circuital law* in a more general form.

As with dielectrics, it is convenient to define a *magnetic susceptibility* χ_m and a *relative permeability* μ_r as follows:

$$\boldsymbol{M} = \chi_m \boldsymbol{H}, \tag{16-20}$$

$$\boldsymbol{B} = \mu_0 \mu_r \boldsymbol{H}, \qquad \mu_r = 1 + \chi_m. \qquad \text{(16-21), (16-22)}$$

Then $\mu = \mu_r \mu_0$ is the *permeability.* In ferromagnetic materials the value of μ_r can vary by orders of magnitude, and even in sign, depending on the value of H and on the previous magnetic history of the sample.

At the interface between two media the normal component of $\boldsymbol{B}$ and the tangential component of $\boldsymbol{H}$ are continuous. There is *refraction of the lines of* $\boldsymbol{B}$ at an interface, the larger angle with the normal being in the medium with the larger permeability.

Ferromagnetic materials spontaneously magnetize over microscopic *domains* that can grow, one at the expense of another, upon application of a magnetic field.

If one plots B as a function of H for an initially unmagnetized sample, B first increases with H along the *magnetization curve.* Over a complete cycle of H one obtains a closed curve called a *hysteresis loop.* The area of the loop is equal to the energy lost per cubic meter of material and per cycle.

*PROBLEMS

16-1. (*16.2*) The number of Bohr magnetons per atom in iron

The magnetization M in iron can contribute as much as 2 teslas to B. If one electron contributes one Bohr magneton, how many electrons per atom, on average, can contribute to M? See Prob. 14-5.

16-2. (*16.2*) Magnetic field of the earth

A sphere of radius R is uniformly magnetized in the direction parallel to a diameter. The external field of such a sphere is similar to that of the earth.

Show that the equivalent surface current density is the same as if the sphere carried a uniform surface charge density σ and rotated at an angular velocity ω such that

$$\sigma \omega R = M.$$

16-3. (*16.2*) Equivalent currents

An iron torus whose major radius is much larger than its minor radius is magnetized in the azimuthal direction with M uniform.

What can you say about α_e?

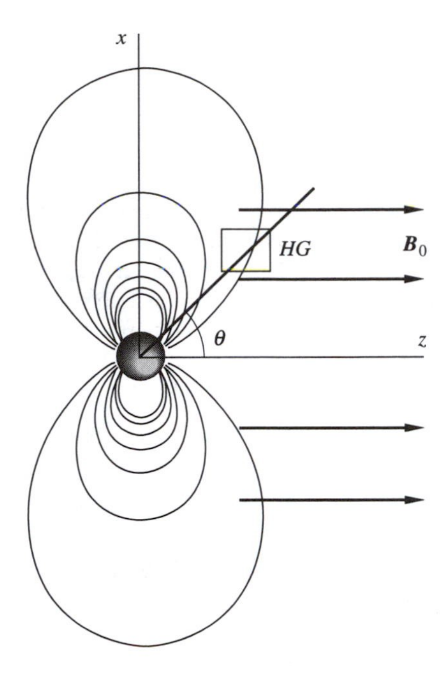

Figure 16-12 Set-up for measuring the magnetization M induced in a small spherical sample of material by a uniform field $\boldsymbol{B}_0$. The Hall generator HG is oriented so as to be sensitive to the x-component of the field originating in the sphere, and not to $\boldsymbol{B}_0$.

16-4. (*16.2*) Measuring M

The following method has been used to measure the magnetization M of a small sphere of material, induced by an applied uniform magnetic field $\boldsymbol{B}_0$ as in Fig. 16-12.

In the figure, HG is a Hall generator† with its current flowing parallel to $\boldsymbol{B}_0$.

Since Hall generators are sensitive only to magnetic fields perpendicular to their current flow, this one measures only the dipole field originating in the small sphere.

If the magnetization is M, the magnetic moment m of the sphere is $(4/3)\pi R^3 M$, and it turns out that the x-component of the dipole field of the sample at HG is

$$\frac{3\mu_0 m}{4\pi}\frac{\sin\theta\cos\theta}{r^3}.$$

Thus, at a given angle θ, this field decreases as $1/r^3$.

At what angle θ will the measured field be largest?

16-5. (*16.2*) Mechanical displacement transducer

One often wishes to obtain a voltage that is proportional to the displacement of an object from a known reference position.

If there are no magnetic materials nearby, one can fix to the object a small permanent magnet and measure its field with a Hall generator,† as in Fig. 16-13. The Hall element is sensitive only to the x-component of the magnetic field, and

$$B_x = \frac{3\mu_0 m}{4\pi}\frac{xz}{(x^2+z^2)^{5/2}}.$$

†See the third Example in Sec. 17.1.

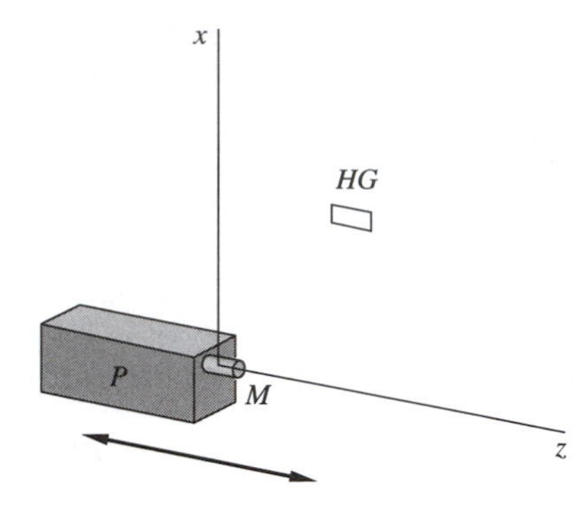

Figure 16-13 Hall generator HG used as a mechanical displacement transducer. The moving part P slides along the z-axis. The Hall generator is oriented so as to monitor the x-component of the magnetic field of the small permanent magnet M fixed to P.

(a) Draw a curve of $(xz)/(x^2 + z^2)^{5/2}$ as a function of z for $x = 100$ millimeters and for $z = -50$ to $+50$ millimeters.

(b) Over what range does B_x deviate by less than 5% from the tangent to the curve at $z = 0$?

The linear region is longer for larger values of x, but B_x decreases rapidly with x.

16-6. (*16.3*) The weber-ampere-turn

Show that one weber-ampere-turn is one joule.

16-7. (*16.3*) Toroidal coil with magnetic core

Compare the fields inside two similar toroidal coils, both carrying a current I, one with a non-magnetic core and the other with a magnetic core.

How are the equivalent currents on the magnetic core oriented with respect to those in the coil?

16-8. (*16.3*) The field of a disk magnet

A disk of iron of radius a and thickness s is magnetized parallel to its axis. Calculate B on the axis, outside the iron.

16-9. (*16.3*) The field inside a tubular magnet

You are asked to design a permanent magnet that would supply a magnetic field in a cylindrical volume about 20 millimeters in length and 20 millimeters in diameter. Someone suggests a tubular magnet, magnetized along its length, that would surround this volume. What do you think of this suggestion?

16-10. (*16.6*) The divergence of $\boldsymbol{H}$ is not always zero

Show that $\boldsymbol{\nabla} \cdot \boldsymbol{H}$ is not zero in a nonhomogeneous magnetic material.

16-11. (*16.6*) The field in a cavity inside magnetic material

Find B and H inside a thin, disk-shaped cavity whose axis is parallel to $\boldsymbol{B}$, inside magnetic material.

16-12. (*16.6*) The free and equivalent volume current densities

Show that, in a homogeneous, isotropic, and linear magnetic material, $\boldsymbol{J}_e = (\mu_r - 1)\boldsymbol{J}_f$.

16-13. (*16.6*) Current-carrying wire along the axis of an iron tube

A wire carrying a current I is situated on the axis of a hollow iron cylinder.

Find H, B, and M in the inner region, in the iron, and in the outer region, as well as the equivalent currents.

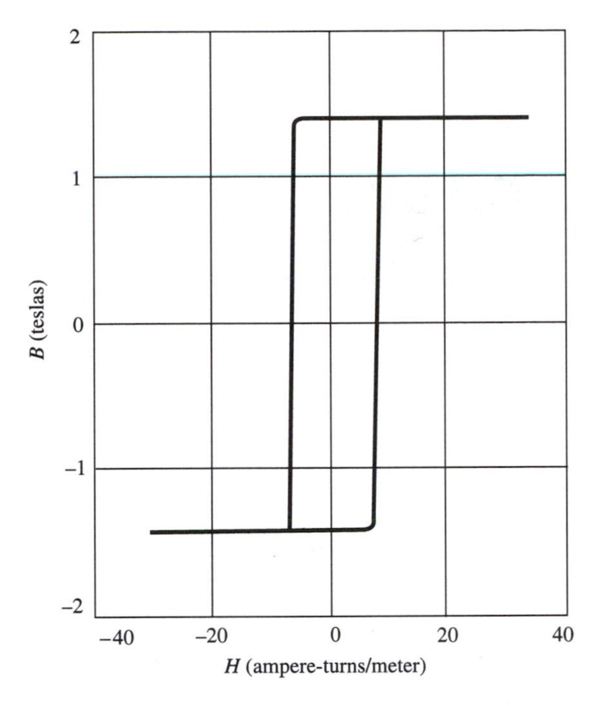

Figure 16-14

16-14. (*16.6*) Magnetization in terms of χ_m and B

Show that in a linear and isotropic magnetic medium,

$$M = \frac{\chi_m B}{\mu_0(1 + \chi_m)}.$$

16-15. (*16.6*) Power loss due to hysteresis

Figure 16-14 shows the hysteresis loop for a nickel-iron alloy called Deltamax.

What is the approximate value of the power dissipation per cubic meter and per cycle when the material is driven to saturation both ways?

16-16. (*16.6*) Correcting the trajectory of a satellite in orbit (?)

This problem concerns both magnetic materials and magnetic forces, which are discussed in Chapter 17.

Someone suggests the following device for exerting a force on a satellite in orbit. A loop of wire fixed to the satellite carries an electric current. Say the loop is rectangular. Since the earth's magnetic field in the satellite is essentially uniform, the net magnetic force exerted on the loop is zero. But if you shield one side with a high-permeability tube, there will be a net magnetic force exerted on the coil.

What is your opinion? Refer to Problem 16-13.

See Problem 18-3 on the magnetic braking of satellites.

CHAPTER

17

MAGNETIC FIELDS IV

Magnetic Forces on Charges and Currents

The main manifestation of the magnetic force in everyday life is the electric motor. But there are magnetic forces everywhere in nature: the magnetic fields of the earth, of the sun, of the solar system, of the stars, of the galaxies, and of Space all have their origin in self-excited dynamos that operate through the $Q\boldsymbol{v} \times \boldsymbol{B}$ force that we are studying here. See the Example on sunspots in Sec. 17.5. The earth's magnetic field is only slightly modified by magnetic ores.

It is sometimes stated in textbooks that the magnetic force can do no work. The electric motor is, of course, proof to the contrary. We shall see here how magnetic forces can be put to work.

We study magnetic forces in two chapters: the present one and Chapter 20. That is because we require the $Q\boldsymbol{v} \times \boldsymbol{B}$ force now, but we must defer a more general discussion of magnetic forces to a later stage.

17.1 THE LORENTZ FORCE

Experiments show that the force exerted on a particle of charge Q moving in a vacuum at an instantaneous velocity $\boldsymbol{v}$ in a region where there exist both an electric and a magnetic field is

$$\boldsymbol{F} = Q[\boldsymbol{E} + (\boldsymbol{v} \times \boldsymbol{B})]. \tag{17-1}$$

This is the *Lorentz force*†. This equation is valid, even if $\boldsymbol{v}$ approaches the speed of light. The variables $\boldsymbol{E}$, $\boldsymbol{B}$, and $\boldsymbol{v}$ can be space- and time-dependent, but they all concern the *same* reference frame.

The term $Q\boldsymbol{v} \times \boldsymbol{B}$ is the *magnetic force*. Observe that the magnetic force is perpendicular to $\boldsymbol{v}$. It can therefore change the direction of $\boldsymbol{v}$, but it cannot alter its magnitude, nor can it alter the kinetic energy of the particle. It can nonetheless do useful work, as we shall see.

EXAMPLE The Pinch Effect

Let us calculate the net outward force on an ion at the *periphery* of an ion beam. First, there is an *outward* electric force because of electrostatic repulsion between the ions. But there is also a magnetic force. Let us calculate both forces. Figure 17-1 shows, schematically, an ion beam.

The ions have a charge Q, a mass m, and a velocity $\boldsymbol{v}$. The beam radius is a, and the beam current is I. Assume that the charge density is uniform. (That is an approximation; the current density is in fact a Gaussian function of the radius.)

We first calculate $\boldsymbol{E}$ and $\boldsymbol{B}$. The linear charge density is I/v. Then, at the periphery,

$$E = \frac{I/v}{2\pi\epsilon_0 a}, \qquad B = \frac{\mu_0 I}{2\pi a}. \tag{17-2}$$

1. *The electric force*. Now refer to Fig. 17-1. An ion of charge Q at the periphery feels a force $Q\boldsymbol{E}$ because of the electrostatic repulsion between ions of the same sign. Note that, if the ions are positive, Q is positive, and $\boldsymbol{E}$ points *outward*, so that the force $Q\boldsymbol{E}$ points *outward*. If the ions are negative, or if we have an electron beam, Q is negative and $\boldsymbol{E}$ points *inward*, so that $Q\boldsymbol{E}$ again points *outward*.

†The "Lorentz" force was discovered by Heaviside in 1889, six years before Lorentz discovered it. See Appendix C.

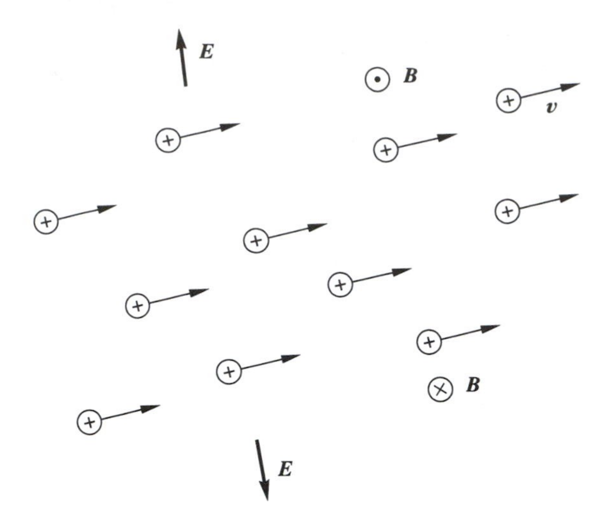

Figure 17-1 Schematic diagram of a positive ion beam, with all the ions moving at the same velocity $\boldsymbol{v}$. $\boldsymbol{E}$ points radially outward, and $\boldsymbol{B}$ is azimuthal.

2. *The magnetic force*. An ion of charge Q at the periphery moves at a speed $\boldsymbol{v}$ in the magnetic field $\boldsymbol{B}$. Thus it is subjected to a force $Q\boldsymbol{v} \times \boldsymbol{B}$. If Q is positive, $\boldsymbol{B}$ points as in the figure and the force points *inward*. If Q is negative, then $\boldsymbol{B}$ changes direction and $Q\boldsymbol{v} \times \boldsymbol{B}$ again points *inward*.
3. *The net force*. So the net *outward* force on an ion at the periphery is

$$F = eE - evB = e\frac{I/v}{2\pi\epsilon_0 a} - ev\frac{\mu_0 I}{2\pi a} \tag{17-3}$$

$$= \frac{eI}{2\pi\epsilon_0 av}\left(1 - \frac{v^2}{c^2}\right). \tag{17-4}$$

We have substituted $1/c^2$ for $\epsilon_0\mu_0$, from Eq. 13-17. So, if $v^2 \ll c^2$, the repulsive electric force predominates.

Note that the electric force at the radius r inside the beam depends only on the charge inside r. Similarly, the magnetic force at r depends only on the current inside r.

If you now cancel the repulsive electric force by adding electrons to a beam of positive charges, say protons, the magnetic force acts alone and the beam *pinches*. This phenomenon is easy to observe in accelerators. Residual gas in the path of the beam ionizes on impact and forms low-energy ions and electrons. Those ions drift away, while the electrons remain trapped in the beam. This is *gas focusing*.

EXAMPLE The van Allen Belts

The van Allen belts illustrated in Fig. 17-2 were discovered by James van Allen (1914–) in 1958. They are clouds of charged particles that surround the earth between a few hundred kilometers and up to about six terrestrial radii. The particles are mostly

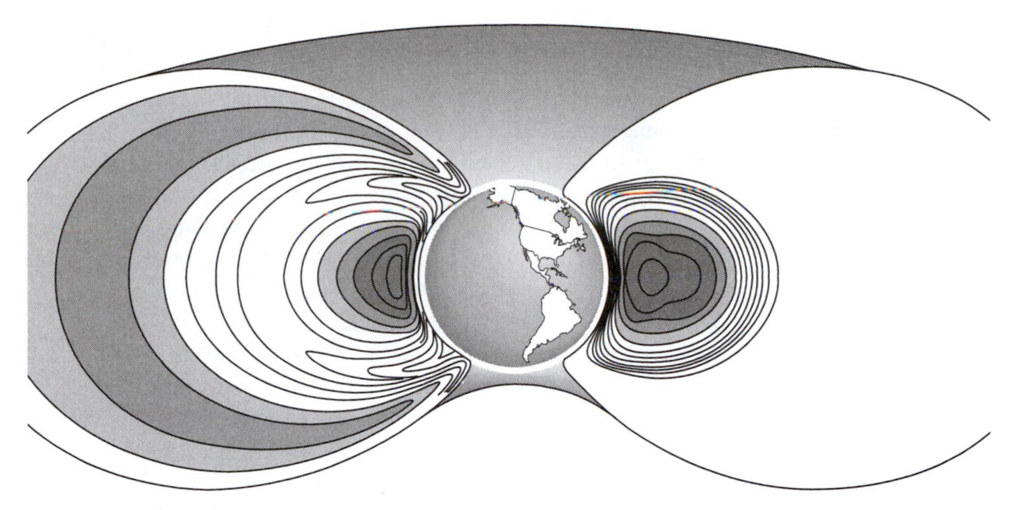

Figure 17-2 The van Allen radiation belts. Right half: protons of energy larger than 15 million electron-volts. Left half: electrons of energy larger than 0.5 million electron volts. Darker shading shows regions of more intense fluxes of trapped particles. (Source, *McGraw-Hill Encyclopedia of Science and Technology,* 8th edition, 1997.)

protons and electrons, but there are also helium, oxygen, and other ions. The protons and electrons originate from the solar wind (see the Example in Sec. 3.4.1) and from the decay of cosmic ray high-energy neutrons.

The particles are trapped in the earth's magnetic field by the $Q\boldsymbol{v} \times \boldsymbol{B}$ force and spiral around field lines. As they approach either the North or the South magnetic pole, B increases, the magnetic field lines become closer and closer, and the particles turn back: the regions of the poles act as *magnetic mirrors*. The particles turn back along more or less the same path, a bit more eastward in the case of the electrons, and a bit more westward in the case of the positive ions. So the particle clouds rotate about the magnetic axis of the earth. The clouds shun the poles, and their axis of symmetry is that of the earth's dipolar magnetic field, which forms an angle of about 11 degrees with respect to the axis of rotation. The positive ion belt is closer in than the electron belt, as in Fig. 17-2.

Since the particles have enough energy to penetrate several centimeters of aluminum, they can affect living organisms and electronic instruments inside satellites, and solar panels.

Other planets that have magnetic fields, for example Jupiter, have van Allen belts.

EXAMPLE The Hall Effect

The Hall effect was discovered by Edwin H. Hall (1855–1938) when he was 24, in the course of working on his PhD dissertation in 1879. The Hall effect is best known for its use in probes for measuring B, but it also has many other uses. The Hall effect is a straightforward application of the Lorentz force, like the magnetohydrodynamic generator of the next example, the electric motor, and so many other devices.

Hall probes have innumerable applications. They can measure B's from about 0.0001 to 14 teslas, and can be used for steady or alternating fields up to about 3 kilohertz. Some probes are as thin as 1 millimeter. Hall probes serve for vehicle detection, and, with a permanent magnet fixed to a moving object, they serve for position sensing, and for angular position sensing. Some probes are submersible and can serve for *borehole logging*, measuring B as a function of depth, as an indication of the nature of the sediments. Triaxial probes measure the three orthogonal components of $\boldsymbol{B}$.

Figures 17-3 (a) and (b) show bars of conducting material connected to voltage sources V_i and carrying longitudinal currents I. External coils apply transverse magnetic fields $\boldsymbol{B}$. In both cases the charge carriers drift longitudinally.

In (a) the charge carriers are holes (which are positive) that drift toward the right. The holes also drift downward because the $Q\boldsymbol{v} \times \boldsymbol{B}$ force points down. This tendency of charge carriers to drift sideways in a transverse magnetic field is called the *Hall effect*. Assume that the voltage V_H is measured with a high-impedance voltmeter. The downward drift of the holes continues until the transverse electric field E_t between the top and bottom electrodes cancels the transverse $\boldsymbol{v} \times \boldsymbol{B}$ field. The voltage V_H is called the *Hall voltage*. Figure (c) shows the six vectors involved.

In (b), the charge carriers are electrons (which are negative) that drift toward the left. The $Q\boldsymbol{v} \times \boldsymbol{B}$ force again points downward, and now the Hall voltage V_H has the opposite sign. Figure 17-3(d) shows the corresponding six vectors.

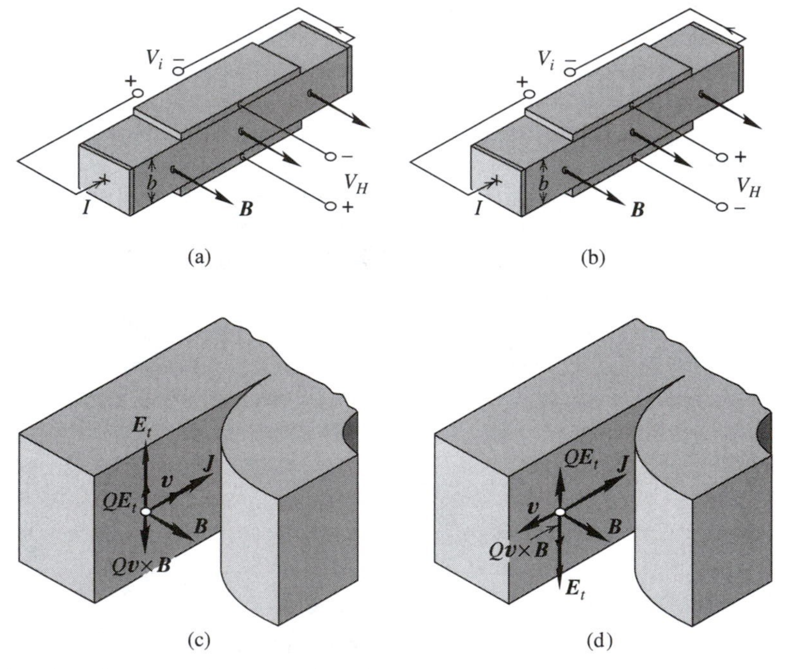

Figure 17-3 The Hall effect in semiconductors. (a) In p-type material the charge carriers are holes, and the Hall voltage is as shown. (b) In n-type material the carriers are electrons, and the Hall voltage has the opposite polarity. (c) In p-type material the transverse forces $Q\boldsymbol{E}_t$ and $Q\boldsymbol{v} \times \boldsymbol{B}$ are equal and opposite. (d) The transverse forces in n-type material.

It is a simple matter to calculate the voltage V_H if we disregard end effects, and assume that the externally applied $\boldsymbol{B}$ is both uniform and much larger than that of the current I of the figure. Once the transverse drift has subsided,

$$Q\frac{V_H}{b} = QvB, \tag{17-5}$$

where b is the sample thickness as in the figure, and v is the longitudinal drift velocity. Then

$$V_H = vBb. \tag{17-6}$$

If the mobility of the carriers is $\mathcal{M}$ (Sec. 4.5.2) and the longitudinal electric field strength is E_{long}, then

$$V_H = E_{\text{long}}\mathcal{M}Bb. \tag{17-7}$$

The voltage is proportional to the mobility.

Note that in the figures the charge carriers tend to drift downward, whether they are positive or negative. Thus the polarity of V_H depends on the sign of the carriers.

The mobilities of semiconductors being larger than those of good conductors by orders of magnitude, Hall devices invariably make use of semiconductors. Some devices are microscopic and form part of integrated circuits.

EXAMPLE The Magnetohydrodynamic Generator

The *magnetohydrodynamic generator* (Messerle, 1995) is an industrial-scale application of the Hall effect. It converts part of the kinetic energy of a hot gas directly into electric energy. Figure 17-4 shows the principle of operation. A hot gas enters on the left at a velocity of the order of 1000 meters/second. It contains a salt such as K_2CO_3 that ionizes readily at high temperature, forming positive ions and electrons. The temperature can range from 2000 to 3000 kelvins, and the conductivity is about 100 siemens/meter. (The conductivity of copper is 5.8×10^7 siemens/meter.) The magnetic field is supplied by superconducting coils.

Positive ions curve downward, electrons curve upward, and the resulting current I flows through the load resistance R. This establishes an electric field $\boldsymbol{E}$ as in Fig. 17-4. The gas remains neutral.

One obvious advantage of the MHD generator is that it comprises no moving parts, except for the gas. Another is that it can operate at such a high input temperature that the overall thermodynamic efficiency

$$\mathscr{E} = \frac{T_{in} - T_{out}}{T_{in}} \tag{17-8}$$

can reach 60% if the hot output gas feeds a conventional turbine generator. The fuel is either coal, oil, or natural gas. The thermodynamic efficiency of conventional thermal plants is about 38%.

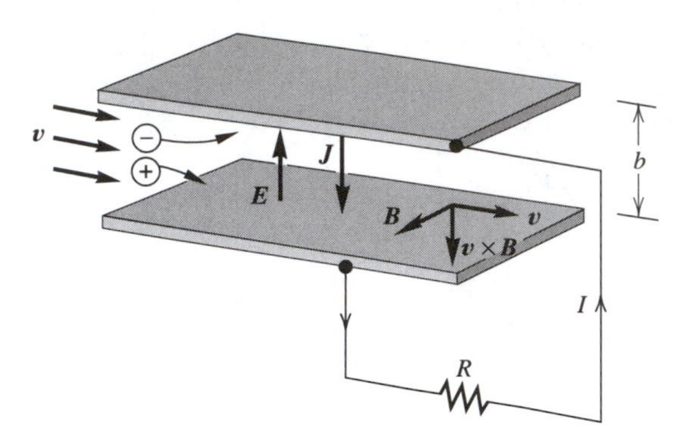

Figure 17-4 Schematic diagram of a magnetohydrodynamic (MHD) generator. Part of the kinetic energy of a very hot gas injected on the left at a velocity v is converted directly into electric energy. The magnetic field $\boldsymbol{B}$ is that of a pair of superconducting coils situated outside the chamber and not shown. The moving ions deflect either up or down, according to their signs.

MHD generators have been the object of much development work, in particular in the former USSR and in the USA, but apparently commercial applications do not yet exist (1999).

In MHD generators, part of the kinetic energy associated with the bulk motion of the gas becomes electric energy in the following way. The magnetic force on a charged particle is normal to the velocity and has no effect on its kinetic energy. The function of the magnetic forces is to compel the positive particles to go to the positive electrode and the negative particles to go to the negative electrode. So the ions and the electrons both move uphill, so to speak, under the action of the forces shown in Fig. 17-5, and slow down. Since the gas pressure is above atmospheric pressure, the mean-free path between collisions is infinitesimal and the charged particles are embedded in the gas. Thus, slowing the particles slows the gas and runs down its bulk kinetic energy.

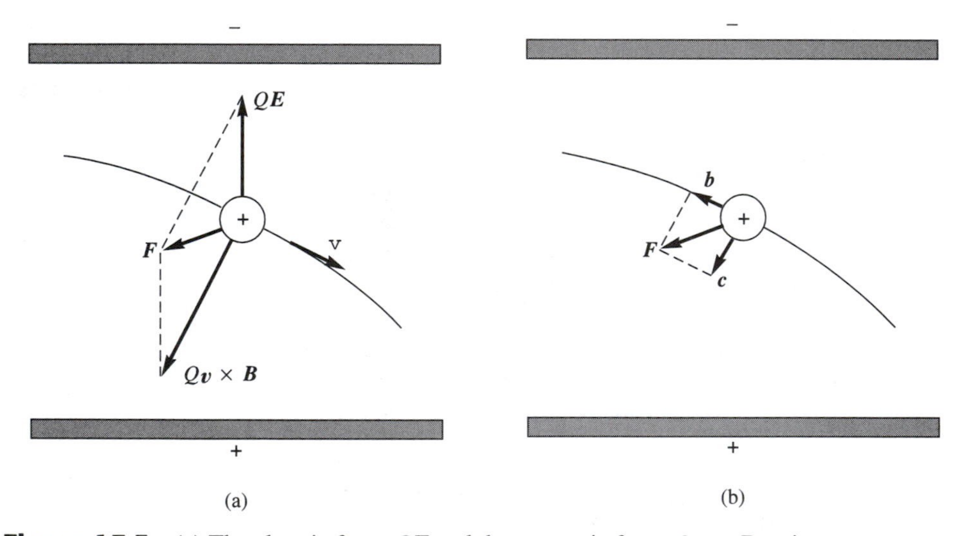

Figure 17-5 (a) The electric force $Q\boldsymbol{E}$ and the magnetic force $Q\boldsymbol{v} \times \boldsymbol{B}$ acting on a positively charged ion in the MHD generator of Fig. 17-4. The sum $\boldsymbol{F}$ of those two forces points downward and to the left. (b) The force $\boldsymbol{F}$ has two components: a tangential braking force $\boldsymbol{b}$ and a normal centripetal force $\boldsymbol{c}$.

Ideally, the charged particles should arrive at the electrodes at zero velocity. They arrive, in fact, at a finite velocity, the electrodes heat up, and only part of the kinetic energy of the *particles* becomes electric energy. An even smaller fraction of the bulk kinetic energy of the *gas* serves to generate electric energy. Also, some of the kinetic energy associated with the transverse motion of the charged particles only increases the random thermal energy of the gas.

Suppose that $\boldsymbol{E}$, $\boldsymbol{B}$, and the particle velocity $\boldsymbol{v}$ are uniform inside the generator. Assume that $\boldsymbol{v}$ is horizontal. These are crude approximations.

The Lorentz force $Q(\boldsymbol{E} + \boldsymbol{v} \times \boldsymbol{B})$ acts on a charge Q as if the electric field strength were $\boldsymbol{E} + \boldsymbol{v} \times \boldsymbol{B}$. So, for a gas of conductivity σ,

$$|\boldsymbol{J}| = |\sigma(\boldsymbol{E} + \boldsymbol{v} \times \boldsymbol{B})| = \sigma(vB - E) = \sigma\left(vB - \frac{V}{b}\right), \tag{17-9}$$

$$I = \mathscr{A}J = \mathscr{A}\sigma\left(vB - \frac{IR}{b}\right), \tag{17-10}$$

where I and R are as in Fig. 17-4, and where $\mathscr{A}$ is the area of one of the electrodes. Solving,

$$I = \frac{vBb}{b/(\sigma\mathscr{A}) + R}. \tag{17-11}$$

Observe that vBb is the open-circuit output voltage and that $b/(\sigma\mathscr{A})$ is the resistance of the gas in the transverse direction.

The output voltage is IR. After cross multiplication,

$$V = IR = vBb - \frac{Ib}{\sigma\mathscr{A}}. \tag{17-12}$$

The voltage V decreases linearly with I.

Now let us look into the efficiency. We are not prepared to calculate the efficiency with which the bulk kinetic energy of the gas becomes electric energy. However, we can compare the Joule losses in the load resistance R to those inside the gas. So let us define the efficiency as

$$\mathscr{E} = \frac{\text{Joule losses in } R}{\text{Joule losses in } R + \text{Joule losses in the gas}} \tag{17-13}$$

$$= \frac{I^2R}{I^2R + I^2b/(\sigma\mathscr{A})} = \frac{R}{R + b/(\sigma\mathscr{A})}. \tag{17-14}$$

From the above expressions for I and for IR,

$$\mathscr{E} = R\frac{I}{vBb} = \frac{IR}{vBb} = 1 - \frac{I}{\sigma\mathscr{A}vB}. \tag{17-15}$$

The efficiency is therefore equal to unity when $I = 0$, or when $R \longrightarrow \infty$. It is equal to zero when $I = \sigma\mathscr{A}vB$, or when $R = 0$, $V = 0$.

17.2 THE MAGNETIC FORCE ON A CURRENT-CARRYING WIRE

A stationary wire of cross section $\mathcal{A}$ carries a current I in a region where there exists a magnetic field $\boldsymbol{B}$ originating elsewhere. The wire contains N charge carriers per cubic meter drifting at a velocity $\boldsymbol{v}$, each one of charge Q.

An element of length $\boldsymbol{dl}$ of the wire contains $\mathcal{A}N dl$ charge carriers. Then the magnetic force on $\boldsymbol{dl}$ is

$$\boldsymbol{dF} = \mathcal{A}N dl Q\boldsymbol{v} \times \boldsymbol{B} = (\mathcal{A}NQv)\,\boldsymbol{dl} \times \boldsymbol{B} = I\boldsymbol{dl} \times \boldsymbol{B}, \tag{17-16}$$

since I is equal to the charge contained in a length v of the wire.

The magnetic force per unit length on a wire bearing a current I is therefore $\boldsymbol{I} \times \boldsymbol{B}$.

Now we have calculated the force on the charge carriers. How is this force transmitted to the wire? As we saw in Sec. 4.5.1, the mean-free-path of conduction electrons between collisions is only of the order of 100 atomic diameters. The electrons thus stumble along, continuously colliding with the crystal lattice and pushing it sideways. So the force exerted on the conduction electrons is also exerted on the wire. The same phenomenon occurs in solid state devices where the mean-free-paths are much longer, and where the charge carriers are positive.

This sideways force can do work. Imagine an idealized case where I is constant and $\boldsymbol{B}$ is both constant and uniform. The wire moves parallel to itself by a distance $\boldsymbol{ds}$. Then the work done by the magnetic force is

$$I(\boldsymbol{dl} \times \boldsymbol{B}) \cdot \boldsymbol{ds}.$$

This simple-minded "thought experiment" illustrates the way in which electric motors work.

The cases of the above MHD generator and of the Faraday disk discussed below show other examples of a magnetic force doing work.

The total magnetic force exerted on a closed circuit C carrying a current I in a magnetic field $\boldsymbol{B}$ is

$$\boldsymbol{F} = I \oint_C \boldsymbol{dl} \times \boldsymbol{B}. \tag{17-17}$$

If $\boldsymbol{B}$ is uniform, the net magnetic force $\boldsymbol{F}$ is zero.

17.3 THE MAGNETIC FORCE BETWEEN TWO CLOSED CIRCUITS

We saw above that the magnetic force exerted on a stationary circuit carrying a current I is I times the line integral of $\boldsymbol{dl} \times \boldsymbol{B}$. Then, applying the Biot-Savart law of Sec. 14.2, the magnetic force exerted *by* a current I_a *on* a current I_b as in Fig. 17-6, is given by

$$\boldsymbol{F}_{ab} = I_b \oint_b \boldsymbol{dl}_b \times \frac{\mu_0}{4\pi} I_a \oint_a \frac{\boldsymbol{dl}_a \times \hat{\boldsymbol{r}}}{r^2} = \frac{\mu_0}{4\pi} I_a I_b \int_a \int_b \boldsymbol{dl}_b \times \frac{\boldsymbol{dl}_a \times \hat{\boldsymbol{r}}}{r^2}, \tag{17-18}$$

where r is the distance between $\boldsymbol{dl}_a$ and $\boldsymbol{dl}_b$, and $\hat{\boldsymbol{r}}$ points from $\boldsymbol{dl}_a$ to $\boldsymbol{dl}_b$.

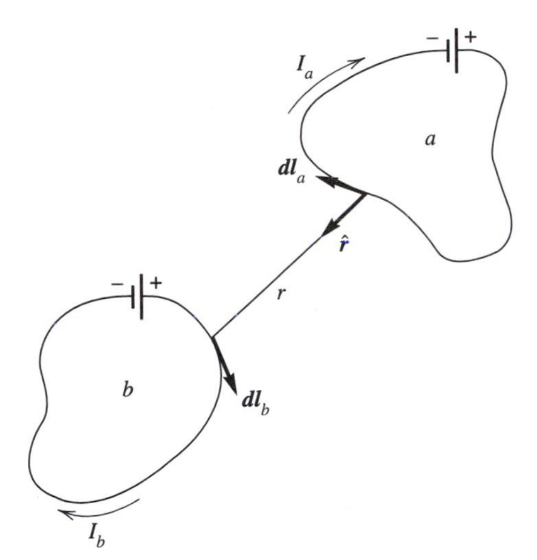

Figure 17-6 Two closed circuits a and b.

The fact that $\boldsymbol{dl}_a$ and $\boldsymbol{dl}_b$ do not play symmetrical roles in this integral is disturbing. The asymmetry appears to indicate that $\boldsymbol{F}_{ab} \neq -\boldsymbol{F}_{ba}$, which is contrary to Newton's law.† That impression is false. We can transform the double line integral to a symmetrical one and show that $\boldsymbol{F}_{ab} = -\boldsymbol{F}_{ba}$ as follows. First, we expand the triple vector product:

$$\boldsymbol{dl}_b \times (\boldsymbol{dl}_a \times \hat{\boldsymbol{r}}) = \boldsymbol{dl}_a(\boldsymbol{dl}_b \cdot \hat{\boldsymbol{r}}) - \hat{\boldsymbol{r}}(\boldsymbol{dl}_a \cdot \boldsymbol{dl}_b). \tag{17-19}$$

Now

$$\oint_a \oint_b \frac{\boldsymbol{dl}_a(\boldsymbol{dl}_b \cdot \hat{\boldsymbol{r}})}{r^2} = \oint_a \boldsymbol{dl}_a \oint_b \frac{\boldsymbol{dl}_b \cdot \hat{\boldsymbol{r}}}{r^2}. \tag{17-20}$$

The second integral on the right is zero for the following reason. It is the ordinary integral of dr/r^2, with identical upper and lower limits, because circuit b is closed, by hypothesis. So the double integral on the left is zero, and

$$\boldsymbol{F}_{ab} = -\frac{\mu_0}{4\pi} I_a I_b \oint_a \oint_b \hat{\boldsymbol{r}} \frac{\boldsymbol{dl}_a \cdot \boldsymbol{dl}_b}{r^2}. \tag{17-21}$$

It follows that $\boldsymbol{F}_{ab} = -\boldsymbol{F}_{ba}$ because $\hat{\boldsymbol{r}}$ points toward the circuit on which the force acts.

The above double line integral is usually difficult to calculate analytically. We shall find more useful expressions in Chap. 20.

†Action is equal to reaction only at speeds much less than the speed of light.

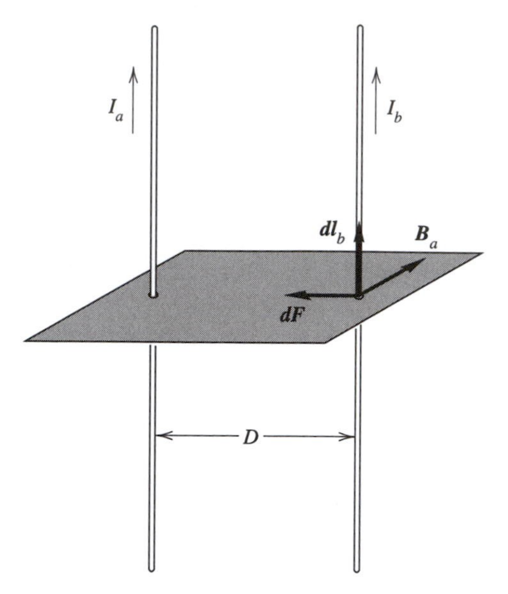

Figure 17-7 Two long parallel wires bearing currents I_a and I_b. The force is attractive when the currents flow in the same direction.

EXAMPLE The Force Between Two Parallel Currents

We can calculate the force per unit length between two long parallel wires bearing currents as in Fig. 17-7, without having to carry out the integration. At the position of I_b, B_a is $\mu_0 I_a/(2\pi D)$ in the direction shown in the figure. See the first example in Sec. 14.4. The force on a unit length of wire b is thus

$$F' = B_a I_b = \frac{\mu_0 I_a I_b}{2\pi D}. \tag{17-22}$$

The force is attractive if the currents flow in the same direction and repulsive otherwise. The force is usually negligible.

See Fig.17-8.

17.3.1 The Definitions of μ_0, the Ampere, the Coulomb, and ϵ_0

As stated previously in Sec. 14.2,

$$\mu_0 \equiv 4\pi \times 10^{-7} \qquad \text{weber/ampere-meter.} \tag{17-23}$$

If $I_a = I_b = I$ amperes, then the force per unit length is

$$F' = \frac{2 \times 10^{-7} I^2}{D} \qquad \text{newtons/meter.} \tag{17-24}$$

This equation defines the *ampere*.

The definition of the *coulomb* follows: it is the charge carried by a current of 1 ampere during 1 second.

So the definitions of μ_0, I, and Q are arbitrary and related.

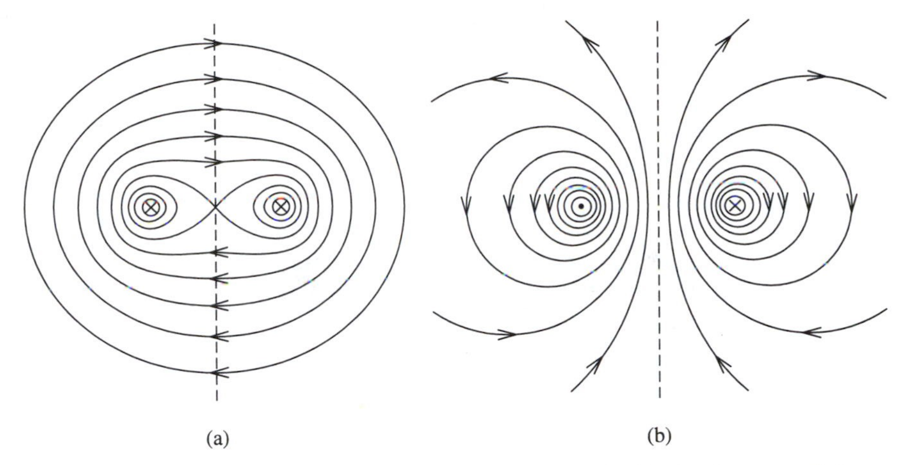

Figure 17-8 (a) Magnetic field lines for the vector sum of the fields of two parallel *currents* flowing in the *same* direction. We have chosen the lines so that the magnetic flux between any two neighboring lines is the same. Lines closer to the wires are not shown because they are too close together. The force between the wires is attractive: magnetic field lines "are under tension." (b) Magnetic field lines for two parallel *currents* flowing in *opposite* directions. The force between the wires is repulsive: magnetic field lines "repel laterally." These lines are identical to the equipotentials of Fig. 3-11 because, in a two-dimensional field, a line of $\boldsymbol{B}$ is a line of constant A (Prob. 14-13), and both V and A vary as 1 over the distance to the wire.

But Coulomb's law relates the force of attraction between two electric charges to their magnitudes, and force is defined in mechanics. Coulomb's law must therefore involve a constant of proportionality whose value must be measured.

We could, in principle, deduce the value of ϵ_0 from the measurement of F, the Q's, and r in Coulomb's law. However, this would not make much sense because none of those measurements can be very accurate. Instead, we use the fact that

$$\epsilon_0 = \frac{1}{\mu_0 c^2}, \tag{17-25}$$

where c, the speed of light, is defined to 9 significant figures. Thus

$$\epsilon_0 = 8.85418782 \times 10^{-12} \qquad \text{farad/meter.} \tag{17-26}$$

17.4 THE MAGNETIC FORCE ON A VOLUME DISTRIBUTION OF CURRENT

Now how can one calculate a magnetic force if the current flows, not through a thin wire, but through a finite volume? Consider a small element of volume of length $\boldsymbol{dl}$ parallel to $\boldsymbol{J}$ and of cross-section $d\mathcal{A}$, as in Fig. 17-9. The magnetic force exerted on the element is

$$\boldsymbol{dF} = (J d\mathcal{A})\boldsymbol{dl} \times \boldsymbol{B} = \boldsymbol{J} \times \boldsymbol{B}\, dv, \tag{17-27}$$

since $J d\mathcal{A}\boldsymbol{dl} = \boldsymbol{J} dv$, as in Sec. 14.2.

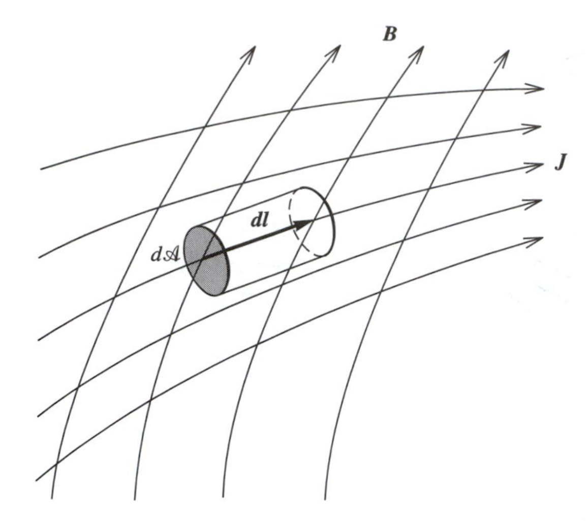

Figure 17-9 Element of volume $dv = dl\,d\mathcal{A}$ in a current distribution $\boldsymbol{J}$.

The force per unit volume is therefore

$$\boldsymbol{F}' = \boldsymbol{J} \times \boldsymbol{B}. \tag{17-28}$$

The total magnetic force on a current distribution that occupies a volume v is thus

$$\boldsymbol{F} = \int_v \boldsymbol{J} \times \boldsymbol{B}\, dv. \tag{17-29}$$

17.4.1 Magnetic Pressure

Magnetic pressure is an important concept, both for the design of solenoids, as we shall see below, and for the study of natural magnetic fields in plasmas such as those in the sun and in Space. See the Example on sunspots below.

A long solenoid has N' turns per meter and carries a current I. So the current per unit length is $N'I$. The winding has a thickness a. Inside the solenoid, from Sec. 15.4, $\boldsymbol{B}$ is axial and uniform, and

$$B = \mu_0 N' I. \tag{17-30}$$

Within the winding, the *average* B has the same value, divided by 2: $\mu_0 N' I/2$.

Now, within the winding, $\boldsymbol{B}$ is axial and $\boldsymbol{J}$ is azimuthal. So the magnetic force density $\boldsymbol{J} \times \boldsymbol{B}$, within the winding, points radially *outward*, as if there existed a gas pressure inside the solenoid.

What is the magnitude of the pressure? The force density F'', per unit *volume*, is given by JB, where

$$J = \frac{N'I}{a} = \frac{B}{\mu_0 a}. \tag{17-31}$$

So the average force density per unit *volume* is

$$F''_{av} = JB_{av} = \frac{B}{\mu_0 a}\frac{B}{2} = \frac{1}{a}\frac{B^2}{2\mu_0}. \tag{17-32}$$

Just inside the winding, the force density F' per unit *area*, or the *magnetic pressure,* is thus

$$p_{mag} = F' = aF''_{av} = \frac{B^2}{2\mu_0}. \tag{17-33}$$

Note that p_{mag} is proportional to the *square* of B. Note also that we have assumed that $B = 0$ outside the solenoid.

Is this a large pressure? Say $B = 1$ tesla. That is a large field, about the field between the pole-pieces of a powerful electromagnet. Then

$$p_{mag} = \frac{1}{2\mu_0} = \frac{1}{2 \times 4\pi \times 10^{-7}} \approx 4 \times 10^5 \text{ pascals} \approx 4 \text{ atmospheres}. \tag{17-34}$$

Keep clear! The solenoid will explode, unless it was designed carefully. More usual magnetic fields of, say, 0.1 tesla or less, pose no problems: for example, if $B \approx 0.1$ tesla, $p_{mag} \approx 0.04$ atmosphere.

At this time (1999), 100-tesla solenoids are being planned both in Toulouse (France) and in Los Alamos (United States). What will the magnetic pressure be? Forty thousand atmospheres! Clearly, the containment of such a huge pressure is a major engineering challenge.

We return to magnetic pressure in Sec. 20.5.

17.5 OHM'S LAW FOR MOVING CONDUCTORS

Conductors that move in magnetic fields are everywhere. There are, of course, electric motors, generators, and a host of other magnetic devices. But also, the liquid conducting shell of the earth's core, and the solar, stellar, and space plasmas convect in their own magnetic fields. Certain features, such as sunspots, are self-excited dynamos that generate magnetic fields, while elsewhere the fluid convects in the ambient field. See the next Examples.

Recall first that a charge Q inside a *stationary* conductor and in a field $\boldsymbol{E}$ is subjected to a force $Q\boldsymbol{E}$, and that $\boldsymbol{J} = \sigma\boldsymbol{E}$. This is Ohm's law (Sec. 4.3).

If now a conductor *moves* at a velocity $\boldsymbol{v}$ in superimposed $\boldsymbol{E}$ and $\boldsymbol{B}$ fields, the Lorentz force on a charge Q inside is

$$\boldsymbol{F} = Q(\boldsymbol{E} + \boldsymbol{v} \times \boldsymbol{B}), \tag{17-35}$$

as in Eq. 17-1, and

$$\boldsymbol{J} = \sigma(\boldsymbol{E} + \boldsymbol{v} \times \boldsymbol{B}). \tag{17-36}$$

This is Ohm's law for moving conductors.

Note the following points. (1) This latter equation applies at a given time and at a given point inside the conductor. All five variables are measured with respect to the *same* fixed frame of reference, and can be functions of the coordinates and of the time. (2) We have here a conduction current density, or a current of free charges. Then we could add a subscript f on $\boldsymbol{J}$. (3) The $\boldsymbol{B}$ is the *net* magnetic field, which is the externally applied field plus the field of $\boldsymbol{J}$. This equation neglects the *convection current*, whose density is

$$\boldsymbol{J}_{conv} = \rho \boldsymbol{v}, \tag{17-37}$$

where ρ is the space charge density of Eq. 17-54, and $\boldsymbol{v}$ is the velocity of the conductor. However, the convection current is negligible as long as $v^2 \ll c^2$ (Lorrain, 1990).

It is usually stated that, if the conductivity σ tends to infinity in Eq.17-36, then $\boldsymbol{E}$ tends to $-(\boldsymbol{v} \times \boldsymbol{B})$. That is incorrect for the simple reason that both $\boldsymbol{E}$ and $\boldsymbol{v} \times \boldsymbol{B}$ are independent of σ (Lorrain, 1990; Lorrain, McTavish, and Lorrain, 1998). Indeed, increasing σ increases $\boldsymbol{J}$ proportionately.

EXAMPLE The Faraday Disk, or Homopolar Generator

The *Faraday disk*, or *homopolar generator* (Corson, 1956; Lorrain, 1990) is a low-voltage, high-current device. It comprises a copper disk that rotates in an axial magnetic field, with contacts on the axle and on the periphery, as in Fig. 17-10. The $\boldsymbol{v} \times \boldsymbol{B}$ field is radial and drives a current between the axle and the periphery of the disk. One application is the generation of the large currents required for purifying metals by electrolysis on an industrial scale. Superconducting coils provide the axial $\boldsymbol{B}$.

We assume that the electric field in the disk is purely radial. This requires contacts all around the axle and around the periphery, which is the usual configuration.

Refer to Fig. 17-11. The $\boldsymbol{v} \times \boldsymbol{B}$ field points *outward*, and $\boldsymbol{E}$ points *inward*. From Eq. 17-36,

$$\boldsymbol{J} = \sigma[E + (\boldsymbol{v} \times \boldsymbol{B})] = \sigma(E + vB)\hat{\boldsymbol{\rho}}, \tag{17-38}$$

where E is a *negative* quantity, as in the figure.

Refer to the lower half of the disk. (1) An electron of charge $-e$ at the radius ρ below the axle has an azimuthal velocity $v = \omega\rho$. (2) It is therefore subjected to the radial *inward* magnetic force $-evB$. (3) The electron is also subjected to the radial *outward* electric force $-eE$ (E is a negative quantity), which is smaller than the inward magnetic force: $|E| < |vB|$. (4) Its *inward* radial drift velocity is therefore $\mathcal{M}(E + vB)$, where $\mathcal{M}$ is the mobility of a conduction electron in copper.† This radial drift is smaller than v by many orders of magnitude. (5) This radial drift gives an azimuthal braking force $-e\mathcal{M}(E + vB)B$ on a conduction electron.

1. The *magnitude* of the azimuthal braking force per unit volume at the radius ρ is thus

†See Sec. 4.5.2. The drift velocity of a charge carrier is equal to $\mathcal{M}$ times the field. Here, the equivalent electric field is $E + vB$.

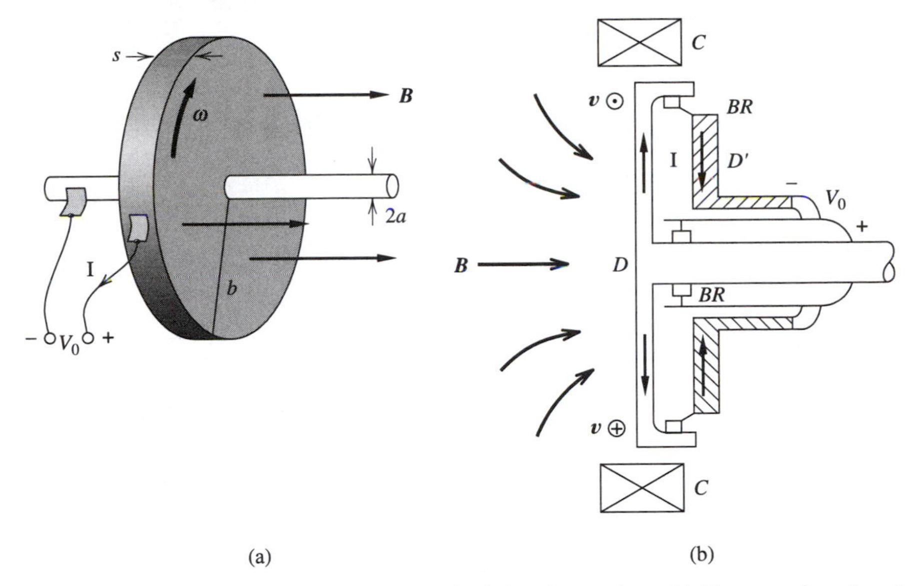

Figure 17-10 Homopolar generator. (a) Principle of operation. (b) Cross section showing the coil C, the disk D, the brush contacts BR, and the terminals V_0. The torque on the rotating disk D is equal and opposite to that on the fixed disk D'. Moreover, the field of the current through D and D' is azimuthal. Thus there is no force on coil C. This is important because C is superconducting and enclosed in a cryostat.

$$F' = eN\mathcal{M}(E + vB)B, \tag{17-39}$$

where N is the number density of the conduction electrons. Now, from Sec. 4.5.2, $Ne\mathcal{M}$ is equal to the conductivity σ, so that

$$F' = \sigma(E + vB)B, \tag{17-40}$$

and $\boldsymbol{F}' = \boldsymbol{J} \times \boldsymbol{B}$, as expected. Recall again that E is a negative quantity.

This force is appreciable, despite the smallness of the drift speed and of $\mathcal{M}$, because of the huge value of the product Ne: in Sec. 4.5.1 we calculated that $Ne \approx 10^{10}$ coulombs per cubic meter in copper!

The braking force acts *on the charge carriers*, but their mean-free-paths between collisions with the crystal lattice is so short, as we saw in Sec. 4.5.1, that the braking force also acts *on the disk.*

Observe that the magnetic braking force does work against whatever device rotates the disk, despite the fact that the force $-e(\boldsymbol{v} \times \boldsymbol{B})$ does not affect the speed of a conduction electron.

2. *The current I.* Call the disk thickness s. Then from Eq. 17-38,

$$I = 2\pi\rho s\sigma(E + vB). \tag{17-41}$$

If $I = 0$, then $|E| = |vB|$. What if $I \neq 0$? Call the load resistance connected between the output terminals R_{load}, and the output voltage V_0. Then

$$V_0 = IR_{load}. \tag{17-42}$$

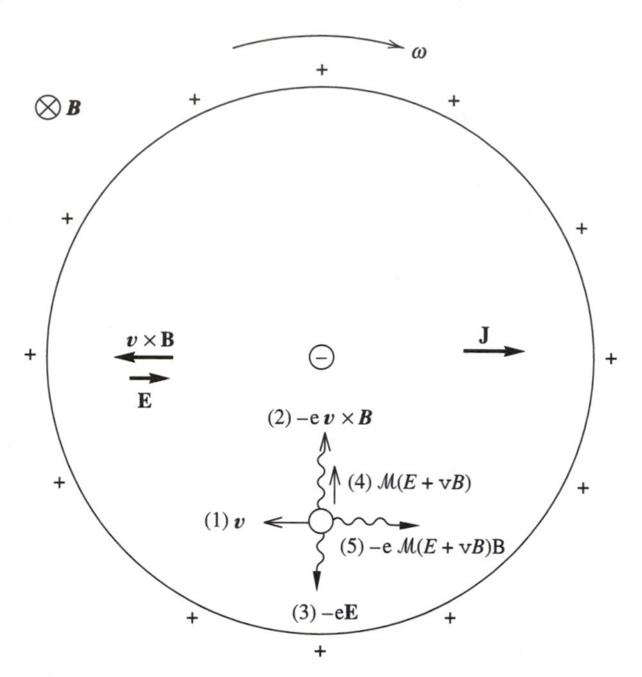

Figure 17-11 Faraday disk: $\boldsymbol{v} \times \boldsymbol{B}$ and $\boldsymbol{E}$ are radial, and $\boldsymbol{B}$ is axial. With the polarities shown, the periphery of the disk is positive, and the axle negative. A conduction electron has an azimuthal velocity $v = \omega r$, the product $\omega r B$ is larger than E, and current flows radially outward. Arrows (1) to (5): the velocity components (straight arrows) and the forces (wavy arrows) on a conduction electron. See the main text.

Call V the potential at the radius ρ on the disk, a the radius of the axle, and b the radius of the disk. From Sec. 3.4, since $\boldsymbol{E} = -\boldsymbol{\nabla} V$ points *inward*, *V increases* with increasing radius and

$$V_0 = \int_b^a \boldsymbol{E} \cdot \boldsymbol{d\rho}. \tag{17-43}$$

Thus

$$IR_{load} = V_0 = \int_b^a \left(-\ \omega \rho B + \frac{I}{2\pi\sigma s \rho} \right) d\rho \tag{17-44}$$

$$= \omega B \left(\frac{b^2 - a^2}{2} \right) - \frac{I}{2\pi\sigma s} \ln \frac{b}{a}. \tag{17-45}$$

Then

$$I\left[R_{load} + \frac{\ln(b/a)}{2\pi\sigma s} \right] = \omega B \left(\frac{b^2 - a^2}{2} \right). \tag{17-46}$$

The second term in the bracket on the left is the resistance of the disk R_{disk} between the radii a and b, as in the Example in Sec. 4.3. So

$$I = \frac{\omega B(b^2 - a^2)}{2(R_{load} + R_{disk})}. \tag{17-47}$$

Let us check. The disk current I is proportional to the angular speed ω of the disk and to B; that is reasonable. Also, $I = 0$ if R_{load} is infinite, or if $b = a$, or if $B = 0$, all of which makes sense.

3. What about the *centrifugal force*? Let us see. The centrifugal force on a conduction electron is $m\omega^2\rho$, while the magnetic force is $e\omega\rho B$. Then

$$\frac{m\omega^2\rho}{e\omega\rho B} = \frac{m}{e}\frac{\omega}{B} = \frac{9.1 \times 10^{-31}}{1.6 \times 10^{-19}}\frac{\omega}{B} \approx 6 \times 10^{-12}\frac{\omega}{B}. \tag{17-48}$$

The ratio is independent of ρ. If, for example, $\omega = 1000$ radians/second and $B = 1$ tesla, the centrifugal force is completely negligible. No sensible combination of ω and B can make the centrifugal force important.

4. *The value of E* is particularly interesting. Rewriting Eq. 17-41,

$$E = -\omega\rho B + \frac{I}{2\pi\sigma s\rho}. \tag{17-49}$$

Let us calculate the divergence of $\boldsymbol{E}$. In cylindrical coordinates†,

$$\boldsymbol{\nabla} \cdot \boldsymbol{E} = \frac{1}{\rho}\frac{\partial}{\partial\rho}(\rho E) = -2\omega B. \tag{17-50}$$

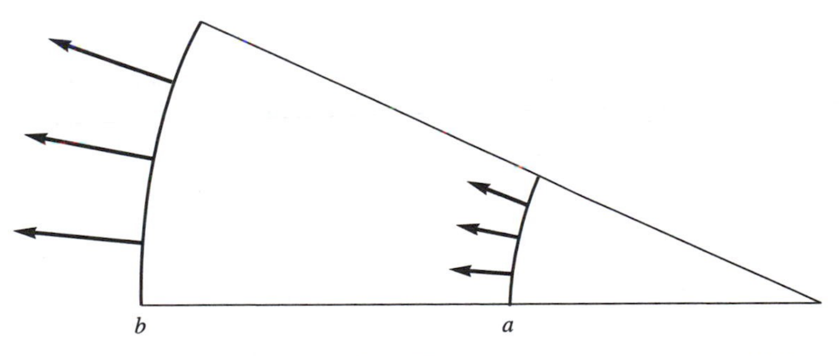

Figure 17-12 Portion of the Faraday disk. The arrows represent the $\boldsymbol{v} \times \boldsymbol{B}$ field, which is proportional to the radial position ρ, and $\boldsymbol{\nabla} \cdot (\boldsymbol{v} \times \boldsymbol{B})$ is positive.

†See the inside of the front cover.

Observe that $\nabla \cdot \boldsymbol{E} = -\nabla \cdot (\boldsymbol{v} \times \boldsymbol{B})$, as we shall see in Sec. 17.6. Now call the charge density $\tilde{Q}$ and recall from Eq. 7-34 that

$$\nabla \cdot \boldsymbol{E} = \frac{\tilde{Q}}{\epsilon_r \epsilon_0}, \text{ so that } \tilde{Q} = -2\epsilon_r \epsilon_0 \omega B: \tag{17-51}$$

the copper disk carries a uniform, steady *volume charge* of density $\tilde{Q}$.

But we saw in Sec. 4.5.4 that, under steady conditions, the *volume* charge density inside a conductor is *zero*! Is there a contradiction? No, because the above $\tilde{Q}$ is a function of $\boldsymbol{v} \times \boldsymbol{B}$, and Sec. 4.5.4 disregards magnetic fields.

Now why is the charge *negative*? Figure 17-12 shows why. Recall that $\boldsymbol{B}$ is uniform over the surface of the disk, and that $v = \omega\rho$. Then $|\boldsymbol{v} \times \boldsymbol{B}|$ increases with increasing radius as in the figure. For the element of volume shown, the flux of $\boldsymbol{v} \times \boldsymbol{B}$ is larger at the outer radius than at the inner radius, there is a net *outward* flux of $\boldsymbol{v} \times \boldsymbol{B}$, and $\nabla \cdot (\boldsymbol{v} \times \boldsymbol{B})$ is positive. So there is an *inward* flux of conduction electrons that soon stops because of the resulting electric field $\boldsymbol{E}$.

Note that the space charge density $\tilde{Q}$ is independent of both the conductivity σ and the current I. It is negative if the angular velocity is in the positive direction with respect to $\boldsymbol{B}$, as above. We return to this important phenomenon in Sec. 17.6.

Homopolar *motors* are built like homopolar generators. Voltage is applied between the terminals in Fig. 17-10(a), and then $|E| > |vB|$. Homopolar motors are also low-voltage, high-current devices, and they usually operate with direct current. They provide high torques at low speeds and are suitable, for example, for ship propulsion. The current is then supplied by a diesel motor-generator through a step-down transformer and a rectifying circuit. The current that flows radially through the disk provides the driving torque. Again, the magnetic force $Q(\boldsymbol{v} \times \boldsymbol{B})$ does work.

EXAMPLE The Sunspot as a Self-Excited Dynamo

†Sunspots are exceedingly complex magnetic structures on the sun. See Fig. 17-13. They were discovered by Galileo (1564–1642), who proved that they were not due to planets rotating in front of the sun, as was believed at the time. The discovery caused a scandal, because how can a perfect body like the sun have spots?

The diameter of the central dark region, called the *umbra*, varies widely, but it is typically 5–10 megameters. (The diameter of the earth is about 13 megameters.) The magnetic field is roughly vertical and, at the center, B is a few tenths of a tesla, which is about the field between the pole-pieces of an electromagnet. There are two types of spot, with $\boldsymbol{B}$ pointing either up or down, and coupled in some unknown way below the surface. Thousands of papers have been written on sunspots, amounting to about 150 per year at this time.

It is observed that the photospheric plasma flows down into the sunspot as in Fig. 17-14, and eventually returns to the surface. The plasma density increases with depth.

†The Example in Sec. 3.4.1 discusses the electrostatic potential at the surface of the sun, and there is an Example on solar coronal loops in Sec. 14.2.

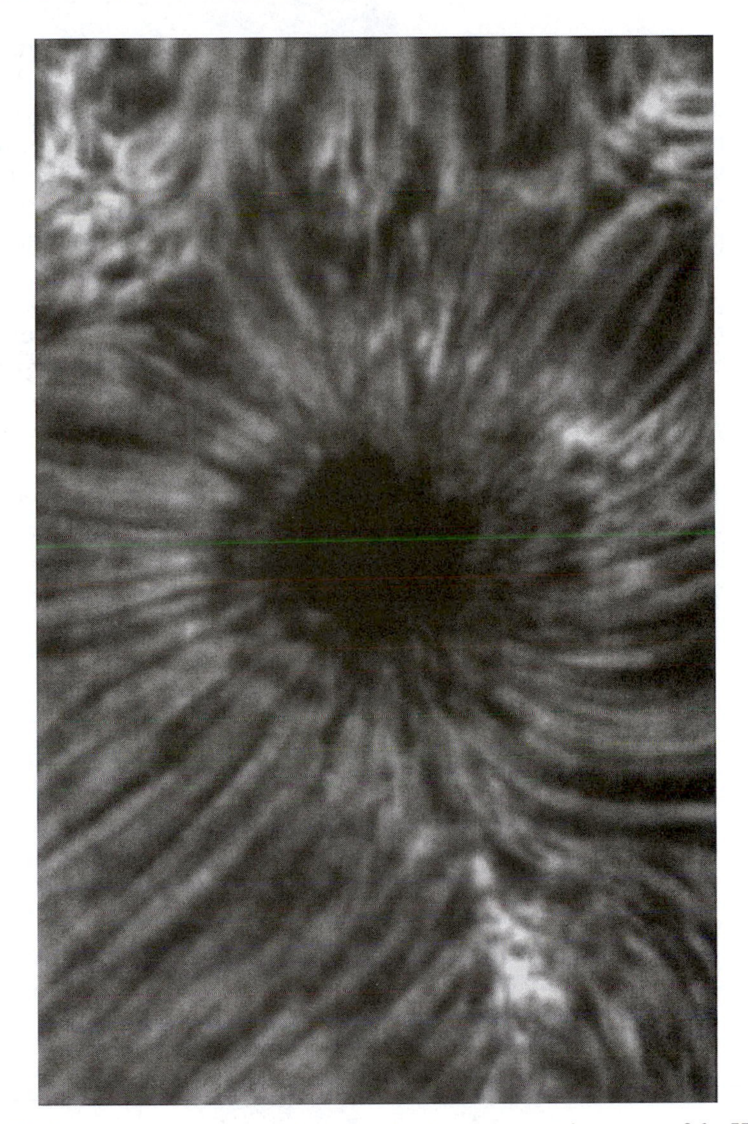

Figure 17-13 A sunspot photographed near the disk center, at the center of the H alpha line ($\lambda = 656.3$ nanometers). The dark central region is the *umbra*, while the striated surrounding ring is the *penumbra* closer in, and the *superpenumbra* farther out. The figure covers an area of 70 by 44 seconds, or about 51 by 32 megameters, one second of arc being equal to 725 kilometers at the sun. For comparison, the diameter of the earth is about 13 megameters. There exists a wide variety of sizes and shapes of sunspots.

The umbra, which is the top of the magnetic flux tube, is far from uniform; a longer exposure reveals a complex structure with bright *umbral dots*, *light bridges*, etc. The bright and dark striations of the penumbra and superpenumbra are still the object of debate.

This image was obtained with the R.B. Dunn solar telescope at the National Solar Observatory at Sacramento Peak NM by Eugenia Christopoulou, University of Patras, Greece, Alexander Georgakilas of the Astronomical Institute, National Observatory of Athens, Greece, and Serge Koutchmy of the Institut d'astrophysique de Paris, Paris, France.

Assume that the field is vertical and points up as in the figure, and that the spot is circular. The inward plasma velocity gives an azimuthal $\boldsymbol{v} \times \boldsymbol{B}$ field that points as in the figure. The magnetic field of the induced current, of density $\boldsymbol{J} = \sigma(\boldsymbol{v} \times \boldsymbol{B})$, is proportional to B, and it points up, in the *same* direction as the assumed seed field. If the seed field points down instead of up, then $\boldsymbol{v} \times \boldsymbol{B}$ changes sign and the induced current again flows in the *same* direction as the seed field. This type of magnetic configuration in a convecting plasma is known as a *magnetic flux tube* (Lorrain and Salingaros, 1993).

A sunspot is thus a *self-excited dynamo* (Lorrain and O. Koutchmy, 1998) that generates a magnetic field that has the same sign as the seed field. There is positive feedback and, at first sight, B increases exponentially with time. Of course, B does not become infinite, but rather reaches an asymptotic value that depends on the power that drives the convection (Lorrain, 1995). Observe that the magnetic force, of density $\boldsymbol{J} \times \boldsymbol{B}$, points radially outward and brakes the inward velocity of the plasma.

The outward magnetic force (Sec. 17.4.1) is countered by the pressure gradient: the gas pressure is *lower* inside the flux tube than outside. Indeed, it has been shown that, ideally, the sum of the magnetic pressure plus the gas pressure is independent of the radius (Lorrain and Salingaros, 1993).

One major question is: Why is the umbra dark, compared to the neighboring photosphere? One explanation is that the pressure is lower, as we just saw, so there are

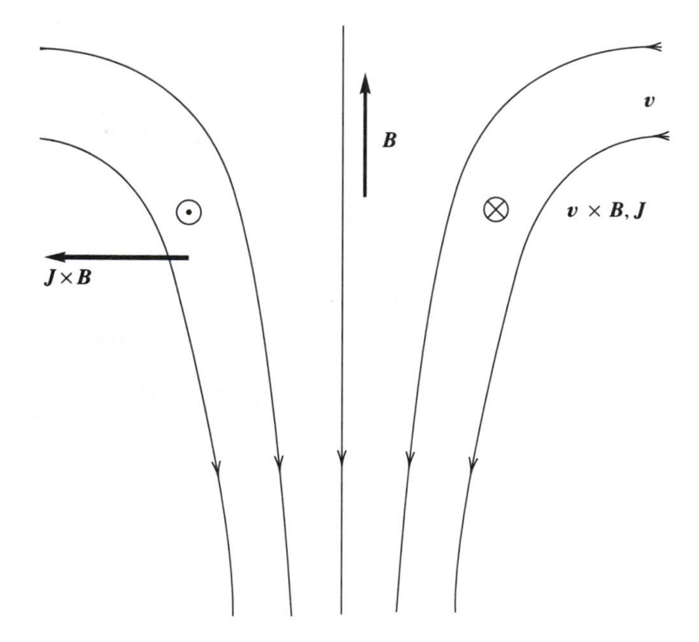

Figure 17-14 Section through a sunspot. The photosphere is roughly at the level of the top of the $\boldsymbol{B}$ arrow. The plasma flows down and its velocity is not only downward, but also inward because the density increases with depth. Assume a seed field $\boldsymbol{B}$ that points upward. Then the $\boldsymbol{v} \times \boldsymbol{B}$ field is azimuthal, in the direction shown. *The magnetic field of this induced current points in the same direction as the seed field.* So there is positive feedback: the larger the field, the faster it grows! The magnetic force density $\boldsymbol{J} \times \boldsymbol{B}$ points outward, as in a solenoid.

fewer excited atoms that can radiate. The azimuthal current heats the gas because the plasma has a finite conductivity, but the Joule power density is negligible (Lorrain and O. Koutchmy, 1998). Various other explanations have been proposed (Zirin, 1988).

See also Lorrain and Koutchmy (1993, 1996).

EXAMPLE Natural Magnetic Fields. The Case of the Earth

The preceding example showed how convection of the solar plasma generates the magnetic fields of sunspots. There certainly exist many other types of self-excited dynamos in convecting, conducting fluids in nature. The general principle is that, given a seed field, the $\boldsymbol{v} \times \boldsymbol{B}$ field of Eq. 17-36 gives a current which, under proper conditions, generates a magnetic field *of the same polarity as the seed field*. The magnetic forces are those of Sec. 17.4.

Magnetic fields in nature are thus of two types: those associated with self-excited dynamos in convecting, conducting fluids, and those associated with magnetic ores.

The earth has a radius of 6.4 megameters, and its core a radius of 3.5 megameters. See Fig. 17-15. The core consists largely of iron, but it is nonmagnetic because its temperature is about 3100 kelvins at its surface, and higher inside. (The *Curie temperature*, above which iron is nonmagnetic, is about 800° C.) From a radius of 3.5 megameters down to 1.2 megameters the core is liquid and its conductivity is thought to be about 3×10^5 siemens/meter, or about $1/200$th of the conductivity of copper at room temperature.

Since the density of the liquid outer core is not uniform because of temperature and chemical gradients, the liquid convects under the combined action of gravitational, centrifugal, and Coriolis forces. It is that convection that generates the earth's main magnetic field (Lorrain, 1993).

There are presumably innumerable self-excited dynamos of many different types in the earth's core, all more or less coupled together. It is certain that the field at the surface of the core is vastly more complex than the one at the surface of the earth. The time dependence is also vastly different: at the surface, we see only the very lowest-frequency components because the earth's crust, and the *mantle*, which is just below, act as conducting shields (Sec. 22.6).

The dipolar component accounts for only 80% of the total magnetic field at the surface. The *South magnetic* pole is close to the *North geographic* pole. So, in the Northern hemisphere, $\boldsymbol{B}$ points North and down; in the southern hemisphere, $\boldsymbol{B}$ points North and up. The axis of symmetry of the dipolar component and the axis of rotation form an angle of about 11°, and the axes are offset: the magnetic field is perpendicular to the surface at 78.1° N, 256.3° E and at 64.9° S, 138.9° E (Lorrain, 1990) (Merrill, McElhinny, and McFadden, 1996; Stacey, 1992). In the past century, the South *magnetic* pole has moved North 1000 kilometers, and it is presently migrating North at the rate of 20 kilometers/year.

The mean magnetic field has decreased by 1% over the past 10 years. At first sight, that is alarming because the Earth's magnetic field acts as a shield against cosmic rays, most charged particles being deflected away by the $e(\boldsymbol{v} \times \boldsymbol{B})$ force. Without this shield, could life be wiped out? Certainly not, because the polarity of the earth's magnetic field has changed maybe 100 times in the past 80 million years, presumably leaving the earth without its shield at every change, and life goes

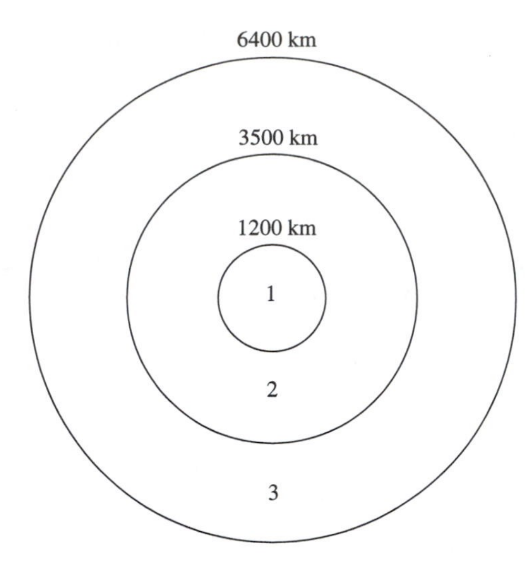

Figure 17-15 Cross-section of the earth. 1. Solid inner core. 2. Liquid outer core. 3. Mantle.

on—but there have been many mass extinctions in the past! See *Physics World* of February 1999.

See Prob. 2-22 concerning the earth as a habitable planet.

17.6 ELECTROSTATIC VOLUME CHARGES INSIDE CONDUCTORS THAT MOVE IN MAGNETIC FIELDS

In Sec. 4.5.4 we saw that, as a rule, electrostatic charges on conductors reside solely on the surface. But recall from Eq. 17-51 that, in the case of the Faraday disk, the net *volume* charge density is *not* zero.

We now show that the volume charge density is in general *not* equal to zero when a conductor moves in a magnetic field (Lorrain, 1990). Surprisingly, this important phenomenon is little-known.

Assume that the conductivity σ is uniform, and take the divergence of Eq. 17-36:

$$\boldsymbol{\nabla} \cdot \boldsymbol{J} = \sigma[\boldsymbol{\nabla} \cdot \boldsymbol{E} + \boldsymbol{\nabla} \cdot (\boldsymbol{v} \times \boldsymbol{B})]. \tag{17-52}$$

On the left we have the divergence of the free current density, which is equal to $-\partial\rho_f/\partial t$, from Eq. 7-10. Assume now a steady state. Then the LHS is equal to zero.

Next, substitute Eq. 7-34 on the right. Then

$$0 = \sigma\left[\frac{\rho_f}{\epsilon_r\epsilon_0} + \boldsymbol{\nabla} \cdot (\boldsymbol{v} \times \boldsymbol{B})\right] \tag{17-53}$$

and

$$\rho_f = -\epsilon_r\epsilon_0 \boldsymbol{\nabla} \cdot (\boldsymbol{v} \times \boldsymbol{B}). \tag{17-54}$$

This is the electric volume charge density in a conductor that moves at a velocity $\boldsymbol{v}$ in a magnetic field $\boldsymbol{B}$. We could also write simply ρ, as in Eq. 17-51, instead of ρ_f. Note that the electric charge density is independent of σ.

If σ is not uniform, or if the steady-state assumption does not apply, then ρ_f is given by more complex expressions, but it is not zero, in general.

The existence of these volume charges results from the fact that the $\boldsymbol{v} \times \boldsymbol{B}$ field sweeps conduction electrons inside. In a region where $\boldsymbol{\nabla} \cdot (\boldsymbol{v} \times \boldsymbol{B})$ is positive, conduction electrons move in, and the region becomes negative. But if the divergence is negative, then some electrons move out and the region becomes positive. So ρ_f and $\boldsymbol{\nabla} \cdot (\boldsymbol{v} \times \boldsymbol{B})$ have opposite signs, as above.

Clearly, $\rho_f = 0$ only under exceptional circumstances. Moreover, surface charges seem inevitable. So it is doubtful that the electrostatic potential V, and hence $\boldsymbol{\nabla} V$, and hence $\boldsymbol{E}$ in Eq. 17-36, can ever be zero in the presence of a $\boldsymbol{v} \times \boldsymbol{B}$ field on a finite-sized conductor.

Electrostatic surface and volume charges play an essential role whenever conductors move in magnetic fields because their $\boldsymbol{E}$ cancels all or part of $\boldsymbol{v} \times \boldsymbol{B}$ in Eq. 17-36.

If $\boldsymbol{E}$ cancels all of $\boldsymbol{v} \times \boldsymbol{B}$, then there is no induced current, no induced magnetic field, and the net field is simply the applied field. Then the magnetic field lines are unaffected by the moving conductor.

But if $\boldsymbol{E}$ cancels only part of $\boldsymbol{v} \times \boldsymbol{B}$, as in the Faraday disk of Sec. 17.5, then there is an induced current and an induced magnetic field that adds to the externally applied field.

17.7 SUMMARY

The force on a charge Q moving at a velocity $\boldsymbol{v}$ in a field $\boldsymbol{E}, \boldsymbol{B}$ is

$$\boldsymbol{F} = Q(\boldsymbol{E} + \boldsymbol{v} \times \boldsymbol{B}). \tag{17-1}$$

This is the *Lorentz force*. The term $Q\boldsymbol{v} \times \boldsymbol{B}$ is the *magnetic force*.

Charge carriers flowing along a conductor situated in a magnetic field tend to drift sideways because of the magnetic force. This is the *Hall effect*.

The magnetic force per unit length on a current-carrying wire is $\boldsymbol{I} \times \boldsymbol{B}$. The total net force on a closed circuit is thus

$$\boldsymbol{F} = I \oint_C d\boldsymbol{l} \times \boldsymbol{B}. \tag{17-17}$$

The magnetic force exerted by a closed circuit a on a closed circuit b is

$$\boldsymbol{F}_{ab} = \frac{\mu_0}{4\pi} I_a I_b \oint_a \oint_b d\boldsymbol{l}_b \times \frac{d\boldsymbol{l}_a \times \hat{\boldsymbol{r}}}{r^2} = -\frac{\mu_0}{4\pi} I_a I_b \oint_a \oint_b \hat{\boldsymbol{r}} \frac{d\boldsymbol{l}_a \cdot d\boldsymbol{l}_b}{r^2}. \tag{17-21}$$

By definition, $\mu_0 \equiv 4\pi \times 10^{-7}$ weber/ampere-meter. Then the force per meter between parallel wires carrying the same current I and separated by a distance D is $2 \times 10^{-7}\, I^2/D$. The force is attractive if the currents flow in the same direction.

The *magnetic force exerted on a volume distribution of current* is

$$\boldsymbol{F} = \int_v \boldsymbol{J} \times \boldsymbol{B}\, dv. \tag{17-29}$$

The *magnetic pressure* on a conductor, with a magnetic field B on one side and zero magnetic field on the other side, is

$$p_{mag} = \frac{B^2}{2\mu_0}. \tag{17-33}$$

The *generalized Ohm's law* applies to a body of conductivity σ that moves at a velocity $\boldsymbol{v}$ in fields $\boldsymbol{E}, \boldsymbol{B}$:

$$\boldsymbol{J} = \sigma[\boldsymbol{E} + (\boldsymbol{v} \times \boldsymbol{B})]. \tag{17-36}$$

All five variables can be functions of the coordinates and of the time.

A conductor that moves at a velocity $\boldsymbol{v}$ in a field $\boldsymbol{B}$ carries an electrostatic *volume* charge of density

$$\rho_f = -\epsilon_r \epsilon_0 \boldsymbol{\nabla} \cdot (\boldsymbol{v} \times \boldsymbol{B}). \tag{17-54}$$

This expression applies to homogeneous conductors under steady conditions.

PROBLEMS

17-1. (*17.1*) Electrons in the Crab nebula†

In the Crab nebula there is a magnetic field of about 2×10^{-8} tesla and electrons whose energy is about 2×10^{14} electronvolts.

(a) Find the radius of gyration. Compare this radius with that of the earth's orbit (see the page inside the back cover).

(b) How long does an electron take to complete one turn, in days?

17-2. (*17.1*) The acceleration of an electron in a field $\boldsymbol{E}, \boldsymbol{B}$†

(a) Show that the equation of motion for a particle of rest mass m_0, charge Q, and velocity $\boldsymbol{v}$ in a field $\boldsymbol{E}, \boldsymbol{B}$ is

$$\gamma m_0 \frac{d\boldsymbol{v}}{dt} = Q\left(\boldsymbol{E} + \boldsymbol{v} \times \boldsymbol{B} - \frac{\boldsymbol{v}}{c^2}\boldsymbol{v} \cdot \boldsymbol{E}\right).$$

If $\boldsymbol{E}$ is zero and $\boldsymbol{B}$ is static,

$$\frac{d\boldsymbol{v}}{dt} = \boldsymbol{v} \times \frac{Q\boldsymbol{B}}{m} = \boldsymbol{v} \times \boldsymbol{\omega}_c,$$

and the electron describes a circle at the angular velocity $\omega_c = QB/m$, which is called the *cyclotron frequency.*

(b) A 12.0 millionelectronvolt electron moves in the positive direction of the z-axis in a field $\boldsymbol{E} = 1.00 \times 10^6 \hat{\boldsymbol{z}}, \boldsymbol{B} = 1.00 \hat{\boldsymbol{x}}$.

Calculate its acceleration. The rest mass of an electron is 5.11×10^5 electron volts.

---

†This problem requires a knowledge of relativity.

17-3. (*17.1*) The motion of a charged particle in uniform and perpendicular $\boldsymbol{E}$ and $\boldsymbol{B}$ fields

A particle of charge Q starts from rest at the origin in a region where $\boldsymbol{E} = E\hat{\mathbf{y}}$ and $\boldsymbol{B} = B\hat{\mathbf{z}}$.

(a) Find two simultaneous differential equations for v_x and v_y.

(b) Find v_x and v_y. Set $\omega_c = BQ/m$. This is the *cyclotron frequency.* Also, set $v_x = 0$, $v_y = 0$ at $t = 0$. There will be two constants of integration.

(c) Find $x(t)$ and $y(t)$. Set $x = 0$, $y = 0$ at $t = 0$.

(d) Describe the motion of the charge.

(e) Plot the trajectory for $\omega_c = 1$, $E/B = 1$.

(f) Show that the particle drifts at the velocity $\boldsymbol{E}\times\boldsymbol{B}/B^2$. Note that this velocity is independent of the nature of the particle and of its energy. In a plasma, charges of both signs drift at the same velocity, and the net drift current is zero.

(g) Calculate the drift velocity of a proton at the equator under the combined actions of gravity and the $\boldsymbol{B}$ of the earth. Assume that $B = 4 \times 10^{-5}$ tesla, in the horizontal direction. Remember that in the region of the *north geographic* pole there is a *south magnetic* pole. Thus, at the equator, $\boldsymbol{B}$ points *north.*

(h) In which direction does an electron drift at the equator?

17-4. (*17.1*) The crossed-field photomultiplier

Figure 17-16 shows the principle of operation of a crossed-field photomultiplier. A sealed and evacuated enclosure contains two parallel plates called *dynodes.* They provide the electric field $\boldsymbol{E}$. An external permanent magnet superimposes the magnetic field $\boldsymbol{B}$.

A photon ejects a low-energy photoelectron. The electron accelerates upward, but the magnetic field deflects it back to the negative dynode. At this point it ejects a few secondary electrons, and the process repeats itself. Eventually, the electrons impinge on the collector C.

Let us find the value of a.

(a) Find the differential equations for v_x and v_y. The trajectory is not circular. You can simplify the calculation by setting $Be/m = \omega_c$, the cyclotron frequency.

(b) Find v_x as a function of y.

(c) You can now find y, and then x, as functions of t. Set $t = 0$, $dx/dt = 0$, and $dy/dt = 0$ at $x = 0$, $y = 0$. You should find that the trajectory is a cycloid.

(d) What is the maximum value of y?

(e) What is the value of a?

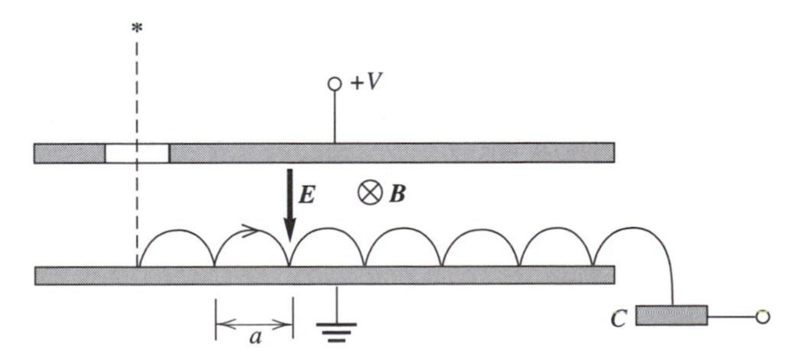

Figure 17-16

17-5. (*17.1*) Fermi acceleration†

Fermi proposed the following mechanism, now called *Fermi acceleration,* to explain the existence of very high-energy particles in space. Imagine a clump of plasma traveling at some velocity $v_c\hat{\boldsymbol{x}}$. The plasma carries a current and thus has a magnetic field. Imagine now a particle traveling in the opposite direction at a velocity $-v_a\hat{\boldsymbol{x}}$, both and being positive quantities. The particle is deflected in the magnetic field and acquires a velocity $v_b\hat{\boldsymbol{x}}$, where v_b is also a positive quantity.

Set $v_c = v_a = c/2$. Calculate the initial and final values of γ.

17-6. (*17.1*) Electromagnetic pumps

Electromagnetic pumps are convenient for pumping highly conducting fluids, for example liquid sodium in certain nuclear reactors.

The conduction current density in a liquid metal of conductivity σ that moves at a velocity $\boldsymbol{v}$ in a field $\boldsymbol{E}$, $\boldsymbol{B}$ is $\boldsymbol{J} = \sigma(\boldsymbol{E} + \boldsymbol{v} \times \boldsymbol{B})$. All quantities are measured with respect to a fixed reference frame. Then the magnetic force per unit volume is $\boldsymbol{F}' = \boldsymbol{J} \times \boldsymbol{B} = \sigma(\boldsymbol{E} + \boldsymbol{v} \times \boldsymbol{B}) \times \boldsymbol{B}$. As a rule, $\boldsymbol{E}$, $\boldsymbol{B}$, and $\boldsymbol{v}$ are orthogonal as in Fig. 17-17.

(a) Show that the force density $\boldsymbol{F}'$ is $\sigma B^2(\boldsymbol{u} - \boldsymbol{v})$, where $\boldsymbol{u} = \boldsymbol{E} \times \boldsymbol{B}/B^2$ and $\boldsymbol{v}$ is the fluid velocity. The magnetic force tries to make $\boldsymbol{v}$ equal to $\boldsymbol{u}$.

(b) Calculate the efficiency, on the assumption that a permanent magnet supplies the magnetic field. Neglect edge and end effects.

17-7. (*17.1*) The Hall effect

Let us investigate the Hall effect more closely. We assume that the charge carriers are electrons of charge $-e$. Their effective mass is m^*. The *effective mass* takes into account the periodic forces exerted on the electrons as they travel through the crystal lattice. As a rule, the effective mass is *smaller* than the mass of an isolated electron.

The force on an electron is $\boldsymbol{F} = -e(\boldsymbol{E} + \boldsymbol{v} \times \boldsymbol{B})$, where $\boldsymbol{E}$ has two components, the applied field E_x and the Hall field E_y. The average drift velocity is

$$\boldsymbol{v} = \frac{\mathcal{M}\boldsymbol{F}}{e} = -\mathcal{M}(\boldsymbol{E} + \boldsymbol{v} \times \boldsymbol{B}),$$

where $\mathcal{M}$ is the mobility (Sec. 4.5.2). The law $\boldsymbol{F} = m^*\boldsymbol{a}$ applies only for collisions with the crystal lattice.

(a) Show that

$$v_x = -\mathcal{M}(E_x + v_y B), \qquad v_y = -\mathcal{M}(E_y - v_x B), \qquad v_z = 0.$$

(b) Show that

$$J_x = Ne\mathcal{M}\frac{E_x - \mathcal{M}E_y B}{1 + \mathcal{M}^2 B^2}, \qquad J_y = Ne\mathcal{M}\frac{E_y + \mathcal{M}E_x B}{1 + \mathcal{M}^2 B^2}.$$

Thus, if $J_y = 0$,

†This problem requires a knowledge of relativity.

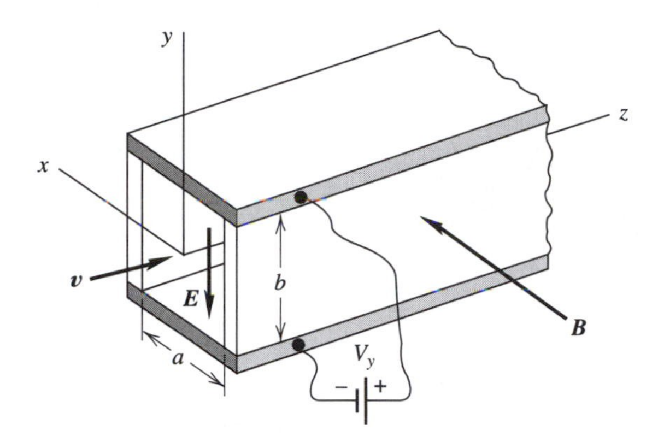

Figure 17-17

$$E_y = -\mathcal{M} E_x B \qquad \text{or} \qquad V_y = \frac{b}{a} \mathcal{M} V_x B.$$

Note that the Hall voltage V_y is proportional to the *product* of the applied voltage V_x and B. The Hall effect is thus useful for multiplying one variable by another.

When it is connected in this way, the Hall element has four terminals and is called a *Hall generator,* or a *Hall probe.*

(c) Calculate V_y for $b = 1$ millimeter, $a = 5$ millimeters, $\mathcal{M} = 7$ meters^2/volt-second (indium antimonide), $V_x = 1$ volt, $B = 10^{-4}$ tesla.

(d) Show that, if $E_y = 0$, then

$$\frac{\Delta R}{R_0} = \mathcal{M}^2 B^2,$$

where R_0 is the resistance of the probe in the x-direction when $B = 0$, and ΔR is the increase in resistance upon application of the magnetic field.

17-8. (*17.1*) The electromagnetic flowmeter

The *electromagnetic flowmeter* is the inverse of the electromagnetic pump (Prob. 17-6). It operates as follows. See Fig 17-18. A conducting fluid flows in a nonconducting tube between the poles of a magnet. Electrodes on either side of the tube and in contact with the fluid measure the $\boldsymbol{v} \times \boldsymbol{B}$ field, and thus the quantity of fluid that flows through the tube per second. This is a Hall effect, except that here ions of both signs move with the fluid in the same direction.

Faraday attempted to measure the velocity of the Thames River in this way in 1832. The magnetic field was, of course, that of the earth.

In the absence of turbulence, the fluid speed in a tube of radius R is of the form $v = v_0(1 - r^2/R^2)$. The $\boldsymbol{v} \times \boldsymbol{B}$ field in the fluid is therefore not uniform. This gives rise to circulating electric currents with $\boldsymbol{J} = \sigma(-\nabla V + \boldsymbol{v} \times \boldsymbol{B})$. The potential V_0 results from the charges that accumulate on the electrodes, and also from the volume charge density ρ.

(a) Sketch a cross section of the tube, showing qualitatively, by means of arrows of various lengths, the magnitude and direction of $\boldsymbol{v} \times \boldsymbol{B}$.

(b) Sketch another cross section, showing the lines of current flow. The current drawn by the electrodes is negligible.

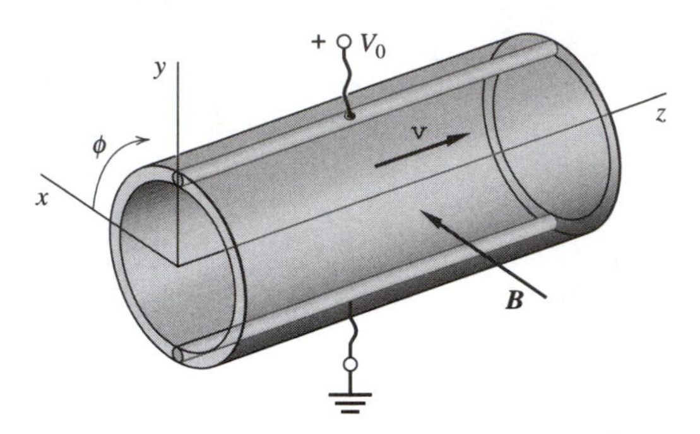

Figure 17-18

(c) Neglect end effects by setting $\partial/\partial z = 0$. Use the fact that $\nabla \cdot \boldsymbol{J} = 0$ to show that

$$\nabla^2 V = B\frac{\partial v}{\partial y} = B\frac{\partial v}{\partial r}\sin\phi.$$

Since this Laplacian is equal to $-\rho/\epsilon_0$, the volume charge density ρ is zero on the axis, where $\partial v/\partial r = 0$ and at $\phi = 0$ and π.

(d) Set $V = V'\sin\phi$, where V' is independent of ϕ, and show that

$$\frac{d^2V'}{dr^2} + \frac{1}{r}\frac{dV'}{dr} - \frac{V'}{r^2} = B\frac{dv}{dr}.$$

(e) You can solve this differential equation as follows: (1) express the left-hand side as a derivative, (2) integrate, (3) multiply both sides by r, (4) express both sides as derivatives, (5) integrate. This will leave you with two constants of integration, one of which is easy to dispose of. You can find the value of the other by remembering that $J_y = 0$ at $x = 0$, $y = R$ if the voltmeter draws zero current.

Note that the output voltage is independent of the conductivity of the fluid, if one assumes that the voltmeter draws zero current.

Find the output voltage as a function of the volume τ of fluid that flows in one second.

(f) We have neglected edge effects in the region where the fluid enters into, and emerges from, the magnetic field. Sketch lines of current flow for these two regions. These currents reduce the output voltage somewhat.

17-9. (*17.1*) Magnetic focusing

There exist many devices that utilize fine beams of charged particles. The cathode-ray tube that is used in television receivers and in oscilloscopes is the best-known example. The electron microscope is another example. In these devices the particle beam is focused and deflected in much the same way as a light beam in an optical instrument.

Beams of charged particles can be focused and deflected by properly shaped *electric* fields. The electron beam in a TV tube is focused with electric fields and deflected with magnetic fields. Beams can be deflected along a circular path in a *magnetic* field. Let us see how they can be focused by a magnetic field.

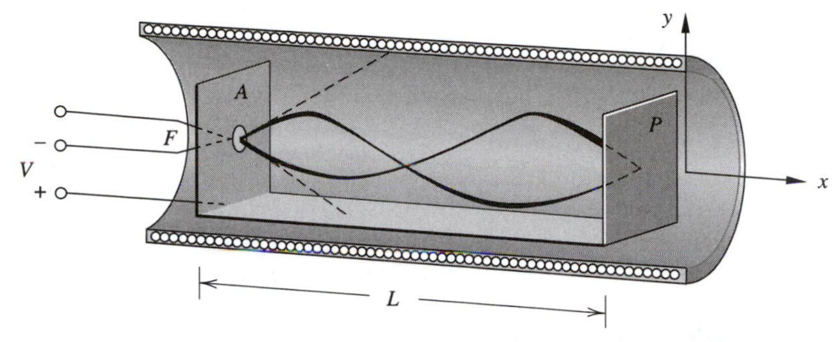

Figure 17-19 Focusing of an electron beam in a uniform magnetic field: F is a heated filament, A is the anode, and the electron beam is focused at P. The accelerating voltage is V. A filament supply of a few volts is connected between the top two terminals. When the solenoid is not energized, the beam diverges as shown by the dashed lines.

Figure 17-19 shows an electron gun situated inside a long solenoid. The electrons that emerge from the hole in the anode have a small transverse velocity component and, if there is no current in the solenoid, they spread out as in the figure. Let us see what happens when we turn on the magnetic field.

Let the velocity components at the hole be v_x and v_y, with $v_x^2 \gg v_y^2$. Then

$$eV = \frac{1}{2}m(v_x^2 + v_y^2) \approx \frac{1}{2}mv_x^2.$$

If $v_y = 0$, the electron continues parallel to the axis. If $v_y \neq 0$, the electron follows a helical path at an angular velocity ω. After one complete turn, it has returned to the axis. So, if B is adjusted correctly, all the electrons will converge at the same point P, as in the figure, and form an image of the hole in the anode.

(a) Show that this will occur if

$$B = 2^{3/2}\pi\frac{(mV/e)^{1/2}}{L}.$$

(b) Calculate the number of ampere-turns per meter IN' in the solenoid, for an accelerating voltage V of 10 kilovolts and a distance L of 0.5 meter.

In actual practice, magnetic focusing is achieved with short coils, and not with solenoids.

17-10. (*17.1*) Dempster mass-spectrometer

Figure 17-20 shows a *Dempster mass-spectrometer*. An ion describes a circle and the centripetal force is $Q\boldsymbol{v} \times \boldsymbol{B}$. The magnetic field is provided by an electromagnet.

(a) Show that

$$m = QR^2B^2/(2V).$$

(b) In one particular experiment, a mass-spectrometer of this type was used for the hydrogen ions H_1^+, H_2^+, H_3^+ with $R = 60.0$ millimeters and $V = 1000$ volts. The

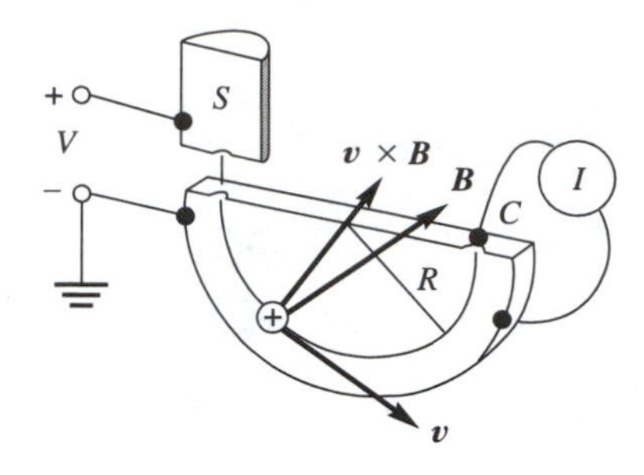

Figure 17-20 Cross-section of a mass-spectrometer of the Dempster type. Ions produced at the source S are collected at C. The ions are accelerated through a difference of potential V and describe a semicircle of radius R. The magnetic field $\boldsymbol{B}$ is perpendicular to the plane of the trajectory.

H_1^+ ion is a proton, H_2^+ is composed of two protons and one electron, and H_3^+ is composed of three protons and two electrons.

Find the values of B for these three ions.

(c) Draw a curve of B as a function of m under the above conditions, from hydrogen to uranium.

Would you use this type of spectrometer for heavy ions? Why?

17-11. (*17.1*) High-temperature plasmas

One method of injecting and trapping ions in a high-temperature plasma is illustrated in Fig.17-21. Molecular ions of deuterium, D_2^+ (two deuterons plus one electron, the deuteron being composed of one proton and one neutron), are injected into a magnetic field and are dissociated into pairs of D^+ ions (deuterons) and electrons in a high-intensity arc. The radius of the trajectory is reduced and the ions are trapped.

(a) Calculate the radius of curvature R for D_2^+ ions having a kinetic energy of 600 kiloelectronvolts in a B of 1.00 tesla.

(b) Calculate R for the D^+ ions produced in the arc. The D^+ ions have one half the kinetic energy of the D_2^+ ions.

17-12. (*17.2*) Magnetic force

Calculate the magnetic force on an arc 50 millimeters long carrying a current of 400 amperes in a direction perpendicular to a uniform $\boldsymbol{B}$ of 5×10^{-2} tesla.

High current *circuit breakers* often comprise coils that generate a magnetic field to blow out the arc that forms when the contacts open.

17-13. (*17.2*) Magnetic force

(a) Find the current density necessary to float a copper wire in the earth's magnetic field at the equator.

Assume a field of 10^{-4} tesla. The density of copper is 8.9×10^3 kilograms per cubic meter.

(b) Will the wire become hot?

The conductivity of copper is 5.80×10^7 siemens/meter.

(c) In what direction must the current flow?

(d) What would happen if the experiment were performed at one of the magnetic poles?

17-14. (*17.2*) Magnetic force

Calculate the force due to the earth's magnetic field on a horizontal wire 100 meters long carrying a current of 50 amperes due north.

Set $B = 5 \times 10^{-5}$ tesla, pointing downward at an angle of 70 degrees with the horizontal.

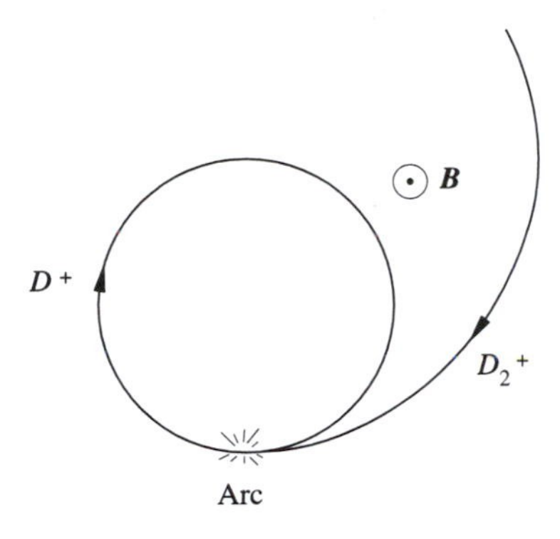

Figure 17-21 Method for injecting and trapping D^+ ions.

17-15. (*17.2*) Magnetic force

Show that the total force on a closed circuit carrying a current I in a uniform magnetic field is zero.

17-16. (*17.4*) The pinch effect on a conductor

We have seen that the magnetic force density on a conductor is $\mathbf{J} \times \mathbf{B}$.

If the force density is enormous, then any solid can be treated as a fluid, and if p is the pressure, $\boldsymbol{\nabla} p = \mathbf{J} \times \mathbf{B}$.

(a) A conducting wire of radius R carries a current I. At the surface, $p = 0$. Show that at the radius r, $p = [\mu_0 I^2/(4\pi^2 R^2)](1 - r^2/R^2)$. The magnetic force compresses the wire.

(b) Calculate the instantaneous pressure on the axis for a current of 30 kiloamperes in a wire 1 millimeter in radius. The wire, of course, vaporizes. Such large currents are obtained by discharging large, low-inductance capacitors.

(c) Show that, with a tubular conductor of inner radius R_1 and outer radius R_2, the pressure in the cavity is given by

$$p = \frac{\mu_0 I^2}{4\pi^2 R_2^2} \frac{1 - (R_1/R_2)^2[1 + 2\ln(R_2/R_1)]}{[1 - (R_1/R_2)^2]^2}.$$

Transient pressures approaching 10^6 atmospheres have been obtained this way.

One type of x-ray source implodes thin, aluminized plastic tubes by discharging 1 megajoule in them. The resulting plasma generates a 150-kilojoule pulse of radiation.

17-17. (*17.5*) The homopolar motor

(a) Find the mechanical power as a function of ω for a homopolar motor, for a given applied voltage V. See Fig. 17-10(a).

You can do this by first writing $P = \omega T = \omega B(b^2 - a^2)I/2 = \omega A I$, where A is a constant, and then expressing I as a function of ω.

(b) Sketch a curve of P as a function of ω.

(c) Show that the mechanical power is maximum at the angular velocity $\omega_{\text{maximum power}} = V/[B(b^2 - a^2)]$.

CHAPTER

18

MAGNETIC FIELDS V

The Faraday Induction Law

The Faraday induction law is, arguably, the most subtle part of electromagnetic theory. This law groups two phenomena that, at first sight, are unrelated: if you move a conductor in a static magnetic field, then there is an induced current (but not always, as we saw in Secs. 17.5 and 17.6), and if a stationary conductor lies in a time-dependent magnetic field, then there is also an induced current. We show that the two phenomena are not really distinct, but the proof requires the use of Relativity.

It is the custom to relate the two phenomena by ascribing the induced current to "flux cutting," but that is an incomplete, although useful, explanation of the law, as we show in the first Example in Sec. 18.3.

This chapter will demonstrate the fundamental importance of the vector potential $\boldsymbol{A}$.

†Here material marked with an asterisk requires relativity.

18.1 THE FARADAY INDUCTION LAW FOR $\boldsymbol{v} \times \boldsymbol{B}$ FIELDS. MOTIONAL ELECTROMOTANCE

Consider a closed circuit C that moves as a whole and distorts in some arbitrary way in a constant, but not necessarily uniform, magnetic field, as in Fig. 18-1. Then, by definition, the *induced,* or *motional, electromotance* is

$$\mathcal{V} = \oint_C (\boldsymbol{v} \times \boldsymbol{B}) \cdot d\boldsymbol{l} = -\oint_C \boldsymbol{B} \cdot (\boldsymbol{v} \times d\boldsymbol{l}). \tag{18-1}$$

The negative sign comes from the fact that we have altered the cyclic order of the terms under the integral sign.

Now $\boldsymbol{v} \times d\boldsymbol{l}$ is the area swept by the element $d\boldsymbol{l}$ in 1 second. Thus $\boldsymbol{B} \cdot (\boldsymbol{v} \times d\boldsymbol{l})$ is the rate at which the magnetic flux linking the circuit increases because of the motion of the element $d\boldsymbol{l}$. Integrating over the complete circuit, we find that the induced electromotance is proportional to the time rate of change of the magnetic flux Φ linking the circuit:

$$\mathcal{V} = -\frac{d\Phi}{dt}. \tag{18-2}$$

The positive directions for $\mathcal{V}$ and for Φ satisfy the right-hand screw rule. The current is the same as if the circuit comprised a battery of voltage $\mathcal{V}$.

This is the *Faraday induction law for* $\boldsymbol{v} \times \boldsymbol{B}$ *fields.* This law is important. As far as our demonstration goes, it applies only to constant $\boldsymbol{B}$s, but it is, in fact, general, as we see in Sec. 18.2. Quite often Φ is difficult to define; then we can integrate $\boldsymbol{v} \times \boldsymbol{B}$ around the circuit to obtain $\mathcal{V}$.

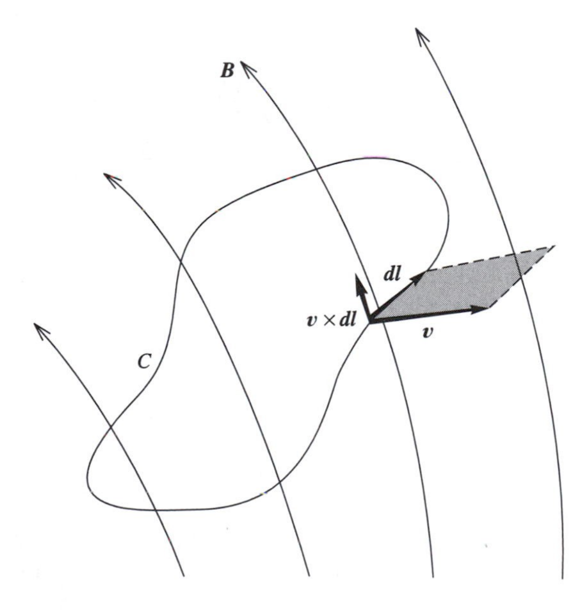

Figure 18-1 Closed circuit C that moves and distorts in some arbitrary way in a constant magnetic field $\boldsymbol{B}$. The element $d\boldsymbol{l}$ moves at a velocity $\boldsymbol{v}$ and sweeps an area $\boldsymbol{v} \times d\boldsymbol{l}$ in 1 second.

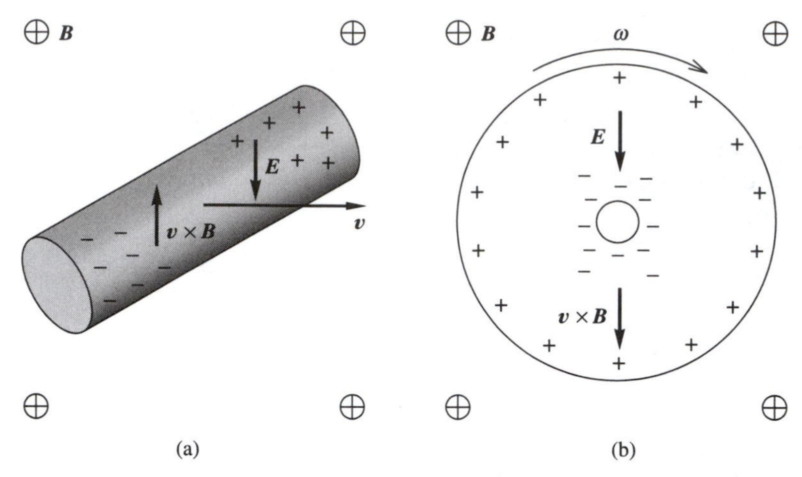

Figure 18-2 (a) Conducting rod moving at a velocity $\boldsymbol{v}$ in a magnetic field $\boldsymbol{B}$. The $\boldsymbol{v} \times \boldsymbol{B}$ field points *upward* and drives conduction electrons *downward*. Current flows until the resulting electric field $\boldsymbol{E}$ exactly cancels the $\boldsymbol{v} \times \boldsymbol{B}$ field at every point. (b) Disk rotating in a magnetic field $\boldsymbol{B}$. The $\boldsymbol{v} \times \boldsymbol{B}$ field points *outward* and drives conduction electrons *inward*. A radial current flows until $\boldsymbol{E}$ cancels $\boldsymbol{v} \times \boldsymbol{B}$. See the Example on the Faraday disk in Sec. 17.5.

If C is open, as in Fig. 18-2, then current flows until the electric field resulting from the accumulations of charge exactly cancels the $\boldsymbol{v} \times \boldsymbol{B}$ field.

There are usually surface charges but, as we saw in Sec. 17.6, there are also volume charges inside a conductor that moves in a magnetic field, unless $\boldsymbol{\nabla} \cdot (\boldsymbol{v} \times \boldsymbol{B}) = 0$.

EXAMPLE A Simple-Minded Generator

An electric generator transforms mechanical energy into electric energy, usually by moving conducting wires in a direction perpendicular to a magnetic field.

The simplest (and most impractical!) type of generator is that of Fig. 18-3. The link slides to the right at a speed v such that $v^2 \ll c^2$, where c is the speed of light, in a constant and uniform $\boldsymbol{B}$ that is normal to the paper. The resistance at the left-hand end of the line is R, and that of the link is R_l. The horizontal wires have zero resistance.

The electromotance is

$$\mathcal{V} = \left| -\frac{d\Phi}{dt} \right| = BDv. \tag{18-3}$$

We have disregarded the magnetic flux resulting from the current I itself. In other words, the resistance R is large. Then

$$I = \frac{BDv}{R + R_l}. \tag{18-4}$$

Observe that sliding the bar to the right increases the linking flux, but the induced current I tends to decrease it. So the induced electromotance drives a current whose

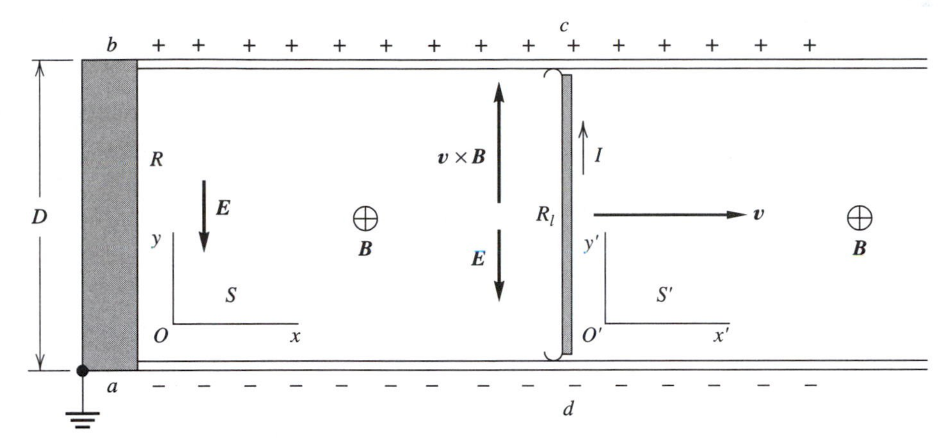

Figure 18-3 Simple-minded electric generator. The magnetic field $\boldsymbol{B}$ is constant and uniform. Sliding the link to the right at the velocity $\boldsymbol{v}$ generates a current I in the direction shown. The load resistance is R, and the link has a resistance R_l.

field opposes a *change* in the net magnetic flux linking the circuit *abcd*. This is *Lenz's law*. If the circuit were superconducting, the enclosed flux would remain constant.

In a fixed reference frame S, the force on a conduction electron of charge Q inside the link is $Q(\boldsymbol{E} + \boldsymbol{v} \times \boldsymbol{B})$. Thus, in the link,

$$\boldsymbol{J} = \sigma(\boldsymbol{E} + \boldsymbol{v} \times \boldsymbol{B}) = \sigma(-\boldsymbol{\nabla} V + \boldsymbol{v} \times \boldsymbol{B}). \tag{18-5}$$

Let us calculate V in the stationary reference frame S.

At b in Fig. 18-3, $V_b = IR$. In either horizontal wire, $\boldsymbol{J} = \sigma \boldsymbol{E}$ is finite. Since $\sigma \longrightarrow \infty$, by hypothesis, then $\boldsymbol{E} = 0$, $\boldsymbol{\nabla} V = 0$, and

$$V_d = V_a = 0, \qquad V_c = V_b = IR. \tag{18-6}$$

Inside R and R_l, with the y-axis as in the figure,

$$V = IR\frac{y}{D} = \frac{vBR}{R + R_l}y. \tag{18-7}$$

The voltage V_c across R_l is IR:

$$V_c = IR = I(R + R_l) - IR_l = vBD - IR_l. \tag{18-8}$$

This means that the motion generates a voltage vBD in the link, while its current causes a voltage drop IR_l.

Suppose we connect a voltmeter across the link as in Fig. 18-4(a). Call its resistance R_v with $R_v \gg R_l$. This hardly affects the current I. What will the reading on the voltmeter be? If the current through the voltmeter is I_v, then it will read a voltage $I_v R_v$, with the polarity shown in the figure.

Now refer to Fig. 18-4(b). Clearly, $I_v R_v = I_l R_l$ and the voltmeter reads the voltage drop $I_l R_l$.

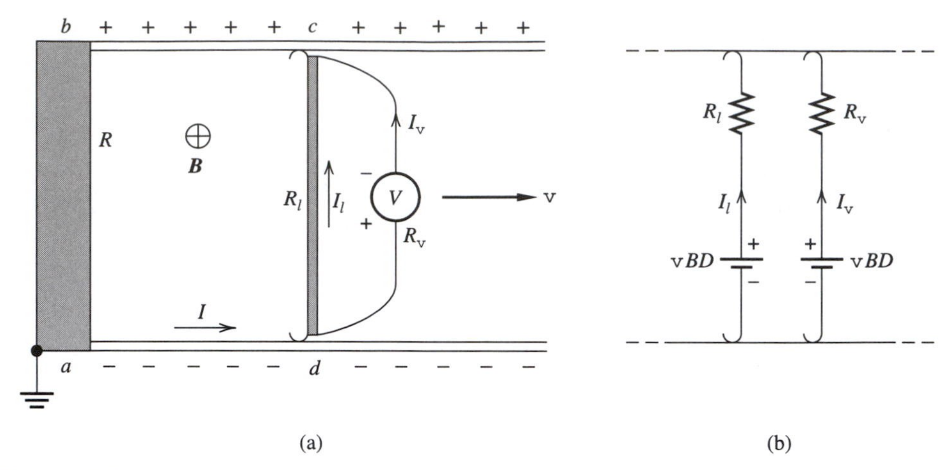

Figure 18-4 (a) Voltmeter V connected across the link in the generator of Fig. 18-3. (b) Equivalent circuit.

All we know about the magnetic field is that $\boldsymbol{B} = -B\hat{\boldsymbol{z}}$; the current distribution that generates $\boldsymbol{B}$ is unspecified. Let us set

$$A_x = -nBy, \qquad A_y = -(n+1)Bx, \tag{18-9}$$

where n is a pure number. It is a simple matter to check that $\boldsymbol{B} = \boldsymbol{\nabla} \times \boldsymbol{A}$.

If $n = 0$, then the currents supplying $\boldsymbol{B}$ are all vertical. If $n = -1$, they are all horizontal. With a solenoid whose axis coincides with the z-axis, $n = -\frac{1}{2}$. Therefore, inside R and R_l,

$$\boldsymbol{E} = -\boldsymbol{\nabla} V \tag{18-10}$$

$$= -\frac{vBR}{R + R_l}\hat{\boldsymbol{y}}. \tag{18-11}$$

The rest of this example requires Relativity. Let us now see what happens inside the link, in its own reference frame S'. We assume that $v^2 \ll c^2$, which makes $\gamma \approx 1$.

From Sec. 13.4,

$$A'_x = A_x - \frac{vV}{c^2} = -nBy - \frac{v^2}{c^2}\frac{BR}{R + R_l}y \approx -nBy, \tag{18-12}$$

$$A'_y = A_y = -(n+1)Bx = -(n+1)Bvt, \tag{18-13}$$

$$V' = V - vA_x = \frac{vBR}{R + R_l}y + vnBy = \left(\frac{R}{R + R_l} + n\right)vBy. \tag{18-14}$$

$$V'_c = \left(\frac{R}{R + R_l} + n\right)vBD. \tag{18-15}$$

Note that the values of A' and of V' depend on the value of n. In other words, they depend on the particular geometry of the coils selected for generating $\boldsymbol{B}$. Observe also the appearance of a $\partial A'/\partial t'$ term in S'.

Now

$$\boldsymbol{E}' = -\boldsymbol{\nabla}' V' - \frac{\partial \boldsymbol{A}'}{\partial t'}, \tag{18-16}$$

where

$$\boldsymbol{\nabla}' = \frac{\partial}{\partial y'}\hat{\mathbf{y}} = \frac{\partial}{\partial y}\hat{\mathbf{y}}, \qquad t' = t - \frac{v}{c^2}x = t - \frac{v^2}{c^2}t \approx t. \tag{18-17}$$

Thus

$$\boldsymbol{E}' = -\frac{\partial V'}{\partial y}\hat{\mathbf{y}} - \frac{\partial \boldsymbol{A}'}{\partial t} = -\frac{\partial V'}{\partial y}\hat{\mathbf{y}} - \frac{\partial A'_y}{\partial t}\hat{\mathbf{y}} \tag{18-18}$$

$$= \left[-\left(\frac{R}{R+R_l} + n\right)vB + (n+1)vB\right]\hat{\mathbf{y}} \tag{18-19}$$

$$= -\left(\frac{R}{R+R_l} - 1\right)vB\hat{\mathbf{y}} = \frac{R_l}{R+R_l}vB\hat{\mathbf{y}}. \tag{18-20}$$

In general, $-\partial \boldsymbol{A}'/\partial t'$ *is not equal to* $\boldsymbol{v} \times \boldsymbol{B}$.

The quantity n has disappeared! We could have expected this because, clearly, $\boldsymbol{E}'$ must be independent of the configuration of the coils that generate the given magnetic field.

We could also have found $\boldsymbol{E}'$ directly, by simply transforming $\boldsymbol{E}$, with $v^2 \ll c^2$:

$$\boldsymbol{E}' = \boldsymbol{E}_{\parallel} + (\boldsymbol{E}_{\perp} + \boldsymbol{v} \times \boldsymbol{B}) = \boldsymbol{E}_{\perp} + \boldsymbol{v} \times \boldsymbol{B} \tag{18-21}$$

$$= \left(-\frac{vBR}{R+R_l} + vB\right)\hat{\mathbf{y}} = \frac{R_l}{R+R_l}vB\hat{\mathbf{y}}, \tag{18-22}$$

as above. This shows that, in the moving reference frame of the link, $\boldsymbol{E}'$ is equal to $\boldsymbol{E}$ plus $\boldsymbol{v} \times \boldsymbol{B}$.

The current I' is equal to I, and the voltmeter reading is the same for an observer in S' as for an observer in S.

EXAMPLE An Alternating-Current Generator

The rectangular loop of Fig. 18-5 rotates at an angular velocity ω in a uniform, constant $\boldsymbol{B}$. We calculate the induced electromotance $\mathcal{V}$, first through $\boldsymbol{v} \times \boldsymbol{B}$ and then through $d\Phi/dt$.

1. Along the right-hand side of the loop,

$$b\boldsymbol{v} \times \boldsymbol{B} = \frac{\omega a}{2}Bb\sin\theta\hat{\mathbf{x}} = \frac{\omega abB}{2}\sin\omega t\hat{\mathbf{x}}. \tag{18-23}$$

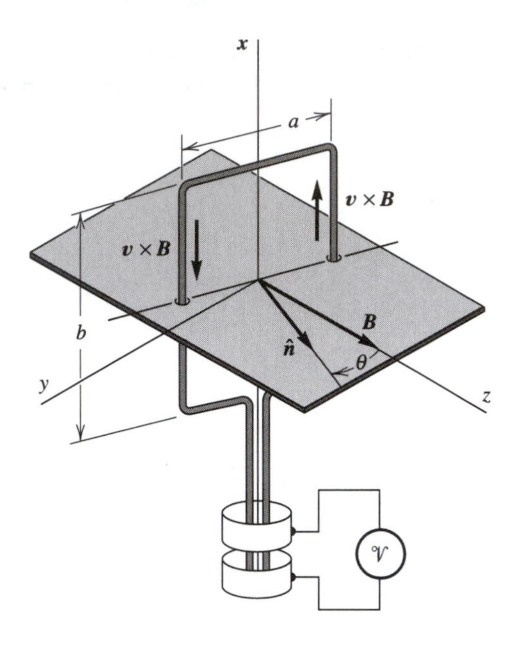

Figure 18-5 Loop rotating in a constant and uniform magnetic field $\boldsymbol{B}$. The slip rings provide contacts between the voltmeter and the loop.

Along the left-hand side we have the same induced electromotance, but directed downward as in the figure. Along the upper and lower sides, $\boldsymbol{v} \times \boldsymbol{B}$ is perpendicular to the wire. This crowds the conduction electrons sideways, thereby increasing the resistance imperceptibly, but $\boldsymbol{v} \times \boldsymbol{B}$ contributes nothing to the electromotance. So

$$\mathcal{V} = abB\omega \sin \omega t. \tag{18-24}$$

Notice that there is zero electromotance when $\omega t = n\pi$, where n is a whole number. Then $\boldsymbol{v}$ and $\boldsymbol{B}$ are parallel, and $\boldsymbol{v} \times \boldsymbol{B}$ is zero.

2. The time rate of change of the magnetic flux gives the same result:

$$\mathcal{V} = -\frac{d\Phi}{dt} = -\frac{d}{dt}(abB \cos \omega t) = abB\omega \sin \omega t. \tag{18-25}$$

18.2 THE FARADAY INDUCTION LAW FOR TIME-DEPENDENT *B*s. THE CURL OF *E*

Imagine now two closed and rigid circuits as in Fig. 18-6. The active circuit a is stationary, while the passive circuit b moves in some arbitrary way, say in the direction of a as in the figure. The current I_a is constant.

From Sec. 18.1, the electromotance induced in circuit b is

$$\mathcal{V} = \oint_b (\boldsymbol{v} \times \boldsymbol{B}) \cdot d\boldsymbol{l} = -\frac{d\Phi}{dt}, \tag{18-26}$$

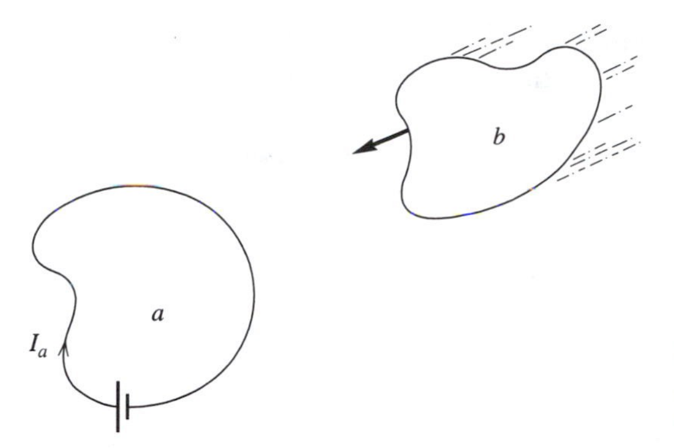

Figure 18-6 Circuit a is active and fixed in position. Circuit b is passive and moves in the field of a.

where Φ is the magnetic flux linking b. This seems trivial, but it is not, because $d\Phi/dt$ would be the same if both circuits were stationary and if I_a changed appropriately. This means that the *Faraday induction law,*

$$\mathcal{V} = -\frac{d\Phi}{dt}, \tag{18-27}$$

applies whether there are moving conductors in a constant $\boldsymbol{B}$, *or stationary conductors in a time-varying* $\boldsymbol{B}$. However, our argument is no more than plausible. A proper demonstration follows at the end of this chapter. It requires relativity. See the next section.

Assuming the correctness of the above result, the electromotance induced in a rigid and stationary circuit C lying in a time-varying magnetic field is

$$\mathcal{V} = \oint_C \boldsymbol{E} \cdot \boldsymbol{dl} = \int_{\mathcal{A}} (\nabla \times \boldsymbol{E}) \cdot \boldsymbol{d\mathcal{A}} = -\frac{d\Phi}{dt} = -\int_{\mathcal{A}} \frac{\partial \boldsymbol{B}}{\partial t} \cdot \boldsymbol{d\mathcal{A}}. \tag{18-28}$$

We have used Stokes's theorem in going from the first to the second integral, $\mathcal{A}$ being an arbitrary surface bounded by C. Also, we have a partial derivative under the last integral sign, to take into account the fact that the magnetic field can be a function of the coordinates as well as of the time. The right-hand screw rule applies.

The path of integration need not lie in conducting material.

Observe that the above equation involves only the integral of $\boldsymbol{E} \cdot \boldsymbol{dl}$. It does *not* give $\boldsymbol{E}$ as a function of the coordinates. Exceptionally, for simple geometries, one can deduce $\boldsymbol{E}$ from the value of $\mathcal{V}$.

Since the surface of area $\mathcal{A}$ chosen for the surface integrals is arbitrary, the equality of the third and last terms above means that

$$\boxed{\nabla \times \boldsymbol{E} = -\frac{\partial \boldsymbol{B}}{\partial t}.} \tag{18-29}$$

This is yet another Maxwell equation. This equation, like the other two (Eqs. 7-16 and 14-25), is valid on the condition that all the variables relate to the same reference frame.

The negative sign in Eqs. 18-27 and 18-29 is important. If Φ points into the paper and increases, then $d\Phi/dt$ points into the paper. Then, according to the right-hand screw rule, the negative sign means that the induced electromotance is counterclockwise: the induced electromotance tends to generate a magnetic field that counters the imposed *change* in flux. Lenz's law always applies.

If a closed circuit comprises N turns, each intercepting the same magnetic flux, then the electromotances add and the net electromotance is N times larger. Then the quantity $N\Phi$ is termed the *flux linkage*:

$$\Lambda = N\Phi \qquad \text{and} \qquad \mathcal{V} = -\frac{d\Lambda}{dt}. \tag{18-30}$$

Of course, the geometry of the circuit and the configuration of the field can be quite complex. Then this equation still applies and the geometric meaning of Λ becomes obscure, but $\mathcal{V}$, and hence Λ, are measurable quantities.

18.3 THE ELECTRIC FIELD STRENGTH E EXPRESSED IN TERMS OF THE POTENTIALS V AND A

An arbitrary, rigid, and stationary closed circuit C lies in a time-dependent $\boldsymbol{B}$. Then, from Sec. 18.2,

$$\oint_C \boldsymbol{E} \cdot d\boldsymbol{l} = -\frac{d}{dt}\int_{\mathcal{A}} \boldsymbol{B} \cdot d\boldsymbol{\mathcal{A}}, \tag{18-31}$$

where $\mathcal{A}$ is the area of any open surface bounded by C.

Now, from Sec. 15.1, we can replace the surface integral on the right by the line integral of the vector potential $\boldsymbol{A}$ around C:

$$\oint_C \boldsymbol{E} \cdot d\boldsymbol{l} = -\frac{d}{dt}\oint_C \boldsymbol{A} \cdot d\boldsymbol{l} = -\oint_C \frac{\partial \boldsymbol{A}}{\partial t} \cdot d\boldsymbol{l}. \tag{18-32}$$

There is no objection to inserting the time derivative under the integral sign, but then it becomes a partial derivative because $\boldsymbol{A}$ is normally a function of the coordinates as well as of the time.

Thus

$$\oint_C \left(\boldsymbol{E} + \frac{\partial \boldsymbol{A}}{\partial t}\right) \cdot d\boldsymbol{l} = 0, \tag{18-33}$$

where C is a closed curve, as stated above. Then, from Sec. 1.8, the expression enclosed in parentheses is equal to the gradient of some function:

$$\boldsymbol{E} + \frac{\partial \boldsymbol{A}}{\partial t} = -\boldsymbol{\nabla} V, \tag{18-34}$$

$$\boldsymbol{E} = -\boldsymbol{\nabla} V - \frac{\partial \boldsymbol{A}}{\partial t}, \tag{18-35}$$

where V is, of course, the electric potential.

So $\boldsymbol{E}$ is the sum of two terms, $-\boldsymbol{\nabla}V$ that results from accumulations of charge, and $-\partial\boldsymbol{A}/\partial t$, if there are time-dependent fields in the given reference frame.

This is a fundamental equation; we shall use it repeatedly. Note that it expresses $\boldsymbol{E}$ itself, and *not* its derivatives or its integral, at a given point, in terms of the derivatives of V and of $\boldsymbol{A}$, at that point. Its magnetic equivalent is

$$\boldsymbol{B} = \boldsymbol{\nabla} \times \boldsymbol{A}$$

of Sec. 14.4.

Note that Eq. 18-35 is a *local* relation: it relates $\boldsymbol{E}$, V, and $\boldsymbol{A}$ *at a given point*. But Eq. 18-31 is *non-local*: it relates *integrals* of $\boldsymbol{E}$ and $\boldsymbol{B}$. Both equations are correct, but the local equation 18-35 is often more useful. See, for example, Lorrain and S. Koutchmy (1998) for a discussion of the field $-\partial\boldsymbol{A}/\partial t$ in the solar atmosphere. Unfortunately, the local relation is much less well known.

The Faraday induction law, in differential form (Eq. 18-29), relates *space* derivatives of $\boldsymbol{E}$ to the *time* derivative of $\boldsymbol{B}$ at a given point.

Observe that $\boldsymbol{\nabla}V$ is a function of V, which depends on the positions of the charges. However, $\partial\boldsymbol{A}/\partial t$ is a function of the time derivative of the current density $\boldsymbol{J}$, and hence of the acceleration of the charges.

The relations

$$\boldsymbol{E} = -\boldsymbol{\nabla}V - \frac{\partial \boldsymbol{A}}{\partial t} \quad \text{and} \quad \boldsymbol{B} = \boldsymbol{\nabla} \times \boldsymbol{A} \tag{18-36}$$

are always valid in any given inertial reference frame.

In a time-dependent $\boldsymbol{B}$, the electromotance induced in a circuit C is

$$\mathcal{V} = -\int_C \frac{\partial \boldsymbol{A}}{\partial t} \cdot d\boldsymbol{l}. \tag{18-37}$$

EXAMPLE: The Mystery of the Secondary on a Long Solenoid

You wind a secondary on a long solenoid, near the center, as in Fig. 18-7. If you vary the current in the solenoid, a voltage $\mathcal{V}$ appears at the terminals of the secondary, as in Eq. 18-2. But how can that be? There is no magnetic field outside the solenoid! A.P. French (1994) asked the question, and several teachers proposed various answers over the following years, but none mentioned the vector potential $\boldsymbol{A}$!

True, there is essentially no $\boldsymbol{B}$-field outside a long solenoid. But there is an $\boldsymbol{A}$-field! The vector $\boldsymbol{A}$ is azimuthal, as in the Example that follows Sec. 15.1. Say the secondary has a single turn of radius a. According to Eq. 15-3, at the radius a, $A = \Phi/(2\pi a)$, where Φ is the magnetic flux in the solenoid, and the electromotance $\mathcal{V}$ induced on the secondary is $-d\Phi/dt$. If the secondary has N turns, then $\mathcal{V} = -Nd\Phi/dt$.

This thought experiment shows that "flux cutting" is not an entirely satisfactory interpretation of the Faraday induction law.

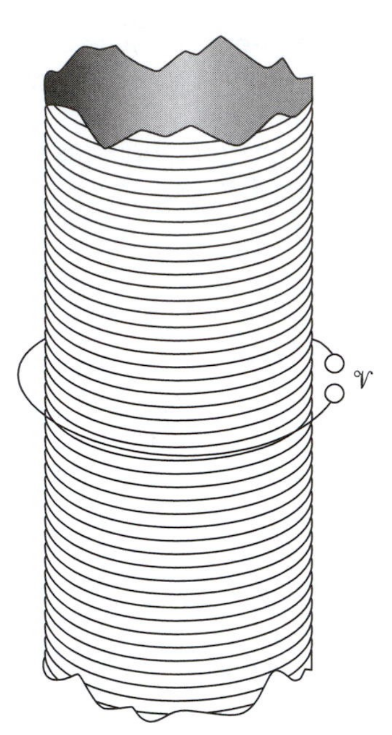

Figure 18-7 Single-turn secondary on a long solenoid.

EXAMPLE Eddy Currents

Imagine a sheet of copper lying inside a solenoid, in a plane perpendicular to the axis. The solenoid carries an alternating current.

According to Lenz's law, the $-\partial \boldsymbol{A}/\partial t$ electric field induces currents in the copper that tend to cancel the changes in the net $\boldsymbol{B}$. These currents are azimuthal because $\boldsymbol{A}$ and $\partial \boldsymbol{A}/\partial t$ are azimuthal (Example in Sec. 15.1).

Currents induced in bulk conductors by changing magnetic fields are termed *eddy currents*.

Eddy currents can be useful. For example, they dissipate energy in various damping mechanisms.

They are harmful in transformers because they cause Joule losses in the core. Transformer cores are usually assembled from thin sheets of transformer iron, called *laminations*, a fraction of a millimeter thick in small units, insulated from each other by a thin layer of oxide. With a solid core the eddy currents would largely cancel changes in magnetic flux. Also, the Joule losses would be excessive.

In audio transformers the iron alloy is sometimes in the form of a powder molded in an insulating binder. The transformer is then said to have a *powdered iron core.*

Ferrites serve at audio frequencies and above, right up to 5 gigahertz, for inductors and transformers, and in recording heads for reading tapes, diskettes, or hard disks. Ferrites are ceramic-type materials composed of oxides of various metals, often without iron. The oxides are first powdered, then molded, and finally heated under pressure. Their magnetic permeabilities, of the order of 100, are frequency-dependent, and their conductivities can be as low as 10^{-3} siemens/meter. The conductivity of transformer iron is about 10^{6} siemens/meter.

EXAMPLE Induced Electromotance in a Rigid Circuit

In the rigid circuit of Fig. 18-8, we assume that (1) the horizontal wires have zero resistance, (2) the resistance R_a is a long distance away from I', and (3) both R_a and R_b are large enough to render the magnetic field of I negligible compared to that of I'. Thus $\boldsymbol{B}$ and $\boldsymbol{A}$ are essentially those of I', and point in the directions shown. The vector potential of the current I' at the position of R_b is $\boldsymbol{A}_b$, and it points down. According to Lenz's law (Sec. 18.2), an increase in I' induces an electromotance and a current I in the counterclockwise direction.

One may ascribe the induced electromotance either to the changing magnetic flux Φ or to the electric field strength $-\partial \boldsymbol{A}/\partial t$ in R_b, with

$$\Phi = \oint_C \boldsymbol{A} \cdot \boldsymbol{dl} \approx A_b D. \tag{18-38}$$

The line integral runs clockwise because of the right-hand screw rule. The induced electromotance is thus

$$\mathcal{V} = -\frac{d\Phi}{dt} = -\frac{dA_b}{dt} D. \tag{18-39}$$

This is counterclockwise if Φ points into the paper and increases, or if I' increases. Then

$$I = \frac{(dA_b/dt)D}{R_a + R_b} \tag{18-40}$$

in the direction shown.

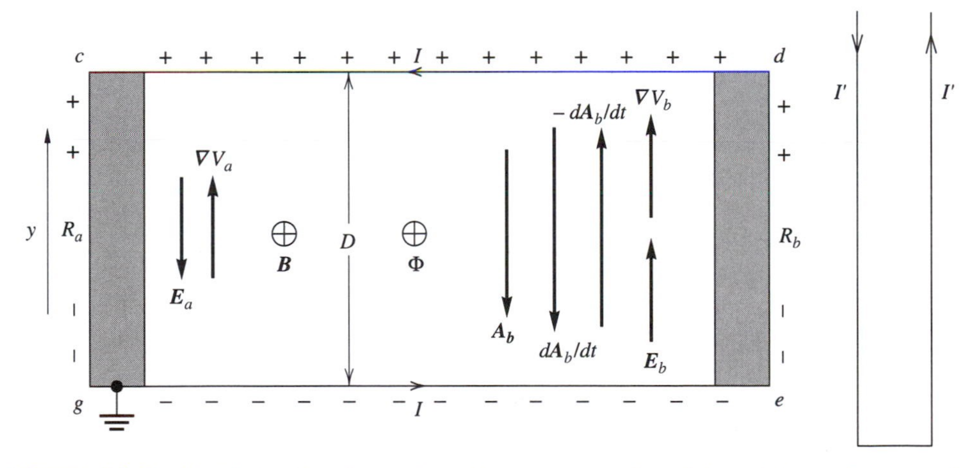

Figure 18-8 Rigid circuit *gcde* terminated by resistances R_a and R_b and situated near a pair of wires carrying a current I'. The lengths of the arrows show the relative magnitudes of $\boldsymbol{E}$, ∇V, and $d\boldsymbol{A}/dt$, with $R_a = R_b$.

Inside the resistance R_a, $A \approx 0$ by hypothesis and

$$\mathbf{E}_a = -\nabla V_a = -\frac{IR_a}{D}\hat{\mathbf{y}}, \qquad V_c = IR_a, \tag{18-41}$$

with $\mathbf{E}_a$ and ∇V_a pointing as in Fig. 18-8.

The potential at e is zero for the following reason. Inside the horizontal wires, $\mathbf{E} = \mathbf{J}/\sigma$, where $\mathbf{J}$ has some finite value and σ is infinite, by hypothesis. So

$$E_x = 0, \qquad -\frac{\partial V}{\partial x} - \frac{\partial A_x}{\partial t} = 0. \tag{18-42}$$

But A_x is everywhere zero because the current I' has essentially no x-component. So

$$\frac{\partial V}{\partial x} = 0, \qquad V_e = 0 \tag{18-43}$$

and

$$V_c = V_d = IR_a. \tag{18-44}$$

This potential on the cd wire results from the accumulation of positive surface charges on the upper half of the circuit, and negative charges on the lower half, as in Fig. 18-8.

Inside R_b, therefore,

$$\nabla V_b = \frac{IR_a}{D}\hat{\mathbf{y}} \tag{18-45}$$

points upward, and

$$\mathbf{E}_b = -\nabla V_b - \frac{\partial \mathbf{A}_b}{\partial t} = \frac{IR_b}{D}\hat{\mathbf{y}} \tag{18-46}$$

also points upward. Of course, $\mathbf{E}$ points in the same direction as $\mathbf{J}$.

If $R_a = R_b$, then

$$|\mathbf{E}_a| = |\mathbf{E}_b|, \qquad \left|\frac{\partial \mathbf{A}_b}{\partial t}\right| = 2|\nabla V_b|. \tag{18-47}$$

18.4 THE $\mathbf{E}$, $-\nabla V$, $-\partial \mathbf{A}/\partial t$, AND $\mathbf{v} \times \mathbf{B}$ FIELDS

In any given inertial reference frame, say S, the equation

$$\mathbf{E} = -\nabla V - \frac{\partial \mathbf{A}}{\partial t} \tag{18-48}$$

always applies.

If a charge Q moves at a velocity $\boldsymbol{v}$ with respect to S, then for an observer on S the force is

$$\boldsymbol{F} = Q(\boldsymbol{E} + \boldsymbol{v} \times \boldsymbol{B}) = Q\left(-\nabla V - \frac{\partial \boldsymbol{A}}{\partial t} + \boldsymbol{v} \times \boldsymbol{B}\right) \tag{18-49}$$

and, in a medium of conductivity σ,

$$\boldsymbol{J} = \sigma\left(-\nabla V - \frac{\partial \boldsymbol{A}}{\partial t} + \boldsymbol{v} \times \boldsymbol{B}\right), \tag{18-50}$$

which is Eq. 18-5 with $\boldsymbol{E}$ replaced by its two components as in Eq. 18-35.

All the variables are measured with respect to the *same* reference frame S. These equations are valid even if v approaches the speed of light.

†For an observer on the moving body, say in reference frame S', the body is at rest and

$$\boldsymbol{F}' = Q\boldsymbol{E}' = Q\left(-\nabla' V' - \frac{\partial \boldsymbol{A}'}{\partial t'}\right). \tag{18-51}$$

*18.5 RELATING THE TWO FORMS OF THE FARADAY INDUCTION LAW

†We found above that if a rigid circuit moves in a constant $\boldsymbol{B}$, then the induced electromotance follows Faraday's induction law, Eq. 18-2. Then we concluded that the same law applies to a stationary circuit lying in a time-dependent $\boldsymbol{B}$.

Passing from one form of the law to the other requires Relativity.

In the remainder of this chapter we call $\boldsymbol{v}$ the velocity of S' with respect to S. We assume that $\boldsymbol{v}$ is constant.

Equation 18-27 refers to the induced electromotance, as measured in the fixed reference frame S. From our experience with relativistic calculations, it is by no means evident that, in the frame S' of the moving circuit,

$$\mathcal{V}' = -\frac{d\Phi'}{dt'}. \tag{18-52}$$

That is, in fact, true because, as we shall see,

$$\mathcal{V}' = \gamma\mathcal{V}, \qquad \Phi' = \Phi, \qquad dt' = \frac{dt}{\gamma}. \tag{18-53}$$

We can immediately accept the equation for dt' for the following reason. For any point in S',

$$t = \gamma\left(t' + \frac{vx'}{c^2}\right), \tag{18-54}$$

---

†Relativity is a prerequisite here.

from Sec. 11.4. Thus, at any given point in S', x' is fixed and

$$dt = \gamma dt'. \tag{18-55}$$

Proving the other two equations takes a bit longer.

*18.5.1 Transformation of a Magnetic Flux

The magnetic flux linking a given closed circuit bounding an area $\mathcal{A}$ is

$$\Phi = \int_{\mathcal{A}} \boldsymbol{B} \cdot d\boldsymbol{\mathcal{A}} \tag{18-56}$$

in a reference frame S and

$$\Phi' = \int_{\mathcal{A}'} \boldsymbol{B}' \cdot d\boldsymbol{\mathcal{A}}' \tag{18-57}$$

in S'.

Of course, the surface of area $\mathcal{A}$ in frame S has a different shape in frame S', because of the Lorentz contraction, and a different area $\mathcal{A}'$. A given element of area (say it is painted red) carries a flux $\boldsymbol{B} \cdot d\boldsymbol{\mathcal{A}}$ in frame S, and $\boldsymbol{B}' \cdot d\boldsymbol{\mathcal{A}}'$ in S'. Thus

$$\frac{d\Phi'}{d\Phi} = \frac{\boldsymbol{B}' \cdot d\boldsymbol{\mathcal{A}}'}{\boldsymbol{B} \cdot d\boldsymbol{\mathcal{A}}}. \tag{18-58}$$

But, from the first example in Sec. 11.4 and from Sec. 13.4, and setting $\boldsymbol{E} = 0$,

$$\boldsymbol{B}' \cdot d\boldsymbol{\mathcal{A}}' = (\boldsymbol{B}_{\parallel} + \gamma \boldsymbol{B}_{\perp}) \cdot \left(d\boldsymbol{\mathcal{A}}_{\parallel} + \frac{d\boldsymbol{\mathcal{A}}_{\perp}}{\gamma} \right) \tag{18-59}$$

$$= \boldsymbol{B}_{\parallel} \cdot d\boldsymbol{\mathcal{A}}_{\parallel} + \boldsymbol{B}_{\perp} \cdot d\boldsymbol{\mathcal{A}}_{\perp} = \boldsymbol{B} \cdot d\boldsymbol{\mathcal{A}}, \tag{18-60}$$

and $\Phi' = \Phi$. *The magnetic flux linking a rigid closed curve is invariant under a Lorentz transformation.*

*18.5.2 Transformation of an Electromotance

Refer again to Fig. 18-6 and call S' the reference frame of a moving rigid circuit. In S',

$$\mathcal{V}' = \oint \boldsymbol{E}' \cdot d\boldsymbol{l}'. \tag{18-61}$$

Since we are only interested in motional electromotance for the moment, we may assume that $\boldsymbol{B}$ is constant, and $\boldsymbol{E}$ is zero in S. Then, from Sec. 13.4,

$$\mathcal{V}' = \oint_C \gamma(\boldsymbol{v} \times \boldsymbol{B}) \cdot d\boldsymbol{l}'. \tag{18-62}$$

Now the vector product is perpendicular to $\boldsymbol{v}$. It is therefore only the perpendicular component of $\boldsymbol{dl'}$ that matters, $\boldsymbol{dl'}_\perp = \boldsymbol{dl}_\perp$, and

$$\mathcal{V}' = \gamma \oint (\boldsymbol{v} \times \boldsymbol{B}) \cdot \boldsymbol{dl} = \gamma \mathcal{V}. \tag{18-63}$$

We have therefore proved all three equations 18-53. As a consequence, we have shown that, under any circumstance, but in a single reference frame, the electromotance induced in a closed circuit associated with a changing magnetic flux is given by

$$\mathcal{V} = -\frac{d\Phi}{dt}. \tag{18-64}$$

The positive directions chosen for $\mathcal{V}$ and for Φ follow the right-hand screw rule.

18.6 FIVE KEY EQUATIONS

It is useful at this stage to group the following equations:

$$\boldsymbol{E} = -\boldsymbol{\nabla} V - \frac{\partial \boldsymbol{A}}{\partial t}, \tag{18-36}$$

$$\boldsymbol{J} = \sigma(\boldsymbol{E} + \boldsymbol{v} \times \boldsymbol{B}), \tag{18-50}$$

$$\boldsymbol{\nabla} \times \boldsymbol{E} = -\frac{\partial \boldsymbol{B}}{\partial t}, \tag{18-29}$$

$$\boldsymbol{B} = \boldsymbol{\nabla} \times \boldsymbol{A}, \tag{14-26, 18-36}$$

$$\boldsymbol{\nabla} \times \boldsymbol{B} = \mu_0 \boldsymbol{J}. \tag{15-17}$$

These equations are general, if $\boldsymbol{J}$ in the last equation includes more than the conduction current, as in Sec. 21.2. All the equations are *local* relations that apply to the field variables and to their derivatives *at a given point.*

In each equation all the terms concern the *same* reference frame.

The integral form of the third equation above is Eq. 18-28, which is non-local because it involves integrals over a curve and over a surface. The integral form of the fifth equation above is Eq. 15-18, which is also non-local.

18.7 SUMMARY

A body moves in some arbitrary fashion in a constant, but not necessarily uniform, magnetic field. At a point fixed to the moving body, a charge Q experiences a force $Q\boldsymbol{v} \times \boldsymbol{B}$. The *motional electromotance* along an arbitrary curve C is

$$\mathcal{V} = \int_C (\boldsymbol{v} \times \boldsymbol{B}) \cdot \boldsymbol{dl}. \tag{18-1}$$

Here $\boldsymbol{v}$ is the velocity of the charge, and $\boldsymbol{B}$ is the magnetic flux density at that point in space.

When the curve C is closed, the motional electromotance is also given by

$$\mathcal{V} = -\frac{d\Phi}{dt}, \tag{18-2}$$

where Φ is the enclosed flux. This is the *Faraday induction law.* The right-hand screw rule applies. This law also applies to a fixed circuit situated in a time-varying magnetic field.

At any point in space, in a given reference frame,

$$\boxed{\nabla \times \boldsymbol{E} = -\frac{\partial \boldsymbol{B}}{\partial t}.} \tag{18-29}$$

This is one of Maxwell's equations.

Lenz's law states that the electromotance induced in a closed circuit tends to oppose *changes* in the magnetic flux linking the circuit.

With a multiturn closed circuit,

$$\mathcal{V} = -\frac{d\Lambda}{dt}, \tag{18-30}$$

where Λ is the *flux linkage.* If an N-turn circuit is linked by a flux Φ, $\Lambda = N\Phi$.

In any given inertial reference frame,

$$\boldsymbol{E} = -\nabla V - \frac{\partial \boldsymbol{A}}{\partial t} \quad \text{and} \quad \boldsymbol{B} = \nabla \times \boldsymbol{A}. \tag{18-36}$$

A charge Q moving at a velocity $\boldsymbol{v}$ in superposed electric and magnetic fields is subjected to a force

$$\boldsymbol{F} = Q(\boldsymbol{E} + \boldsymbol{v} \times \boldsymbol{B}) = Q\left(-\nabla V - \frac{\partial \boldsymbol{A}}{\partial t} + \boldsymbol{v} \times \boldsymbol{B}\right), \tag{18-49}$$

and, in a medium of conductivity σ,

$$\boldsymbol{J} = \sigma(\boldsymbol{E} + \boldsymbol{v} \times \boldsymbol{B}) = \sigma\left(-\nabla V - \frac{\partial \boldsymbol{A}}{\partial t} + \boldsymbol{v} \times \boldsymbol{B}\right). \tag{18-50}$$

PROBLEMS

18-1. (*18.1*) The thought experiment of Fig. 18-3

Show that there is conservation of energy in the thought experiment of Fig. 18-3.

18-2. (*18.1*) Tides and the magnetic field of the earth

Discuss how tides affect the magnetic field of the earth by considering the case of a river flowing into the sea in the east-to-west direction in the northern hemisphere. Remember that the magnetic pole situated at the *north geographic* pole is a *south magnetic* pole. The vector $\boldsymbol{B}$ points *downward* in the *northern* hemisphere.

18-3. (*18.1*) The magnetic braking force on a satellite

A natural satellite whose diameter is 10^4 meters moves at a velocity of 1 kilometer/second in the direction normal to the magnetic field of a planet in a region where $B = 10^{-7}$ tesla. The satellite has an appreciable conductivity.

(a) The satellite moves in a perfect vacuum. What happens?

(b) The ambient gas has a density of the order of 10^{10} particles per cubic meter, the particles being either electrons or singly charged ions. Each half of the satellite collects particles of the correct sign in sweeping through space. Calculate the order of magnitude of the current.

(c) Calculate the order of magnitude of the braking force.

(d) Someone suggests that this current could provide power for an artificial satellite traveling in the same field at the same velocity. Inversely, a current in the opposite direction could serve to propel the satellite. What is your opinion?

Artificial satellite speeds range from about 4 to 8 kilometers/second, and $\boldsymbol{v} \times \boldsymbol{B}$ in the ionosphere and magnetosphere ranges from about 100 microvolts/meter to 320 millivolts/meter.

18-4. (*18.1*) Boat testing tank

A carriage runs on rails on either side of a long tank of water equipped for testing boat models. The rails are 3.0 meters apart and the carriage has a maximum speed of 20 meters per second.

(a) Calculate the maximum voltage between the rails if the vertical component of the earth's magnetic field is 2.0×10^{-5} tesla.

(b) What would be the voltage if the tank were situated at the magnetic equator?

18-5. (*18.1*) A bar magnet is pulled through a conducting ring at a constant velocity as in Fig. 18-9.

Sketch curves of (a) the magnetic flux Φ, (b) the current I, (c) the power dissipated in the ring, as functions of the time.

Use the positive directions for Φ and I shown in the figure.

This phenomenon has been used to measure the speeds of projectiles. The projectile, with a tiny permanent magnet inserted into its nose, is made to pass in succession through two coils, separated by a distance of about 100 millimeters. The time delay between the pulses is a measure of the speed. The method has been used up to speeds of 5 kilometers/second.

18-6. (*18.2*) Induced electromotance

A loop of wire is situated in a time-dependent magnetic field with

$$B = 1.00 \times 10^{-2} \cos(2\pi \times 60)t$$

perpendicular to the plane of the loop.

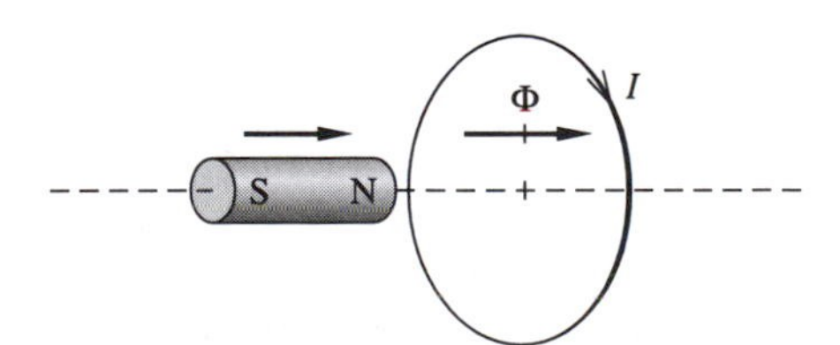

Figure 18-9

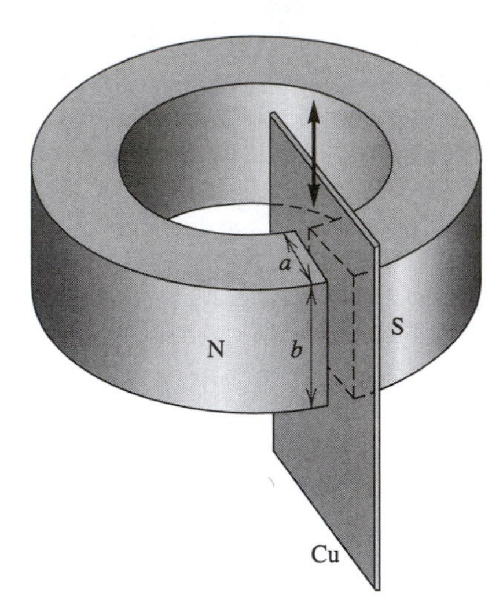

Figure 18-10

Calculate the induced electromotance in a 100-turn square loop 100 millimeters on the side.

18-7. (*18.3*) Eddy-current damping

Figure 18-10 shows one common type of eddy-current damper. Motion of the copper plate in the field of the permanent magnet induces currents that tend to oppose the motion, according to Lenz's law. Joule losses in the plate dissipate its kinetic energy.

Dampers of this general type are used mostly, but not exclusively, in low-power devices such as watt-hour meters and balances. As you will see, the braking force is proportional to the speed, as in a viscous fluid.

(a) Explain qualitatively, but in greater detail, the origin of the braking force.

(b) Could you design an automobile speedometer that uses eddy currents?

(c) Say B is uniform over the pole face. The path followed by the current is complex; set $R \approx 3a/(\sigma bs)$. This quantity is of the order of 3. The plate has a thickness s and a conductivity σ. Calculate the current.

(d) Calculate the braking force F. This is proportional to the conductivity. So the plate should be either copper or aluminum. An even better solution is to use an iron plate faced with copper.

(e) Calculate the power IV dissipated in the plate. This should be equal to Fv.

(f) Estimate the value of B required in eddy-current disk brakes for a small bus. The conductivity of copper is 5.8×10^7 siemens/meter. Estimate the power dissipated at each wheel. Why does a vehicle equipped with eddy-current brakes still need conventional brakes?

In mountainous regions some buses and trucks are equipped with dynamos that brake by generating electric power that is dissipated in a large resistance on the roof.

18-8. (*18.2*) Detecting flaws in metal tubing

Figure 18-11 shows the principle of operation of a device for detecting flaws in metal tubing, or rod. The coils a provide a large *gradient* of magnetic field along the axis, as in Prob. 14-10. Coil b is connected to a monitor. The tubing T moves at a con-

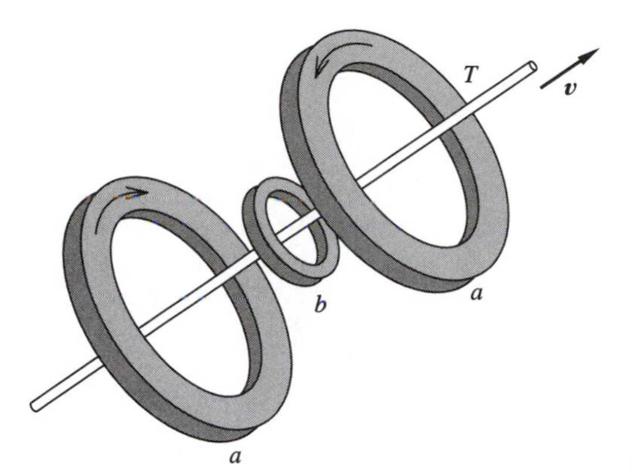

Figure 18-11

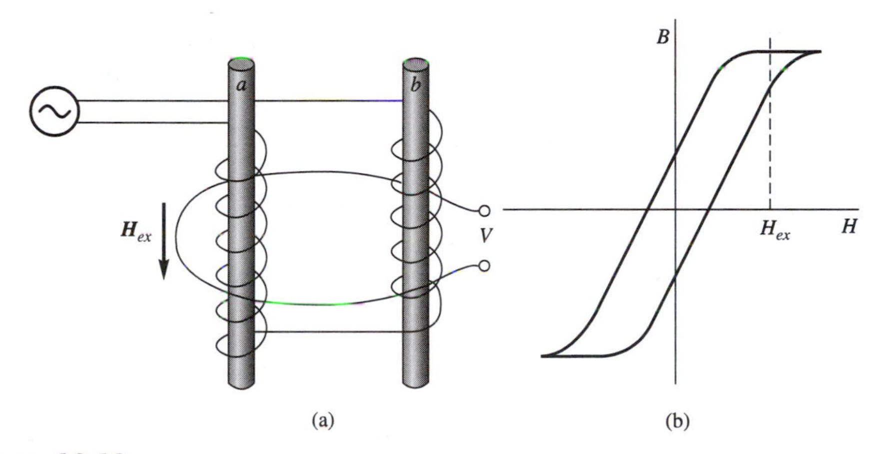

Figure 18-12

constant velocity $\boldsymbol{v}$ along the axis of symmetry. A voltage appears across coil b when a flaw passes through. Explain.

18-9. (*18.2*) The flux-gate magnetometer†

A *magnetometer* measures B. One common type is the *flux-gate magnetometer*, which puts to use the hysteresis curve. There exist many forms, one of which is shown in Fig. 18-12(a). The two rods are made of a ferromagnetic material such as a ferrite, whose hysteresis curve is shown in Fig. 18-12(b). The twin coils are in series and are wound as in the figure so as to magnetize the rods in opposite directions. The current through these coils is sufficient to carry the material through a complete hysteresis loop.

In the absence of an external field H_{ex}, the magnetic fluxes through the rods cancel, and $V = 0$.

(a) Sketch $\Phi(t)$ and $V(t)$ for each rod when $H_{ex} = 0$.

†This problem requires Sec. 16.6.

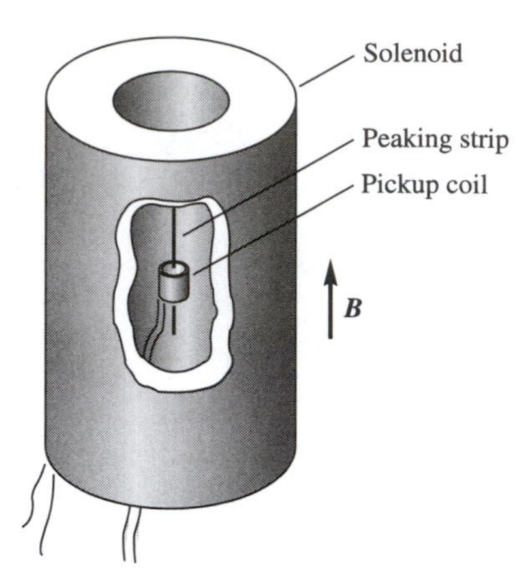

Figure 18-13

(b) Sketch the same quantities for $H_{ex} \neq 0$. You will notice that if the oscillator operates at a frequency f, the fundamental frequency of V is $2f$. This facilitates the measurement because the detector can be made to reject the frequency f.

Flux-gate magnetometers can measure fields down to a few nanoteslas.

18-10. (*18.2*) The peaking strip†

A *peaking strip* serves to measure B. It consists of a fine wire of Permalloy (see below) oriented in the direction of $\mathbf{B}$ with a small pickup coil of a few thousand turns near the center, on the axis of a solenoid, as in Fig. 18-13.

To measure the ambient B, the solenoid carries a direct current that just cancels B, plus a small alternating current. Then the H on the axis of the solenoid is that of the alternating current, and the strip goes through a hysteresis loop at every cycle.

With molybdenum Permalloy the hysteresis loop is approximately rectangular, and the voltage induced in the small coil has two sharp peaks, one positive and one negative, which can be observed on an oscilloscope.

When the oscilloscope sweep is synchronized with the alternating current in the solenoid, the two peaks are symmetric if the time-averaged H on the axis of the solenoid is zero. Then the steady field of the solenoid exactly cancels the ambient B and the current in the solenoid is a measure of B.

The peaking strip has a rather limited range of applications. (1) The solenoid has to be at least about 10 centimeters long because it must be at least a few times longer than the strip, to avoid excessive end effects. But the length of the strip must be much larger than its diameter, again to reduce end effects, and a decrease in the strip cross-section decreases the signal proportionately. (2) The ambient B cannot be larger than a few hundredths of a tesla, for otherwise the power dissipated in the solenoid becomes excessive. (3) If one measures B in the neighborhood of a pole-piece, the field of the solenoid alters the permeability of the iron locally.

Calculate the peak voltage induced in the pickup coil under the following conditions: strip diameter, 25 micrometers; number of turns in the pickup coil, 1000; maxi-

†This problem requires Sec. 16.6.

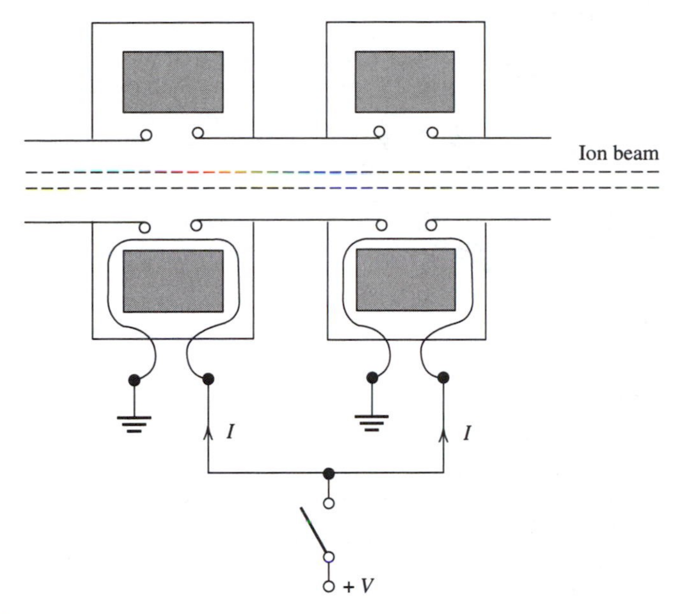

Figure 18-14

mum value of μ_r, 75,000; frequency, 60 hertz; amplitude of the alternating H, 7 ampere-turns/meter.

18-11. (*18.2*) Measuring a resistivity without contacts

It is useful to be able to measure the resistivity of a sample without having to cement contacts to it. One method involves placing a disk of the material inside a solenoid carrying an alternating current, with the two axes parallel, and measuring the power absorbed by the disk. The disk has a radius a, a thickness s, and a conductivity σ. The magnetic field is uniform, and $B = B_m \cos \omega t$. We neglect the magnetic field of the induced currents. We therefore restrict ourselves to low-conductivity materials.

Find the relation between σ and the average dissipated power P.

18-12. (*18.3*) The induction linear accelerator

Figure 18-14 shows a schematic diagram of a section of an induction linear accelerator. It consists of a series of ferrite toroids linked by the ion beam and by one-turn loops that carry large pulsed currents.

One such accelerator comprises 200 toroids and accelerates a 10-kiloampere pulsed electron beam to 50 megaelectronvolts. Its total length is 80 meters, and the current pulses are 70 nanometers wide.

Explain its operation qualitatively.

18-13. (*18.3*) A magnetometer that uses eddy currents

Figure 18-15 shows the principle of operation of a magnetometer that can measure magnetic fields as small as 10^{-8} tesla and up to 10^{-2} tesla. The aluminum plate P turns on the axis AA at the angular velocity ω in the ambient field B_0 that we wish to measure. The fluctuating eddy currents induced in P produce a fluctuating magnetic flux through the fixed coil C, which has N turns, and the voltage V is a measure of B_0.

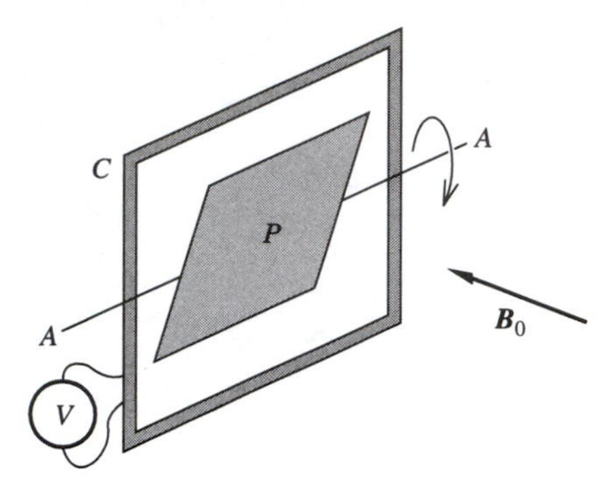

Figure 18-15

The plate is 10 millimeters square and is cemented inside the plastic rotor of a small air turbine that operates at 1000 revolutions/second. The only metallic parts are the plate and the coil.

An exact calculation of V as a function of geometry, of ω, and of B_0 would be difficult. But this is unnecessary because we can calibrate the instrument with Helmholtz coils (Prob. 14-9).

(a) How does V vary with B_0 and with ω? Set $\omega t = 0$ when the plate lies in the plane of C.

(b) What is the frequency of V?

18-14. (*18.3*) Electric conduits

The U.S. National Electrical Code rules that both conductors of a circuit operating on alternating current, if enclosed in a metallic conduit, must be run in the same conduit.

Let us suppose that we have a single conductor carrying an alternating current and enclosed within a conducting tube.

Show that both $\mathbf{A}$ and $\partial \mathbf{A}/\partial t$ are longitudinal in the tube.

So, with a single wire, a longitudinal current is induced in the tube. This causes a needless power loss, and may even cause sparking at faulty joints. With two conductors carrying equal and opposite currents, both $\mathbf{A}$ and $\partial \mathbf{A}/\partial t$ are essentially zero in the conduit.

CHAPTER

19

*MAGNETIC FIELDS VI†

Mutual and Self-Inductance

This chapter concerns the electromotance induced in a circuit when its magnetic flux linkage changes. The change in flux can occur either because of a change in the current flowing through the circuit itself or in currents flowing elsewhere, or because of a change in the geometry of the circuits.

*19.1 MUTUAL INDUCTANCE *M*

In Fig. 19-1 the active circuit a carries a current I_a. The magnitude of the flux Φ_{ab} that originates in a and links b is

$$\Phi_{ab} = \oint_b \mathbf{A}_a \cdot d\mathbf{l}_b = \oint_b \left(\frac{\mu_0 I_a}{4\pi} \oint_a \frac{d\mathbf{l}_a}{r} \right) \cdot d\mathbf{l}_b, \tag{19-1}$$

where r is the distance between the elements $d\mathbf{l}_a$ and $d\mathbf{l}_b$. Thus

$$\Phi_{ab} = \frac{\mu_0 I_a}{4\pi} \oint_a \oint_b \frac{d\mathbf{l}_a \cdot d\mathbf{l}_b}{r} = M_{ab} I_a, \tag{19-2}$$

†This whole chapter can be omitted without losing continuity. For a more extensive discussion of electric circuits, see Lorrain, Corson, and Lorrain (1988).

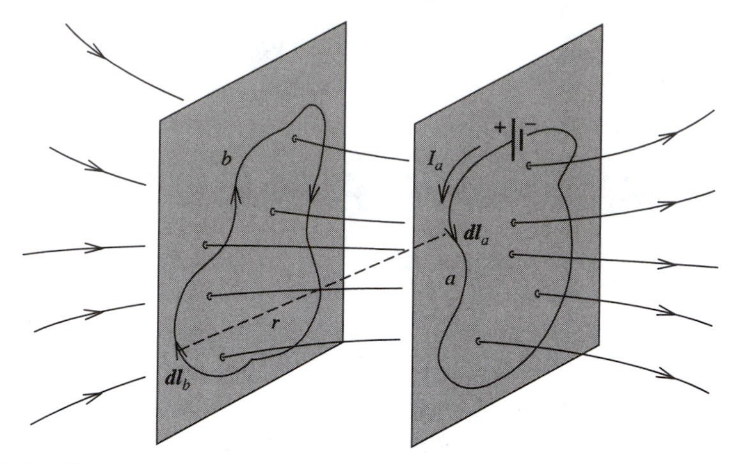

Figure 19-1 The active circuit a bears a current I_a that increases. Part of its magnetic flux links the passive circuit b. The electromotance $\mathcal{V}_b$, induced in b is in the direction shown.

where

$$M_{ab} = \frac{\mu_0}{4\pi}\oint_a \oint_b \frac{\mathbf{dl}_a \cdot \mathbf{dl}_b}{r} \tag{19-3}$$

is the *mutual inductance* between the two circuits. Mutual inductance is expressed in webers/ampere, or in *henrys*. This is the *Neumann equation*.

If the geometry is more complex than that of Fig. 19-1, the above reasoning still applies, except that the flux Φ_{ab} becomes the flux linkage Λ_{ab} (Secs. 15.1, 18.2), and

$$\Lambda_{ab} = M_{ab} I_a. \tag{19-4}$$

The electromotance induced in b by a change in I_a is

$$\mathcal{V}_b = -\frac{d\Lambda_{ab}}{dt} = -M_{ab}\frac{dI_a}{dt} - I_a\frac{dM_{ab}}{dt}. \tag{19-5}$$

The mutual inductance is usually constant. Then the last term on the right vanishes.

Therefore the mutual inductance between two circuits is 1 henry if a current changing at the rate of 1 ampere/second in one circuit induces an electromotance of 1 volt in the other.†

If current I_a is sinusoidal, then

$$\mathcal{V}_b = -j\omega M_{ab} I_a. \tag{19-6}$$

†As a rule, the sign of M is immaterial. There are occasions, however, when phases matter, and then one requires the sign of M. Also, the directions of forces and torques between circuits depend on the sign of M, as we shall see in Chap. 20. The sign of M is defined as follows. First select arbitrary positive directions for the currents in the two circuits. Then M is positive if a positive current in a gives in b a flux linkage of the same sign as that resulting from a positive current in b.

A device comprising two circuits designed to possess mutual inductance is termed a *mutual inductor*, or a *transformer.*

The Neumann equation is seldom useful, because the double integral is difficult to evaluate, even for simple geometries. This is not a matter for concern, because mutual inductances are easily measured. One can also calculate M_{ab} from the ratio $\mathcal{V}_b/(dI_a/dt)$.

The Neumann equation is nonetheless interesting. It shows that mutual inductance depends solely on the geometry of the system. We had a similar situation with respect to capacitance.

Also, we can interchange the subscripts in the Neumann equation without altering the mutual inductance. Therefore

$$M_{ab} = M_{ba} = M. \tag{19-7}$$

This is surprising, because the circuits can have different shapes and different numbers of turns. That is,

$$\text{if} \quad \mathcal{V}_b = -M\frac{dI_a}{dt}, \quad \text{then} \quad \mathcal{V}_a = -M\frac{dI_b}{dt}. \tag{19-8}$$

*EXAMPLE The Mutual Inductance Between Two Coaxial Solenoids

In the third example in Sec. 14.2 we found that, inside a long solenoid with N' turns per meter and bearing a current I (ignoring end effects),

$$B = \mu_0 N' I. \tag{19-9}$$

We add a second winding over the solenoid, as in Fig. 19-2, and we assume that both windings are long compared to their common diameter, in order to render end effects negligible.

First, we assume that solenoid a, of radius R and number of turns N_a, bears a current I_a. The magnetic flux of a that links solenoid b, of the same radius and with N_b turns, is then, from the third example in Sec. 14.2,

$$\Phi_{ab} = \pi R^2 \mu_0 \frac{N_a}{a} I_a, \tag{19-10}$$

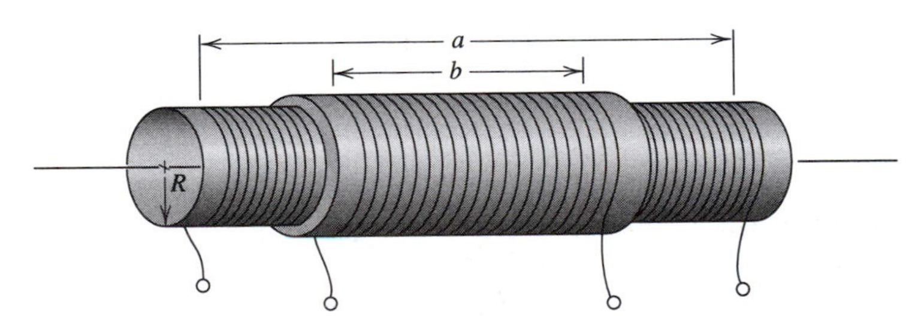

Figure 19-2 Two coaxial solenoids. We have shown different radii for clarity, but coil b is wound directly over a.

and

$$M_{ab} = \frac{N_b \Phi_{ab}}{I_a} = \frac{\mu_0 \pi R^2 N_a N_b}{a}. \tag{19-11}$$

Alternatively, we assume a current I_b in solenoid b. Then

$$\Phi_{ba} = \pi R^2 \mu_0 \frac{N_b}{b} I_b. \tag{19-12}$$

This flux links only $(b/a)N_a$ turns of coil a, since B falls rapidly to zero beyond the end of a long solenoid. Thus

$$M_{ba} = \frac{b}{a} \frac{N_a \Phi_{ba}}{I_b} = \frac{\mu_0 \pi R^2 N_a N_b}{a}, \tag{19-13}$$

and $M_{ab} = M_{ba}$, as expected.

It is paradoxical that a varying current in the inner solenoid should induce an electromotance in the outer one, since $B \approx 0$ outside a long solenoid. The reason for this is that the induced $\mathbf{E}$ is equal to minus the time derivative of $\mathbf{A}$, *not* $\mathbf{B}$, and $\mathbf{A}$ does not vanish outside a long solenoid, even though $\nabla \times \mathbf{A}$ is zero. See the first Example in Sec. 18.3.

*19.2 SELF-INDUCTANCE L

A circuit carrying a current I is, of course, linked by its own magnetic flux, as in Fig. 19-3. The ratio

$$L = \frac{\Lambda}{I} \tag{19-14}$$

is termed the *self-inductance* of the circuit. As for mutual inductance, self-inductance depends solely on the geometry of the circuit and is measured in henrys. Self-inductance is always positive.

If the current I in a self-inductance L changes, then a voltage

$$\mathcal{V} = -\frac{d\Lambda}{dt} = -L\frac{dI}{dt} \tag{19-15}$$

appears between its terminals. The negative sign means that this voltage opposes the change in current. See Fig. 19-3.

Therefore the self-inductance of a zero-resistance circuit is 1 henry if the current increases at the rate of 1 ampere/second when the difference of potential applied between the terminals is 1 volt.

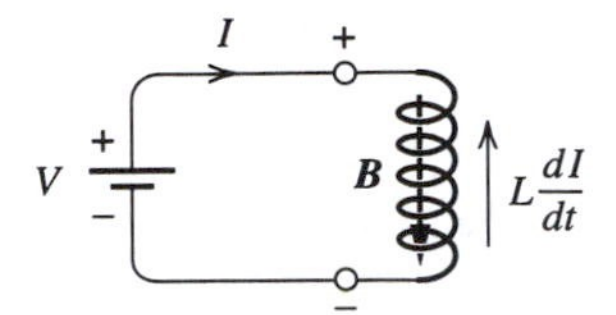

Figure 19-3 Circuit linked by its own flux. If I increases, the induced voltage $-LdI/dt$ across L opposes I.

If both I and L are time-dependent, then

$$\mathcal{V} = -\frac{d\Lambda}{dt} = -\frac{d}{dt}(LI) = -L\frac{dI}{dt} - I\frac{dL}{dt}. \tag{19-16}$$

As a rule, L is constant.

One can calculate a self-inductance, at least in principle, from the Neumann equation (Sec.19.1), with both line integrals running over the same circuit. If the conductor cross section is infinitely small, then the linking flux and the self-inductance turn out to be infinite. This arises from the fact that B tends to infinity in the immediate neighborhood of the wire. The region where B tends to infinity is itself infinitely small, but the flux tends to infinity logarithmically. With currents distributed over an infinitely thin *surface,* as in the next example, B remains finite and L is also finite.

Self-inductance is easily measured. A circuit designed to possess self-inductance is called an *inductor.*

*EXAMPLE The Self-Inductance of a Long Solenoid

If l is the length of a long solenoid, N the number of turns, and R its radius, then

$$L = \frac{\Lambda}{I} = \frac{N\Phi}{I} = \frac{N\pi R^2\mu_0(N/l)I}{I} = \frac{\mu_0 N^2\pi R^2}{l}. \tag{19-17}$$

The self-inductance of a short solenoid is smaller by a factor that is a function of the ratio R/l.

*EXAMPLE The Self-Inductance of a Toroidal Coil

The magnetic flux density inside the air-core toroidal coil of Fig.19-4 follows from the circuital law (Sec. 16.3):

$$B = \frac{\mu_0 NI}{2\pi r}. \tag{19-18}$$

Thus

$$\Phi = \frac{\mu_0 NI}{2\pi}\int_{a-b/2}^{a+b/2}\frac{b}{r}dr = \frac{\mu_0 NIb}{2\pi}\ln\frac{2a+b}{2a-b}, \tag{19-19}$$

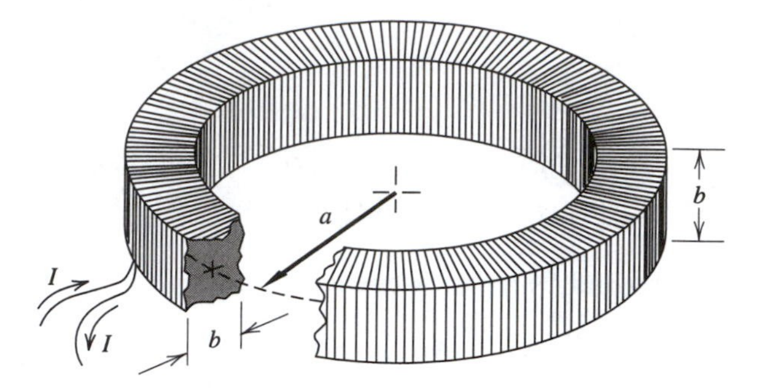

Figure 19-4 Toroidal coil of square cross section.

$$L = \frac{N\Phi}{I} = \frac{\mu_0 N^2 b}{2\pi} \ln \frac{2a + b}{2a - b}. \tag{19-20}$$

If the relative permeability of the core is μ_r, then the self-inductance is μ_r times as large.

---

*19.3 TIME CONSTANT AND IMPEDANCE Z OF AN INDUCTOR

In Secs. 6.5 and 6.6 we discussed time constants and impedances in relation with capacitors. What about inductors?

1. *Time constant.* If you apply a fixed voltage V to an inductor of inductance L and resistance R as in Fig. 19-5(a), then

$$V = IR + L\frac{dI}{dt}, \quad \text{or} \quad \frac{dI}{dt} + \frac{R}{L}I = \frac{V}{L} \tag{19-21}$$

and

$$I = \frac{V}{R} + I_0 \exp\left(-\frac{t}{L/R}\right), \tag{19-22}$$

where I_0 is a constant of integration.

If $I = 0$ at $t = 0$, then $I_0 = -V/R$ and

$$I = \frac{V}{R}\left[1 - \exp\left(-\frac{t}{L/R}\right)\right]. \tag{19-23}$$

The current I builds up from zero with a *time constant* L/R seconds, and eventually reaches its asymptotic value V/R.

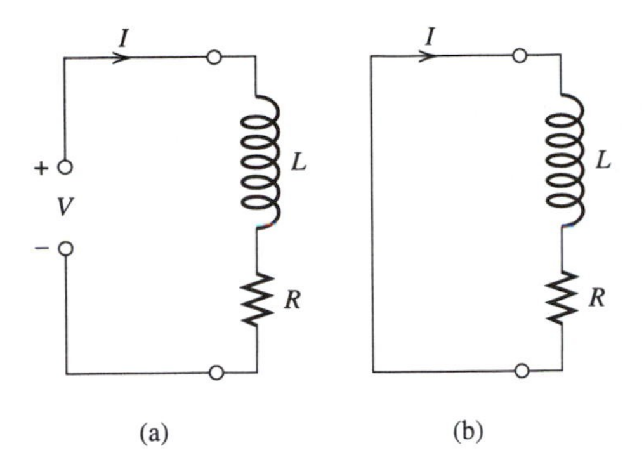

Figure 19-5

On the other hand, if a closed circuit comprising an inductance L and a resistance R as in Fig. 19-5(b) carries an initial current I_0, then Eq. 19-22 applies again with $V = 0$ and

$$I = I_0 \exp\left(-\frac{t}{L/R}\right). \tag{19-24}$$

The current now decays with the same time constant L/R.

2. *Impedance*. If you apply an alternating voltage $V_m \exp j\omega t$ on an inductor R, L, then, from Eq. 19-21,

$$V = V_m \exp j\omega t = IR + Lj\omega I = (R + j\omega L)I, \tag{19-25}$$

and

$$I = \frac{V}{R + j\omega L} = \frac{V}{Z}, \tag{19-26}$$

where $Z = R + j\omega L$ is the *impedance* of the inductor. The imaginary part $j\omega L$ is proportional to the frequency f because, by definition, $\omega = 2\pi f$.†

The Kirchhoff current and voltage laws of Secs. 4.4 and 6.6.1 apply to circuits comprising any combination of resistances, capacitances, or inductances and fed by alternating currents or voltages. One simply has to use phasors and impedances. Complex impedances in series and in parallel behave like resistances in series and in parallel.

†Increasing the frequency increases the ratio $\omega L/R$, which lessens the relative importance of R. However, this brings in two other phenomena. (1) Since there is a voltage difference between turns, the inductor also acts as a capacitor and there is a *stray capacitance* between the terminals. See Prob. 19-9. The impedance of that stray capacitance decreases with increasing frequency. (2) Increasing the frequency increases R because the *skin effect* (Sec. 22.6) makes the current flow closer and closer to the surface of the wire.

*19.4 SUMMARY

The *mutual inductance* M between two circuits a and b is given by the *Neumann equation*

$$M_{ab} = \frac{\mu_0}{4\pi} \oint_a \oint_b \frac{\mathbf{dl}_a \cdot \mathbf{dl}_b}{r}, \tag{19-3}$$

where the line integrals run around each circuit, and r is the distance between the elements $\mathbf{dl}_a$ and $\mathbf{dl}_b$. Because of the symmetry of the integral,

$$M_{ab} = M_{ba} = M. \tag{19-7}$$

If Λ_{ab} is the flux that originates in a and links b, then

$$\Lambda_{ab} = M_{ab} I_a. \tag{19-4}$$

The electromotance induced in b by the current in a is

$$\mathcal{V}_b = -\frac{d\Lambda_{ab}}{dt} = -M\frac{dI_a}{dt} - I_a\frac{dM}{dt}. \tag{19-5}$$

The *self-inductance* of a circuit is

$$L = \frac{\Lambda}{I}, \tag{19-14}$$

where Λ is the flux linkage when the current is I. The impedance of an ideal inductor is $j\omega L$.

If you apply a fixed voltage V to a circuit comprising a resistance R and an inductance L in series, the current increases with a *time constant* L/R:

$$I = \frac{V}{R}\left[1 - \exp\left(-\frac{t}{L/R}\right)\right]. \tag{19-23}$$

If a closed circuit has a resistance R in series with an inductance L and carries an initial current I_0, then the current decays with the same time constant:

$$I = I_0 \exp\left(-\frac{t}{L/R}\right). \tag{19-24}$$

The impedance Z of an inductor of inductance L and resistance R is $R + j\omega L$.

*PROBLEMS

19-1. (*19.1*) The mutual inductance between a toroid and an axial wire

A long straight wire lies along the axis of a toroid of N turns, major radius a, and square cross section of side b, with $a \gg b$.

Calculate the mutual inductance (a) assuming a current I in the wire and (b) assuming a current I in the toroid.

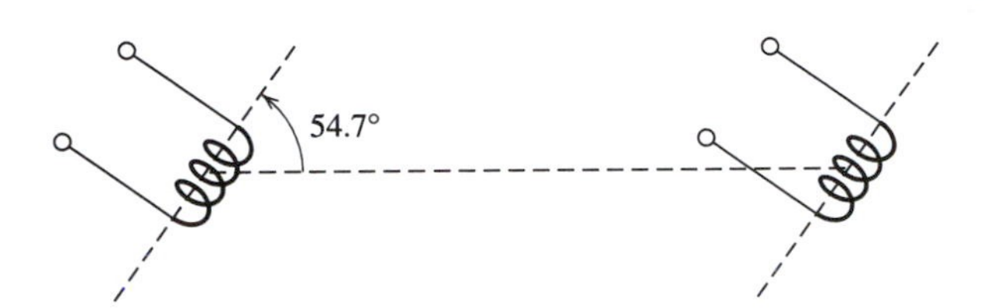

Figure 19-6

19-2. *(19.1)* The mutual inductance between a straight wire and a loop

A loop of wire of radius R is centered at a distance $2R$ from a long straight wire. The wire is in the plane of the loop.

Calculate the mutual inductance.

19-3. *(19.1)* A zero-mutual-inductance magnetic dipole pair

A certain device for geophysical exploration comprises two short coils in the position shown in Fig. 19-6. The manufacturer states that the mutual inductance is zero. Is that true?

19-4. *(19.1)* Current transformer

Figure 19-7 shows a *side-look current transformer* for measuring large current pulses. Show that for a single-turn coil

$$V = \frac{\mu_0 a}{\pi} \ln\left(\frac{b+a}{b-a}\right) \frac{dI}{dt}.$$

The current $I(t)$ follows by integrating V.

19-5. *(19.1)* Mutual inductance

(a) Show that the mutual inductance between two coaxial coils, separated by a distance z as in Fig. 19-8, one of radius a and N_a turns, and the other of radius $b \ll a$ and N_b turns is

$$\frac{\pi \mu_0 N_a N_b a^2 b^2}{2(a^2 + z^2)^{3/2}}.$$

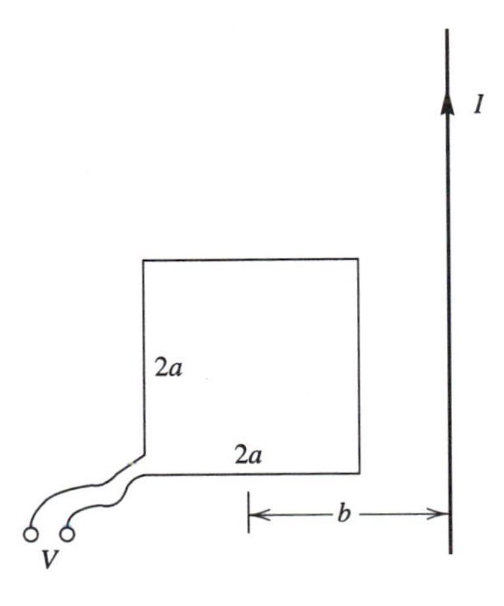

Figure 19-7

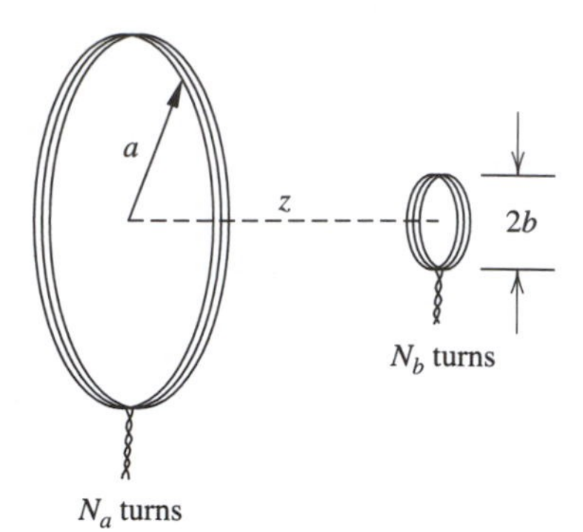

Figure 19-8

You can calculate this in one of two ways. You can either assume a current in coil a and calculate the flux through coil b, or assume a current in b and calculate the flux through a. From Sec. 19.1, M is the same, either way. Now, if the current is in a, you know how to calculate the field at b, from the second Example in Sec.14.2, since $b \ll a$. However, if the current is in b, then the calculation is vastly more difficult because you have to calculate $\boldsymbol{B}$ away from the axis. So you should calculate the magnetic flux through coil b due to a current through coil a.

(b) How does the mutual inductance vary when the small coil is rotated around the vertical axis?

(c) What if one rotated the large coil around a vertical axis? Would the mutual inductance vary in the same way?

19-6. (*19.2*) Induced currents

An azimuthal current is induced in a conducting tube of length l, average radius a, and thickness b, with $b \ll a$.

(a) Show that its resistance and inductance are given by

$$R = 2\pi a/(\sigma b l), \qquad L = \mu_0 \pi a^2 / l.$$

(b) Calculate L, R, L/R for a copper tube ($\sigma = 5.8 \times 10^7$ siemens/meter), 1 meter long, 10 millimeters in diameter, and with a wall thickness of 1 millimeter.

19-7. (*19.2*) A conducting shield for fluctuating magnetic fields

It is often necessary to shield instruments from stray magnetic fields. If the only disturbing field is that of the earth, then one can set up a pair of Helmholtz coils (Prob. 14-9) to oppose the earth's field. If the field is static but not uniform, then one must use a shield made of high-permeability material. Multiple shields, one inside the other, are better than a single thick shield.

Could a conducting enclosure be a good shield against fluctuating magnetic fields? The answer is yes, as we shall see, but only at quite high frequencies.

Imagine a simple situation where the external magnetic field B_{ex} is uniform, with $B_{ex} = B_{ex,m} \exp j\omega t$. The shield is a long tube, parallel to the lines of $\boldsymbol{B}$, a few times longer than its diameter $2a$, and a few times longer than the shielded region.

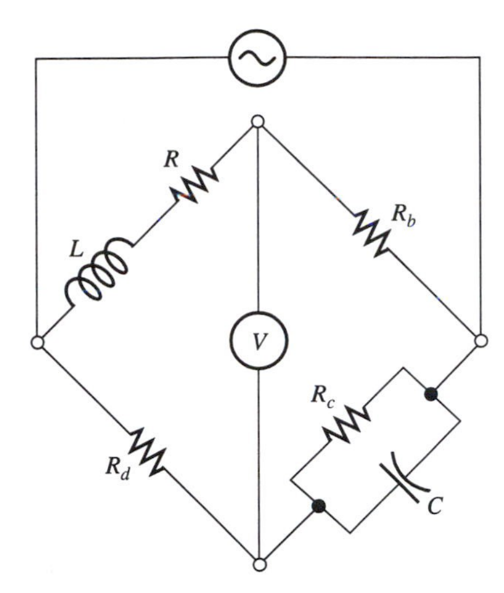

Figure 19-9

We assume that the current induced in the shield is uniformly distributed throughout its thickness b. In other words, we disregard the skin effect (Sec. 22.6). If this assumption is not valid, then the shielding is better than our calculation would indicate.

(a) Calculate the resistance R' of the tube, per unit length, in the azimuthal direction. Calculate L.

(b) Let $B_{in} = B_{in,m} \exp j\omega t$ be the value of B inside the tube, away from the ends. Find the ratio $B_{in,m}/B_{ex,m}$.

(c) Show that, if the skin effect is negligible, this ratio cannot be smaller than 0.5. A conducting enclosure therefore acts as a shield mostly through the skin effect.

19-8. (*19.3*) The Maxwell bridge

Figure 19-9 shows a *Maxwell bridge*. This circuit serves to measure the inductance L and the resistance R of an inductor. One adjusts the values of R_b, R_c, R_d, and C until V equals zero.

Find L and R in terms of the other components.

19-9. (*19.3*) The impedance Z of a real inductor

Figure 19-10 shows the equivalent circuit of a real inductor, where R, L, and C depend on the geometry of the coil and on the nature of the materials used. The wire has a finite resistance R, and there is a stray capacitance C between the turns.

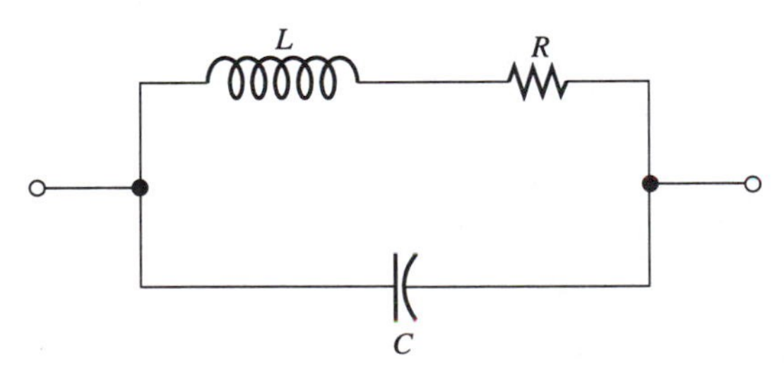

Figure 19-10

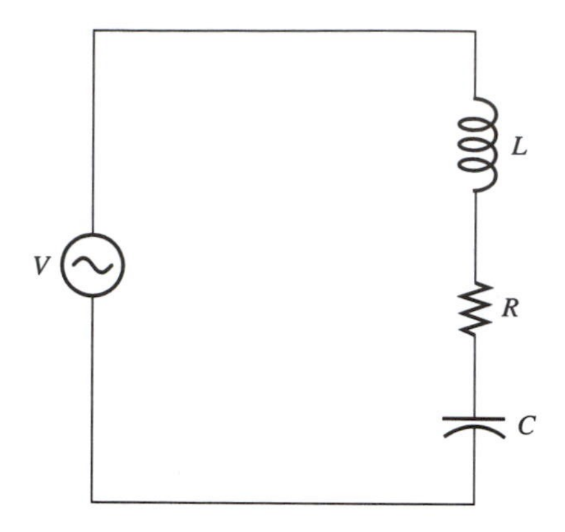

Figure 19-11

Set $Z = R' + jX$ and calculate R' and X. Disregard the fact that R' depends on the frequency because of the skin effect (Sec. 22.6).

If the frequency $f = \omega/(2\pi)$ changes, the point (R', X) in the complex plane moves. Plot the locus of Z in the complex plane with $R = 1000$ ohms, $L = 100$ millihenrys, and $C = 100$ picofarads, for frequencies around 1 megahertz.

19-10. (*19.3*) Series resonance

Figure 19-11 shows a source of alternating current that feeds a series *LRC* circuit. The circuit has two important parameters:

$$\omega_r = \frac{1}{(LC)^{1/2}}, \quad \text{and} \quad Q = \frac{\omega_r L}{R} = \frac{1}{R}\left(\frac{L}{C}\right)^{1/2},$$

where $\omega_r/(2\pi)$ is loosely called the *resonance frequency*, and Q is the *Quality factor*, or simply the Q of the circuit.

Calculate $Z = R + jX$ as a function of ω. Recall that impedances in series and in parallel are treated like resistances in series and in parallel.

Figure 19-12 shows (a) the ratio $R/|Z|$ and (b) the phase angle $\phi = \arctan X/R$ as functions of $\omega' = \omega/\omega_r$ for $Q = 1, 3, 10$, and 30.

At resonance, the imaginary component is zero and $Z = R$.

By definition, *the width of the resonance peak* is the difference Δf between the frequencies for which

$$R/|Z| = 1/2^{1/2} = 0.7071, \quad \text{and} \quad \Delta f = f_r/Q.$$

The higher the Q, the sharper the resonance peak.

Also, the phase angle, for this circuit,

$$\phi = \arctan\left[Q\left(\omega' - \frac{1}{\omega'}\right)\right],$$

varies from $-\pi/2$ to $\pi/2$.

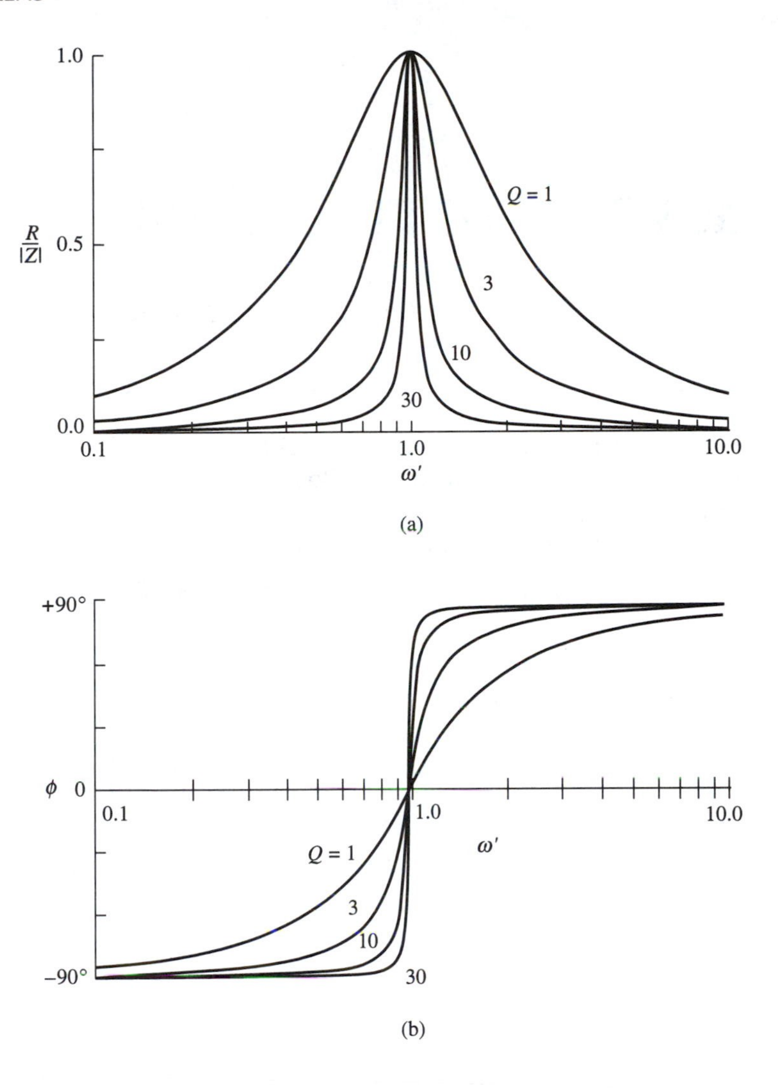

Figure 19-12

CHAPTER

20

MAGNETIC FIELDS VII

Magnetic Energy and Macroscopic Magnetic Forces

This is the last chapter dealing specifically with magnetic fields. We find several expressions for the magnetic energy stored in a field, and then deduce the forces exerted on a current-carrying body situated in a magnetic field that originates elsewhere. We assume that the currents arise from the motion of free charges, and a low frequency.

20.1 ENERGY STORAGE IN AN INDUCTIVE CIRCUIT

The circuit of Fig. 20-1 will illustrate the storage of magnetic energy in an inductive circuit. The wire has a uniform cross section $\mathcal{A}$ and a uniform conductivity σ. There are no ferromagnetic materials in the field, and the circuit is rigid and stationary. We assume that $\epsilon_r = 1$ in the wire.

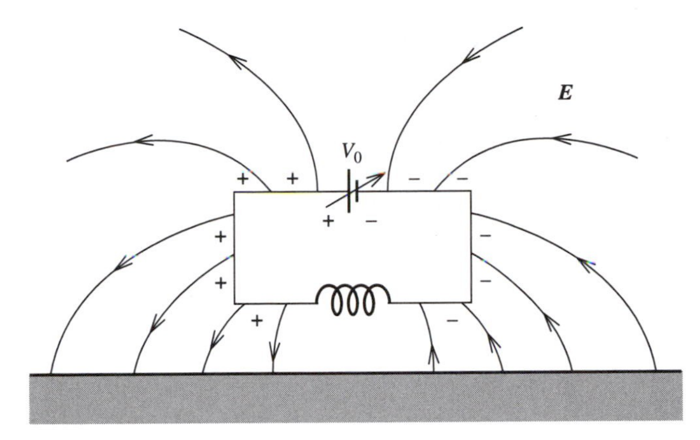

Figure 20-1 Coil fed by a battery.

1. The current is constant. Inside the wire, the volume charge density is zero and

$$E = |\nabla V| = \frac{J}{\sigma}. \tag{20-1}$$

The lines of $\boldsymbol{E}$, inside, follow the wire, parallel to its axis.

At the surface of the wire, $\boldsymbol{E}$ has both a normal and a tangential component. If the wire is situated in a vacuum, the free surface charge density is $\epsilon_0 E_n$, where E_n is the normal component of $\boldsymbol{E}$ just outside. If it lies in a dielectric, then the free surface charge density is $D_n = \epsilon_r \epsilon_0 E_n$.

At any point in space, $\boldsymbol{E}$ is proportional to V_0. Of course, $\boldsymbol{E}$ also depends on the geometry of the circuit and on the neighboring objects.

In principle, one could calculate V and $\boldsymbol{E}$ everywhere from the surface charge density all around the circuit, source included, and on the neighboring bodies.

If l is the length of the wire, $\mathcal{A}$ its cross section, and R its resistance $l/(\sigma\mathcal{A})$, then

$$l|\nabla V| = lE = \frac{lJ}{\sigma} = \frac{lI}{\mathcal{A}\sigma} = IR = V_0. \tag{20-2}$$

Part of the power supplied by the source serves to establish the magnetic field, and the rest dissipates as heat.

2. Now set the source voltage V_0 at zero and increase it slowly. Then

$$\boldsymbol{E} = -\nabla V - \frac{\partial \boldsymbol{A}}{\partial t}. \tag{20-3}$$

The wire now lies in its own $-\partial \boldsymbol{A}/\partial t$ field.

The relation $\boldsymbol{J} = \sigma \boldsymbol{E}$ still applies, and both J and E are uniform throughout the wire, as previously. This means that the surface charges distribute themselves so as to maintain a uniform E inside, despite the presence of the $\partial \boldsymbol{A}/\partial t$ term.

Inside the wires that go from the source to the coil, $\boldsymbol{A}$ is weak and $\boldsymbol{E} \approx -\boldsymbol{\nabla} V$. Inside the coil wire, we have the same $\boldsymbol{E}$, but it comes partly from $-\boldsymbol{\nabla} V$, which points in the direction of $\boldsymbol{J}$, and partly from $-\partial \boldsymbol{A}/\partial t$, which points in the opposite direction because $\boldsymbol{A}$ points in the direction of $\boldsymbol{J}$ and increases.

Now we have just seen that $lE = IR$. Integrating $\boldsymbol{E}$ inside the wire from the positive to the negative terminal of the source,

$$\int_+^- \left(-\boldsymbol{\nabla} V - \frac{\partial \boldsymbol{A}}{\partial t} \right) \cdot \boldsymbol{dl} = IR, \tag{20-4}$$

$$-\int_+^- \boldsymbol{\nabla} V \cdot \boldsymbol{dl} - \int_+^- \frac{\partial \boldsymbol{A}}{\partial t} \cdot \boldsymbol{dl} = IR. \tag{20-5}$$

Since the first integral in the latter equation equals V_0,

$$V_0 = IR + \int_+^- \frac{\partial \boldsymbol{A}}{\partial t} \cdot \boldsymbol{dl} = IR + \frac{d}{dt} \int_+^- \boldsymbol{A} \cdot \boldsymbol{dl} = IR + \frac{d\Lambda}{dt}. \tag{20-6}$$

We have used the fact that the line integral of $\boldsymbol{A}$ is equal to the flux linkage Λ (Sec. 15.1).

At any instant the power supplied by the source is

$$IV_0 = I^2 R + I \frac{d\Lambda}{dt}. \tag{20-7}$$

The first term on the right is the power dissipated as heat, and the second is the rate of increase of the magnetic energy.

Thus, if $\mathscr{E}_m$ is the stored magnetic energy at a given instant,

$$\frac{d\mathscr{E}_m}{dt} = I \frac{d\Lambda}{dt} = I \frac{d(LI)}{dt} = \frac{1}{2} L \frac{dI^2}{dt}. \tag{20-8}$$

The inductance L is a constant because we have assumed that the circuit is rigid and that there are no ferromagnetic materials in the vicinity. Clearly, $\mathscr{E}_m = 0$ when $I = 0$. Then

$$\mathscr{E}_m = \frac{LI^2}{2} = \frac{I\Lambda}{2}. \tag{20-9}$$

If we have two circuits a and b, then

$$\mathscr{E}_{ma} + \mathscr{E}_{mb} = \frac{1}{2}(I_a \Lambda_a + I_b \Lambda_b) \tag{20-10}$$

$$= \frac{1}{2}[I_a(L_a I_a + M I_b) + I_b(L_b I_b + M I_a)] \tag{20-11}$$

$$= \frac{1}{2} L_a I_a^2 + \frac{1}{2} L_b I_b^2 + M I_a I_b. \tag{20-12}$$

EXAMPLE The Long Solenoid

The magnetic energy stored in a long solenoid follows from the value of the self-inductance that we found in the first Example in Sec. 19.2:

$$\mathscr{E}_m = \frac{LI^2}{2} = \frac{\pi \mu_0 N^2 R^2 I^2}{2l}. \tag{20-13}$$

20.2 MAGNETIC ENERGY DENSITY

We can rewrite the time derivative of $\mathscr{E}_m$ as follows. The self-inductance L is constant. Since $\Lambda = LI$,

$$\frac{d\mathscr{E}_m}{dt} = \frac{1}{2}\left(I\frac{d\Lambda}{dt} + \Lambda\frac{dI}{dt}\right) = I\frac{d\Lambda}{dt} = I\oint_C \frac{\partial \boldsymbol{A}}{\partial t} \cdot d\boldsymbol{l}$$

$$= \int_{\mathscr{A}} \boldsymbol{J} \cdot d\boldsymbol{\mathscr{A}} \oint_C \frac{\partial \boldsymbol{A}}{\partial t} \cdot d\boldsymbol{l} \tag{20-14}$$

$$= \int_v \boldsymbol{J} \cdot \frac{\partial \boldsymbol{A}}{\partial t}\, dv, \tag{20-15}$$

where $\mathscr{A}$ is the cross-sectional area of the wire, C is the curve defined by the wire, and v is the volume of the wire. The volume v is finite.

Note in passing that $\boldsymbol{J} \cdot (\partial \boldsymbol{A}/\partial t)dv$ is the work done in moving the conduction charges situated in the element of volume dv against the electric field $\partial \boldsymbol{A}/\partial t$ during 1 second.

Now consider the identity

$$\frac{d}{dt}\int_v \boldsymbol{J} \cdot \boldsymbol{A}\, dv \equiv \int_v \boldsymbol{J} \cdot \frac{\partial \boldsymbol{A}}{\partial t}\, dv + \int_v \frac{\partial \boldsymbol{J}}{\partial t} \cdot \boldsymbol{A}\, dv. \tag{20-16}$$

The first integral on the right is equal to $Id\Lambda/dt$, from Eq. 20-15, and to $d\mathscr{E}_m/dt$, also from Eq. 20-15. Similarly, the second integral is equal to $\Lambda dI/dt$, which is the same as $Id\Lambda/dt$ if the inductance is constant. So the term on the left is equal to $2Id\Lambda/dt$,

$$2\frac{d\mathscr{E}_m}{dt} = \frac{d}{dt}\int_v \boldsymbol{J} \cdot \boldsymbol{A} dv \tag{20-17}$$

and

$$\mathscr{E}_m = \frac{1}{2}\int_v \boldsymbol{J} \cdot \boldsymbol{A} dv \tag{20-18}$$

where, again, v is the volume of the conductor. This equation applies only if the source is of finite size.

If there are several circuits, then the integral runs over all of them and the vector potential in one circuit is the sum of the $\boldsymbol{A}$s of all the circuits. One can add to $\boldsymbol{A}$ any quantity whose curl is zero without affecting this integral.

The magnetic energy density at a point can therefore be taken to be

$$\mathscr{E}'_m = \frac{1}{2}\boldsymbol{J}\cdot\boldsymbol{A}. \tag{20-19}$$

To express the magnetic energy in terms of $\boldsymbol{H}$ and $\boldsymbol{B}$, we use Eq. 20-9 and apply it to the loop of Fig. 20-2. The loop lies in a homogeneous, isotropic, linear, and stationary (HILS) magnetic medium. This excludes ferromagnetic media. From Ampère's circuital law,

$$I = \oint_{C'} H\, dl, \tag{20-20}$$

where C' is any line of $\boldsymbol{H}$.

Also, let $\mathscr{A}$ be the area of any open surface bounded by the loop C and orthogonal to the lines of $\boldsymbol{H}$ and of $\boldsymbol{B}$. Then

$$\Lambda = \Phi = \int_{\mathscr{A}} \boldsymbol{B}\cdot d\boldsymbol{\mathscr{A}} \tag{20-21}$$

and, from Eq. 20-9,

$$\mathscr{E}_m = \frac{1}{2}I\Lambda = \frac{1}{2}\oint_{C'} H\, dl \int_{\mathscr{A}} B\, d\mathscr{A}. \tag{20-22}$$

Now, the lines of $\boldsymbol{H}$ and the set of open surfaces define a coordinate system in which $\boldsymbol{dl}\cdot d\boldsymbol{\mathscr{A}}$ is an element of volume with $\boldsymbol{dl}$ and $d\boldsymbol{\mathscr{A}}$ both parallel to $\boldsymbol{H}$. Also, for each element $\boldsymbol{dl}$ along the chosen line of $\boldsymbol{H}$, one integrates over all the corresponding sur-

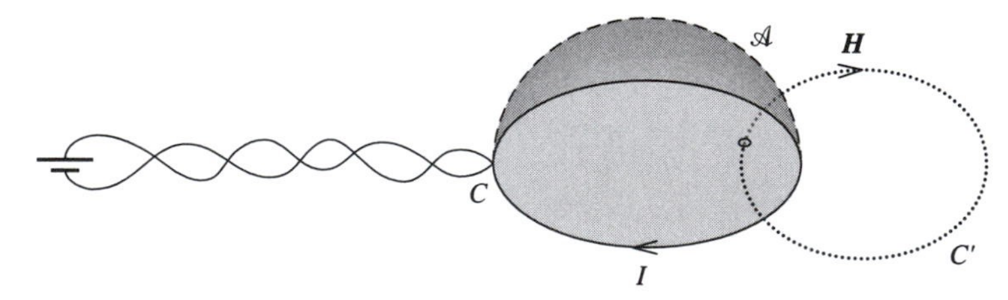

Figure 20-2 Single-turn loop of wire C bearing a current I. The *dotted* line is a typical line of $\boldsymbol{H}$. The open surface, of area $\mathscr{A}$, is bounded by C, and it is everywhere orthogonal to $\boldsymbol{H}$.

face. Since the field extends to infinity, this double integral is the volume integral of $\boldsymbol{H} \cdot \boldsymbol{B}$ over all space, and

$$\mathscr{E}_m = \frac{1}{2}\int_\infty \boldsymbol{H} \cdot \boldsymbol{B}\, dv. \tag{20-23}$$

The *magnetic energy density* in nonferromagnetic media is thus

$$\mathscr{E}'_m = \frac{\boldsymbol{H} \cdot \boldsymbol{B}}{2} = \frac{B^2}{2\mu_0} = \frac{\mu_0 H^2}{2}. \tag{20-24}$$

The magnetic energy density varies as B^2. Thus, after superimposing several fields, the total field energy is *not* equal to the sum of the individual energies. See Eq. 20-12.

Compare this section with Sec. 6.2.

EXAMPLE The Long Solenoid

Neglecting end effects, we find that the magnetic energy stored in the field of a long solenoid in air is

$$\mathscr{E}_m = \frac{1}{2\mu_0}(\mu_0 N'I)^2 \pi R^2 l = \frac{\pi R^2 \mu_0 N^2 I^2}{2l}, \tag{20-25}$$

as in the Example in Sec. 20.1.

20.3 SELF-INDUCTANCE OF A VOLUME DISTRIBUTION OF CURRENT

We saw in Sec. 19.2 that the self-inductance of a circuit comprising infinitely thin wires is infinite. A *real* circuit comprises conductors of finite cross section and its self-inductance L, by definition, is proportional to the stored energy:

$$\tfrac{1}{2}LI^2 = \mathscr{E}_m = \frac{1}{2\mu_0}\int_\infty B^2 dv. \tag{20-26}$$

Thus

$$L = \frac{1}{\mu_0 I^2}\int_\infty B^2\, dv. \tag{20-27}$$

EXAMPLE The Coaxial Line

Assume that the frequency is low enough to ensure that the currents spread uniformly throughout the cross sections of the conductors, and neglect end effects.

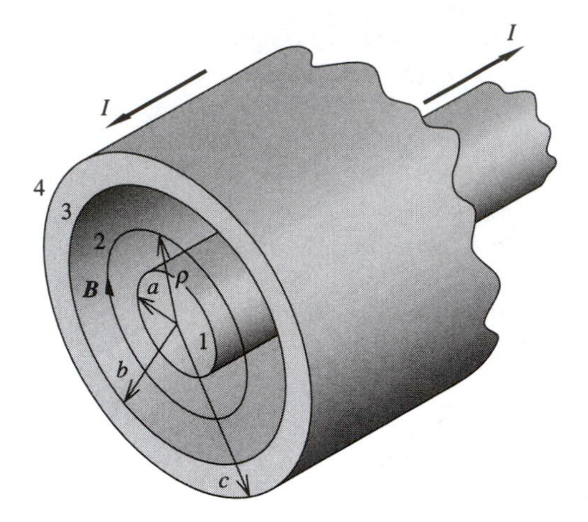

Figure 20-3 Coaxial line.

We calculate successively the magnetic energies per unit length of the line in regions 1, 2, 3, 4 as in Fig. 20-3, and then we set the sum equal to $L'I^2/2$, where L' is the self-inductance per meter. Lines of $\boldsymbol{B}$ are circles centered on the axis.

1. Region 1. From Ampère's circuital law, Secs. 15.4 and 16.3, at the radius $\rho \le a$

$$2\pi\rho B_1 = \mu_0 I \frac{\pi\rho^2}{\pi a^2}, \tag{20-28}$$

and the magnetic energy per unit length is

$$\mathscr{E}'_{m1} = \frac{1}{2\mu_0}\int_0^a \left(\frac{\mu_0 I\rho}{2\pi a^2}\right)^2 2\pi\rho \, d\rho = \frac{\mu_0}{16\pi}I^2. \tag{20-29}$$

2. Region 2, where $a \le \rho \le b$. Here

$$B_2 = \frac{\mu_0 I}{2\pi\rho}, \qquad \mathscr{E}'_{m2} = \frac{\mu_0 I^2}{4\pi}\ln\frac{b}{a}. \tag{20-30}$$

3. Region 3. We require the net current that flows within a circular path with $b \le \rho \le c$. This is I minus that part of the current in the outer conductor that flows between radii b and ρ. Thus

$$B_3 = \frac{\mu_0 I}{2\pi\rho}\left(1 - \frac{\rho^2 - b^2}{c^2 - b^2}\right) = \frac{\mu_0 I}{2\pi\rho}\,\frac{c^2 - \rho^2}{c^2 - b^2}, \tag{20-31}$$

$$\mathscr{E}'_{m3} = \frac{\mu_0 I^2}{4\pi}\left[\frac{c^4}{(c^2 - b^2)^2}\ln\frac{c}{b} - \frac{3c^2 - b^2}{4(c^2 - b^2)}\right]. \tag{20-32}$$

4. Region 4. From Ampère's circuital law, there is zero field in this region.
Finally, the self-inductance per unit length of the coaxial line is

$$L' = \mu_0 \left\{ \frac{1}{8\pi} + \frac{1}{2\pi} \ln \frac{b}{a} + \frac{1}{2\pi} \left[\frac{c^4}{(c^2 - b^2)^2} \ln \frac{c}{b} - \frac{3c^2 - b^2}{4(c^2 - b^2)} \right] \right\}. \tag{20-33}$$

The second term between the braces comes from $\mathscr{E}'_2$ and is normally the most important.

20.4 THE FORCE AND THE TORQUE BETWEEN TWO CURRENT-CARRYING CIRCUITS

Within a single isolated circuit one has magnetic forces, because the current in one part flows in the magnetic field of the rest of the circuit. For example, if the circuit is a simple loop, then the magnetic force on the wire tends to expand the loop.

We found an integral for the force between *two* current-carrying circuits in Sec. 17.3. However, as we noted at the time, the integral is difficult to evaluate.

Here we express the force and the torque between two circuits in terms of their mutual inductance. Mutual inductance is just as difficult to calculate as the force, but it is easy to measure, much more so than the force itself. In the process, we shall find that, whenever one changes the geometry of a circuit or of a pair of circuits, precisely one-half of the energy furnished by the source, exclusive of Joule and other losses, becomes magnetic energy, and the other half performs mechanical work.

We consider two circuits carrying currents I_a and I_b in the same direction, as in Fig. 20-4. The magnetic force is such that the loops tend to move toward each other. They are fixed in position by opposing mechanical forces. All materials are nonmagnetic.

Now assume a small virtual translation (Sec. 6.8) of one circuit, without any rotation. Since there is conservation of energy, the energy expended by the sources is equal to the increase in magnetic energy plus the mechanical work done. The displacement takes place slowly so as to avoid taking kinetic energy into account.

To simplify the calculation, we assume that the currents are constant. This assumption will not affect our result. We had a similar situation in electrostatics.

Loop b moves a distance dr toward loop a. Only M changes and, from Eq. 20-12, the magnetic energy increases by

$$d\mathscr{E}_m = I_a I_b \, dM = I_a \, d\Lambda_{ba} = I_b \, d\Lambda_{ab}, \tag{20-34}$$

Λ_{ba} being the flux originating in b and linking a, and similarly for Λ_{ab}. Since M is positive (Sec. 19.1) and increases, $d\mathscr{E}_m$ is positive.

Now consider the extra energy supplied by the sources. In loop b, the linking flux increases and the induced electromotance tends to generate a magnetic field that op-

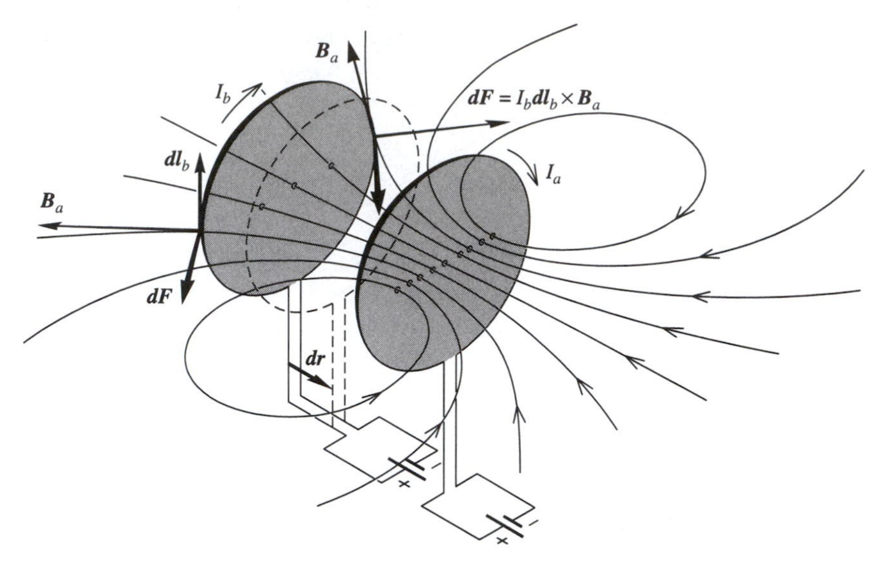

Figure 20-4 Two parallel loops bearing currents I_a and I_b, with typical lines of $\boldsymbol{B}$ originating in loop a. The element of force $\boldsymbol{dF}$ possesses a component in the direction of a, and $\boldsymbol{F}$ is attractive.

poses this increase. Therefore the electromotance induced in b tends to oppose I_b. To keep that current constant, its source supplies the extra voltage $d\Lambda_{ab}/dt$ and the extra energy

$$d\mathscr{E}_{sb} = I_b \frac{d\Lambda_{ab}}{dt} dt = I_b d\Lambda_{ab} = I_b I_a dM. \tag{20-35}$$

By symmetry, $d\mathscr{E}_{sa}$ is the same, and the extra energy supplied by the two sources is

$$d\mathscr{E}_s = 2I_a I_b dM, \tag{20-36}$$

which is exactly twice the increase in magnetic energy. The remainder has gone into mechanical work. In other words, the energy supplied by the sources divides equally between mechanical energy and magnetic energy. See the Example in Sec. 6.8.

If $\boldsymbol{F}_{ab}$ is the force that coil a exerts on coil b, then the mechanical work done is

$$\boldsymbol{F}_{ab} \cdot \boldsymbol{dr} = I_a I_b dM. \tag{20-37}$$

Since the quantity on the right is positive, $\boldsymbol{F}_{ab}$ points toward coil a, like $\boldsymbol{dr}$, which is correct.

The x-component of the force is

$$F_{abx} = I_a I_b \frac{\partial M}{\partial x}, \tag{20-38}$$

where dx is the x-component of $\boldsymbol{dr}$. This equation applies to any pair of circuits.

This is an alternate expression for the force that we found in Sec. 17.3. We can show that the two expressions are equal as follows. Let circuit b move as a whole, without rotating, parallel to the x-axis. Then

$$F_{abx} = -I_a I_b \frac{\partial}{\partial x}\left(\frac{\mu_0}{4\pi}\oint_a \oint_b \frac{\boldsymbol{dl}_a \cdot \boldsymbol{dl}_b}{r}\right), \tag{20-39}$$

where r is the distance from $\boldsymbol{dl}_a$ to $\boldsymbol{dl}_b$. The derivative with respect to x acts only on the $1/r$ term under the integral sign because the vectors $\boldsymbol{dl}$ are not affected by a *translation* of circuit b. Thus

$$F_{abx} = -\frac{\mu_0}{4\pi} I_a I_b \oint_a \oint_b x \frac{\boldsymbol{dl}_a \cdot \boldsymbol{dl}_b}{r^3} \tag{20-40}$$

and, more generally,

$$\boldsymbol{F}_{ab} = -\frac{\mu_0}{4\pi} I_a I_b \oint_a \oint_b \boldsymbol{r} \frac{\boldsymbol{dl}_a \cdot \boldsymbol{dl}_b}{r^3} = -\frac{\mu_0}{4\pi} I_a I_b \oint_a \oint_b \hat{\boldsymbol{r}} \frac{\boldsymbol{dl}_a \cdot \boldsymbol{dl}_b}{r^2}, \tag{20-41}$$

as in Sec. 17.3.

Since the term on the right in Eq. 20-37 is the increase in magnetic energy, we could also write that

$$\boldsymbol{F}_{ab} \cdot \boldsymbol{dr} = d\mathscr{E}_m, \tag{20-42}$$

remembering that the force pulls in the direction that *increases* the magnetic energy. Also,

$$F_{abx} = \frac{\partial \mathscr{E}_m}{\partial x}. \tag{20-43}$$

By analogy a circuit a that forms an angle θ with another circuit b exerts on b a torque

$$T = I_a I_b \frac{\partial M}{\partial \theta} = \frac{\partial \mathscr{E}_m}{\partial \theta}. \tag{20-44}$$

The torque tends to *increase* both the mutual inductance M and the magnetic energy $\mathscr{E}_m$.

EXAMPLE The Force Between Two Long Coaxial Solenoids

Figure 20-5 shows two coaxial solenoids, one of which extends a distance l inside the other. The mutual inductance is positive. The net force is axial, and it is attractive, as can be seen from Fig. 20-6. Remember that the force between two parallel currents flowing in the same direction is attractive.

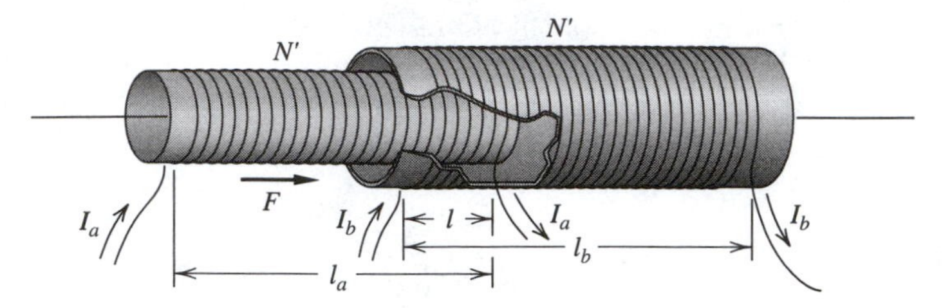

Figure 20-5 Two coaxial solenoids of approximately equal diameters. The force $\boldsymbol{F}$ is attractive when the currents flow in the same direction. The solenoids have the same number of turns per meter N'.

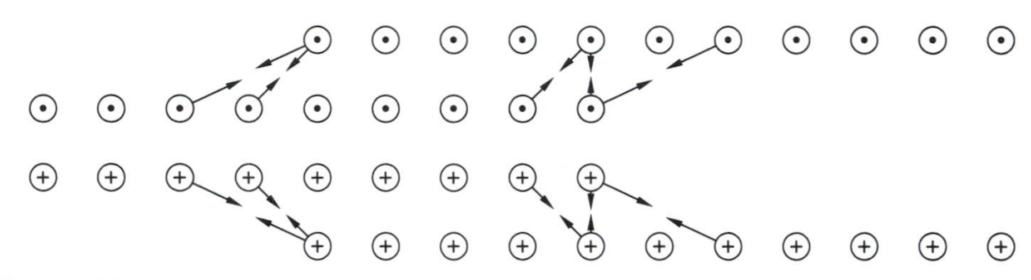

Figure 20-6 Section through part of the solenoids of Fig. 20-5. There is a force of attraction at the ends of the solenoids.

We cannot perform a rigorous calculation of the force, because it clearly depends on the very end effect that we have disregarded until now. However, we can find an approximate expression by applying the formulas that we found above.

We assume that the field of a solenoid stops abruptly at the end. Then

$$M = \frac{N' l \Phi_{ba}}{I_b} = \frac{N' l \Phi_{ab}}{I_a}. \tag{20-45}$$

For a cross-sectional area $\mathcal{A}$,

$$M = \frac{(N'l)(\mu_0 N' I_b \mathcal{A})}{I_b} = \mu_0 N'^2 \mathcal{A} l, \tag{20-46}$$

$$F = I_a I_b \frac{\partial M}{\partial l} = \mu_0 N'^2 \mathcal{A} I_a I_b. \tag{20-47}$$

The force is attractive because M increases with l.

Now let us calculate the force from $\partial \mathscr{E}_m / \partial l$:

$$\mathscr{E}_m = \frac{1}{2\mu_0} \int_\infty B^2 dv = \frac{1}{2\mu_0} [B_a^2 (l_a - l) + B_b^2 (l_b - l) + (B_a + B_b)^2 l] \mathcal{A} \tag{20-48}$$

$$= \frac{1}{2\mu_0} (B_a^2 l_a + B_b^2 l_b + 2 B_a B_b l) \mathcal{A}, \tag{20-49}$$

where $B_a = \mu_0 N' I_a$ originates in solenoid a, and similarly for b. Then

$$F = \frac{d\mathscr{E}_m}{dl} = \frac{B_a B_b \mathscr{A}}{\mu_0} = \mu_0 N'^2 \mathscr{A} I_a I_b. \tag{20-50}$$

Observe that the force would be zero if $\mathscr{E}_m$ were proportional to B.

EXAMPLE Magnetic Torque on a Current Loop

A rectangular loop of wire b carrying a current I_b lies in a uniform $\boldsymbol{B}$ in air, as in Fig. 20-7. We calculate the torque at the angle θ.

1. The simplest procedure here is to calculate the torque from the magnetic force $I_b\, \boldsymbol{dl} \times \boldsymbol{B}$ on the element $\boldsymbol{dl}$ of the wire as in Sec. 17.2:

$$T = 2(a/2)\sin\theta\, B I_b b = B I_b\, \mathscr{A} \sin\theta, \tag{20-51}$$

where $\mathscr{A}$ is the area of the loop. The torque acts in the positive direction of θ.

2. Now let us use Eq. 20-44. Let I_a be the unknown current that provides the field $\boldsymbol{B}$. Then

$$M = -\frac{B\mathscr{A}\cos\theta}{I_a}. \tag{20-52}$$

We require a negative sign here because the flux $B\mathscr{A}\cos\theta$ links the loop in the direction opposite to that of the flux of a current I_b flowing in the loop as in the figure. Then

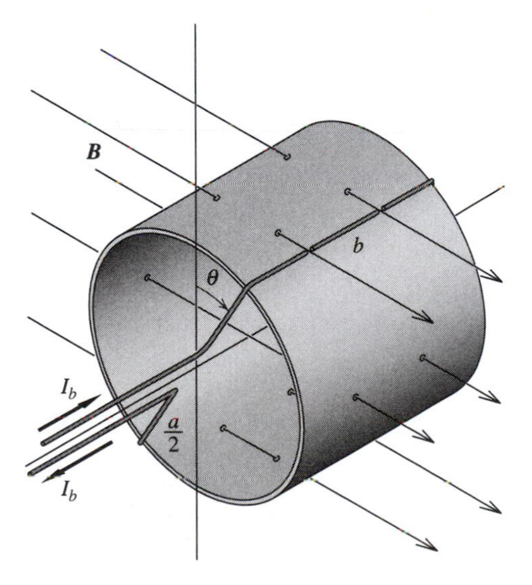

Figure 20-7 Loop carrying a current I_b in a constant $\boldsymbol{B}$. The magnetic torque tends to turn the loop in the direction of the θ arrow.

$$T = I_a I_b \frac{\partial M}{\partial \theta} = -I_b \frac{d}{d\theta}(B\mathcal{A}\cos\theta) = BI_b\mathcal{A}\sin\theta. \tag{20-53}$$

The positive sign means that the torque tends to increase θ and hence to increase M.

20.5 MAGNETIC PRESSURE

We return to the concept of *magnetic pressure* discussed in Sec. 17.4.1, now using a different approach.

If current flows through a conducting sheet, then it is appropriate to think in terms of *magnetic pressure.* Imagine a conducting sheet, in air, carrying $\boldsymbol{\alpha}$ amperes/meter and situated in a uniform tangential magnetic field $\boldsymbol{B}/2$ originating in currents flowing elsewhere, as in Fig. 20-8(a), with $\boldsymbol{\alpha}$ normal to $\boldsymbol{B}$. The force per unit area is $\boldsymbol{\alpha} \times \boldsymbol{B}/2$.

Now increase $\boldsymbol{\alpha}$ until its field cancels the ambient field on one side and doubles it on the other, as in Fig 20-8(b). Then, from Prob. 15-4, $\alpha' = B/\mu_0$, and the force per unit area, or the pressure, is

$$p = \frac{B^2}{2\mu_0}. \tag{20-54}$$

This applies to any current sheet with zero $\boldsymbol{B}$ on one side.

The pressure is equal to the energy density, as with electric fields (Sec. 6.7) This pressure pushes the current sheet *away* from the field. On the field side, the lines of B are parallel to the sheet and repel laterally.

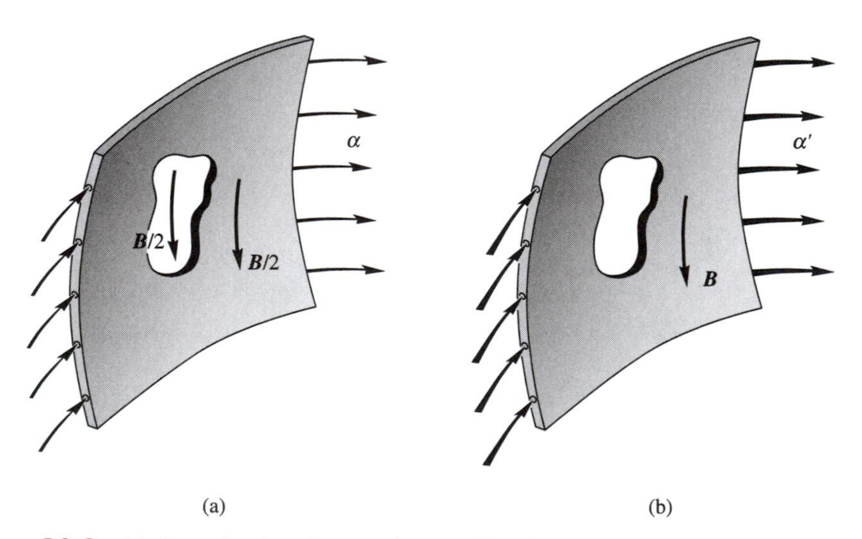

Figure 20-8 (a) A conducting sheet carries a negligible surface current density of $\boldsymbol{\alpha}$ amperes/meter of width and lies in a uniform magnetic field $\boldsymbol{B}/2$ originating elsewhere. (b) The surface current density $\boldsymbol{\alpha}'$ is now such that the total $\boldsymbol{B}$ on the near side is twice what it was, while the total $\boldsymbol{B}$ on the far side is zero.

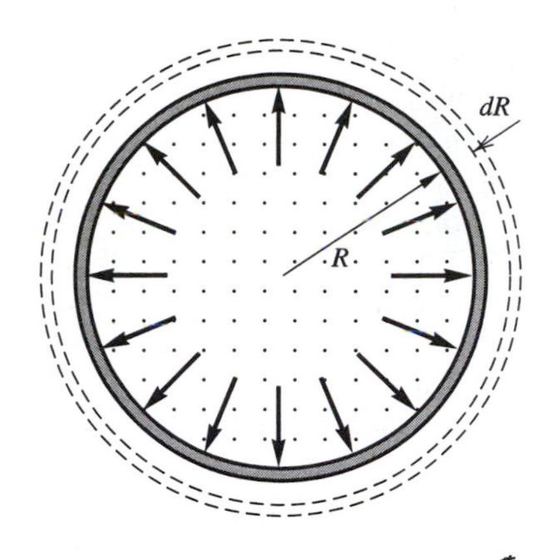

Figure 20-9 End view of a solenoid. The dots represent lines of $\boldsymbol{B}$, normal to the paper. The arrows show how the magnetic pressure pushes the winding away from the field.

On the contrary, electric "pressure" pulls a sheet of charge *into* the field because electric field lines "are under tension." For example, the plates of a dielectric-insulated parallel-plate capacitor squeeze the dielectric. See Secs. 6.7 and 6.8.

EXAMPLE Magnetic Pressure Inside a Long Solenoid

Figure 20-9 shows an end view of a solenoid. We use the method of virtual work (Sec. 6.8) to show that the magnetic pressure is $B^2/(2\mu_0)$. Imagine that the current remains constant, while the magnetic pressure increases the radius from R to $R + dR$. Then, for a solenoid of length l, the magnetic energy increases by

$$d\mathscr{E}_m = \frac{B^2}{2\mu_0} l 2\pi R \, dR. \tag{20-55}$$

This is equal to the mechanical work performed by the magnetic pressure p over an area $2\pi Rl$ over the virtual displacement dR, and

$$p = \frac{B^2}{2\mu_0}. \tag{20-56}$$

20.6 MAGNETIC FORCES ON CONDUCTORS THAT MOVE IN MAGNETIC FIELDS

The magnetic force density at a point inside a conductor that moves at a velocity $\boldsymbol{v}$ in a magnetic field $\boldsymbol{B}$ is

$$\boldsymbol{F}'_{mag} = \boldsymbol{J} \times \boldsymbol{B} = \sigma[\boldsymbol{E} + (\boldsymbol{v} \times \boldsymbol{B})] \times \boldsymbol{B} \tag{20-57}$$

$$= \sigma[\boldsymbol{E} \times \boldsymbol{B} + (\boldsymbol{v}_\perp \times \boldsymbol{B}) \times \boldsymbol{B}] \tag{20-58}$$

$$= \sigma(\boldsymbol{E} \times \boldsymbol{B} - B^2 \boldsymbol{v}_\perp), \tag{20-59}$$

where $\boldsymbol{v}_\perp$ is the velocity component of the moving conductor that is perpendicular to the local $\boldsymbol{B}$. We have used Ohm's law for moving conductors of Sec. 17.5.

The magnetic force density on a moving conductor therefore has two components. The first component is perpendicular to $\boldsymbol{B}$, and can point in any direction with respect to the local fluid velocity $\boldsymbol{v}$. The second component is proportional to $-\boldsymbol{v}_\perp$ and is analogous to a viscous braking force: it tries to cancel the velocity component that is perpendicular to $\boldsymbol{B}$. That second component is proportional to the local B, squared.

In a convecting, conducting fluid, such as the liquid part of the Earth's core, magnetic forces expend energy in disturbing the flow pattern. They not only affect the velocity $\boldsymbol{v}$; they also affect $\boldsymbol{v} \times \boldsymbol{B}$, and thus the electric charge density ρ of Sec. 17.6. Hence they affect the electric potential V and its gradient ∇V, which in turn alters the current density $\boldsymbol{J}$, and hence $\boldsymbol{B}$. Also, a change in $\boldsymbol{J}$ alters the stored magnetic energy and the Joule losses! So all these variables are tightly woven together.

In the absence of an electric field, the magnetic force reduces to the "viscous" term.

20.7 SUMMARY

When a current I flows in a circuit of self-inductance L, the *magnetic energy* stored in the field can be expressed in various ways:

$$\mathscr{E}_m = \tfrac{1}{2}LI^2 = \tfrac{1}{2}I\Lambda = \frac{1}{2}\int_\infty \boldsymbol{H} \cdot \boldsymbol{B}\, dv, \qquad (20\text{-}9),\ (20\text{-}23)$$

where Λ is the magnetic flux linkage and where the integration runs over all space. Also, for a finite current distribution,

$$\mathscr{E}_m = \frac{1}{2}\int_v \boldsymbol{J} \cdot \boldsymbol{A}\, dv, \qquad (20\text{-}18)$$

where v is the volume occupied by the current.

Thus the *energy density* can be taken to be

$$\mathscr{E}'_m = \frac{1}{2}\boldsymbol{H} \cdot \boldsymbol{B} = \frac{B^2}{2\mu_0}. \qquad (20\text{-}24)$$

We define the *self-inductance* of a real circuit comprising currents distributed over a finite volume in terms of the magnetic energy stored in the field:

$$L = \frac{1}{\mu_0 I^2}\int_\infty B^2\, dv. \qquad (20\text{-}27)$$

The x-component of the *force* exerted on a circuit b situated in the field of another circuit a is

$$F_{abx} = I_a I_b \frac{\partial M}{\partial x} = \frac{\partial \mathscr{E}_m}{\partial x}. \qquad (20\text{-}38), (20\text{-}43)$$

The force pulls in the direction that increases both M and the magnetic energy.

Similarly, the *torque* is given by

$$T = I_a I_b \frac{\partial M}{\partial \theta} = \frac{\partial \mathscr{E}_m}{\partial \theta}. \qquad (20\text{-}44)$$

If the current flowing through a conducting sheet situated in a magnetic field is such that the magnetic flux density is $\boldsymbol{B}$ on one side and zero on the other, then the *magnetic pressure* on the sheet is $B^2/(2\mu_0)$. The pressure tends to push the sheet *away* from the field.

The magnetic force density on a conductor that moves at a velocity $\boldsymbol{v}$ in a field $\mathbf{E}, \mathbf{B}$ is

$$\boldsymbol{F}'_{mag} = \sigma(\boldsymbol{E} \times \boldsymbol{B} - B^2 \boldsymbol{v}_\perp). \qquad (20\text{-}59)$$

PROBLEMS

20-1. (*20.1*) The average stored energies in capacitors and in inductors

The average stored energies in capacitors and in inductors are $CV^2/2$ and $LI^2/2$, respectively, where V and I are rms values.

Show that, for an angular frequency ω,

$$\mathscr{E}_{C,av} = \frac{I^2}{2\omega^2 C}, \qquad \mathscr{E}_{L,av} = \frac{V^2}{2\omega^2 L}.$$

20-2. (*20.1*) Pulsed magnetic fields

Extremely high magnetic fields can be obtained by discharging a capacitor through a low-inductance coil. The capacitor leads must of course have a low-inductance. Such fast capacitors cost approximately five dollars per joule of stored-energy capacity.

(a) Estimate the cost of a capacitor that could store an energy equal to that of a 100-tesla magnetic field occupying a volume of one liter.

(b) Estimate the cost of the electricity required to charge the capacitors.

20-3. (*20.2*) Energy storage

Compare the energies per unit volume in (a) a magnetic field of 1.0 tesla and (b) an electrostatic field of 10^6 volts/meter.

20-4. (*20.3*) The inductance of a coaxial line is slightly frequency-dependent

High-frequency currents do not penetrate a conductor as do low-frequency currents. This is the skin effect (Sec. 22.6). Does the self-inductance of a coaxial line increase or decrease with frequency?

20-5. (*20.4*) The electromagnetic levitation of high-speed tracked vehicles

The suspension and the propulsion of tracked vehicles become major problems at speeds above about 300 kilometers/hour. Wheels are then impractical because vehicle vibration, track damage, and power loss become excessive. The tractive force also deteriorates with increasing speed.

An air cushion provides a satisfactory suspension at high speeds, but it consumes a large amount of power. Propulsion then requires either a propeller or a linear electric motor, with the stator in the track.

It is also possible to support a vehicle by means of magnetic forces, and several methods have been developed. In one of these, superconducting coils in the vehicle generate a magnetic field that extends down into the track, which is a sheet of aluminum. At rest and at low speeds, the vehicle uses wheels. As the speed increases, the eddy currents induced in the track by the traveling magnetic field exert a force of repulsion on the currents in the vehicle coils, and the vehicle flies about 10 centimeters above the track. There are, of course, problems of stability. Also, the suspension is not lossless because there are Joule losses in the track.

Let us consider a simplified form of levitation. A pair of parallel and coaxial coils of radius R and N turns are separated by a distance D. The lower coil simulates the track. For $D \approx 0.1R$, the mutual inductance is given by $N^2\{2.154 - 12.04[(D/R) - 0.1]\}R$ microhenrys.

(a) Calculate the number of ampere-turns required in each coil to support a mass of 1 metric ton when $R = 1$ meter.

(b) Draw a sketch showing the two coils and lines of $\mathbf{B}$. Can you explain the force of repulsion qualitatively?

20-6. (*20.5*) The force between two parallel bus bars of finite cross section.

Two parallel bus bars have equal circular cross sections and carry equal currents I. The currents are equally distributed over the cross sections.

Show, without any calculation, that the force is the same as if the bus bars were thin wires.

20-7. (*20.5*) A superconducting power transmission line

A superconducting DC power transmission line has been proposed that would carry 100 gigawatts of power at 200 kilovolts over 1000 kilometers. The conductors would have a diameter of 25 millimeters and be separated by a center-to-center distance of 50 millimeters.

(a) Calculate the magnetic force per meter. See the previous problem. It is clearly preferable to use a coaxial line.

(b) Calculate the stored energy in kilowatt-hours. The self-inductance per meter is $[\mu_0/(4\pi)][1 + 4\ln(D/R)]$.

20-8. (*20.5*) Large-scale energy storage in inductors and in capacitors

Much work has been done on the large-scale storage of energy in inductors, for public utilities. One author proposes a huge, underground, cryogenized inductor that would operate at a field of 14 teslas.

(a) Calculate the energy density in kilowatt-hours/meter^3.

(b) Calculate the magnetic pressure in atmospheres, (1 atmosphere $\approx 10^5$ pascals).

(c) It seems more reasonable to store energy in a capacitor, because a capacitor need not be cryogenized, and because the force points inward, not outward as in an inductor. Calculate the energy density in kilowatt-hours/meter^3 if $\epsilon_r = 3$, and the dielectric strength is 1.5×10^8 volts /meter.
Gasoline can store over 100 kilowatt-hours/meter^3, and flywheels over 200.

20-9. (*20.5*) The mechanical work performed by mechanical forces on an isolated, active, and deformable circuit

We showed in Sec. 20.4 that, if one active circuit moves with respect to another, the mechanical work performed by the sources is equal to the *increase* in magnetic energy if the currents are maintained constant. Hence the force between two active circuits is given by the rate of increase of magnetic energy.

Show that, if the geometry of an isolated active circuit changes, the energy supplied by the sources divides in the same way. Assume again that the current is constant. It follows, on this assumption, that the force on an element of an active circuit is equal to the rate of increase of magnetic energy.

20-10. (*20.5*) The axial compression force on a solenoid

(a) Show qualitatively, in two different ways, that the turns of a solenoid tend to squeeze together lengthwise.
(b) Calculate the axial compression force on a long solenoid.

20-11. (*20.5*) Magnetic pressure

(a) Show that magnetic pressure is about $4B^2$ atmospheres. One atmosphere is about 10^5 pascals. The magnetic flux density rarely exceeds 1 tesla.
(b) Find the corresponding expression for an electric field. The maximum electric field strength that can be maintained in air at normal temperature and pressure is about 3×10^6 volts/meter.
(c) Discuss the similarities and differences between magnetic pressure and electric force per unit area.

20-12. (*20.5*) Magnetic pressure

It is possible to attain very high pressures by discharging a large capacitor through a hollow wire.

(a) Show that, for a thin tube of radius R carrying a current I, the inward magnetic pressure is

$$p_m = \mu_0 I^2/(8\pi^2 R^2).$$

(b) Calculate the pressure for a current of 30 kiloamperes in a tube one millimeter in diameter.
One atmosphere is approximately 10^5 pascals.

20-13. (*20.5*) Magnetic shutter

Magnetic fields can perform mechanical tasks that require a high power level for a very short time. For example, magnetic pressure can crush a light aluminum tube that acts as a shutter to turn off a beam of light or of soft x-rays. The tube is placed inside a coil, parallel to the axis. When the coil is suddenly connected to a large capacitor, the change in flux induces a large current in the tube, which collapses under the magnetic pressure.

Let us calculate the pressure. If the current I in the solenoid increases gradually from zero to some large value, the induced current is small and the magnetic pressure is negligible. Let us assume that dI/dt in the coil is so large that the induced current in the tube maintains zero magnetic field inside it. Then there is a magnetic field B only in the annular region between the solenoid and the conducting tube.

(a) Calculate the pressure on the tube in atmospheres at 1 tesla.

(b) What would be the pressure if the conducting tube were parallel to the axis but off the axis?

20-14. (*20.5*) Flux compression

Flux compression is one method of obtaining large magnetic fields. For example, one can insert a light conducting tube in the field B_0 of a solenoid and then implode the tube by means of an annular explosive charge situated between the tube and the solenoid. Currents flow in the tube, and the magnetic pressure builds up until it is equal to the external gas pressure. The solenoid is fed by a constant-current source.

Neglect the mechanical, thermal, and acoustic energies, and neglect end effects. Let a be the initial radius, and b the final radius of the tube.

(a) Show that, if the tube radius shrinks very rapidly, B inside increases to approximately $B_0(a^2/b^2)$. For example, if $B_0 = 10$ teslas and if $a/b = 10$, then $B = 1000$ teslas.

(b) Calculate the surface current density on the tube at the very end of the compression.

(c) Calculate the increase of magnetic energy, the energy fed into the source during the compression, and the explosive energy required. The tube is 200 millimeters long and its initial radius is 50 millimeters.

20-15. (*20.5*) The torque on a current-carrying coil

(a) Show that a current-carrying coil tends to orient itself in a magnetic field in such a way that the total magnetic flux linking the coil is *maximum.*

(b) Show that the torque on the coil is $\boldsymbol{m} \times \boldsymbol{B}$, where $\boldsymbol{m}$ is the magnetic moment of the coil and $\boldsymbol{B}$ is the magnetic flux density when the current in the coil is zero.

A cylindrical permanent magnet behaves in the same way.

20-16. (*20.5*) High-gradient magnetic separation

It is possible to separate magnetic particles in suspension in a fluid by passing the mixture through steel wool subjected to a strong magnetic field. The magnetic particles cling to the steel wires where the field gradient is large. Arrays of fine steel wires normal to $\boldsymbol{B}$ are also used.

With a field of the order of several teslas supplied by superconducting coils, the separation occurs even with materials that are only slightly magnetic. The method is also applicable in air for removing magnetic particles, say in pulverized coal.

Let us see how a small magnetic dipole behaves in a nonuniform $\boldsymbol{B}$. The dipole first orients itself. Then, as we shall see, it tends to move in the direction in which the applied $\boldsymbol{B}$ increases.

Figure 20-10 shows a small current loop of radius R that is already oriented in a field $\boldsymbol{B}$ that increases symmetrically about the positive direction of the z-axis.

(a) Show, without any calculation, that the magnetic force points to the right. Note that this force tends to *increase* the linking flux.

(b) Show that $F = 2\pi RIB_\rho$, where B_ρ is the component of $\boldsymbol{B}$ that is normal to the z-axis.

(c) Now consider a small volume of thickness Δz, as in the figure. Use the fact that the net outward flux of $\boldsymbol{B}$ is zero to find B_ρ and F.

(d) Calculate the force from the rate of increase of magnetic energy.

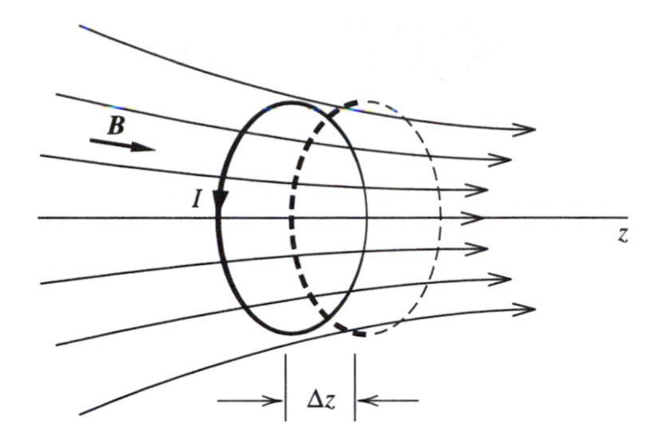

Figure 20-10

CHAPTER

21

THE MAXWELL EQUATIONS

This chapter concerns the four fundamental equations of electromagnetism that bear the name of James Clerk Maxwell (1831-1879), but that should rather be called after both Maxwell and Heaviside (1850-1925); see Appendix C.

21.1 THE EQUATION FOR THE CURL OF *B*

In Sec. 15.4 we found that, at a given point in space where the current density is $\boldsymbol{J}$,

$$\nabla \times \boldsymbol{B} = \mu_0 \boldsymbol{J}. \tag{21-1}$$

However, we warned the reader that the relation is valid only for static fields. So there must be a term missing in this equation.

Now recall Sec. 13.2, where we found the $\boldsymbol{E}$ and $\boldsymbol{B}$ fields at a point P near a charge Q that moves at a velocity $\boldsymbol{\mathcal{V}}$, as in Fig. 13-2. At P, there is both an $\boldsymbol{E}$ and a $\boldsymbol{B}$ field:

$$\boldsymbol{E} = \frac{Q\hat{\boldsymbol{r}}}{4\pi\epsilon_0\gamma^2 r^2(1-\beta^2\sin^2\theta)^{3/2}}, \tag{21-2}$$

$$\boldsymbol{B} = \frac{\mu_0 Q\boldsymbol{\mathcal{V}} \times \hat{\boldsymbol{r}}}{4\pi\gamma^2 r^2(1-\beta^2\sin^2\theta)^{3/2}}. \tag{21-3}$$

The vector $\boldsymbol{E}$ points outward in the direction of the unit vector $\hat{\boldsymbol{r}}$, while $\boldsymbol{B}$ is azimuthal, as in Figs. 13-2 and 13-3.

Clearly, the above $\boldsymbol{E}$ and $\boldsymbol{B}$ fields are closely related, mathematically. Indeed, you can show in Problem 21-1 that, in this specific case,

$$\nabla \times \boldsymbol{B} = \epsilon_0 \mu_0 \frac{\partial \boldsymbol{E}}{\partial t}. \tag{21-4}$$

Equations 21-1 and 21-4 are complementary, and the complete Maxwell equation has two terms on the right:

$$\boxed{\nabla \times \boldsymbol{B} = \mu_0 \left(\boldsymbol{J} + \epsilon_0 \frac{\partial \boldsymbol{E}}{\partial t} \right).} \tag{21-5}$$

The term $\epsilon_0 \partial \boldsymbol{E}/\partial t$ is the *displacement current density* at the point P, in a vacuum. See also Sec. 7.10.

See Bork (1963) for the story of $\epsilon_0 \partial \boldsymbol{E}/\partial t$.

21.2 THE MAXWELL EQUATIONS IN DIFFERENTIAL FORM

Let us group the four Maxwell equations together. We found them successively in Secs. 3.7, 18.2, 14.3, and 21.1 above:

$$\nabla \cdot \boldsymbol{E} = \frac{\rho}{\epsilon_0}, \tag{21-6}$$

$$\nabla \times \boldsymbol{E} + \frac{\partial \boldsymbol{B}}{\partial t} = 0, \tag{21-7}$$

$$\nabla \cdot \boldsymbol{B} = 0, \tag{21-8}$$

$$\nabla \times \boldsymbol{B} - \epsilon_0 \mu_0 \frac{\partial \boldsymbol{E}}{\partial t} = \mu_0 \boldsymbol{J}. \tag{21-9}$$

We have written the field terms on the left and the source terms on the right, but that is deceptive, because $\boldsymbol{E}$, $\boldsymbol{B}$, ρ, and $\boldsymbol{J}$ are all closely related together.

The above equations are general in that the media can be nonhomogeneous, nonlinear, and nonisotropic. However, they apply only to media that are stationary with respect to the coordinate axes. For an extensive discussion of moving media, see Penfield and Haus (1967). Also, the coordinate axes must not accelerate, and they must not rotate.

These are the four fundamental equations of electromagnetism. They form a set of partial differential equations, relating certain time and space *derivatives* at a point to the charge and current densities at that point.

Equation 21-8 applies on the condition that there do not exist magnetic monopoles (Sec. 14.1).

As usual,

$\boldsymbol{E}$ is the electric field strength, in volts/meter;

$\rho = \rho_f + \rho_b$ is the total electric volume charge density, in coulombs/meter^3;

ρ_f is the free volume charge density;

$\rho_b = -\nabla \cdot \boldsymbol{P}$ is the bound volume charge density;

$\boldsymbol{P}$ is the electric polarization, in coulombs/meter^2;

$\boldsymbol{B}$ is the magnetic flux density, in teslas;

$\boldsymbol{J} = \boldsymbol{J}_f + \partial \boldsymbol{P}/\partial t + \nabla \times \boldsymbol{M}$ is the total current density, in amperes/meter^2;†

$\boldsymbol{J}_f$ is the current density that results from the motion of free charges;

$\partial \boldsymbol{P}/\partial t$ is the polarization current density in a dielectric;

$\nabla \times \boldsymbol{M}$ is the equivalent current density in magnetized matter;

$\boldsymbol{M}$ is the magnetization, in amperes/meter;

ϵ_0 is the permittivity of free space, about 8.85×10^{-12} farad/meter;
and

$\mu_0 \equiv 4\pi \times 10^{-7}$ henry/meter is the permeability of free space.

Note that the Maxwell equations 21-6 to 21-9 are *linear:* they contain neither products nor powers of the variables or of their derivatives. It follows that, if the field $\boldsymbol{E}_1$, $\boldsymbol{B}_1$ satisfies the Maxwell equations when $\rho = \rho_1$ and $\boldsymbol{J} = \boldsymbol{J}_1$, and if the field $\boldsymbol{E}_2$, $\boldsymbol{B}_2$ corresponds similarly to $\rho = \rho_2$, $\boldsymbol{J} = \boldsymbol{J}_2$, then the total field is $\boldsymbol{E}_1 + \boldsymbol{E}_2$, $\boldsymbol{B}_1 + \boldsymbol{B}_2$ when $\rho = \rho_1 + \rho_2$, $\boldsymbol{J} = \boldsymbol{J}_1 + \boldsymbol{J}_2$. This is the *principle of superposition* in its general form (Secs. 3.3 and 14.2). In other words, each charge distribution ρ, and each current distribution $\boldsymbol{J}$ acts independently of all the others.

See alternate forms of the Maxwell equations on the inside of the back cover.

In linear media, $\boldsymbol{P}$ and $\boldsymbol{M}$ are proportional, respectively, to $\boldsymbol{E}$ and to $\boldsymbol{B}$, and the principle of superposition applies.

In nonlinear media, $\boldsymbol{P}$ and $\boldsymbol{M}$ are complicated functions of $\boldsymbol{E}$ and $\boldsymbol{B}$, and the principle of superposition does not apply to ρ_f and $\boldsymbol{J}_f$. The principle of superposition continues to apply to ρ and $\boldsymbol{J}$.

All materials become nonlinear at high field strengths.

21.3 THE MAXWELL EQUATIONS IN INTEGRAL FORM

We now deduce the integral form of the Maxwell equations.

Integrating Eq. 21-6 over a finite volume v, and then applying the divergence theorem, we find the integral form of Gauss's law (Sec. 7.5):

$$\int_{\mathcal{A}} \boldsymbol{E} \cdot d\boldsymbol{\mathcal{A}} = \frac{1}{\epsilon_0}\int_v \rho\, dv = \frac{Q}{\epsilon_0}, \tag{21-10}$$

where $\mathcal{A}$ is the area of the surface bounding the volume v, and Q is the total charge enclosed within v. See Fig. 21-1.

†Until now we were solely concerned with free current densities, and we used $\boldsymbol{J}$ instead of $\boldsymbol{J}_f$ to simplify the notation.

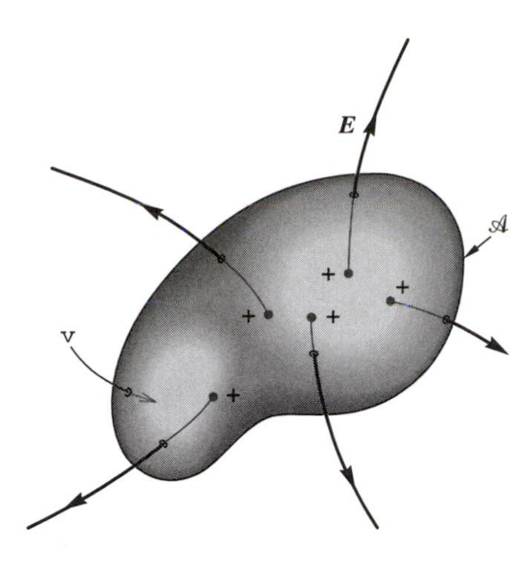

Figure 21-1 Lines of $\boldsymbol{E}$ emerging from a volume containing a net charge Q. The outward flux of $\boldsymbol{E}$ is equal to Q/ϵ_0.

Equation 21-7 is the differential form of the Faraday induction law for time-dependent magnetic fields. Integrating over an open surface of area $\mathcal{A}$ bounded by the curve C gives the integral form, as in Sec. 18.2:

$$\oint_C \boldsymbol{E} \cdot d\boldsymbol{l} = -\frac{d}{dt}\int_{\mathcal{A}} \boldsymbol{B} \cdot d\boldsymbol{\mathcal{A}} = -\frac{d\Phi}{dt}, \tag{21-11}$$

where Φ is the linking flux, as in Fig. 21-2. The electromotance induced around a closed curve C is equal to minus the time derivative of the flux linkage. The positive directions for Φ and around C satisfy the right-hand screw convention.

Equation 21-8 says that the net outward flux of $\boldsymbol{B}$ through any closed surface is zero, as in Fig. 21-3:

$$\int_{\mathcal{A}} \boldsymbol{B} \cdot d\boldsymbol{\mathcal{A}} = 0. \tag{21-12}$$

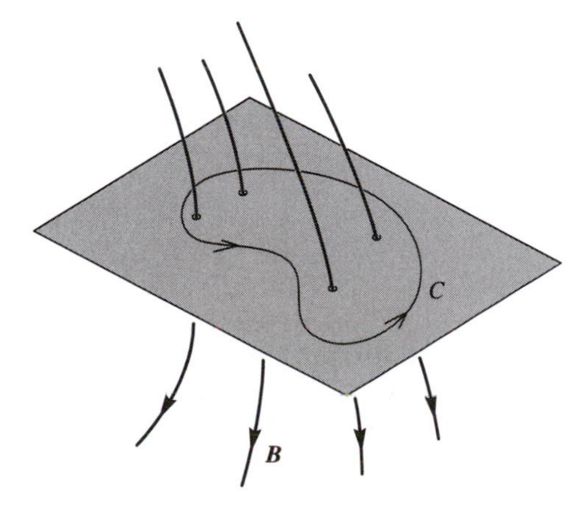

Figure 21-2 If the magnetic flux linking C *increases*, it induces an electromotance around C in the direction of the arrow. The electromotance points in the same direction if $\boldsymbol{B}$ points upward and *decreases*.

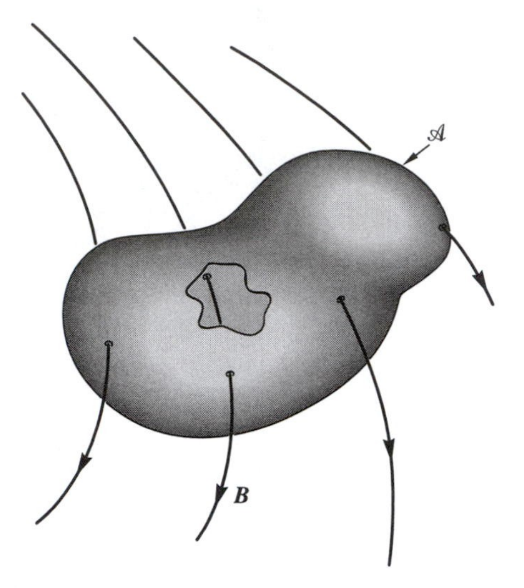

Figure 21-3 Lines of $\boldsymbol{B}$ passing through a closed surface. The net outward flux of $\boldsymbol{B}$ is equal to zero.

Finally, integrating Eq. 21-9 over the closed curve C yields

$$\oint_C \boldsymbol{B} \cdot d\boldsymbol{l} = \mu_0 \int_{\mathcal{A}} \left(\boldsymbol{J} + \epsilon_0 \frac{\partial \boldsymbol{E}}{\partial t} \right) \cdot d\boldsymbol{\mathcal{A}}. \tag{21-13}$$

We found two less general forms of this law in Secs. 15.4 and 16.3. The closed curve C bounds an open surface of area $\mathcal{A}$ through which flows a current of density $\boldsymbol{J} + \epsilon_0 \partial \boldsymbol{E}/\partial t$, as in Fig. 21-4.

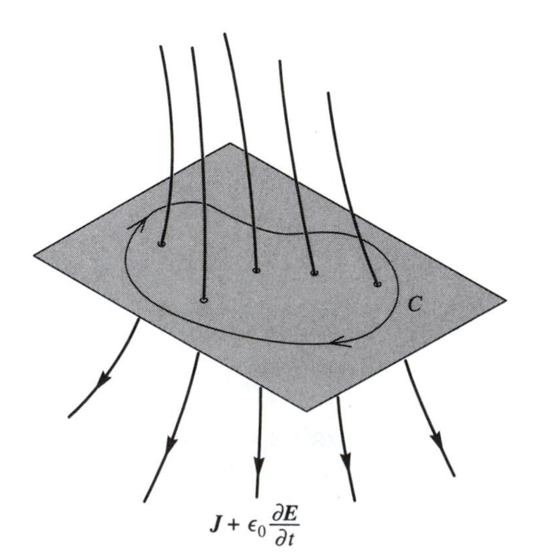

Figure 21-4 The line integral of $\boldsymbol{B} \cdot d\boldsymbol{l}$ around C is positive if the integration runs in the direction of the arrows with a current density $\boldsymbol{J} + \epsilon_0 \partial \boldsymbol{E}/\partial t$ pointing downward.

21.4 THE LAW OF CONSERVATION OF CHARGE

In Sec. 4.2 we saw that there is conservation of *free* charge. At that time we used $\boldsymbol{J}$ for the current density of free charges, instead of $\boldsymbol{J}_f$. Is there a more general law? We need the divergence of $\boldsymbol{J}$, as defined in Sec. 21-2, for the next section. First,

$$\boldsymbol{\nabla} \cdot \boldsymbol{J} = \boldsymbol{\nabla} \cdot \left(\boldsymbol{J}_f + \frac{\partial \boldsymbol{P}}{\partial t} + \boldsymbol{\nabla} \times \boldsymbol{M} \right) = \boldsymbol{\nabla} \cdot \boldsymbol{J}_f + \frac{\partial}{\partial t}(\boldsymbol{\nabla} \cdot \boldsymbol{P}), \tag{21-14}$$

since the divergence of a curl is zero. Thus

$$\boldsymbol{\nabla} \cdot \boldsymbol{J} = -\frac{\partial \rho_f}{\partial t} - \frac{\partial \rho_b}{\partial t} = -\frac{\partial(\rho_f + \rho_b)}{\partial t} = -\frac{\partial \rho}{\partial t}. \tag{21-15}$$

This is the more general form of the *law of conservation of charge*.

21.5 THE MAXWELL EQUATIONS ARE REDUNDANT!

Surprise! There are really only two independent Maxwell equations, not four!

The equations for $\boldsymbol{\nabla} \times \boldsymbol{E}$ and for $\boldsymbol{\nabla} \cdot \boldsymbol{B}$ form the first pair of Maxwell's equations, while the equations for $\boldsymbol{\nabla} \times \boldsymbol{B}$ and for $\boldsymbol{\nabla} \cdot \boldsymbol{E}$ form the second pair.

The equations of the first pair are related as follows. Take the divergence of Eq. 21-7, remembering that the divergence of a curl is zero. Then

$$\boldsymbol{\nabla} \cdot \frac{\partial \boldsymbol{B}}{\partial t} = 0, \qquad \text{or} \qquad \frac{\partial}{\partial t} \boldsymbol{\nabla} \cdot \boldsymbol{B} = 0. \tag{21-16}$$

So, at any point in space, $\boldsymbol{\nabla} \cdot \boldsymbol{B}$ is constant. Then we can set $\boldsymbol{\nabla} \cdot \boldsymbol{B} = 0$ everywhere and for all time *if we assume* that, for each point in space, $\boldsymbol{\nabla} \cdot \boldsymbol{B} = 0$ *at some time,* in the past, at present, or in the future. With this assumption, Eq. 21-8 follows from Eq. 21-7.

The equations of the second pair are also related. Take the divergence of Eq. 21-9 and apply the law of conservation of charge, Eq. 21-15,

$$\epsilon_0 \boldsymbol{\nabla} \cdot \frac{\partial \boldsymbol{E}}{\partial t} = -\boldsymbol{\nabla} \cdot \boldsymbol{J} = \frac{\partial \rho}{\partial t}, \tag{21-17}$$

$$\frac{\partial}{\partial t}(\boldsymbol{\nabla} \cdot \boldsymbol{E}) = \frac{\partial}{\partial t}\left(\frac{\rho}{\epsilon_0}\right), \quad \boldsymbol{\nabla} \cdot \boldsymbol{E} = \frac{\rho}{\epsilon_0} + C. \tag{21-18}$$

The constant of integration C can be a function of the coordinates. *If we now assume* that, at every point in space, *at some time,* $\boldsymbol{\nabla} \cdot \boldsymbol{E}$ and ρ are simultaneously equal to zero, then $C = 0$ and we have Eq. 21-6.

So the equations for $\boldsymbol{\nabla} \times \boldsymbol{B}$ and for $\boldsymbol{\nabla} \cdot \boldsymbol{E}$ are also related, and the four Maxwell equations are equivalent to only two independent equations.

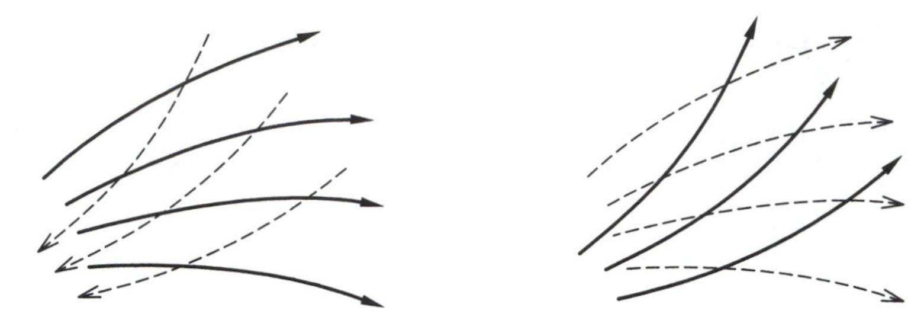

Figure 21-5 Pair of dual fields. Lines of $\boldsymbol{E}$ are solid, and lines of $\boldsymbol{H}$ dashed.

21.6 DUALITY

This section is interesting, heuristically: if you are familiar with one field, then you can guess the possibility of a different one.

Imagine a field $\boldsymbol{E}$, $\boldsymbol{B}$ that satisfies the Maxwell equations, with $\rho_f = 0$, $\boldsymbol{J}_f = 0$ in a given region. The medium is homogeneous, isotropic, linear, and stationary (HILS). Now imagine a different field

$$\boldsymbol{E}' = -K\boldsymbol{B} = -K\mu\boldsymbol{H}, \tag{21-19}$$

$$\boldsymbol{H}' = +K\boldsymbol{D} = +K\epsilon\boldsymbol{E}, \tag{21-20}$$

where the constant K has the dimensions of a speed and is independent of the space and time coordinates. This other field *also* satisfies the Maxwell equations, as you can check by substitution.

Figure 21-5 illustrates this duality property of electromagnetic fields. One field is said to be the *dual*, or the *dual field* of the other. If one field can exist, then its dual can also exist.

Even though the two fields are closely related through the above equations, and even though they both satisfy the Maxwell equations, they are entirely different. For example, the Lorentz force of Sec. 17.1 for one field bears no relation to the Lorentz force for the other.

EXAMPLE The Fields of Electric and Magnetic Dipoles

In Sec. 5.1 we found that, in the field of an electric dipole of moment $\boldsymbol{p}$,

$$\boldsymbol{E} = \frac{p}{4\pi\epsilon_0 r^3}(2\cos\theta\,\hat{\boldsymbol{r}} + \sin\theta\,\hat{\boldsymbol{\theta}}), \tag{21-21}$$

in spherical coordinates.

Later, in the third example in Sec. 14.4, we showed that, in the field of a magnetic dipole of moment $\boldsymbol{m}$, again in spherical coordinates,

$$\boldsymbol{B} = \frac{\mu_0 m}{4\pi r^3}(2\cos\theta\,\hat{\boldsymbol{r}} + \sin\theta\,\hat{\boldsymbol{\theta}}), \tag{21-22}$$

or

$$\boldsymbol{H} = \frac{m}{4\pi r^3}(2\cos\theta\hat{\boldsymbol{r}} + \sin\theta\hat{\boldsymbol{\theta}}). \tag{21-23}$$

We have set $\epsilon_r = 1$, $\mu_r = 1$. If the field of the electric dipole is the unprimed field and that of the magnetic dipole the primed field, then $K = m/p$.

21.7 THE LORENTZ CONDITION

The so-called *Lorentz condition* relates the vector potential $\boldsymbol{A}$ to the electric potential V:

$$\nabla \cdot \boldsymbol{A} + \epsilon_0\mu_0\frac{\partial V}{\partial t} = 0, \tag{21-24}$$

with

$$V = \frac{1}{4\pi\epsilon_0}\int_{v'}\frac{\rho}{r}dv', \tag{21-25}$$

$$\boldsymbol{A} = \frac{\mu_0}{4\pi}\int_{v'}\frac{\boldsymbol{J}}{r}dv', \tag{21-26}$$

as in Secs. 3.4.1 and 14.4. These are the potentials at the field point $P(x,y,z)$, $\rho = \rho_f + \rho_b$ is the total volume charge density at the source point $P'(x', y', z')$, dv' is the element of volume $dx'dy'dz'$ at P', r is the distance between P and P', and $\boldsymbol{J}$, defined as in Sec. 21.2, is the total current density at P'. The volume v' encloses all the charges and all the currents.

The Lorentz condition follows from the law of conservation of charge (Sec. 21.4), which links ρ to $\boldsymbol{J}$ (Lorrain, 1967; Lorrain, Corson, and Lorrain, 1988).

Is this relation important? It is! Think of an antenna that radiates an electromagnetic wave. You wish to calculate its $\boldsymbol{E}$ and $\boldsymbol{B}$ fields. So you need to know *six* quantities: the three components of $\boldsymbol{E}$ and the three components of $\boldsymbol{B}$. But wait! In Secs. 18.3 and 14.4, we found that

$$\boldsymbol{E} = -\nabla V - \frac{\partial \boldsymbol{A}}{\partial t}, \quad \boldsymbol{B} = \nabla \times \boldsymbol{A}. \tag{21-27}$$

So you really need only *four* quantities, the three components of $\boldsymbol{A}$ and the single component of V to find $\boldsymbol{E}$ and $\boldsymbol{B}$.

The Lorentz condition 21-24 is a further simplification. If the field is a sinusoidal function of the time, then $\partial V/\partial t$ is just a multiple of V: $\partial V/\partial t = j\omega V$. So, if you can find the *three* components of $\boldsymbol{A}$, the Lorentz condition provides V, and you can deduce all *six* components of $\boldsymbol{E}$ and $\boldsymbol{B}$!

Of course, it is always easy to calculate the gradient, or the divergence, or the curl of a known function.

21.8 SUMMARY

The Maxwell equations, in differential form, are as follows:

$$\boldsymbol{\nabla} \cdot \boldsymbol{E} = \frac{\rho}{\epsilon_0}, \tag{21-6}$$

$$\boldsymbol{\nabla} \times \boldsymbol{E} + \frac{\partial \boldsymbol{B}}{\partial t} = 0, \tag{21-7}$$

$$\boldsymbol{\nabla} \cdot \boldsymbol{B} = 0, \tag{21-8}$$

$$\boldsymbol{\nabla} \times \boldsymbol{B} - \epsilon_0 \mu_0 \frac{\partial \boldsymbol{E}}{\partial t} = \mu_0 \boldsymbol{J}. \tag{21-9}$$

The various terms are defined in Sec. 21.2, and other forms of the Maxwell equations are given on the back cover. The integral form is more meaningful, intuitively, but less useful, as a rule.

The equation for the divergence of $\boldsymbol{B}$ follows from the one for the curl of $\boldsymbol{E}$. Similarly, the equation for the divergence of $\boldsymbol{E}$ follows from the one for the curl of $\boldsymbol{B}$.

According to the *law of conservation of charge*,

$$\boldsymbol{\nabla} \cdot \boldsymbol{J} = -\frac{\partial \rho}{\partial t}. \tag{21-15}$$

If

$$\boldsymbol{E}' = -K\boldsymbol{B} = -K\mu\boldsymbol{H}, \tag{21-19}$$

$$\boldsymbol{H}' = +K\boldsymbol{D} = +K\epsilon\boldsymbol{E}, \tag{21-20}$$

where K has the dimensions of a speed and is independent of the space and time coordinates, the primed and the unprimed fields are called *dual* fields. Both fields satisfy the Maxwell equations. The fields of the electric and magnetic dipoles are examples of such dual fields.

The *Lorentz condition*

$$\boldsymbol{\nabla} \cdot \boldsymbol{A} + \epsilon_0 \mu_0 \frac{\partial V}{\partial t} = 0 \tag{21-24}$$

follows from the conservation of charge, with

$$V = \frac{1}{4\pi\epsilon_0} \int_{v'} \frac{\rho}{r} dv', \tag{21-25}$$

$$A = \frac{\mu_0}{4\pi}\int_{v'} \frac{J}{r} dv'. \tag{21-26}$$

Thanks to this relation, it is possible to calculate the six components of $\boldsymbol{E}$ and $\boldsymbol{B}$, for a given charge and current distribution, from only the three components of $\boldsymbol{A}$.

PROBLEMS

21-1. (*21.1*) The Maxwell equation for the curl of $\boldsymbol{B}$

Show that Eq. 21-4 applies to the fields of Eqs. 21-2 and 21-3.

(a) First do the non-relativistic case, with $\beta = 0$, $\gamma = 1$.

Disregard the factors $Q/(4\pi\epsilon_0)$ for $\boldsymbol{E}$ and $\mu_0 Q/(4\pi)$ for $\boldsymbol{B}$, and call the new fields $\boldsymbol{E}'$ and $\boldsymbol{B}'$. You should find after simplification that

$$(\nabla \times \boldsymbol{B}')_\rho = \frac{\partial E'_\rho}{\partial t} = \frac{3\mathcal{V}\rho(z - \mathcal{V}t)}{(\rho^2 + z^2 - 2z\mathcal{V}t + \mathcal{V}^2t^2)^{5/2}},$$

$$(\nabla \times \boldsymbol{B}')_z = \frac{\partial E'_z}{\partial t} = \frac{\mathcal{V}(-\rho^2 + 2z^2 - 4z\mathcal{V}t + 2\mathcal{V}^2t^2)}{(\rho^2 + z^2 - 2z\mathcal{V}t + \mathcal{V}^2t^2)^{5/2}}.$$

(b) If you have studied Chapters 11-13 on Relativity, do the relativistic case, with $\gamma = 2$, $\beta = 0.75$. Make sure that you use *cylindrical* coordinates, and refer to Fig. 21-6. You had better use math software here. You should find that

$$(\nabla \times \boldsymbol{B}')_\rho = \frac{\partial E'_\rho}{\partial t} = \frac{96\mathcal{V}\rho(z - \mathcal{V}t)}{(\rho^2 + 4z^2 - 8z\mathcal{V}t + 4\mathcal{V}^2t^2)^{5/2}},$$

$$(\nabla \times \boldsymbol{B}')_z = \frac{\partial E'_z}{\partial t} = \frac{8\mathcal{V}(-\rho^2 + 8z^2 - 16\mathcal{V}t + 8\mathcal{V}^2t^2)}{(\rho^2 + 4z^2 - 8z\mathcal{V}t + 4\mathcal{V}^2t^2)^{5/2}}.$$

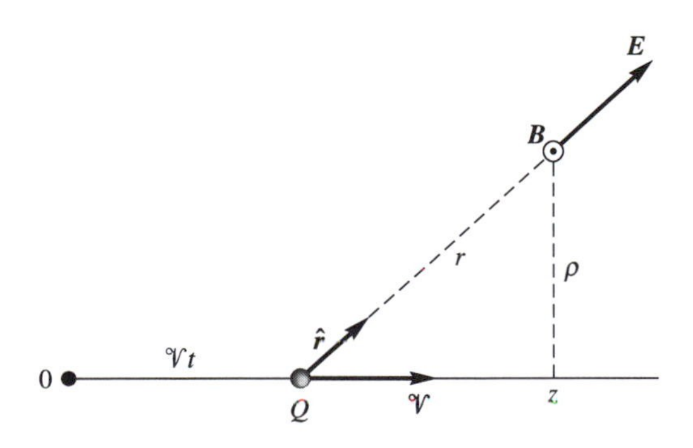

Figure 21-6

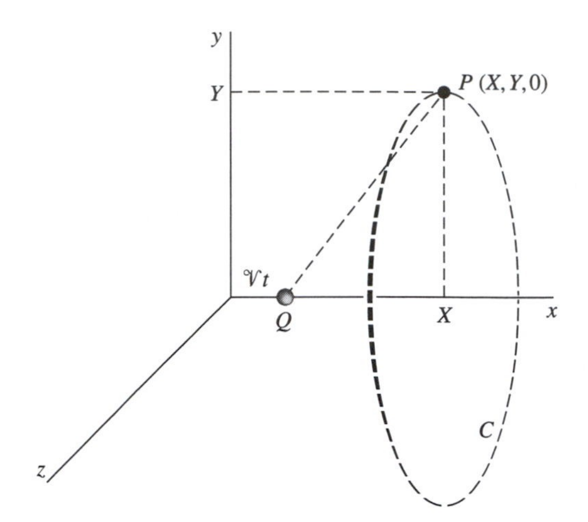

Figure 21-7

21-2. (*21.1*) The magnetic field of a point charge that moves at a constant velocity

Figure 21-7 shows a point charge Q that travels along the x-axis at a velocity $\mathcal{V}\hat{\boldsymbol{x}}$. Its position at time t is $(\mathcal{V}t, 0, 0)$.

(a) Find $\boldsymbol{B}$ at a point $P(X, Y, 0)$, not at the origin, at the instant that the charge passes through the origin. The particle travels in a vacuum, and $\mathcal{V}^2 \ll c^2$, where c is the speed of light.

(b) If you have studied Chap. 13, compare your result with that of Sec. 13.2.

(c) Sketch the value of

$$\frac{d}{dt}\int_{\mathcal{A}} \boldsymbol{D} \cdot d\boldsymbol{\mathcal{A}},$$

where $\mathcal{A}$ is the area of the spherical segment of radius Y, as a function of $\mathcal{V}t$. When the charge is just to the left of $\mathcal{V}t = X$, the flux of $\boldsymbol{D}$ through $\mathcal{A}$ points to the right and is $+\ Q/2$. Immediately afterward, the flux points to the left and is $-Q/2$, so there is a discontinuity in this curve.

(d) The integral of $\boldsymbol{H} \cdot \boldsymbol{dl}$ around the circle shown in the figure is equal to $2\pi YB/\mu_0$. Calculate the integral of this quantity over time from minus infinity to plus infinity. Set $X = 0$ to simplify the calculation. Explain your result.

21-3. (21.2) Express the following quantities in terms of kilograms, meters, seconds, and amperes:

joule, watt,

coulomb, volt, ohm, siemens, farad,

weber, tesla, henry.

21-4. (*21.2*) Magnetic monopoles

If monopoles exist, then Maxwell's equations require two more terms, to take into account magnetic charges and magnetic currents. It is the custom to write the equations in the following form:

$$\nabla \cdot \boldsymbol{D} = \rho, \qquad \nabla \times \boldsymbol{E} = -\frac{\partial \boldsymbol{B}}{\partial t} - \boldsymbol{J}^*, \qquad \nabla \cdot \boldsymbol{B} = \rho^*, \qquad \nabla \times \boldsymbol{H} = \frac{\partial \boldsymbol{D}}{\partial t} + \boldsymbol{J},$$

where ρ^* is the magnetic charge density, expressed in webers/meter ^{3}, and $\boldsymbol{J}^*$ is the magnetic current density, in webers/second-meter ^{2}.

(a) Show that

$$\nabla \cdot \boldsymbol{J}^* = -\frac{\partial \rho^*}{\partial t}.$$

This is the equation of conservation for magnetic monopoles.

(b) Show that, by analogy with electric fields, near a point magnetic charge Q^*,

$$\boldsymbol{B} = \frac{Q^*}{4\pi r^2}\hat{\boldsymbol{r}}, \qquad \boldsymbol{H} = \frac{Q^*}{4\pi\mu_0 r^2}\hat{\boldsymbol{r}}.$$

(c) Calculate the energy acquired by a magnetic monopole that accelerates over a distance of 1000 kilometers in the earth's magnetic field ($\approx 10^{-5}$ tesla).

(d) Magnetic monopoles go through a loop of copper wire. Show that the induced electromotance is equal to minus the magnetic current, with the right-hand screw convention. This is one method of detecting magnetic monopoles.

21-5. (*21.2*) A transformation that leaves Maxwell's equations unchanged

Show that Maxwell's equations for free space are unchanged by the transformation

$$\boldsymbol{E}' = a\boldsymbol{E} + bc\boldsymbol{B}, \qquad \boldsymbol{B}' = -\left(\frac{b}{c}\right)\boldsymbol{E} + a\boldsymbol{B},$$

where a and b are constants and c is the speed of light.

21-6. (21.2) Another transformation that leaves Maxwell's equations unchanged

Show that Maxwell's equations for free space are unchanged by the transformation

$$\boldsymbol{E}' = \boldsymbol{E}\cos\theta + c\boldsymbol{B}\sin\theta, \qquad \boldsymbol{B}' = -\frac{\boldsymbol{E}}{c}\sin\theta + \boldsymbol{B}\cos\theta.$$

The transformation $\boldsymbol{E}' = -K\boldsymbol{B}$, $\boldsymbol{H}' = K\boldsymbol{D}$ of Sec 21.6 and the transformation $\boldsymbol{E}' = -\boldsymbol{E}$, $\boldsymbol{B}' = -\boldsymbol{B}$ are special cases corresponding to $\theta = \pi/2$ and $\theta = \pi$, respectively.

21-7. (*21.2*) The skin effect

As we shall see in Secs. 22.6 and 23.4, a high-frequency field does not penetrate significantly into the body of a good conductor. Also, both $\boldsymbol{E}$ and $\boldsymbol{B}$ inside are approximately tangent to the surface. Let the z-axis be normal to the surface, pointing outward, with $\boldsymbol{E}$ in the direction of the x-axis and $\boldsymbol{B}$ in the direction of the y-axis. Set $\partial/\partial x = 0$, $\partial/\partial y = 0$.

(a) Show that, inside, $\partial E/\partial z = -\partial B/\partial t$.

(b) Show that, just outside the conductor, $\boldsymbol{B}$ is tangent, or nearly so.

(c) Let the current density near the surface be $\boldsymbol{\alpha}$ amperes/meter. Show that, just outside the conductor, $\boldsymbol{B} = \mu_0\boldsymbol{\alpha} \times \hat{\boldsymbol{z}}$.

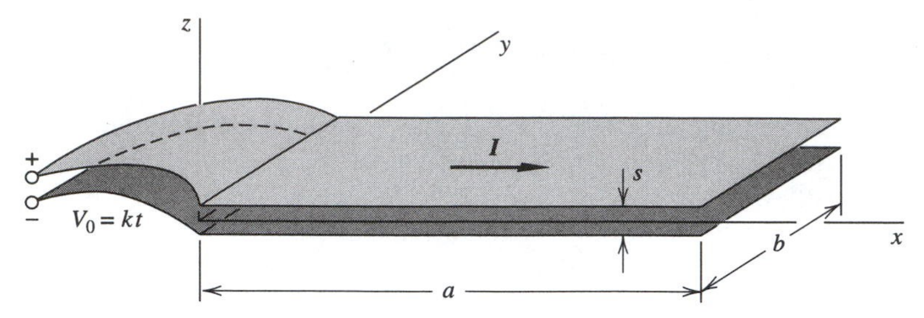

Figure 21-8

(d) Does this last result depend on how the current varies with depth inside the conductor?

21-8. (*21.2*) Parallel-plate capacitor fed at one end by a time-dependent source

Figure 21-8 shows a parallel-plate capacitor connected at one end to a source whose voltage increases slowly and linearly with time: $dV_0/dt = k$. Edge effects are negligible: $a \gg s$, $b \gg s$.

(a) Find the current I as a function of x.
(b) Find $\boldsymbol{B}$ inside in two different ways. Find $\boldsymbol{A}$ inside.
(c) Find $\boldsymbol{B}$ and $\boldsymbol{A}$ outside. The capacitor plates are thin.
(d) Draw a large cross section of the capacitor in the midplane parallel to the xz-plane, showing $\boldsymbol{I}$ and the vectors $\boldsymbol{A}, \boldsymbol{B}, \boldsymbol{\nabla}\times\boldsymbol{B}, \boldsymbol{E}, \partial\boldsymbol{E}/\partial t$ near both ends. Use arrows of different sizes to indicate qualitatively how these vectors vary with x and with z.

21-9. (*21.2*) Superconductivity

A superconductor offers zero resistance to the motion of superconducting charge carriers. These are pairs of electrons that move as a unit.

(a) If there are N such carriers per cubic meter, of mass m' and charge e', show that

$$\boldsymbol{E} = \frac{m'}{Ne'^2}\frac{d\boldsymbol{J}}{dt}.$$

This is the *first London equation*. Note that $\boldsymbol{E}$ is zero only if $\boldsymbol{J}$ is constant.

(b) Set $K = m'/(Ne'^2)$. Show that in an alternating field

$$\sigma = \frac{1}{j\omega K} = -\frac{j}{\omega K}.$$

(c) Show that

$$\boldsymbol{\nabla}\times\left(K\frac{\partial \boldsymbol{J}}{\partial t}\right) = -\frac{\partial \boldsymbol{B}}{\partial t}.$$

The equation

$$\boldsymbol{\nabla}\times K\boldsymbol{J} = -\boldsymbol{B}$$

is the *second London equation*. It does *not* follow mathematically from the first.

(d) Show that under steady-state conditions

$$\nabla^2 \boldsymbol{B} = \frac{\mu_0}{K}\boldsymbol{B}.$$

In one dimension, this means that

$$\frac{d^2\boldsymbol{B}}{dx^2} = \frac{\mu_0}{K}\boldsymbol{B},$$

or that

$$\boldsymbol{B} = \boldsymbol{B}_0 \exp\left[-\left(\frac{\mu_0}{K}\right)^{1/2} x\right],$$

where $(K/\mu_0)^{1/2}$ is the *depth of penetration* of the field.

(e) Calculate the value of the depth of penetration, setting m' equal to twice the mass of an electron, $e' = 2e$, and $N = 10^{29}$ meter^{-3}. The depth of penetration is, in fact, a few times larger.

(f) Much beyond the depth of penetration, $\boldsymbol{E} = 0, \boldsymbol{J} = 0, \boldsymbol{B} = 0$.

Show that just outside a superconductor, $\boldsymbol{B}$ is tangential to the surface and $\boldsymbol{H}$ is equal in magnitude to the surface current density.

21-10. (*21.3*) The magnetic field of a leaky capacitor

A charged capacitor whose electrodes are parallel and circular lies in a large volume of dielectric that is slightly conducting, as in Fig. 21.9. The capacitor discharges.

(a) Calculate the value of the ratio $J_f/(\partial D/\partial t)$ at any point in the dielectric in terms of the resistance and the capacitance.

(b) Show that this ratio is equal to –1 at the surface of an electrode.

(c) Show that $\boldsymbol{B}$ is zero everywhere in the dielectric. This means that the magnetic field of the conduction and polarization (*not* displacement) currents in the fringing field exactly cancels the magnetic field of the conduction and polarization currents in the region between the plates.

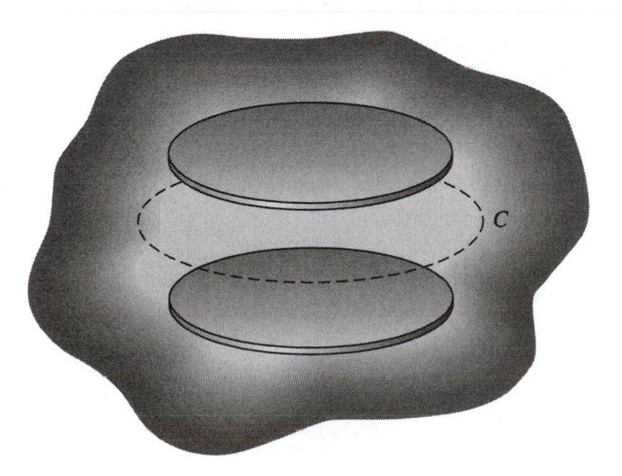

Figure 21-9

(d) Show that $\boldsymbol{B}$ is also zero for electrodes of any shape.

If the dielectric occupies only part of the field of a capacitor, say the region between the plates of a parallel-plate capacitor, then the value of the ratio calculated under (a) applies. However, at the surface of an electrode, this ratio is not equal to -1 because charge migrates from the outer surface of an electrode to the inner surface, where it leaks out. Then, in the dielectric, $|\boldsymbol{J}_f| > |\partial \boldsymbol{D}/\partial t|$, $\boldsymbol{J}_f - \partial \boldsymbol{D}/\partial t$ points in the direction of $\boldsymbol{J}_f$ and thus of $\boldsymbol{E}$, and there is an azimuthal magnetic field.

21-11. *(21.7)* The continuous-creation theory and Maxwell's equations

Imagine an expanding, spherically symmetrical universe in which there is continuous creation of charge at the rate of q coulombs/meter^3-second. Creation of electric charge occurs through the creation of hydrogen atoms carrying a slight excess charge ye as in Prob. 3-22.

The rate of mass creation Q is proportional to q: $Q = [m/(ye)]q$, where m is the mass of the proton, the universe being mostly hydrogen.

(a) By symmetry, the vector potential can only be radial.

Show that under steady-state conditions the current density $\boldsymbol{J}$ is everywhere zero, according to Maxwell's equations.

Lyttleton and Bondi (see Prob. 3-22) suggested that, if continuous creation does exist, then Maxwell's equations must be modified as follows:

$$\boldsymbol{\nabla} \times \boldsymbol{B} = \mu_0 \boldsymbol{J} + \frac{1}{c^2}\frac{\partial \boldsymbol{E}}{\partial t} - \left[\frac{1}{l^2}\boldsymbol{A}\right], \qquad \boldsymbol{\nabla} \cdot \boldsymbol{E} = \frac{\rho}{\epsilon_0} - \left[\frac{1}{l^2}V\right],$$

where the new terms are enclosed in brackets. The quantities V and $\boldsymbol{A}$ are the usual scalar and vector potentials:

$$\boldsymbol{E} = -\boldsymbol{\nabla} V - \frac{\partial \boldsymbol{A}}{\partial t}, \qquad \boldsymbol{B} = \boldsymbol{\nabla} \times \boldsymbol{A}.$$

The other two equations of Maxwell for $\boldsymbol{\nabla} \times \boldsymbol{E}$ and $\boldsymbol{\nabla} \cdot \boldsymbol{B}$ remain unchanged. Lyttleton and Bondi suggested that the constant l, which has the dimensions of a length, would be of the order of the radius of the universe. The new terms would therefore be negligible in all but cosmological problems.

(b) If these modified Maxwell equations are correct, are V and $\boldsymbol{A}$ measurable, in principle? Remember that, with the above equations for $\boldsymbol{E}$ and $\boldsymbol{B}$ in terms of V and $\boldsymbol{A}$, only the rates of change of V and $\boldsymbol{A}$ determine $\boldsymbol{E}$ and $\boldsymbol{B}$.

(c) Write out the equation for the conservation of the total charge (Sec. 21.4).

(d) Would the Lorentz condition still be valid?

(e) Now set $\boldsymbol{A} = A'\boldsymbol{r}$, where A' is a constant, and assume V to be constant. Show that $\boldsymbol{B} = 0$, $\boldsymbol{E} = 0$, $\boldsymbol{J} = (q/3)\boldsymbol{r}$, $\rho = \epsilon_0 V/l^2$.

Assuming that the velocity of the outward flow of matter is the same as that of the charge, namely $\boldsymbol{J}/\rho$, it follows that the radial velocity is proportional to $\boldsymbol{r}$, which is consistent with the linear velocity-distance relation observed by astronomers: $v = r/T$, where $T \approx 3 \times 10^{17}$ seconds is the *Hubble constant.*

(f) Show that $\rho = qT/3$.

(g) Now the space-charge density ρ is $\eta ye/m$, where η, the mass density of the universe, is about 10^{-26} kilogram/meter^3.

Show that, if this theory is correct, then $Q \approx 1/(2 \times 10^{16})$ hydrogen atom/meter^3-second.

CHAPTER

22

ELECTROMAGNETIC WAVES I

In Free Space, in Nonconductors, and in Conductors

In this chapter we study the propagation of plane electromagnetic waves in free space, in nonconductors, and in conductors. We shall study reflection and refraction in Chapter 23, then guided waves in Chapter 24, and the radiation of electromagnetic waves in Chapter 25.

If you are not familiar with the use of phasors, then you must read Chapter 2 now.

22.1 THE ELECTROMAGNETIC SPECTRUM

Maxwell's equations impose no limit on the frequency of electromagnetic waves. The known spectrum extends continuously from the long radio waves to the very high energy gamma rays of cosmic radiation, as in Fig. 22-1. In the former, the frequencies are of the order of 100 hertz and the wavelengths about 3 megameters; in the latter, the frequencies are of the order of 10^{24} hertz and the wavelengths less than one femtometer (10^{-15} meter).

The known frequencies and wavelengths thus cover a range of about 22 orders of magnitude. Radio waves, heat waves, light, x-rays, and gamma rays are all electromagnetic waves, although the sources and the detectors, as well as the modes of interaction with matter, vary widely as the frequency changes by orders of magnitude.

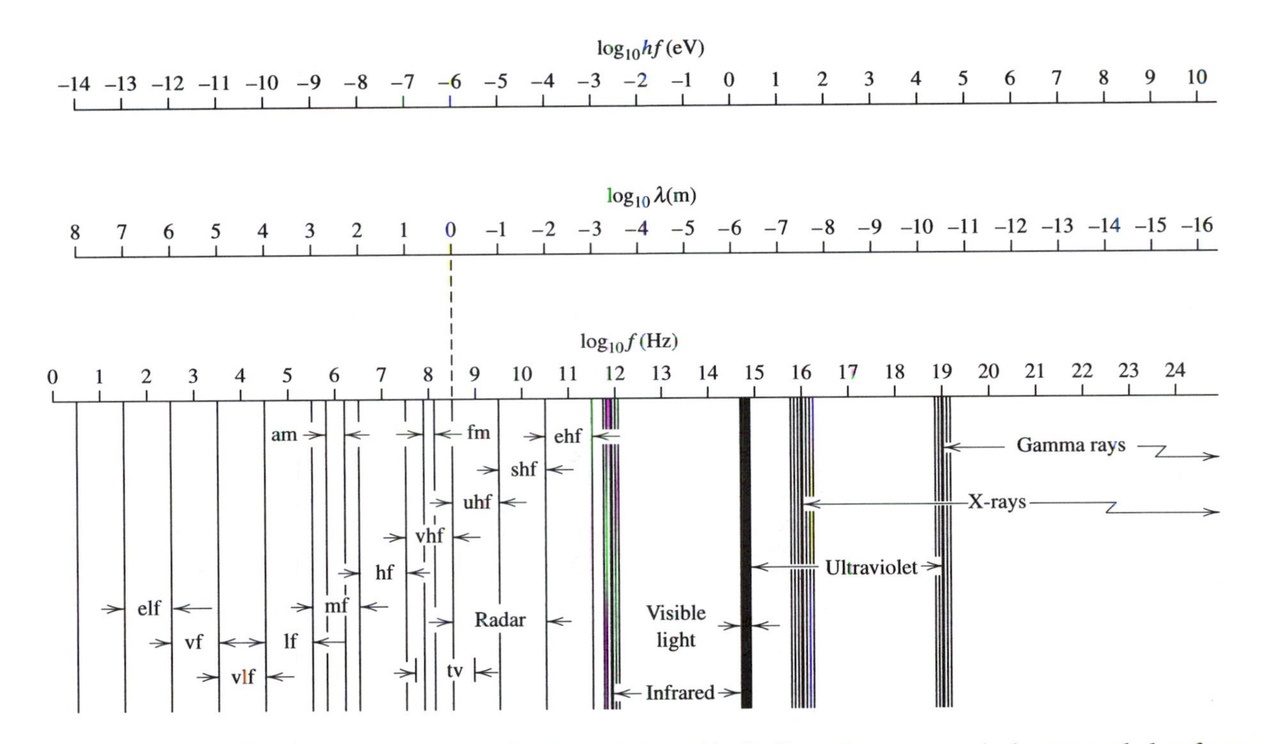

Figure 22-1 The spectrum of electromagnetic waves. The abbreviations elf, vf, vlf, . . . mean, respectively, extremely low frequency, voice frequency, very low frequency, low frequency, medium frequency, high frequency, very high frequency, ultrahigh frequency, super high frequency, and extremely high frequency. The limits indicated by the shaded regions are approximate. The energy hf, where h is Planck's constant (6.63×10^{-34} joule-second) and f is the frequency, is that of a photon or quantum of radiation.

Many experiments demonstrate the fundamental identity of all these waves. In free space, they are all transverse, and they all travel at the same speed. For example, simultaneous radio and optical observations on stars show that the speed of propagation is the same, within experimental error, for wavelengths differing by more than six orders of magnitude.

Before discussing electromagnetic waves, we must first study waves briefly.

22.2 WAVES

A disturbance can propagate as a wave. For example, if you whistle, the air pressure between your lips fluctuates, and the disturbance travels outward in all directions.

Now consider a sound wave and call the air pressure p. In the simplest case, the wave travels along the z-axis, the pressure is independent of the x- and y-coordinates, and there is no attenuation. At $z = 0$, set

$$p = p_m \cos \omega t. \tag{22-1}$$

Then the phase of the disturbance at $z = 0$ is ωt. What is the phase at z? At z, the phase lags by the time it takes for the wave to travel out to the distance z at the constant speed v, or z/v. So, at z, the disturbance is the same as at the source, at the previous time $t - (z/v)$:

$$p = p_m \cos\left[\omega\left(t - \frac{z}{v}\right)\right]. \tag{22-2}$$

This equation defines an *unattenuated plane sinusoidal wave*. Here the *wave fronts* are normal to the z-axis. Figure 22-2 shows p as a function of t at a given z, and p as a function of z at a given t. The *wavelength* λ is the distance over which the phase changes by 2π, the *period* T is the time it takes for the phase to change by 2π, and the *frequency* $f = 1/T$. Then $\lambda = v/f$.

It is usually more convenient to write

$$p = p_m \cos(\omega t - kz), \tag{22-3}$$

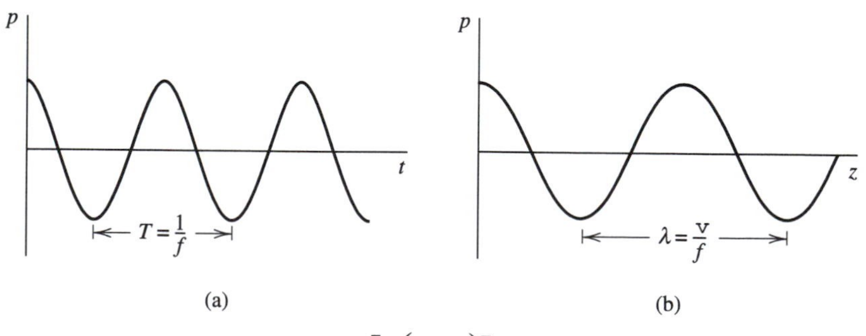

Figure 22-2 The quantity $p = p_m \cos\left[\omega\left(t - \frac{z}{v}\right)\right]$ (a) as a function of t at a given z, and (b) as a function of z at a given t.

where

$$k = \frac{\omega}{v} = \frac{2\pi f}{v} = \frac{2\pi}{\lambda} = \frac{1}{\lambda\!\!\!^{-}} \tag{22-4}$$

is the *wave number*. (This quantity should rather be called the *circular* wave number because it is 2π times larger than the wave number that is used in optics.) The quantity $\lambda\!\!\!^{-}$ (pronounced "lambda bar") is the *radian length*, or the distance over which the phase changes by one radian.

In phasor notation (Chap. 2),

$$p = p_m \exp j(\omega t - kz). \tag{22-5}$$

You can check that the differential equation that corresponds to both Eqs. 22-3 and 22-5 is

$$\frac{\partial^2 p}{\partial z^2} - \frac{1}{v^2}\frac{\partial^2 p}{\partial t^2} = 0. \tag{22-6}$$

For an unattenuated wave in three dimensions, the first term on the left in Eq. 22-6 must be replaced by three second derivatives, with respect to x, y, z, and it becomes a Laplacian:

$$\nabla^2 p - \frac{1}{v^2}\frac{\partial^2 p}{\partial t^2} = 0. \tag{22-7}$$

If a plane wave propagates in an arbitrary direction, then the vector wave number $\boldsymbol{k}$ points in the direction of propagation and

$$p = p_m \exp j(\omega t - \boldsymbol{k} \cdot \boldsymbol{r}). \tag{22-8}$$

Now air pressure is a scalar quantity, like a mass density: at a given point, it does not have a specific direction. What if the quantity propagated in the wave is a vector quantity, like an electric field $\boldsymbol{E}$, for example? Then, in Cartesian coordinates, we have three similar equations, one for the x-component E_x, one for E_y, and one for E_z:

$$\nabla^2 E_x - \frac{1}{v^2}\frac{\partial^2 E_x}{\partial t^2} = 0, \tag{22-9}$$

$$\nabla^2 E_y - \frac{1}{v^2}\frac{\partial^2 E_y}{\partial t^2} = 0, \tag{22-10}$$

$$\nabla^2 E_z - \frac{1}{v^2}\frac{\partial^2 E_z}{\partial t^2} = 0. \tag{22-11}$$

If we now multiply the first equation by $\hat{x}$, the second by $\hat{y}$, the third by $\hat{z}$, and then add the three equations, we find that

$$\nabla^2 \mathbf{E} - \frac{1}{v^2}\frac{\partial^2 \mathbf{E}}{\partial t^2} = 0. \tag{22-12}$$

This is the *wave equation for* **E** in three dimensions, assuming zero attenuation.

22.3 ELECTROMAGNETIC WAVES IN FREE SPACE

By definition, *free space* is an infinite region devoid of matter.

Is the above wave equation for **E** consistent with the Maxwell equations? Refer to Sec. 21.2. In free space, the space charge density ρ and the current density $\mathbf{J}$ are both zero, so that Maxwell's equations simplify to

$$\nabla \cdot \mathbf{E} = 0, \tag{22-13}$$

$$\nabla \times \mathbf{E} + \frac{\partial \mathbf{B}}{\partial t} = 0, \tag{22-14}$$

$$\nabla \cdot \mathbf{B} = 0, \tag{22-15}$$

$$\nabla \times \mathbf{B} - \epsilon_0 \mu_0 \frac{\partial \mathbf{E}}{\partial t} = 0. \tag{22-16}$$

Now take the curl of Eq. 22-14, and refer to Item 5 in the Table of Vector Definitions, Identities, and Theorems at the beginning of this book:

$$\nabla \times \nabla \times \mathbf{E} = -\nabla^2 \mathbf{E} + \nabla(\nabla \cdot \mathbf{E}). \tag{22-17}$$

The last term on the right is equal to zero. Then

$$-\nabla^2 \mathbf{E} + \nabla \times \frac{\partial \mathbf{B}}{\partial t} = 0. \tag{22-18}$$

Now invert the order of the curl and of the time derivative in the last term:

$$-\nabla^2 \mathbf{E} + \frac{\partial}{\partial t} \nabla \times \mathbf{B} = 0. \tag{22-19}$$

Finally, substitute Eq. 22-16:

$$\nabla^2 \mathbf{E} - \epsilon_0 \mu_0 \frac{\partial^2 \mathbf{E}}{\partial t^2} = 0. \tag{22-20}$$

This is the *wave equation for* **E** in free space. Comparing now with Eq. 22-12, we see that *the speed of propagation of a plane electromagnetic wave in free space* is given by

$$c = \frac{1}{(\epsilon_0 \mu_0)^{1/2}}. \tag{22-21}$$

This equation is remarkable because it links three basic constants of electromagnetism: the speed of light c, the permittivity of free space ϵ_0, from the Coulomb force law, and the permeability of free space μ_0, from the magnetic force law.

Since, by definition, $\mu_0 \equiv 4\pi \times 10^{-7}$, the value of ϵ_0 follows from the experimental value of c, as in Sec. 17.3.1:

$$c = 2.99792458 \times 10^8 \text{ meters/second,} \tag{22-22}$$

$$\epsilon_0 = \frac{1}{\mu_0 c^2} = 8.85418782 \times 10^{-12} \text{ farad/meter.} \tag{22-23}$$

Is there a corresponding wave equation for the magnetic flux density $\boldsymbol{B}$? There is, and it is mathematically identical:

$$\nabla^2 \boldsymbol{B} - \epsilon_0 \mu_0 \frac{\partial^2 \boldsymbol{B}}{\partial t^2} = 0. \tag{22-24}$$

Until now we have used the magnetic field strength $\boldsymbol{H}$ only when discussing magnetic materials in Chap. 16. But in discussing electromagnetic waves we had better use $\boldsymbol{H}$ rather than $\boldsymbol{B}$ for two good reasons: as we shall see, $\boldsymbol{E} \times \boldsymbol{H}$ is the power density in an electromagnetic wave, and E/H is the characteristic impedance of the medium of propagation. These two concepts have great practical importance.

The wave equation for $\boldsymbol{H}$ in free space follows immediately from Eq. 22-24 by substituting $\mu_0 \boldsymbol{H}$ for $\boldsymbol{B}$. So, in free space,

$$\nabla^2 \boldsymbol{E} - \epsilon_0 \mu_0 \frac{\partial^2 \boldsymbol{E}}{\partial t^2} = 0, \qquad \nabla^2 \boldsymbol{H} - \epsilon_0 \mu_0 \frac{\partial^2 \boldsymbol{H}}{\partial t^2} = 0. \tag{22-25}$$

Assume now a plane sinusoidal wave traveling in the positive direction of the z-axis. Assume also that the $\boldsymbol{E}$-vectors are all parallel to a given direction: the wave is *linearly polarized*. By definition, the *plane of polarization* is parallel to $\boldsymbol{E}$. If the wave is not linearly polarized, then it is the sum of linearly polarized waves.

In a linearly polarized wave traveling in the positive direction of the z-axis, $\boldsymbol{E}$ and $\boldsymbol{H}$ are of the form

$$\boldsymbol{E} = \boldsymbol{E}_m \exp j(\omega t - kz), \qquad \boldsymbol{H} = \boldsymbol{H}_m \exp j(\omega t - kz), \tag{22-26}$$

where $\boldsymbol{E}_m$ and $\boldsymbol{H}_m$ are vectors that are independent of the time and of the coordinates. In free space, $k = k_0$, with

$$k_0 = \frac{1}{\lambda\!\!\!^{-}_0} = \frac{\omega}{c} = \omega(\epsilon_0 \mu_0)^{1/2}. \tag{22-27}$$

For this particular field, in phasor notation,

$$\frac{\partial}{\partial t} = j\omega \quad \text{and} \quad \nabla = \frac{\partial}{\partial x}\hat{x} + \frac{\partial}{\partial y}\hat{y} + \frac{\partial}{\partial z}\hat{z} = -jk\hat{z}. \tag{22-28}$$

Then Maxwell's equations 22-13 to 22-16, with $\boldsymbol{B} = \mu_0 \boldsymbol{H}$, reduce to

$$-jk_0\hat{z} \cdot \boldsymbol{E} = 0, \quad -jk_0\hat{z} \times \boldsymbol{E} = -j\omega\mu_0\boldsymbol{H}, \tag{22-29}$$

$$-jk_0\hat{z} \cdot \boldsymbol{H} = 0, \quad -jk_0\hat{z} \times \boldsymbol{H} = j\omega\epsilon_0\boldsymbol{E}, \tag{22-30}$$

and then to

$$\hat{z} \cdot \boldsymbol{E} = 0, \qquad \boldsymbol{H} = \frac{k_0}{\omega\mu_0}\hat{z} \times \boldsymbol{E} = \frac{\boldsymbol{k}_0 \times \boldsymbol{E}}{\omega\mu_0}, \tag{22-31}$$

$$\hat{z} \cdot \boldsymbol{H} = 0, \qquad \boldsymbol{E} = -\frac{\boldsymbol{k}_0 \times \boldsymbol{H}}{\omega\epsilon_0}. \tag{22-32}$$

The $\hat{z} \cdot \boldsymbol{E}$ and $\hat{z} \cdot \boldsymbol{H}$ equations show that $\boldsymbol{E}$ and $\boldsymbol{H}$ are perpendicular to the direction of propagation $\hat{z}$, or transverse. Also, because of the $\boldsymbol{H}$ and $\boldsymbol{E}$ equations, the $\boldsymbol{E}$ and $\boldsymbol{H}$ vectors are mutually perpendicular, as in Fig. 22-3.

The *Poynting vector* $\boldsymbol{E} \times \boldsymbol{H}$ is important: it points in the direction of propagation, and we shall see in the next section that it gives the power density in the wave.

The ratio E/H is the *characteristic impedance*† of the medium of propagation:

$$Z_0 = \frac{E}{H} = \frac{k_0}{\omega\epsilon_0} = \frac{\omega\mu_0}{k_0}. \tag{22-33}$$

Then

$$Z_0 = \frac{E}{H} = \frac{\omega\mu_0}{k_0} = \frac{\mu_0}{(\epsilon_0\mu_0)^{1/2}} = \left(\frac{\mu_0}{\epsilon_0}\right)^{1/2}, \tag{22-34}$$

$$= 3.767303 \times 10^2 \approx 377 \text{ ohms}. \tag{22-35}$$

Since the ratio E/H is real, the $\boldsymbol{E}$ and $\boldsymbol{H}$ vectors are in phase.

Also, since $B = \mu_0 H$ in free space,

$$\frac{E}{B} = \frac{1}{(\epsilon_0\mu_0)^{1/2}} = c, \qquad \text{or} \qquad E = Bc, \tag{22-36}$$

a simple relation to remember.

The electric and magnetic energy densities are equal:

$$\frac{\epsilon_0 E^2/2}{\mu_0 H^2/2} = \frac{\epsilon_0}{\mu_0}\left(\frac{\mu_0}{\epsilon_0}\right) = 1. \tag{22-37}$$

†Does the ratio E/H really have the dimensions of an impedance V/I? Well, E has the dimensions of V/L, where L is a length and, from Eq. 16-16, H/L has the dimensions of J, or of I/L^2, so that H has the dimensions of I/L, and E/H has indeed the same dimensions as V/I.

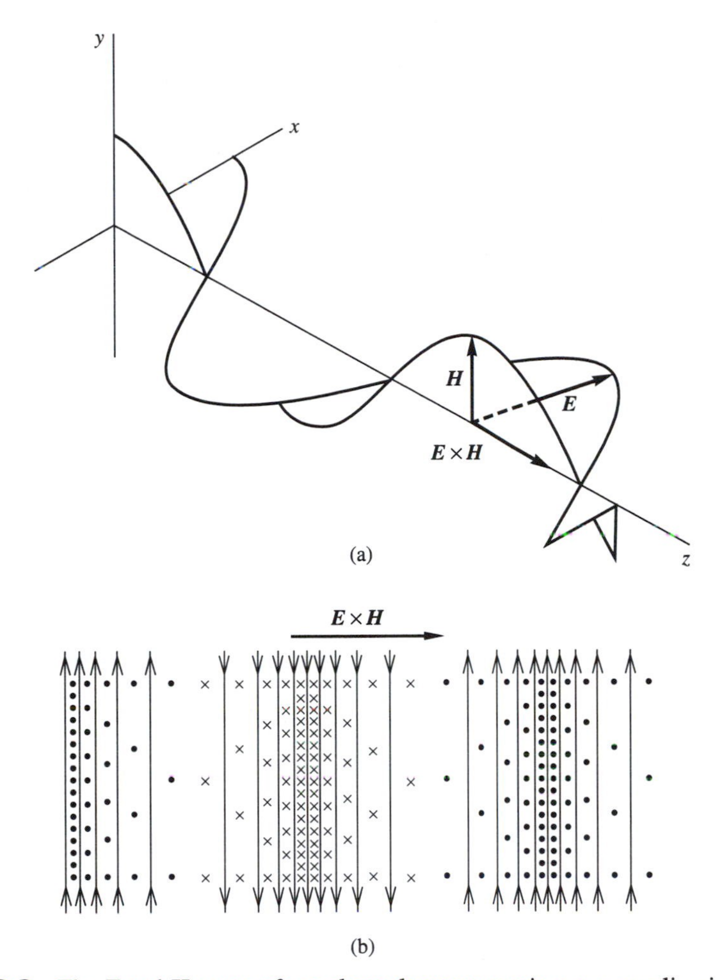

Figure 22-3 The $\boldsymbol{E}$ and $\boldsymbol{H}$ vectors for a plane electromagnetic wave traveling in free space in the positive direction along the z-axis. (a) The fields $\boldsymbol{E}$ and $\boldsymbol{H}$ as functions of z at a particular moment. The two vectors are orthogonal and in phase. (b) Lines of $\boldsymbol{E}$ (arrows), as seen when looking down on the xz-plane. The dots represent lines of $\boldsymbol{H}$ coming out of the paper, and the crosses lines going into the paper. The vector $\boldsymbol{E} \times \boldsymbol{H}$ points everywhere in the direction of propagation.

At any instant the total energy density fluctuates with z as in Fig. 22-4, and its time-averaged value at any point is

$$\mathscr{E}'_{av} = \frac{\epsilon_0 E_{\rm rms}^2}{2} + \frac{\mu_0 H_{\rm rms}^2}{2} = \epsilon_0 E_{\rm rms}^2 = \mu_0 H_{\rm rms}^2. \tag{22-38}$$

Without the phasor notation,

$$\boldsymbol{E} = \boldsymbol{E}_m \cos(\omega t - kz), \quad \boldsymbol{H} = \boldsymbol{H}_m \cos(\omega t - kz), \tag{22-39}$$

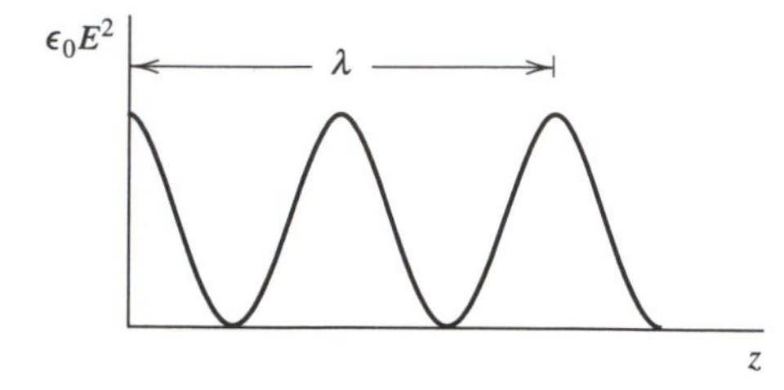

Figure 22-4 The total energy density $\epsilon_0 E^2$, or $\mu_0 H^2$, as a function of z, at $t = 0$, for a plane wave traveling along the z-axis in free space.

with

$$E_m / H_m \approx 377 \text{ ohms}. \tag{22-40}$$

22.4 THE POYNTING THEOREM

In Sec. 22.3 we found that the Poynting vector $\boldsymbol{E} \times \boldsymbol{H}$ points in the direction of propagation. We now show that, more generally, the Poynting vector gives the power density in an electromagnetic field. It thus has great practical importance. See Lorrain (1982, 1984, 1987).

First, we have the vector identity

$$\boldsymbol{\nabla} \cdot (\boldsymbol{E} \times \boldsymbol{H}) = \boldsymbol{H} \cdot (\boldsymbol{\nabla} \times \boldsymbol{E}) - \boldsymbol{E} \cdot (\boldsymbol{\nabla} \times \boldsymbol{H}). \tag{22-41}$$

In free space, Eqs. 22-13 to 22-16 apply and

$$\boldsymbol{\nabla} \cdot (\boldsymbol{E} \times \boldsymbol{H}) = -\boldsymbol{H} \cdot \mu_0 \frac{\partial \boldsymbol{H}}{\partial t} - \boldsymbol{E} \cdot \epsilon_0 \frac{\partial \boldsymbol{E}}{\partial t} \tag{22-42}$$

$$= -\frac{\partial}{\partial t}\left(\frac{\epsilon_0 E^2}{2} + \frac{\mu_0 H^2}{2}\right). \tag{22-43}$$

Now integrate over a finite volume v of area $\mathcal{A}$, and then apply the divergence theorem on the left:

$$\int_{\mathcal{A}} (\boldsymbol{E} \times \boldsymbol{H}) \cdot d\boldsymbol{\mathcal{A}} = -\frac{\partial}{\partial t} \int_v \left(\frac{\epsilon_0 E^2}{2} + \frac{\mu_0 H^2}{2}\right) dv. \tag{22-44}$$

The term on the right, with its negative sign, gives the decrease in the total electromagnetic energy inside the volume v, per unit time. Then the integral on the left must give the rate at which electromagnetic energy flows *out* of v. This is the *Poynting theorem*; it is simply a statement of the conservation of energy in an electromagnetic field.

Then the *Poynting vector*

$$\mathcal{S} = \boldsymbol{E} \times \boldsymbol{H} \tag{22-45}$$

gives the power per unit area flowing in the wave.

The time-averaged Poynting vector is (Sec. 2.4)

$$\mathcal{S}_{av} = \frac{1}{2}\mathrm{Re}(\boldsymbol{E} \times \boldsymbol{H}^*) \tag{22-46}$$

and, for a plane wave propagating in the $\hat{z}$ direction in free space, with $\boldsymbol{E}$ and $\boldsymbol{H}$ as in Eq. 22-26, the time-averaged power density is

$$\mathcal{S}_{av} = \frac{1}{2}E_m H_m \hat{z} \tag{22-47}$$

$$= \frac{1}{2}\left(\frac{\epsilon_0}{\mu_0}\right)^{1/2} E_m^2 \hat{z} = c\epsilon_0 E_{\mathrm{rms}}^2 \hat{z}. \tag{22-48}$$

So $\mathcal{S}_{av}$ is the time-averaged total energy density $\epsilon_0 E_{\mathrm{rms}}^2$, multiplied by the speed of light c. Also,

$$\mathcal{S}_{av} = \frac{E_{\mathrm{rms}}^2}{Z_0}\hat{z} \approx \frac{E_{\mathrm{rms}}^2}{377}\hat{z} \quad \text{watts/meter}^2. \tag{22-49}$$

Abandoning the phasor notation for a moment,

$$\boldsymbol{E} = \boldsymbol{E}_m \cos(\omega t - k_0 z), \quad \boldsymbol{H} = \boldsymbol{H}_m \cos(\omega t - k_0 z), \tag{22-50}$$

and the magnitude of the Poynting vector is

$$|\mathcal{S}| = |\boldsymbol{E} \times \boldsymbol{H}| = E_m H_m \cos^2(\omega t - k_0 z). \tag{22-51}$$

The electromagnetic energy travels in "lumps," one-half wavelength long, as in Fig. 22-4.

EXAMPLE The $\boldsymbol{E}$ and $\boldsymbol{B}$ Fields in a Laser Beam in Air

Most lasers operate at powers of the order of milliwatts. But some lasers are vastly more powerful. One of them supplies a pulsed focused beam whose peak power, at the focus, is 10^{24} watts/meter^2! Thus

$$\mathcal{S} = 10^{24} \text{ watts/meter}^2, \tag{22-52}$$

$$E_{\mathrm{rms}} = (377 \times 10^{24})^{1/2} = 2 \times 10^{13} \text{ volts/meter}. \tag{22-53}$$

This E is *enormous*: 2000 volts over the diameter of an atom ($\approx 10^{-10}$ meter)! Air breaks down at fields of about 3×10^6 volts/meter. Also,

$$B_{\mathrm{rms}} = E_{\mathrm{rms}}/c = 6.5 \times 10^4 \text{ teslas}, \tag{22-54}$$

or about 65,000 times the field between the pole pieces of a powerful electromagnet.

22.5 ELECTROMAGNETIC WAVES IN NONCONDUCTORS

With nonconductors, $\epsilon_r \neq 1$, $\mu_r \neq 1$, and our discussion of electromagnetic waves in free space applies, if we simply replace ϵ_0 by $\epsilon = \epsilon_r \epsilon_0$ and μ_0 by $\mu = \mu_r \mu_0$. We disregard ferromagnetic materials, for which μ_r is undefinable.

First, instead of Eq. 22-20, we now have

$$\nabla^2 \boldsymbol{E} - \epsilon\mu \frac{\partial^2 \boldsymbol{E}}{\partial t^2} = 0 \tag{22-55}$$

and

$$v = \frac{1}{(\epsilon\mu)^{1/2}} = \frac{c}{(\epsilon_r \mu_r)^{1/2}} = \frac{c}{n}, \qquad n = (\epsilon_r \mu_r)^{1/2}, \tag{22-56}$$

where n is the *index of refraction*. So the speed of propagation v of an electromagnetic wave in a nonconductor is lower than in free space by the factor n.† Glasses have indices of refraction of about 1.5.

As a rule, tables of n apply to optical frequencies ($\approx 10^{15}$ hertz), whereas tables of ϵ_r, and of μ_r apply to much lower frequencies, at best up to about 10^{10} hertz. Since ϵ_r, μ_r, and n are all frequency-dependent, values drawn from such tables do not satisfy the above relation.

Also, from Eq. 22-33,

$$k = \frac{1}{\lambda\!\!\!^{-}} = \omega(\epsilon\mu)^{1/2}, \tag{22-57}$$

$$Z = \frac{E}{H} = \frac{k}{\omega\epsilon} = \frac{\omega\mu}{k} = \left(\frac{\mu}{\epsilon}\right)^{1/2} \approx \left(\frac{\mu_r}{\epsilon_r}\right)^{1/2} 377 \text{ ohms}. \tag{22-58}$$

The characteristic impedance Z is again real, and the vectors $\boldsymbol{E}$ and $\boldsymbol{H}$ are in phase.

The electric and magnetic energy densities are again equal:

$$\frac{\epsilon E^2/2}{\mu H^2/2} = \frac{\epsilon}{\mu}\left(\frac{\mu}{\epsilon}\right) = 1, \tag{22-59}$$

and the total energy density fluctuates again as in Fig. 22-4.

†Is this in contradiction of Sec. 12.6, where we wrote that photons always travel at the speed of light c? No. The propagation of light in matter is more complex than it seems because the wave that one observes results from the superposition of the direct wave coming from the source, and of the very large number of waves originating in the individual oscillating dipoles in the medium. These waves are all associated with photons that travel at the speed c. It is lucky that such a complex phenomenon can be analyzed simply in terms of an index of refraction.

The Poynting vector points again in the direction of propagation and

$$\mathscr{S}_{av} = \left(\frac{\epsilon}{\mu}\right)^{1/2} E_{\rm rms}^2 \hat{z} = v\epsilon E_{\rm rms}^2 \hat{z} \tag{22-60}$$

$$= \left(\frac{\epsilon_r}{\mu_r}\right)^{1/2} \frac{E_{\rm rms}^2}{377} \hat{z} \text{ watts/meter}^2. \tag{22-61}$$

The time-averaged Poynting vector is again equal to the wave speed multiplied by the time-averaged total energy density.

EXAMPLE The $\boldsymbol{E}$ and $\boldsymbol{B}$ Fields in a Laser Beam in Glass

There is no point in referring here to the laser beam of Sec. 22.4 because such a beam would instantly vaporize glass. Say we have a 1.0-milliwatt beam with a diameter of 1.0 millimeter in glass whose index of refraction is 1.5. Then

$$\mathscr{S} = \frac{10^{-3}}{\pi \times (5 \times 10^{-4})^2} = 1.3 \times 10^3 \text{ watts/meter}^2, \tag{22-62}$$

$$E_{\rm rms} = \left[\left(\frac{\mu}{\epsilon}\right)^{1/2} \mathscr{S}\right]^{1/2} = \left[\left(\frac{4\pi \times 10^{-7}}{1.5^2 \times 8.85 \times 10^{-12}}\right)^{1/2} \times 1.3 \times 10^3\right]^{1/2} \tag{22-63}$$

$$= 5.7 \times 10^2 \text{ volts/meter}, \tag{22-64}$$

$$B_{\rm rms} = \mu_0 H_{\rm rms} = \mu_0 \left(\frac{\epsilon}{\mu_0}\right)^{1/2} E_{\rm rms} = (\epsilon\mu_0)^{1/2} E_{\rm rms} \tag{22-65}$$

$$= (1.5^2 \times 8.85 \times 10^{-12} \times 4\pi \times 10^{-7})^{1/2} \times 5.7 \times 10^2 \tag{22-66}$$

$$= 2.9 \times 10^{-6} \text{ tesla}. \tag{22-67}$$

22.6 ELECTROMAGNETIC WAVES IN CONDUCTORS

Propagation in stationary conductors is more complex. Refer to the Maxwell equations of Sec. 21.2. We can set $\rho = 0$ but, now, $\boldsymbol{J} = \sigma \boldsymbol{E}$, where σ is the conductivity. We also replace ϵ_0 by $\epsilon = \epsilon_r \epsilon_0$, where ϵ_r is presumably of the order of 3 in conductors. We set $\boldsymbol{B} = \mu \boldsymbol{H}$, where $\mu = \mu_r \mu_0$. We exclude ferromagnetic materials. Then

$$\nabla \cdot \boldsymbol{E} = 0, \tag{22-68}$$

$$\nabla \times \boldsymbol{E} + \mu \frac{\partial \boldsymbol{H}}{\partial t} = 0, \tag{22-69}$$

$$\nabla \cdot \boldsymbol{H} = 0, \tag{22-70}$$

$$\nabla \times \boldsymbol{H} - \epsilon \frac{\partial \boldsymbol{E}}{\partial t} = \sigma \boldsymbol{E}. \tag{22-71}$$

Recall now from Eq. 22-28 that, in phasor notation,

$$\frac{\partial}{\partial t} = j\omega, \quad \text{and} \quad \nabla = -jk\,\hat{z}. \tag{22-72}$$

So the Maxwell equations become

$$-jk\,\hat{z} \cdot \boldsymbol{E} = 0, \quad -jk\,\hat{z} \times \boldsymbol{E} = -j\omega\mu\boldsymbol{H}, \tag{22-73}$$

$$-jk\,\hat{z} \cdot \boldsymbol{H} = 0, \quad -jk\,\hat{z} \times \boldsymbol{H} = \sigma\boldsymbol{E} + j\omega\epsilon\boldsymbol{E}, \tag{22-74}$$

or

$$\hat{z} \cdot \boldsymbol{E} = 0, \quad \boldsymbol{H} = \frac{k}{\omega\mu}\hat{z} \times \boldsymbol{E}, \tag{22-75}$$

$$\hat{z} \cdot \boldsymbol{H} = 0, \quad \boldsymbol{E} = -\frac{k}{\omega\epsilon - j\sigma}\hat{z} \times \boldsymbol{H}. \tag{22-76}$$

As in nonconductors, $\boldsymbol{E}$ and $\boldsymbol{H}$ are transverse and orthogonal, and $\boldsymbol{E} \times \boldsymbol{H}$ points in the direction of propagation.

The characteristic impedance is now complex:

$$Z = \frac{E}{H} = \frac{k}{\omega\epsilon - j\sigma} = \frac{\omega\mu}{k}, \tag{22-77}$$

so E and H are *not* in phase. We return to the characteristic impedance and to the relative phases of E and H below.

Let us calculate k:

$$k^2 = \omega^2\epsilon\mu - j\omega\sigma\mu = \omega^2\epsilon\mu\left(1 - j\frac{\sigma}{\omega\epsilon}\right). \tag{22-78}$$

The σ term accounts for Joule losses and attenuation.

In a good conductor, by definition, $\sigma/(\omega\epsilon) \gg 1$. The conduction current density $\sigma\boldsymbol{E}$ is therefore much larger than the displacement current density of Sec. 7.10:

$$\sigma\boldsymbol{E} \gg j\omega\epsilon\boldsymbol{E} = \frac{\partial \boldsymbol{D}}{\partial t}. \tag{22-79}$$

Note that σ and ϵ_r are both functions of the frequency, especially at optical and x-ray frequencies, so that the ratio $\sigma/(\omega\epsilon)$ does *not* decrease indefinitely with increasing frequency.

We need the value of k, so we need the square root of a complex number. In the general case, that requires a bit of algebra but, here, things are easy because the imaginary part of k^2 is much larger than unity. So

$$k \approx \left[\omega^2 \epsilon\mu \frac{\sigma}{\omega\epsilon}(-j)\right]^{1/2} = [\omega\mu\sigma(-j)]^{1/2}. \tag{22-80}$$

But what is the square root of $-j$? In the complex plane, $-j$ has a modulus of 1 and an argument of $-\pi/2$ so, in polar form,

$$-j = \exp\left(-j\frac{\pi}{2}\right), \qquad \text{and} \qquad (-j)^{1/2} = \exp\left(-j\frac{\pi}{4}\right). \tag{22-81}$$

So

$$k = (\omega\sigma\mu)^{1/2}\exp(-j\pi/4). \tag{22-82}$$

Equation 22-77 now tells us that $\boldsymbol{H}$ lags $\boldsymbol{E}$ by 45 degrees:

$$Z = \frac{E}{H} = \frac{\omega\mu}{k} = \frac{\omega\mu}{(\omega\sigma\mu)^{1/2}\exp(-j\pi/4)} = \left(\frac{\omega\mu}{\sigma}\right)^{1/2}\exp(j\pi/4). \tag{22-83}$$

The characteristic impedance of a good conductor is proportional to $(f/\sigma)^{1/2}$; recall that, in nonconductors, Z is independent of the frequency.

It is the custom to set

$$k = \beta - j\alpha. \tag{22-84}$$

Here,

$$\alpha = \beta = \left(\frac{\omega\mu\sigma}{2}\right)^{1/2}. \tag{22-85}$$

Both α and β are positive. Then

$$\boldsymbol{E} = \boldsymbol{E}_m\exp(-\alpha z)\exp j(\omega t - \beta z), \tag{22-86}$$

$$\boldsymbol{H} = \boldsymbol{H}_m\exp(-\alpha z)\exp j(\omega t - \beta z). \tag{22-87}$$

Note that

$$\beta = \frac{2\pi}{\lambda} = \frac{1}{\lambda\!\!\!^{-}}, \qquad \text{and that} \qquad \alpha = \frac{1}{\delta}, \quad \lambda = 2\pi\delta. \tag{22-88}$$

where δ is the *skin depth*, or the *attenuation distance*, the distance over which the amplitude of the wave decreases by a factor of $e = 2.71828$. This is the *skin effect.*

Now taking into account the phase difference between $\boldsymbol{E}$ and $\boldsymbol{H}$,

$$\boldsymbol{E} = \boldsymbol{E}_m\exp\left[j\left(\omega t - \frac{z}{\delta}\right) - \frac{z}{\delta}\right], \tag{22-89}$$

$$\boldsymbol{H} = \left(\frac{\sigma}{\omega\mu}\right)^{1/2}\boldsymbol{E}_m\left[j\left(\omega t - \frac{z}{\delta} - \frac{\pi}{4}\right) - \frac{z}{\delta}\right]. \tag{22-90}$$

The wave propagates in the positive direction of the z-axis. If $\boldsymbol{E}$ is parallel to the x-axis, then $\boldsymbol{H}$ is parallel to the y-axis and

$$\boldsymbol{E} = E_m \exp\left[j\left(\omega t - \frac{z}{\delta}\right) - \frac{z}{\delta}\right]\hat{\boldsymbol{x}}, \tag{22-91}$$

$$\boldsymbol{H} = \left(\frac{\sigma}{\omega\mu}\right)^{1/2} E_m\left[j\left(\omega t - \frac{z}{\delta} - \frac{\pi}{4}\right) - \frac{z}{\delta}\right]\hat{\boldsymbol{y}}, \tag{22-92}$$

$$|Z| = \left|\frac{\boldsymbol{E}}{\boldsymbol{H}}\right| = \left(\frac{\omega\mu}{\sigma}\right)^{1/2}. \tag{22-93}$$

What is the speed of propagation v? The above value of β supplies the answer because $\beta = \omega/v$:

$$v = \frac{\omega}{\beta} = \omega\left(\frac{2}{\omega\mu\sigma}\right)^{1/2} = \left(\frac{2\omega}{\mu\sigma}\right)^{1/2}. \tag{22-94}$$

The propagation of an electromagnetic wave in a good conductor is peculiar in that the amplitude of the wave decreases by a factor of e in only one radian length $\lambdabar = \lambda/(2\pi)$, and by a factor of $(1/e)^{2\pi} \approx 2 \times 10^{-3}$ over one wavelength! The attenuation is so large that the wave is hardly discernible. See Fig. 22-5.

Good conductors are therefore opaque to light, except in the form of extremely thin films. It does not follow, however, that substances that are nonconducting at low frequencies are transparent at optical frequencies.

You can show that the energy density is mostly magnetic.

We leave the calculation of the Poynting vector in a conductor as a problem at the end of this chapter. Note that both E and H are proportional to $\exp(-\alpha z)$, so that $\mathcal{S}$ is proportional to $\exp(-2\alpha z)$ and the power density decreases *very* rapidly with distance.

The Poynting vector is again equal to the time-averaged total energy density multiplied by the wave speed.

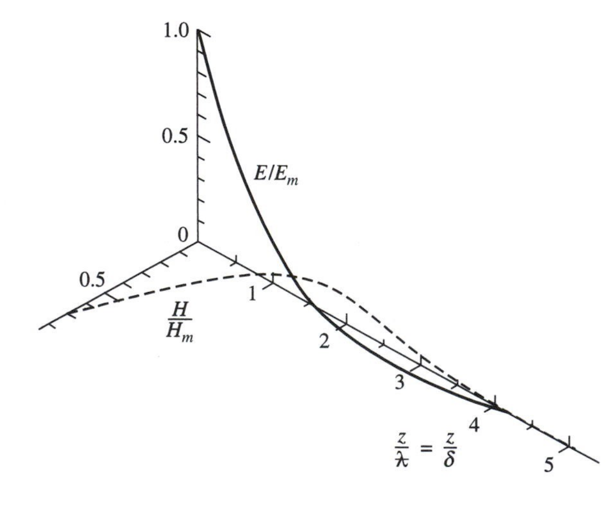

Figure 22-5 The ratios E/E_m and H/H_m at $t = 0$ as functions of $z/\lambdabar$ for an electromagnetic wave propagating in a good conductor.

EXAMPLE Propagation in Copper at 1 Megahertz

The conductivity of copper is 5.8×10^7 siemens/meter. Then, at 1 megahertz,

$$\frac{\sigma}{\omega\epsilon} \approx \frac{\sigma}{\omega\epsilon_0} = \frac{5.80 \times 10^7}{2\pi \times 10^6 \times 8.85 \times 10^{-12}} \approx 10^{12}, \tag{22-95}$$

and the skin depth

$$\delta = \frac{1}{\alpha} = \left(\frac{2}{5.8 \times 10^7 \times 2\pi \times 10^6 \times 4\pi \times 10^{-7}}\right)^{1/2} \tag{22-96}$$

$$= 66 \quad \text{micrometers.} \tag{22-97}$$

A one-megahertz electromagnetic wave in air has a wavelength of $c/f = 3 \times 10^8/10^6 = 300$ meters. What is the wavelength in copper? From Eq. 22-88, $\lambda = 2\pi\delta$ and the wavelength is about 0.4 millimeter, or nearly one million times less. The wave speed in copper is correspondingly low:

$$v = \omega\lambda\!\!\!\!- = 415 \text{ meters/second,} \tag{22-98}$$

which is about 10 times *less* than the speed of *sound* in copper (3.6×10^3 meters/second).

The characteristic impedance of copper at one megahertz is

$$|Z| = \left(\frac{\mu_0\omega}{\sigma}\right)^{1/2} = \left(\frac{4\pi \times 10^{-7} \times 2\pi \times 10^6}{5.8 \times 10^7}\right)^{1/2} = 3.7 \times 10^{-4} \text{ ohm,} \tag{22-99}$$

or about one million times less than the characteristic impedance of free-space, which is 377 ohms.

22.7 SUMMARY

The known electromagnetic spectrum extends from about 100 hertz ($\lambda = 3 \times 10^6$ meters) to about 10^{24} hertz ($\lambda = 3 \times 10^{-16}$ meter).

The *wave equation* for a vector quantity $\boldsymbol{E}$ that propagates unattenuated at the speed v is

$$\nabla^2\boldsymbol{E} - \frac{1}{v^2}\frac{\partial^2\boldsymbol{E}}{\partial t^2} = 0. \tag{22-12}$$

In studying waves one refers to the vectors $\boldsymbol{E}$ and $\boldsymbol{H}$, rather than to $\boldsymbol{E}$ and $\boldsymbol{B}$ because of the importance of the characteristic impedance $Z = E/H$ and of the Poynting vector $\boldsymbol{E} \times \boldsymbol{H}$.

In the case of a plane wave propagating in the positive direction of the z-axis, in phasor notation,

$$\frac{\partial}{\partial t} = j\omega, \qquad \text{and} \qquad \nabla = -jk\,\hat{z}. \tag{22-28}$$

Linearly polarized, plane electromagnetic waves traveling either in free space, or in a HILS nonconductor or conductor have the following properties.

1. The vectors $\boldsymbol{E}$ and $\boldsymbol{H}$ are *transverse* and *orthogonal.*
2. The *Poynting vector* $\boldsymbol{E} \times \boldsymbol{H}$ points in the direction of propagation.
3. Its magnitude, averaged over time, is equal to the power density in the wave:

$$\mathcal{S}_{av} = \frac{1}{2}\text{Re}(\boldsymbol{E} \times \boldsymbol{H}^*) \quad \text{watts/meter}^2. \tag{22-46}$$

4. The magnitude of the Poynting vector is also equal to the time-averaged total energy density multiplied by the wave speed.
5. The *characteristic impedance* of the medium of propagation is

$$Z = E/H. \tag{22-34, 58, 77, 83}$$

6. The *Poynting theorem* states that there is conservation of energy in an electromagnetic field, with the power density in the wave given by the Poynting vector.

In *free space,*

$$c = \frac{1}{(\epsilon_0 \mu_0)^{1/2}} = 2.99792458 \times 10^8 \text{ meters/second,} \tag{22-21, 22}$$

$$Z_0 \approx 377 \text{ ohms,} \tag{22-35}$$

$$E = Bc. \tag{22-36}$$

The electric and magnetic energy densities are equal and

$$\mathcal{S}_{av} \approx \frac{E_{\text{rms}}^2}{377}\hat{z} \quad \text{watts/meter}^2. \tag{22-49}$$

In *nonconductors,*

$$v = \frac{c}{(\epsilon_r \mu_r)^{1/2}} = \frac{c}{n} \quad \text{meters/second,} \tag{22-56}$$

where n is the *index of refraction,* and

$$Z \approx \left(\frac{\mu_r}{\epsilon_r}\right)^{1/2} 377 \quad \text{ohms.} \tag{22-58}$$

The electric and magnetic energy densities are once more equal and

$$\mathcal{S}_{av} \approx \left(\frac{\epsilon_r}{\mu_r}\right)^{1/2} \frac{E_{\rm rms}^2}{377}\,\hat{z} \quad \text{watts/meter}^2. \tag{22-61}$$

In *conductors*, the wave speed

$$v = \left(\frac{2\omega}{\mu\sigma}\right)^{1/2} \tag{22-94}$$

is very low, and so is the characteristic impedance

$$Z = \left(\frac{\omega\mu}{\sigma}\right)^{1/2} \exp(j\pi/4). \tag{22-83}$$

The wave is highly attenuated: the amplitudes of both E and H decrease by a factor of e over one radian length $\lambdabar = \lambda/(2\pi)$, which is the *attenuation distance*, or *skin depth* δ. Most of the energy in the field is magnetic.

PROBLEMS

22-1. (*22.3*) Loop antenna

A 30-megahertz plane electromagnetic wave propagates in free space, and its peak E is 100 millivolts/meter.

Calculate the peak voltage induced in a 1.00-square-meter, 10-turn receiving loop oriented so that its plane forms an angle of 60 degrees with the magnetic vector.

22-2. (*22.4*) The earth as a habitable planet

The power density of the solar radiation is about one kilowatt/meter^2 at the surface of the earth, and 1.4 kilowatts/meter^2 above the atmosphere. So you feel the heat when walking in the sun, even though it is 150 million kilometers away!

(a) Calculate the magnitude of the Poynting vector at the surface of the sun.

(b) How large is an area on the surface of the sun that radiates one gigawatt? That is the electric power generated by a large nuclear station.

(c) Calculate $E_{\rm rms}$ and $H_{\rm rms}$ at the surface of the sun.

(d) It is interesting to reflect upon the fact that conditions at the surface of the earth are just right to sustain life.

As we saw in the third example of Sec. 17.5, the magnetic field of the earth shields us against most of the charged particles of the cosmic radiation. It is the convection in the earth's liquid outer core that generates the magnetic field. If the earth were solid, like the moon, it would not have a magnetic field, cosmic radiation would be much more intense, and living beings would be subjected to many more mutations.

Back to solar radiation. If the earth were either closer in or farther out, the temperature would be either too high or too low. Or if the earth's orbit were much more elliptical than it is, with the same average distance, then the temperature would be too high for part of the year, and too low the rest of the time. The orbit is in fact nearly a perfect circle, the ratio of the maximum distance to the minimum distance being 1.0001395.

What percentage change in the radius of the earth's orbit could we tolerate if the radiative power input could vary by $\pm 20\%$ without too much harm?

(e) The rotation of the earth about its own axis is important. If the earth rotated about the sun in the same way as the moon rotates around the earth, with the same hemisphere always facing the sun, then the average power input on that face would double, and the surface temperature would be unbearable. The other hemisphere would be cold and no more habitable than the moon. Thanks to rotation, nearly all the surface can sustain life.

But the angular velocity must be large enough! What if days were much longer?

(f) If the earth spun at twice its present angular velocity, how much would you weigh?

What if it spun faster and faster? That would require an externally applied torque!

(g) There is also the problem of the angle between the earth's spin axis and the ecliptic (the plane of the orbit). That angle is 66.5 degrees. It is because this angle is not 90 degrees that there are seasons.

What would happen if the angle were 0 degrees? The orientation of the earth's spin axis remains essentially fixed in space, over the years.

(h) The gravitational force of attraction between the sun and the earth is proportional to $1/r^2$. What would happen if that force were not exactly proportional to $1/r^2$?

(i) How does g at the surface vary with the earth's radius, for a given average density?

If the radius of the earth were twice as large, with the same density, by what factor would your weight change?

The value of g is important: if g were, say, 10 times larger than it is, all living beings would have to be designed differently, or life might be impossible.

(j) Assume an atmosphere of oxygen at a temperature of zero degrees. For what minimum value of g will an oxygen molecule remain in the neighborhood of the earth? Assume that all the molecules have the same speed.

You can probably think of many other parameters that must have just about the right values to render our planet habitable.

To date (2000), astronomers have found no other planetary system that could sustain life. See *Physics World,* May 1999, p. 46, Jakosky (1998), and *Physics Today,* October 1999, p. 367.

22-3. (*22.4*) The Poynting vector in a wave

A circularly polarized wave results from the superposition of two waves that are (a) of the same frequency and amplitude, (b) plane-polarized in perpendicular directions, and (c) 90 degrees out of phase.

Show that the average value of the Poynting vector for such a wave is equal to the sum of the average values of the Poynting vectors of the two components.

22-4. (*22.4*) The Poynting vector in a wave

In a particular electromagnetic wave in air, $E_{rms} = 20$ volts/meter. The wave is absorbed by a conducting sheet having a surface mass density of 10^{-3} kilogram/meter^2, and a specific heat capacity of 400 joules/kilogram-kelvin.

Calculate the rate at which the temperature rises, assuming that there are no heat losses and no re-radiation.

22-5. (*22.4*) Solar energy

Calculate the area required to generate one megawatt of electric power from solar energy at the surface of the earth, assuming an average efficiency over one year of 2%. The efficiency of solar cells is about 15% at normal incidence.

22-6. (*22.4*) The Poynting vector in the field of a resistive wire carrying a current

A long, straight wire of radius a and resistance R' ohms/meter carries a current I.

(a) Calculate the Poynting vector at the surface, and explain.

(b) Calculate the Poynting vector both outside and inside the wire. Explain.

22-7. (*22.4*) The Poynting vector in a capacitor

A thin, air-insulated parallel-plate capacitor has circular plates of radius R, separated by a distance s. A constant current I charges the plates through thin wires along the axis of symmetry.

(a) Find the value of E between the plates as a function of the time. Assume a uniform E. Show the direction of $\boldsymbol{E}$ on a figure.

(b) The magnetic field is the sum of two terms, H_w, related to the current in the wire, and H_p, related to the current in the plates. The latter current deposits charges on the inside surfaces of the plates.

Find H_w, H_p, and H. Use cylindrical coordinates with the z-axis along the wire and in the direction of the current. To calculate H_p, apply Ampère's circuital law to each plate. You should find that the magnetic fields tend to infinity as $\rho \longrightarrow 0$. This is simply because we have assumed infinitely thin wires and plates. Show the directions of $\boldsymbol{H}_w$, $\boldsymbol{H}_p$, and $\boldsymbol{H}$ on your figure.

(c) Do $\boldsymbol{E}$ and $\boldsymbol{H}$ satisfy Maxwell's equations? You should find that one of our assumptions is incorrect.

(d) Find $\boldsymbol{E} \times \boldsymbol{H}$.

(e) Find the electric and magnetic energies inside a radius ρ.

You should find that the magnetic energy is negligible if $\rho^2/t^2 \ll c^2$. This condition applies because we have assumed that the capacitor charges up slowly. If it charged very quickly, then there would be a wave of $\boldsymbol{E}$ and $\boldsymbol{H}$ in the capacitor, $\boldsymbol{E}$ would not be uniform, and the above calculation would be invalid.

(f) Now relate the Poynting vector at ρ to the electric energy inside ρ.

(g) Draw a sketch showing $\boldsymbol{E}$, $\boldsymbol{H}$, and $\boldsymbol{E} \times \boldsymbol{H}$ vectors at various points inside and around the capacitor.

22-8. (*22.4*) The Poynting vector in a solenoid

A long solenoid carries a current I.

(a) The current increases. Sketch a longitudinal cross-section showing the direction of the current, and show $\boldsymbol{A}$, $-\partial \boldsymbol{A}/\partial t$, $\boldsymbol{H}$, and $\boldsymbol{E} \times \boldsymbol{H}$. Why is the Poynting vector oriented in that way?

(b) Repeat with a decreasing current.

22-9. (*22.4*) The Poynting vector in an induction motor

In an induction motor, the stator generates a magnetic field that is perpendicular to, and rotates about, the axis of symmetry. The rotor is a cylinder of laminated iron,

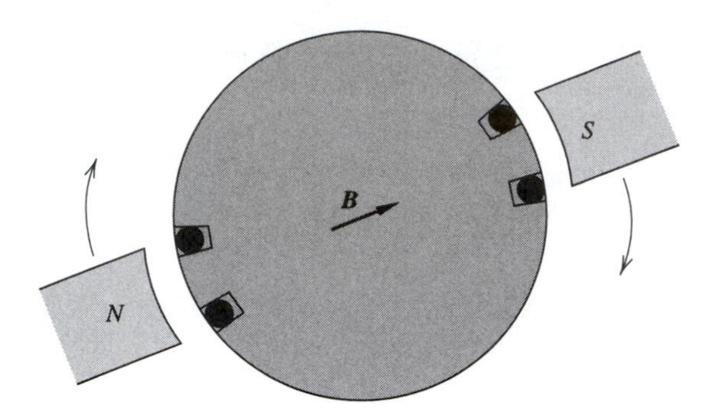

Figure 22-6

with copper bars parallel to the axis and set in grooves in the cylindrical surface. Copper rings at each end of the rotor connect all the copper bars. The rotor is *not* connected to the source of electric current that feeds the motor.

As we shall see, the rotor tends to follow the rotating magnetic field $\boldsymbol{B}$. Figure 22-6 shows the principle of operation. To simplify the analysis, we suppose that the rotor is stationary and that a rotating electromagnet, represented here by its poles N and S, provides the rotating magnetic field.

(a) Draw a larger figure with wide air gaps, showing the direction of the induced currents in the bars and the direction of $\boldsymbol{E}$ in the air gaps.
(b) The current in the rotor generates a magnetic field. Add arrows showing the direction of *that* $\boldsymbol{H}$, inside the rotor and in the air gaps.
(c) Now show Poynting vectors $\boldsymbol{E} \times \boldsymbol{H}$ in the air gaps. The field feeds power into the rotor.
(d) Now draw another figure showing the currents in the bars and a line of $\boldsymbol{B}$ for the sum of the two magnetic fields.
(e) Show the direction of the magnetic forces on the bars.

22-10. (*22.4*) The Poynting vector in a transformer

Figure 22-7 shows, in simplified form, a cross section of a transformer secondary. The field $\boldsymbol{B}$ inside the core C, and the leakage field $\boldsymbol{H}_l$ outside, both result from the currents in the windings and from the equivalent currents in the core. The secondary winding is W. Assume that $\boldsymbol{B}$ and $\boldsymbol{H}_l$ increase. See Lorrain (1984).

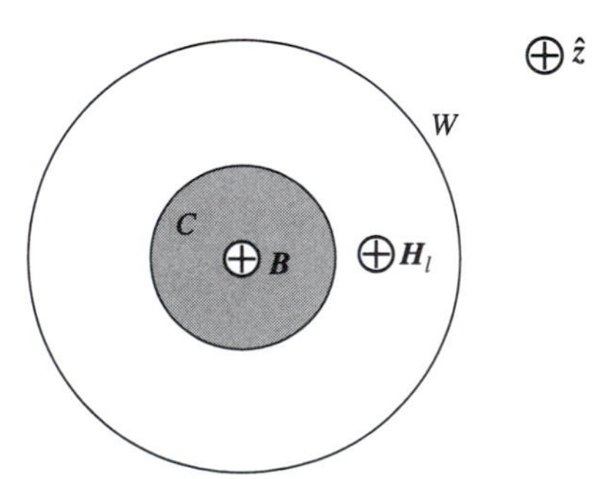

Figure 22-7

(a) Draw a larger figure and show, at one point between C and W, vectors $\boldsymbol{A}$, $\partial \boldsymbol{A}/\partial t$, and $\boldsymbol{E} = -\partial \boldsymbol{A}/\partial t$, disregarding the current in W. Show a vector $\boldsymbol{E}$ at one point outside W.

(b) Assume that the impedance of the secondary is a pure resistance R. Show the direction of the current I in W.

(c) Show the direction of its field $\boldsymbol{H}$ at points between C and W, and outside W.

(d) Show vectors $\boldsymbol{E} \times \boldsymbol{H}$.

(e) How would the directions of the $\boldsymbol{E} \times \boldsymbol{H}$ vectors be affected if $\boldsymbol{B}$ and $\boldsymbol{H}_l$ decreased?

(f) What is the time-averaged value of a vector $\boldsymbol{E} \times \boldsymbol{H}_l$?

(g) Now let us calculate the power flow into the secondary. Assume that the secondary is a long solenoid of N turns and of length L. Disregard $\boldsymbol{H}_l$ and set $\Phi = \Phi_m \exp j\omega t$ in the core. Integrate the Poynting vector over a cylindrical surface situated between the core and the winding, and show that the power flowing into the winding is $(N\omega\Phi_{\rm rms})^2/R = V_{\rm rms}^2/R$, where V is the voltage induced in the secondary winding.

22-11. (*22.4*) Energy and power in a proton beam

Figure 22-8 shows a highly simplified diagram of a proton accelerator. A gas discharge within the source S ionizes hydrogen gas to produce protons. Some of the protons emerge through a hole and are focused into a beam B of radius R_1 inside a conducting tube of radius R_2. The source is at a potential V, and the target is grounded.

To avoid needless complications, we assume that the charge density in the beam is uniform. We also assume that the speed of the protons is much less than $c: v^2 \ll c^2$.

Calculate, in terms of the current I and the speed v:

(a) the electric energy per meter inside the tube $\mathscr{E}'_e$;

(b) the magnetic energy per meter $\mathscr{E}'_m$;

(c) the energy flux associated with the Poynting vector P_P;

(d) the kinetic power P_k, or the flux of kinetic energy, disregarding P_P.

The existence of this Poynting vector is interesting. Because of the radial $\boldsymbol{E}$, the voltage inside the beam is slightly positive. So the protons are not accelerated to the full voltage V, and the kinetic energy in the beam is slightly lower than VI. Most of the power flows down the tube as kinetic energy, and the rest flows as electromagnetic energy. The total power at any point along the tube and on the target is VI.

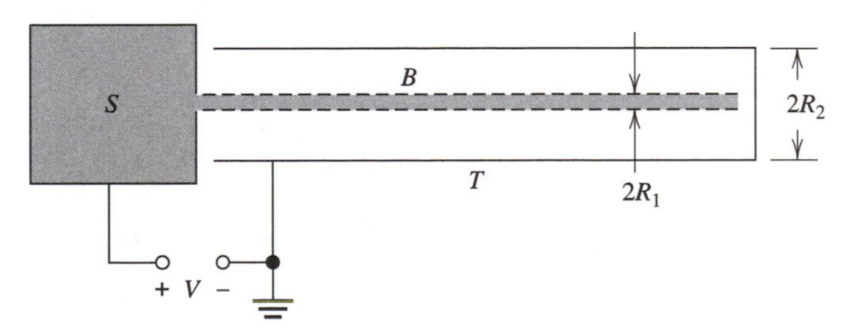

Figure 22-8

(e) Find the numerical values of these quantities for a 1.00-milliampere, 1.00 mega-electronvolt proton beam, with $R_1 = 1.00$ millimeter and $R_2 = 50.0$ millimeters.

22-12. (*22.6*) Induction heating

It is common practice to heat conductors by subjecting them to an alternating magnetic field. The changing flux induces *eddy currents* that heat the conductor by the Joule effect (Sec. 18.3). The method is called *induction heating*. It is used extensively for melting metals and for hardening and forging steel.

The power used ranges from watts to megawatts, and the frequencies used from 60 hertz to several hundred kilohertz.

Induction heating has the advantage of convenience and of not contaminating the metal with combustion gases. It even permits heating a conductor enclosed in a vacuum enclosure. Induction heating has another major advantage. At high frequencies, currents flow near the surface of a conductor. This is the *skin effect*. So, by choosing the frequency correctly, one can apply a brief heat treatment down to a known depth. This is particularly important because there are numerous purposes for which one requires steel parts with a hard skin and a soft core: the hard skin resists abrasion and the soft core reduces breakage. Plowshares are heat-treated in this way.

Induction furnaces are used for melting metals. They consist of large crucibles with capacities ranging up to 30 tons, thermally insulated, and surrounded by current-carrying coils. Operation is usually started with part of the load already molten.

Let us consider Fig. 22-9(a), where a rod of radius a, length L, and conductivity σ is placed inside a solenoid having N' turns/meter and carrying an alternating current

$$I = I_0 \cos \omega t.$$

As usual, we neglect end effects.

We also assume that the frequency is low enough to avoid the complications due to the skin effect. This assumption is well satisfied at 60 hertz with a rod of graphite ($\sigma = 1.0 \times 10^5$ siemens/meter) having a radius of 60 millimeters. In other words, the magnetic flux density of the induced currents is negligible. We also assume that the conductor is nonmagnetic.

This is admittedly a highly simplified illustration of induction heating. Moreover, the power dissipated in the graphite is only a few watts, which is absurdly small for such a large piece. This should nonetheless be a useful exercise on induced currents.

Consider a ring of radius r, thickness dr, and length L inside the conductor, as in Fig. 22-9(b).

(a) Show that the electromotance induced in the ring is

$$\mu_0 \pi r^2 \omega N' I_0 \sin \omega t.$$

(b) Show that, for a current flowing in the azimuthal direction, the ring has a resistance

$$R = 2\pi r/(\sigma L\, dr).$$

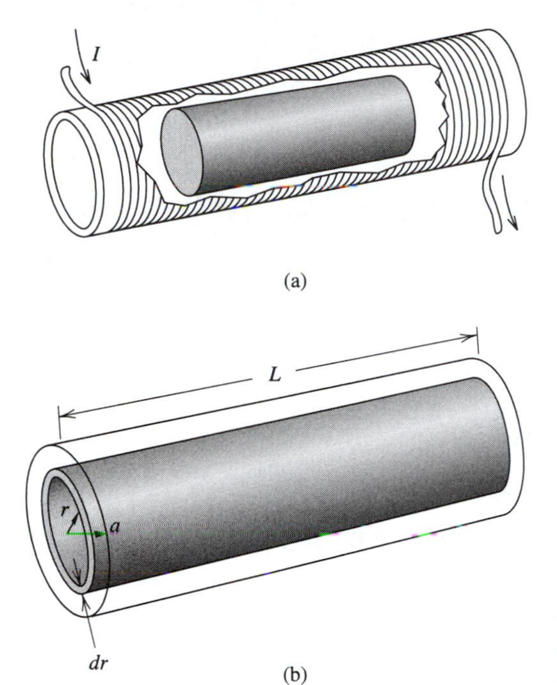

Figure 22-9 (a) Heating a conducting cylinder by induction. (b) Ring inside the cylinder of figure (a).

(c) Show that the average power dissipated in the ring is

$$\frac{1}{4}(\mu_0 \omega N' I_0)^2 \pi \sigma L r^3 dr.$$

(d) Show that the total average power dissipated in the cylinder is

$$\frac{1}{16}(\mu_0 \omega N' I_0)^2 \pi \sigma L a^4.$$

(e) Calculate the power dissipated in the graphite rod, setting $L = 1$ meter, $I_0 = 20$ amperes, $N' = 5000$ turns/meter.

CHAPTER

23

ELECTROMAGNETIC WAVES II

Reflection and Refraction

In general, a wave that encounters a discontinuity in its medium of propagation gives rise to both a reflected wave and a transmitted, or refracted, wave. For example, a sound wave incident on a wall gives both a wave that comes back into the room, and another one that proceeds into the wall.

In Chapter 22 we studied the propagation of electromagnetic waves in unbounded media. We now investigate the behavior of a wave that is incident on the interface between two media, as in Fig. 23-1. As we shall see, there are cases where there is *no reflected wave* (Sec. 23.3.1 on the Brewster angle), and cases where there is *no transmitted wave* (Sec. 23.3.2 on total reflection).

23.1 THE LAWS OF REFLECTION AND SNELL'S LAW OF REFRACTION

The laws of reflection and of refraction are usually taught at the high-school level. Our purpose in this section is to deduce them anew, this time from wave theory.

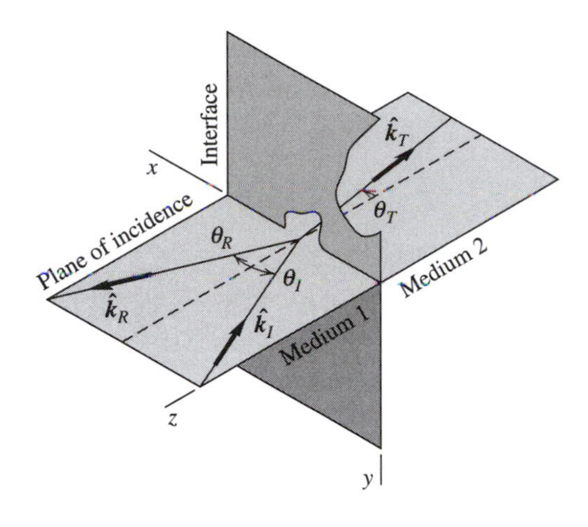

Figure 23-1 An electromagnetic wave incident on the interface between medium 1 and medium 2 gives both a reflected and a transmitted wave. The vectors $\hat{\boldsymbol{k}}$ are normal to their respective wave fronts and point in the direction of propagation. The angles θ_I, θ_R, θ_T are, respectively, the angles of incidence, reflection, and refraction.

The incident and reflected waves are in medium 1, and the transmitted wave is in medium 2, as in Fig. 23-1. Medium 1 is either a vacuum or a nonconductor, and medium 2 either a vacuum, or a nonconductor, or a conductor.

The following assumptions will simplify our calculations. (a) Both media are homogeneous, isotropic, linear, and stationary (HILS). (b) They extend to infinity on either side of the interface; this excludes multiple reflections. (c) The interface is thin and planar. (d) The incident wave is plane. (e) The origin lies in the interface, as in the figure.

Then the incident wave is of the form

$$\boldsymbol{E}_I = \boldsymbol{E}_{Im} \exp j(\omega t - \boldsymbol{k}_I \cdot \boldsymbol{r}), \tag{23-1}$$

where $\boldsymbol{E}_{Im}$ is real, by hypothesis, and where the vector wave number $\boldsymbol{k}_I$ points along a ray in the incident wave. This expression describes an unattenuated plane wave that extends throughout all time and space, but it applies only in medium 1.

The magnitude of $\boldsymbol{k}_I$ is $k_I = n_1 k_0 = n_1/\lambda\!\!\!^{-}_0$, n_1 being the index of refraction of medium 1, and $\lambda\!\!\!^{-}_0$ being the radian length of a wave of the same frequency in a vacuum.

We can expect the reflected and transmitted waves to be of the form

$$\boldsymbol{E}_R = \boldsymbol{E}_{Rm} \exp j(\omega t - \boldsymbol{k}_R \cdot \boldsymbol{r}), \tag{23-2}$$

$$\boldsymbol{E}_T = \boldsymbol{E}_{Tm} \exp j(\omega t - \boldsymbol{k}_T \cdot \boldsymbol{r}). \tag{23-3}$$

The three waves are of the same frequency, and they must be identical functions of position on the interface.

Now let $\boldsymbol{r}_{int}$ designate a point on the interface. The origin also lies in the interface. Let $\boldsymbol{k}_I$ lie in the plane $y = 0$, as in the figure. The three waves must be identical functions of position on the interface, so that

$$\boldsymbol{k}_I \cdot \boldsymbol{r}_{int} = \boldsymbol{k}_R \cdot \mathbf{r}_{int} = \boldsymbol{k}_T \cdot \boldsymbol{r}_{int}. \tag{23-4}$$

The projections of the three $\boldsymbol{k}$'s on the interface must therefore be equal. Since the y-component of $\boldsymbol{k}_I$ is zero, both $\boldsymbol{k}_R$ and $\boldsymbol{k}_T$ have zero y-components, and the three rays shown in Fig. 23-1 are *coplanar.* The plane defined by the three rays shown in the figure is the *plane of incidence.*

Also, the x-components of $\boldsymbol{k}_I$ and of $\boldsymbol{k}_R$ are equal,

$$k_1 \sin\theta_I = k_1 \sin\theta_R \tag{23-5}$$

and $\theta_R = \theta_I$: the angle of reflection is equal to the angle of incidence.

These are the two *laws of reflection*: the incident and reflected rays are coplanar, and the angle of reflection is equal to the angle of incidence.

Now we also have that

$$k_1 \sin\theta_I = k_2 \sin\theta_T. \tag{23-6}$$

Then

$$n_1 \sin\theta_I = n_2 \sin\theta_T, \tag{23-7}$$

which is *Snell's law*: when an electromagnetic wave crosses an interface, there is conservation of the product $n \sin\theta$.

The laws of reflection and Snell's law of refraction are general. They apply to any two HILS media, whether conducting or not.

But $(n_1/n_2)\sin\theta_I$ can conceivably be larger than unity. Then $\sin\theta_T > 1$! Absurd! Well, no. See Sec. 23.3.2 on total reflection.

23.2 THE FRESNEL EQUATIONS

We now require relations between $\boldsymbol{E}_{Im}$, $\boldsymbol{E}_{Rm}$, and $\boldsymbol{E}_{Tm}$ that will ensure continuity of the tangential components of $\boldsymbol{E}$ and of $\boldsymbol{H}$ at the interface. Assume that the waves are plane-polarized. The $\boldsymbol{E}$-vector of the incident wave can point in any direction perpendicular to $\boldsymbol{k}_I$.

Then, at the interface,

$$E_{Ix} + E_{Rx} = E_{Tx}, \qquad E_{Iy} + E_{Ry} = E_{Ty}, \tag{23-8}$$

$$H_{Ix} + H_{Rx} = H_{Tx}, \qquad H_{Iy} + H_{Ry} = H_{Ty}. \tag{23-9}$$

We consider successively incident waves polarized with their $\boldsymbol{E}$ vectors normal to, and then parallel to, the plane of incidence. See Figs. 23-2 and 23-3. Observe that the two figures agree at normal incidence. We utilize the continuity of E_y and H_x in Fig. 23-2, and the continuity of E_x and H_y in Fig. 23-3.

With $\boldsymbol{E}$ normal to the plane of incidence, the $\boldsymbol{E}$ and $\boldsymbol{H}$ vectors of the incident wave point as in Fig. 23-2. Since both media are isotropic, the $\boldsymbol{E}$ vectors of the other two waves are also normal to the plane of incidence. That is because the electrons in both media oscillate in the direction normal to the plane of incidence, and reradiate waves polarized with $\boldsymbol{E}$ normal to the plane of incidence.

With the $\boldsymbol{E}$ and $\boldsymbol{H}$ vectors oriented as in Figure 23-2, the Poynting vectors $\boldsymbol{E} \times \boldsymbol{H}$ point in the direction of propagation, or in the direction of the $\boldsymbol{k}$ vectors.

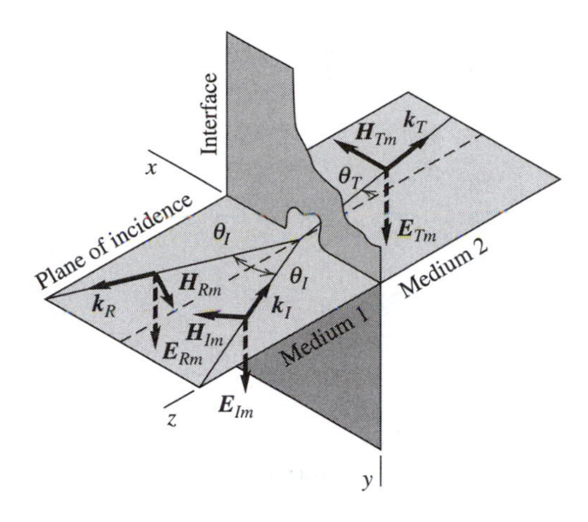

Figure 23-2 The incident, reflected, and transmitted waves for an incident wave polarized with its $\boldsymbol{E}$ field *normal* to the plane of incidence. The arrows show the directions in which the vectors are taken to be positive *at the interface*. The vectors $\boldsymbol{E} \times \boldsymbol{H}$ point everywhere in the direction of propagation, or in the direction of the $\boldsymbol{k}$ vectors.

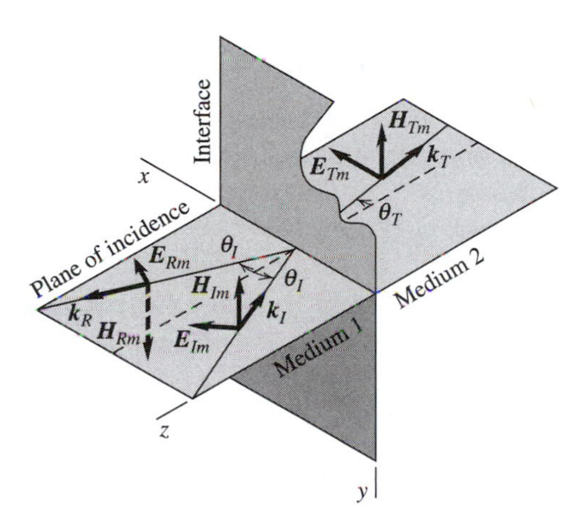

Figure 23-3 This figure is similar to Fig. 23-2, except that now the $\boldsymbol{E}$ fields are all *parallel* to the plane of incidence.

Because of the continuity of the tangential component of $\boldsymbol{E}$ at the interface,

$$E_{Im} + E_{Rm} = E_{Tm}. \tag{23-10}$$

Similarly, because of the continuity of the tangential component of $\boldsymbol{H}$,

$$H_{Im} \cos \theta_I - H_{Rm} \cos \theta_I = H_{Tm} \cos \theta_T, \tag{23-11}$$

or, from Secs. 22.5 and 22.6,

$$\frac{(E_{Im} - E_{Rm}) \cos \theta_I}{Z_1} = \frac{E_{Tm} \cos \theta_T}{Z_2}. \tag{23-12}$$

Solving,

$$\left(\frac{E_{Rm}}{E_{Im}}\right)_\perp = \frac{Z_2 \cos \theta_I - Z_1 \cos \theta_T}{Z_2 \cos \theta_I + Z_1 \cos \theta_T}, \tag{23-13}$$

$$\left(\frac{E_{Tm}}{E_{Im}}\right)_\perp = \frac{2Z_2 \cos\theta_I}{Z_2\cos\theta_I + Z_1\cos\theta_T}. \tag{23-14}$$

The subscript $\perp$ indicates that the $\boldsymbol{E}$ vectors are perpendicular to the plane of incidence. These are two of *Fresnel's equations*.

Similarly, with $\boldsymbol{E}$ parallel to the plane of incidence as in Fig. 23-3, we find the other two Fresnel equations:

$$\left(\frac{E_{Rm}}{E_{Im}}\right)_\parallel = \frac{Z_2\cos\theta_T - Z_1\cos\theta_I}{Z_2\cos\theta_T + Z_1\cos\theta_I}, \tag{23-15}$$

$$\left(\frac{E_{Tm}}{E_{Im}}\right)_\parallel = \frac{2Z_2 \cos\theta_I}{Z_2\cos\theta_T + Z_1\cos\theta_I}. \tag{23-16}$$

See Lorrain (1985).

23.3 REFLECTION AND REFRACTION AT THE INTERFACE BETWEEN TWO NONCONDUCTORS

The subscripts $\perp$ and $\parallel$ refer to the orientation of the $\boldsymbol{E}$ vector of the incident wave with respect to the plane of incidence.

With nonconductors on both sides of the interface, the laws of reflection and Snell's law of Sec. 23.1 apply and

$$Z_1 = c\mu_0/n_1, \qquad Z_2 = c\mu_0/n_2, \tag{23-17}$$

$$\left(\frac{E_{Rm}}{E_{Im}}\right)_\perp = \frac{(n_1/n_2)\cos\theta_I - \cos\theta_T}{(n_1/n_2)\cos\theta_I + \cos\theta_T}, \tag{23-18}$$

$$\left(\frac{E_{Tm}}{E_{Im}}\right)_\perp = \frac{2(n_1/n_2)\cos\theta_I}{(n_1/n_2)\cos\theta_I + \cos\theta_T}. \tag{23-19}$$

Since

$$n_1/n_2 = \sin\theta_T/\sin\theta_I, \tag{23-20}$$

you can show that the first ratio, Eq. 23-18, is proportional to $\sin(\theta_T - \theta_I)$. So that ratio can be either positive or negative, but it cannot be zero. The second ratio, Eq. 23-19, is always positive. See Figs. 23-4 and 23-5.

Also,

$$\left(\frac{E_{Rm}}{E_{Im}}\right)_\parallel = \frac{-\cos\theta_I + (n_1/n_2)\cos\theta_T}{\cos\theta_I + (n_1/n_2)\cos\theta_T}, \tag{23-21}$$

$$\left(\frac{E_{Tm}}{E_{Im}}\right)_\parallel = \frac{2(n_1/n_2)\cos\theta_I}{\cos\theta_I + (n_1/n_2)\cos\theta_T}. \tag{23-22}$$

In that case the first ratio can be positive, negative, or zero, but the second ratio is always positive. See Fig. 23-6.

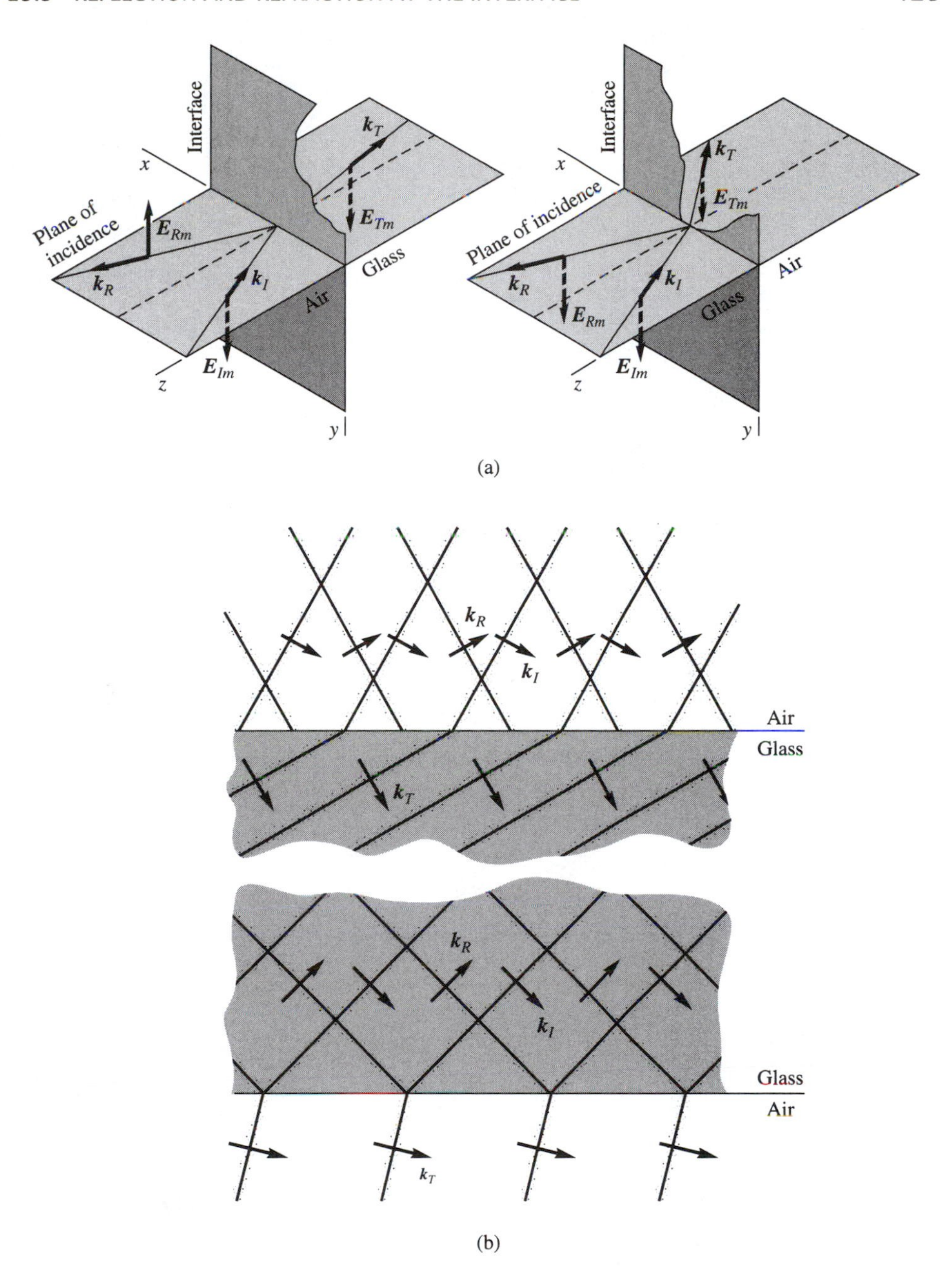

Figure 23-4a, b (a) The relative phases, at the interface, of the $\boldsymbol{E}$'s in the reflected and transmitted waves for $n_1 < n_2$ and for $n_1 > n_2$, with $\boldsymbol{E}_{Im}$ *normal* to the plane of incidence. On the left, the reflected wave is π radians out of phase with respect to the incident wave. The transmitted wave is in phase, in both instances. (b) "Crests" of the field $\boldsymbol{E}$ at some particular time. The crests are one wavelength apart and travel in the directions of the arrows. Note the phase shift of π upon reflection from a glass surface. Note also the interference pattern that results from the superposition of the incident and reflected waves.

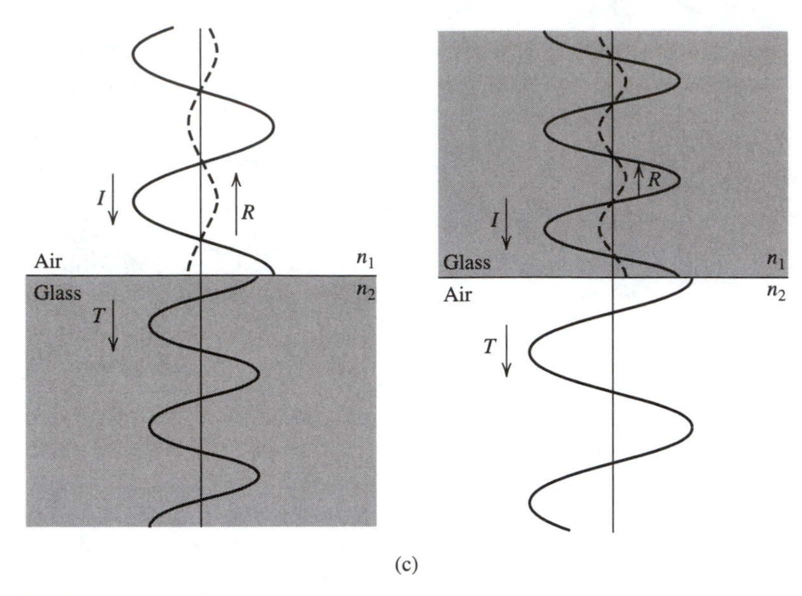

Figure 23-4c (c) The $\boldsymbol{E}$ vectors at a given instant in the incident, reflected, and transmitted waves at normal incidence on a glass-air interface. On the right, E_{Tm} is larger than E_{Im}. However, conservation of energy still applies.

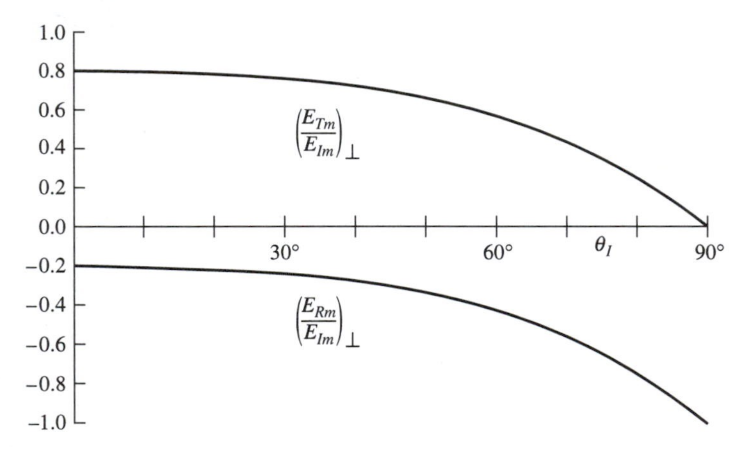

Figure 23-5 Reflection and refraction when $n_1/n_2 = 1/1.5$, for example, when light falls on a glass of $n = 1.5$. The $\boldsymbol{E}$ field is *normal* to the plane of incidence.

23.3.1 The Brewster Angle

The *Brewster angle* is the angle of incidence for which the right-hand side of Eq. 23-21 is equal to zero. For that particular angle of incidence *there is no reflected wave!* Note that there is a Brewster angle only if the incident wave is polarized with its $\boldsymbol{E}$ vector *parallel* to the plane of incidence. Then

$$\cos\theta_I = (n_1/n_2)\cos\theta_T. \tag{23-23}$$

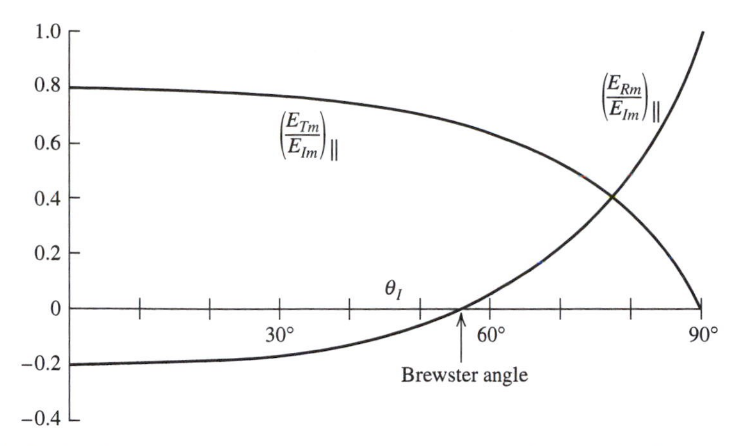

Figure 23-6 Reflection and refraction when $n_1/n_2 = 1/1.5$, as in Fig. 23-5, but with ***E*** *parallel* to the plane of incidence.

Again applying Snell's law,

$$\sin\theta_I \cos\theta_I = \sin\theta_T \cos\theta_T, \qquad \text{or} \qquad \sin(2\theta_I) = \sin(2\theta_T). \tag{23-24}$$

The first solution that comes to mind is $\theta_I = \theta_R$, but that is forbidden by Snell's law, because $n_1 \neq n_2$. The other solution is

$$2\theta_T = \pi - 2\theta_{IB}, \qquad \text{or} \qquad \theta_{IB} + \theta_T = \pi/2: \tag{23-25}$$

the incident and transmitted rays are perpendicular. Then

$$\frac{n_1}{n_2} = \frac{\sin\theta_T}{\sin\theta_{IB}} = \frac{\sin(\pi/2 - \theta_{IB})}{\sin\theta_{IB}} = \cot\theta_{IB}. \tag{23-26}$$

That is the condition that defines the Brewster angle (see Fig. 23-7), for light incident in air on a glass whose index of refraction is 1.5, $\theta_{IB} = 56.3°$.

The Brewster angle is commonly used to measure the index of refraction of a substance by reflecting a ray of light from its surface. The measurement can be accurate to five significant figures.

A plane wave incident on a *plate* of glass at the Brewster angle meets the second interface also at its Brewster angle. So there is no reflection, either at the first or at the second interface.

EXAMPLE The Helium-Neon Laser

There exist many types of gas laser, but the Helium-Neon laser is by far the most common (Waynant and Ediger, 2000). See Fig. 23-8. The source $V \approx 1000$ volts, excites a discharge through the gas in the glass tube, which has an inner diameter of 1 or 2 millimeters, and which can be as long as one meter. The gas is a mixture of 85% Helium and 15% Neon, at a total pressure of about 130 pascals (1 millimeter of mercury), and the wavelength is 632.8 nanometers (red). The beam power is of the order of a milliwatt.

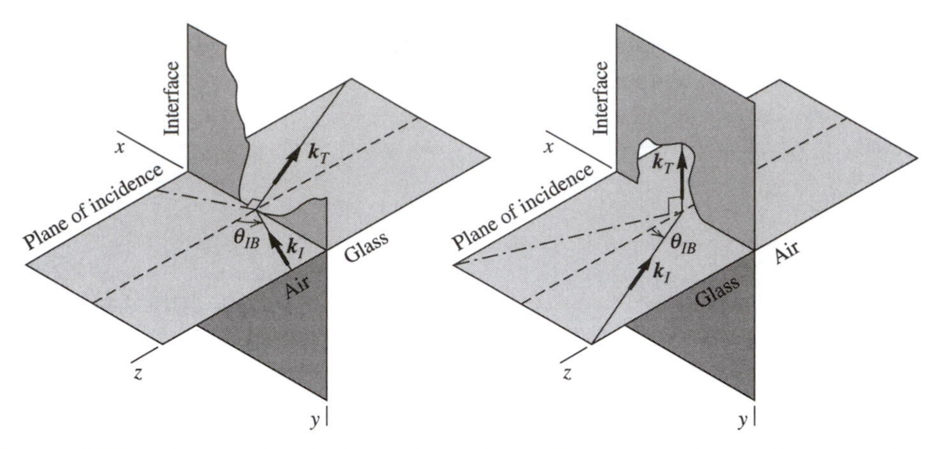

Figure 23-7 If the angle of incidence is equal to the Brewster angle, and if $\boldsymbol{E}$ lies in the plane of incidence, there is no reflected wave. The missing reflected ray is at 90° to the transmitted ray.

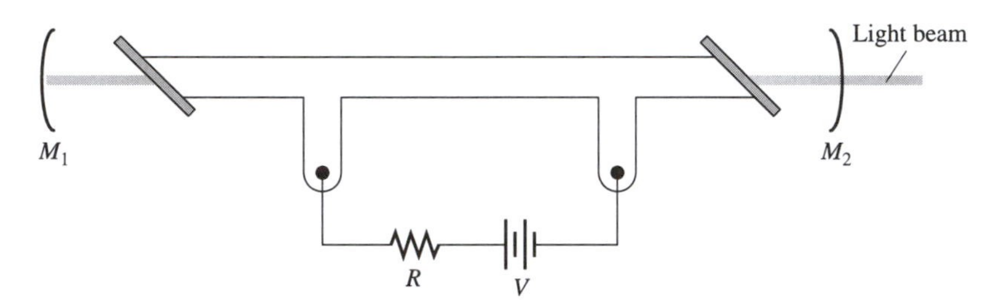

Figure 23-8 Schematic diagram of a Helium-Neon Laser. Note the two Brewster-angle windows.

Let us see how the laser generates a light beam, or *lases*. Forget about the Brewster-angle windows for the moment. The mirrors M_1 and M_2 are slightly concave, and light oscillates back and forth between them. The discharge establishes a *population inversion* in the gas, making the density of higher-energy states larger than the density of lower-energy states. This permits *stimulated emission*, whereby an incident photon stimulates an excited state to decay to a lower-energy state, generating a second photon of the same energy and momentum. There is amplification and the gas acts as a *resonator*; the light intensity increases up to a stable level. The light beam emerges through M_2, which is semi-transparent.

If the light beam need not be polarized, then the mirrors are sealed directly to the ends of the tube.

But, to obtain a polarized beam, one must use at least one Brewster-angle window. These windows are near-transparent for light polarized with its $\boldsymbol{E}$-vector parallel to the plane of incidence (Fig. 23-6), or parallel to the plane of the paper. They do not therefore hinder the amplification for that polarization, but they are poor reflectors for the other polarization. So the light beam that emerges through M_2 is polarized with its $\boldsymbol{E}$-vector parallel to the paper.

EXAMPLE Measuring the Relative Permittivity of the Moon's Surface at Radio Frequencies

The nature of the moon's surface can be inferred, to some extent, from the value of its relative permittivity $\epsilon_r = n^2$ (Sec. 22.5) at radio frequencies. The Brewster angle can serve to measure this quantity in the following way.

If a radio wave originating from a satellite in lunar orbit illuminates the moon, the reflection observed on the earth is similar to the reflection of sunlight from the surface of a lake: most of the light comes from the regions that happen to be correctly oriented for specular reflection. The surface of the moon thus glistens over an area of the order of 100 kilometers in diameter, the area depending on the height of the satellite above the surface of the moon and on the roughness of the surface.

A detector on the earth receives both the reflected wave and a direct wave from the satellite, but it is possible to discriminate between the two by using the fact that the Doppler effect makes the two radio frequencies slightly different. A plot of the intensity of the reflected wave as a function of the angle of incidence when the $\boldsymbol{E}$ vector lies in the plane of incidence shows zero reflection at the Brewster angle.

In one such measurement, performed at a frequency of 140 megahertz, the Brewster angle was $60 \pm 1°$ in the mare northwest of Hanstein. This gives an ϵ_r of 3.0 ± 0.2.

It is possible to perform similar measurements at other points on the surface of the moon because of the relative motions of the three bodies involved, namely the satellite, the moon, and the earth.

23.3.2 Total Reflection

Total reflection is of great practical importance because, as we shall see in the next chapter, it is thanks to total reflection that optical fibers can transmit light signals.

As we noted in Sec. 23.1, if $(n_1/n_2) \sin \theta_I > 1$, then $\sin \theta_T > 1$! That inequality follows from Snell's law, and it applies, whatever the polarization of the incident wave, and whatever the angle of incidence. But how can the sine of an angle be larger than unity? The explanation is a long story: total reflection is a very complex phenomenon and we limit ourselves, here, to a brief qualitative discussion and to the calculation of the phase shift on reflection. Readers who wish to study total reflection in greater depth should consult Lorrain, Corson, and Lorrain (1988).

First, note that total reflection occurs *only* if the index of refraction of medium 1 is *larger* than the index of refraction of medium 2: only if $n_1 > n_2$. See Fig. 23-4a, second part, where the incident ray is in glass, and the transmitted ray is in air. Recall Fig. 23-1 for the definitions of θ_I and θ_T. Note that, with $n_1 > n_2$, Snell's law says that θ_T is larger than θ_I.

Say you have a laser beam. You start with $\theta_I = 0$: the beam is perpendicular to the surface. You slowly rotate the laser to increase θ_I. At first, the reflected and transmitted waves behave normally: the laws of reflection, Snell's law, and the Fresnel equations apply. The angle θ_T increases gradually until it becomes equal to $90°$. You have then reached the *critical angle of incidence*, for which

$$\sin \theta_{Ic} = n_2/n_1. \tag{23-27}$$

For a glass with $n_1 = 1.5$, and with $n_2 = 1$ (air), $\theta_{Ic} = 41.8°$. The critical angle of incidence is independent of the polarization of the incident beam.

You continue to increase the angle of incidence. What happens? The angle θ_T cannot become larger than 90°! What happens is that, from then on, the incident beam is *totally reflected*. Here is an important point: the reflection is *lossless*.

Here is another important point: there is a phase shift Φ on reflection. Let us calculate the phase angle $\Phi_\perp$; we shall need it in Sec. 24.3 on optical fibers. The angle Φ depends on the polarization.

The Fresnel equations of Secs. 23.2 and 23.3 are general; they apply even to total reflection. Thus, with the $\boldsymbol{E}$ vector perpendicular to the plane of incidence, from Sec. 23.3,

$$\left(\frac{E_{Rm}}{E_{Im}}\right)_\perp = \frac{(n_1/n_2)\cos\theta_I - \cos\theta_T}{(n_1/n_2)\cos\theta_I + \cos\theta_T}. \tag{23-28}$$

Fine, but what do we do about $\cos\theta_T$? According to Snell's law, $\theta_T > 90°$, which seems absurd, from Fig. 23-1! Never mind. Apply Snell's law:

$$\cos\theta_T = (1 - \sin^2\theta_T)^{1/2} = [1 - (n_1/n_2)^2 \sin^2\theta_I]^{1/2} \tag{23-29}$$

$$= j[(n_1/n_2)^2 \sin^2\theta_I - 1]^{1/2}. \tag{23-30}$$

So $\cos\theta_T$ is imaginary! Of course there is no problem with $\sin\theta_I$ because θ_I is an ordinary angle: it is real and it can vary from zero to 90°

So

$$\left(\frac{E_{Rm}}{E_{Im}}\right)_\perp = \frac{(n_1/n_2)\cos\theta_I + j[(n_1/n_2)^2 \sin^2\theta_I - 1]^{1/2}}{(n_1/n_2)\cos\theta_I - j[(n_1/n_2)^2 \sin^2\theta_I - 1]^{1/2}} = \exp j\Phi_\perp, \tag{23-31}$$

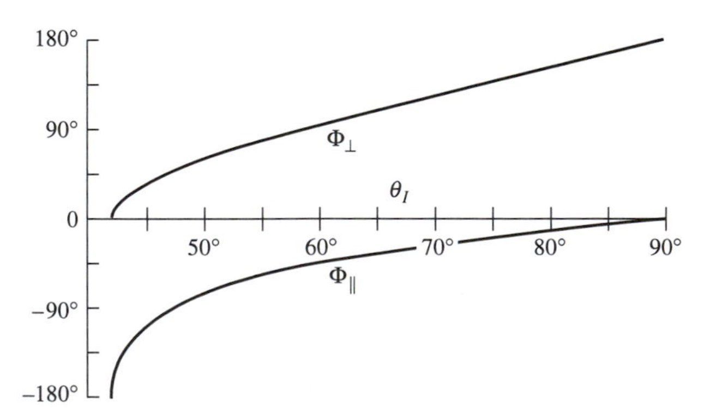

Figure 23-9 The phases $\Phi_\perp$ and $\Phi_\parallel$ of the reflected wave with respect to the incident wave at a point on the interface, for total reflection when the $\boldsymbol{E}$ of the incident wave is normal and parallel to the plane of incidence. The ratio n_1/n_2 is equal to 1.50.

with

$$\Phi_\perp = 2 \arctan \frac{[(n_1/n_2)^2 \sin^2 \theta_I - 1]^{1/2}}{(n_1/n_2)\cos\theta_I}. \tag{23-32}$$

Figure 23-9 shows the phase shifts $\Phi_\perp$ and $\Phi_\parallel$ as functions of the angle of incidence θ_I for $n_1/n_2 = 1.50$. With the $\boldsymbol{E}$ vector of the incident wave perpendicular to the plane of incidence, the reflected wave *leads* the incident wave by an angle that increases from zero at the critical angle of incidence up to 180° at an angle of incidence of 90°.

EXAMPLE Light Emission from a Cathode-Ray Tube

In a cathode-ray tube, the electron beam generates light in a fluorescent coating deposited on the back of the tube face. A given point in the fluorescent material radiates in all directions, but a thin layer of aluminum, as in Fig. 23-10, doubles the light output. Even then, most of the light stays trapped inside the glass by total reflection and travels back to the gun end of the tube.

What fraction F of the light comes out through the tube face? This is easy to calculate if we assume that, inside the cone of angle θ_{Ic}, all the light crosses the glass-air interface. This is not a bad approximation. Then F is the solid angle corresponding to θ_{Ic}, divided by 2π (not 4π, because of the mirror). Since the cone defines a solid angle equal to the area of the spherical segment, shown as a dashed line in the figure, divided by R^2,

$$F = \frac{1}{2\pi}\int_0^{\theta_{Ic}} \frac{2\pi R \sin\theta \, R d\theta}{R^2} = 1 - \cos\theta_{Ic}. \tag{23-33}$$

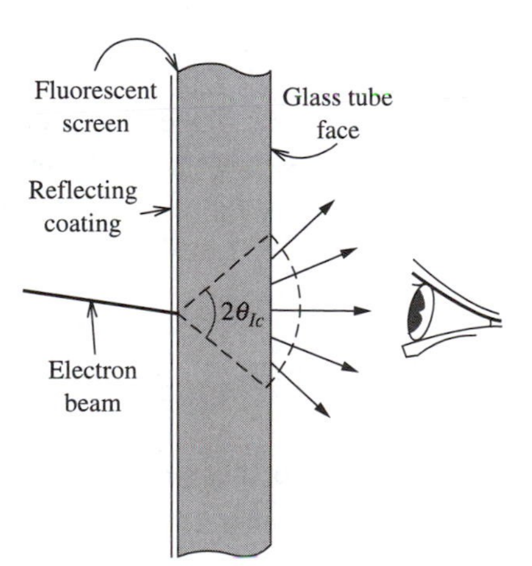

Figure 23-10 Section through the face of a cathode-ray tube.

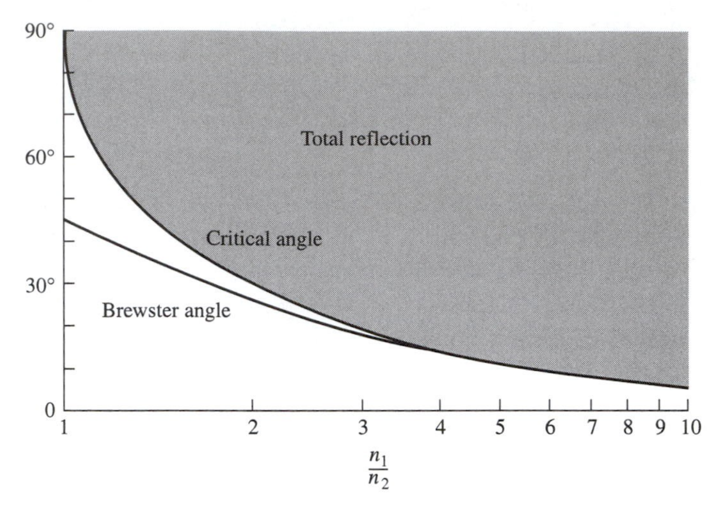

Figure 23-11 The critical angle and the Brewster angle as functions of the ratio n_1/n_2. The incident wave is linearly polarized with the $\boldsymbol{E}$ vector *parallel* to the plane of incidence for the Brewster angle curve.

For a glass whose index of refraction is 1.5, $\theta_{Ic} = 41.8°$ and $F = 0.255$. The fraction F is, in fact, even smaller because of our approximation.

EXAMPLE The Critical Angle and the Brewster Angle

The critical angle is somewhat larger than the Brewster angle (Sec. 23.3.1). For example, again for light propagating inside a glass with an index of refraction n_1 of 1.5, the wave is totally transmitted into the air at a glass-air interface when the angle of incidence is the Brewster angle, 33.7°, and it is totally reflected back into the glass beyond the critical angle of 41.8°.

Figure 23-11 shows these two angles as functions of the ratio n_1/n_2. For large values of n_1/n_2, that is, for light incident in a relatively "dense" medium, θ_{Ic} is nearly equal to θ_{IB}. For media with more similar indices of refraction, the Brewster angle approaches 45°, whereas the critical angle approaches 90°.

For a wave polarized with its $\boldsymbol{E}$ vector parallel to the plane of incidence, the amplitude of the reflected wave changes rapidly when the angle of incidence lies between the Brewster angle and the critical angle. This peculiar behavior of the reflected and transmitted waves could be useful for measuring small angular displacements.

23.4 REFLECTION AND REFRACTION AT THE SURFACE OF A GOOD CONDUCTOR

The incident and reflected waves are in medium 1, which is a nonconductor, but now medium 2 is a conductor (Sec. 22.6). The laws of Sec. 23.1 again apply: the three rays are coplanar and, from Snell's law,

$$n_1 \sin\theta_I = n_2 \sin\theta_T. \tag{23-34}$$

Since $n_2 \gg n_1$, $\sin\theta_T \approx 0$ and $\theta_T \approx 0$: the transmitted wave penetrates into the conductor along the normal to the interface, for any angle of incidence. The transmitted wave is highly attenuated, as in Sec. 22.6.

From the Fresnel equations (Sec. 23.2), again with $n_2 \gg n_1$,

$$\frac{E_{Rm}}{E_{Im}} \approx -1, \tag{23-35}$$

whatever the polarization of the incident wave, and for any angle of incidence: reflection is near-perfect, but not as good as with total reflection. We shall need this result in Sec. 24.3 on the microstrip line. The net $\boldsymbol{E}$ at the interface is approximately zero, as in Fig. 23-12.

EXAMPLE Communicating with Submarines at Sea

For shore-to-ship communication with the submarine antenna submerged, the efficiency is very low, first because of the large coefficient of reflection at the surface of the sea and, second, because of the high attenuation in seawater. The attenuation in seawater is about 172 decibels/meter at 20 megahertz, 5.5 at 20 kilohertz, and 0.33 at 75 hertz. One solution is to operate at low frequencies (about 75 hertz and 17 to 25 kilohertz) and very high power, with huge transmitting antennas, many kilometers on the side.

Another solution for shore-to-ship communication is to modulate a laser beam emitted by a satellite, seawater being quite transparent to blue-green light. Remember that our discussion on the propagation of electromagnetic waves in conductors disregards atomic and molecular phenomena and is valid only up to roughly 1 gigahertz. Optical frequencies are of the order of 10^{15} hertz.

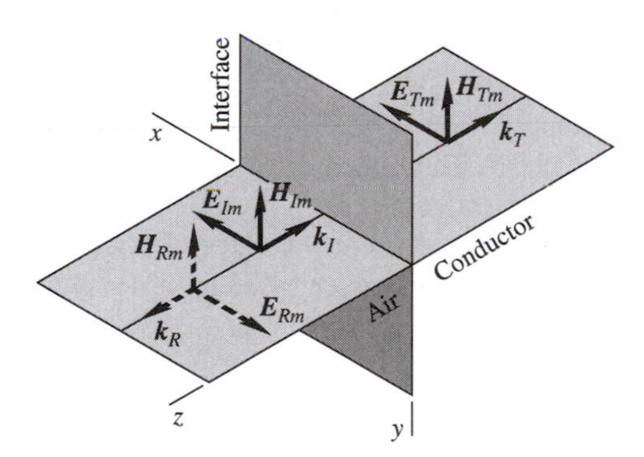

Figure 23-12 Reflection at normal incidence from the surface of a good conductor: $E_{Rm} \approx -E_{Im}$ and $E_{Tm} \ll E_{Im}$.

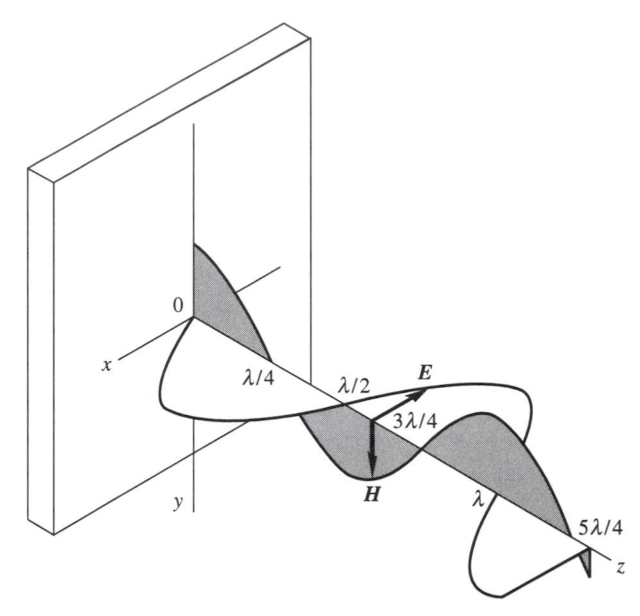

Figure 23-13 The standing wave pattern for reflection at normal incidence on a good conductor, at a particular time. Nodes of $\boldsymbol{E}$ and of $\boldsymbol{H}$ are spaced $\lambda/4$ apart.

Ship-to-shore communication at low frequencies is impossible with long radio waves because a submarine can neither supply the required power nor deploy a long enough antenna. Two-way communication takes place at a few megahertz with the submarine antenna projecting above the water.

EXAMPLE Standing Waves at Normal Incidence on a Good Conductor

Figure 23-12 shows the incident, reflected, and transmitted waves. Since the direction of propagation of the reflected wave is opposite to that of the incident wave, and since $\boldsymbol{E} \times \boldsymbol{H}$ points in the direction of propagation, the $\boldsymbol{H}$ of the reflected wave is in phase with that of the incident wave at the interface, as in the figure. At the reflecting surface, the electric fields nearly cancel and there is a node of $\boldsymbol{E}$; the magnetic fields add, and there is a loop of $\boldsymbol{H}$, as in Fig. 23-13. The nodes of $\boldsymbol{E}$ and $\boldsymbol{H}$ are thus one-quarter wavelength apart. The energy density is uniform.

A similar situation exists for reflection from any surface. Either the $\boldsymbol{E}$ or the $\boldsymbol{H}$ vector must change direction on reflection, in order to change the direction of the Poynting vector $\boldsymbol{E} \times \boldsymbol{H}$.

23.5 SUMMARY

At the plane interface between two homogeneous, isotropic, linear, and stationary (HILS) media, the incident and reflected rays are coplanar, and the angle of reflection is equal to the angle of incidence. These are the *laws of reflection*.

Snell's law states that

$$n_1 \sin\theta_I = n_2 \sin\theta_T, \tag{23-7}$$

where n_1 is the index of refraction of the first medium, and n_2 that of the second medium. *Fresnel's equations* are as follows:

$$\left(\frac{E_{Rm}}{E_{Im}}\right)_\perp = \frac{(n_1/n_2)\cos\theta_I - \cos\theta_T}{(n_1/n_2)\cos\theta_I + \cos\theta_T}, \tag{23-18}$$

$$\left(\frac{E_{Tm}}{E_{Im}}\right)_\perp = \frac{2(n_1/n_2)\cos\theta_I}{(n_1/n_2)\cos\theta_I + \cos\theta_T}, \tag{23-19}$$

$$\left(\frac{E_{Rm}}{E_{Im}}\right)_\parallel = \frac{-\cos\theta_I + (n_1/n_2)\cos\theta_T}{\cos\theta_I + (n_1/n_2)\cos\theta_T}, \tag{23-21}$$

$$\left(\frac{E_{Tm}}{E_{Im}}\right)_\parallel = \frac{2(n_1/n_2)\cos\theta_I}{\cos\theta_I + (n_1/n_2)\cos\theta_T}. \tag{23-22}$$

At the *Brewster angle* of incidence θ_{IB},

$$\frac{n_1}{n_2} = \cot\theta_{IB}, \tag{23-26}$$

and there is no reflected wave if $\boldsymbol{E}$ lies in the plane of incidence.

Total reflection at the interface between two dielectrics occurs at angles of incidence larger than the *critical angle*

$$\sin\theta_{Ic} = n_1/n_2. \tag{23-27}$$

Then the reflection is lossless, and the reflected wave is phase-shifted with respect to the incident wave as in Fig. 23-9.

Reflection at the surface of a good conductor is slightly lossy. The transmitted wave is weak, highly damped, and penetrates in the direction perpendicular to the surface.

PROBLEMS

23-1. *(23.3)* Reflection and refraction at the surface of a dense medium

Write down Fresnel's equations for the case where $n_2 \gg n_1$.

You will find that, if the $\boldsymbol{E}$ vector of the incident wave is parallel to the plane of incidence, then the amplitude of the reflected wave is independent of the angle of incidence!

For what range of θ_I is this correct?

23-2. *(23.3)* A method for measuring an index of refraction

Set

$$p = \left(\frac{E_{Rm}}{E_{Im}}\right)_\parallel, \quad s = \left(\frac{E_{Rm}}{E_{Im}}\right)_\perp.$$

Show that, with a laser beam incident at 45° in air on a medium of index of refraction n,

$$n^2 = \frac{(1-p)(1-s)}{(1+p)(1+s)}.$$

Here, both p and s are negative, from Sec. 23.3. In practice, instruments measure, not p and s, but a beam *power.* So p and s are both equal to *minus* the square root of the ratio of reflected to incident power.

See Lorrain (1985).

23-3. (*23.3.1*) The Brewster angle

Calculate the Brewster angle for each of the following cases:

(a) light incident on a glass whose index of refraction is 1.6,

(b) light emerging from the same type of glass,

(c) a radio-frequency wave incident on water ($n = 9$ at radio frequencies). This case concerns communication with submarines.

23-4. (*23.3*) Reflection at normal incidence on a water surface

A 60-watt light bulb is situated one meter above a water surface.

Calculate the root-mean-square values of E and H for the incident, reflected, and transmitted waves, directly below the bulb. Assume that all the power is dissipated as electromagnetic radiation. The index of refraction of water in the visible spectrum is 1.33.

23-5. (*23.3*) Antireflection coatings for photographic lenses and solar cells

There are instances where the reflection from a dielectric surface must be close to zero. Obvious examples are photographic lenses and solar cells. In complex optical instruments, the loss can be important; moreover, stray reflections reduce contrast in the image.

Clearly, the way to eliminate the reflected wave is by interference. Coating the dielectic with another type of dielectric provides two reflected waves, at the two faces of the coating, that can cancel. The situation is, however, complicated by the presence of multiple reflections in the film. It turns out that, taking multiple reflections into account, there is no reflected wave in air at normal incidence ($n_1 = 1$)when a dielectric n_3 is coated with a quarter-wavelength of dielectric $n_2 = n_3^{1/2}$. Good-quality lenses are coated with magnesium fluoride ($n = 1.38$ at 550 nanometers).

A silicon solar cell has an index of refraction of 3.9 at 600 nanometers. Calculate the thickness and index of refraction of a coating that would eliminate reflection at that wavelength.

23-6. (*23.3.2*) Scintillation particle detector

Figure 23-14 shows one type of *scintillation particle detector.* A scintillator S, usually made out of a single crystal of sodium iodide or of a suitable transparent plastic embedded in a reflector R, emits light when it is traversed by an ionizing particle such as an electron. A photomultiplier PM detects the emitted light.

The scintillator has an index of refraction n_1 and is fixed to the face of the photomultiplier with a cement C of index $n_2 < n_1$. Light is emitted in all directions in the scintillator, but only a fraction F reaches the photomultiplier.

(a) Calculate F as a function of n_2/n_1, assuming that transmission is perfect for angles of incidence smaller than the critical angle, and that the scintillator is surrounded by a nonreflecting substance.

(b) Plot F for values of n_2/n_1 ranging from 0.1 to 1.0.

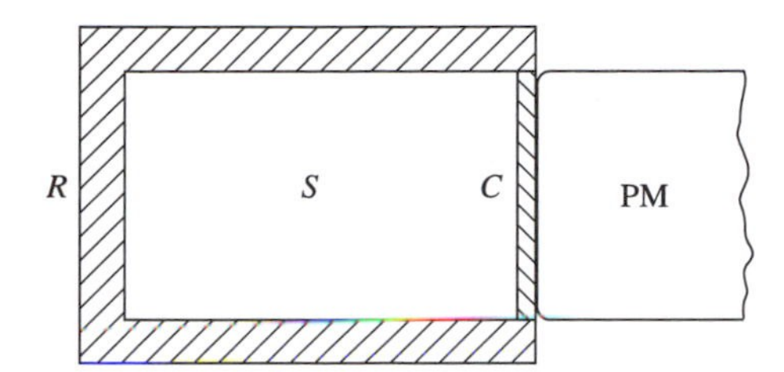

Figure 23-14

23-7. (*23.3.2*) Total reflection in light-emitting diodes

In *light-emitting diodes* (LEDs), radiation occurs in a junction plane within a semiconductor whose index of refraction is quite large. For example, with GaAsP, $n = 3.5$. Total reflection at the semiconductor-air interface limits the efficiency of LEDs to a few percent.

(a) Calculate the critical angle.

(b) Assume that the face of the semiconductor is flat and parallel to the junction. The index of refraction is n.

Calculate the fraction F of the light emitted at the source that reaches the surface at an angle smaller than the critical angle. Show that $F \approx 1/(4n^2)$.

23-8. (*23.4*) Reflection from a good conductor

Draw two figures similar to those of Fig. 23-4, showing $\boldsymbol{E}$ and $\boldsymbol{H}$ for an electromagnetic wave incident on a good conductor. You will, of course, have to exaggerate the values of E_m and of λ in the conductor. Be sure to show the phases correctly. Show x-, y-, z-axes on both figures to relate one with the other.

23-9. (*23.4*) Reflection from a good conductor

Show that, for a good nonmagnetic conductor in air,

$$\text{(a)} \left|\frac{E_{Rm}}{E_{Im}}\right|_{\perp} \approx 1 - \frac{\delta}{\lambda\!\!\!^{-}_0}\cos\theta_I, \qquad \text{(b)} \left|\frac{E_{Rm}}{E_{Im}}\right|_{\parallel} \approx 1 - \frac{\delta}{\lambda\!\!\!^{-}_0\cos\theta_I}.$$

This latter relation is not valid at grazing incidence, where $\cos\theta_I$ tends to zero.

A good conductor is a better reflector when $\boldsymbol{E}$ is normal to the plane of incidence. High-quality metallic reflectors have coefficients of reflection of about 90% near normal incidence in the visible, with unpolarized light.

23-10. (*23.4*) Cutting steel plate with a laser beam

Figure 23-15 shows a laser beam cutting a steel plate,

(a) Why does the beam cut at a faster rate when the $\boldsymbol{E}$ vector lies in the plane of the paper than when it is perpendicular?

(b) Roughly what percentage of the beam power serves to heat the steel in the former case?

(c) Can you explain why the kerf is narrower and more even when $\boldsymbol{E}$ is in the plane of the paper?

If the required kerf is not straight, then the laser should rotate to keep the $\boldsymbol{E}$ vector of the beam parallel to the path. A simpler solution is to use a circularly polarized beam.

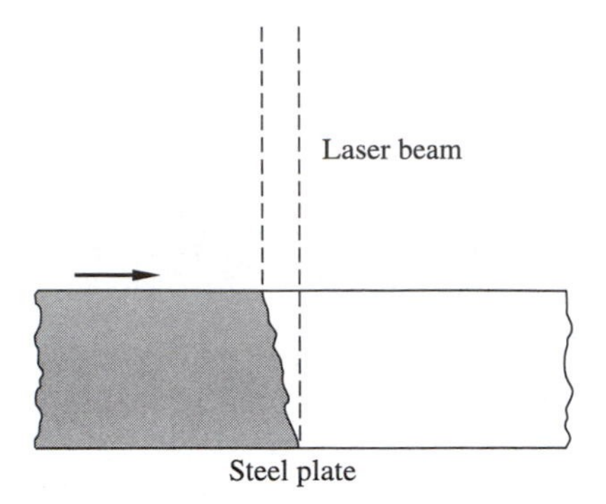

Figure 23-15

23-11. (*23.4*) The standing wave at normal incidence on a good conductor

Show that the electromagnetic energy density in a plane standing wave at normal incidence on a good conductor is uniform.

23-12. (*23.4*) Multiple reflections in a dielectric plate backed by a conductor

An electromagnetic wave falls at an angle of θ_I on a slab of dielectric that is backed by a good conductor.

Under what condition is there a single reflected wave?

CHAPTER

24

ELECTROMAGNETIC WAVES III

The Coaxial and Microstrip Lines. Optical Waveguides

In Chapter 22 we studied the propagation of electromagnetic waves in unbounded media. Then, in Chapter 23, we investigated reflection and refraction at the interface between two media.

We now study how electromagnetic waves can be guided in prescribed directions by *waveguides*. We first study three types of metallic waveguides. The parallel-wire line is the most common and the simplest. The coaxial and microstrip lines are similar, in a way, and also quite simple. Next we briefly study hollow metallic waveguides, which are more sophisticated. Optical fibers will be our next, and last, type of guide. Optical fibers? Well, not quite. The axisymmetry of optical fibers makes them relatively abstruse, and we shall study, instead, planar optical guides, which share essentially the same Physics, and which are used in integrated optical circuits.

These are the main types of waveguide that have survived. Several other types were abandoned years ago.

24.1 PARALLEL-WIRE LINES

Parallel wires, as in house wiring, act as waveguides, although their wave properties are not observable because the wavelength at 60 or 120 hertz is so enormous. (How many kilometers?) Figure 24-1 shows the electric and magnetic fields of a pair of

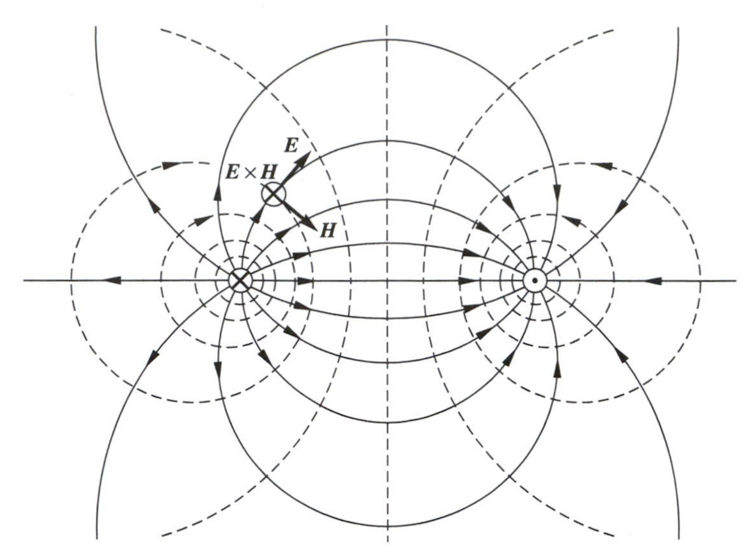

Figure 24-1 Electric field lines (solid) and magnetic field lines (dashed) for a parallel-wire line. The left-hand wire is positive and carries a current that flows into the paper, while the right-hand wire is negative and carries a current that comes out of the paper. The line feeds a resistance at the far end. The vectors $\boldsymbol{E}$ and $\boldsymbol{H}$ are tangent to their field lines, and the Poynting vector $\boldsymbol{E} \times \boldsymbol{H}$ points into the paper. Looks familiar? It is! Refer to Fig. 3-11, which shows the field lines and the equipotentials near a pair of charged wires (no current). The electric field lines are of course the same in the two figures, but the magnetic field lines have the same shape as the equipotentials of Fig. 3-11! See Fig. 17-8.

wires of opposite polarities, carrying currents in opposite directions. Both $\boldsymbol{E}$ and $\boldsymbol{H}$ are transverse and the wave is *Transverse Electric and Magnetic*, or *TEM*. Power flows in the direction of the Poynting vector, into the paper.

Cable pairs were long used in telephony. They comprise two insulated wires, twisted together to reduce their magnetic field, and hence their inductance. The twist also reduces "cross-talk" between neighboring pairs. Cables comprised as many as 2,700 pairs! See Sec. 24.5.

24.2 COAXIAL LINES

Coaxial lines serve to interconnect electronic equipment and, for many years, they served for long-distance telephony. Figure 24-2 shows a portion of a coaxial line. The wave propagates in the annular region between the conductors, and there is zero field outside. The medium of propagation is either a low-loss dielectric, usually polyethylene or Teflon, or air with the inner conductor supported by spacers. There exist a wide variety of coaxial lines, from tiny flexible cables with braided outside conductors up to 15- or 20-centimeter diameter copper pipes capable of carrying hundreds of kilowatts.

A coaxial line operates perfectly well with steady voltages and currents, and then its behavior is trivial. Here, we are interested in the use of coaxial lines with alternat-

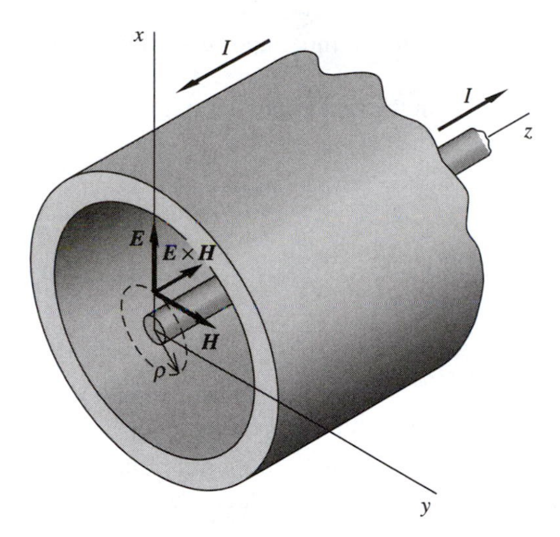

Figure 24-2 The $\boldsymbol{E}$, $\boldsymbol{H}$, and $\boldsymbol{E} \times \boldsymbol{H}$ vectors inside a coaxial line.

ing voltages and currents. Of what frequency? Our analysis is valid at any frequency, as long as the wavelength is large compared to the dimensions of the annular region. For example, if the annular region has an outer diameter of 1 centimeter, then the wavelength of the transmitted signal should be much larger than 1 centimeter, and the frequency much lower than $c/10^{-2}$, or 30 gigahertz. At higher frequencies, the field becomes complex and the wave speed strongly dependent on the frequency, which distorts non-sinusoidal signals. At TV frequencies, the wavelength is about one meter, and both parallel-wire lines and coaxial lines are perfectly adequate.

We assume a wave that travels to the right in Fig. 24-2. We shall see below how the line must be terminated at the far end to eliminate reflection.

There are electric charges and electric currents on the outer surface of the inner conductor and on the inner surface of the outer conductor. These charges and currents travel with the wave. Figure 24-2 shows an $\boldsymbol{E}$ vector pointing outward, at an instant when the inner conductor is positive with respect to the outer conductor. With the $\boldsymbol{H}$ vector pointing as in the figure, the current in the inner conductor flows to the right. An equal current flows in the opposite direction in the outer conductor.

This is again a Transverse Electric and Magnetic, or *TEM* wave.

For an observer outside the coaxial line, there is no net charge, no net current, and $\boldsymbol{E} = 0$, $\boldsymbol{H} = 0$: there is no field outside a coaxial line. That is a highly convenient feature. But flexible coax with a braided outer conductor can have an appreciable leakage field.

The electric field is radial. So, from Gauss's law (Sec. 3.6), $\boldsymbol{E}$ varies inversely as the radius ρ. Set

$$\boldsymbol{E} = \frac{E_m}{\rho} \exp j(\omega t - kz)\hat{\boldsymbol{\rho}}, \tag{24-1}$$

where $k = \omega/v = 1/\lambdabar$, v being the speed of propagation. We have assumed that there is zero attenuation. Observe that this equation describes a *wave*: $\boldsymbol{E}$ depends on both z and t, as in Chapter 22.

We use subscripts 1 and 2 to designate the inner and outer radii of the annular region and, as usual, $\epsilon = \epsilon_r \epsilon_0$.

If Q_1' is the linear charge density on the inner conductor, then

$$E = \frac{Q_1'}{2\pi\epsilon\rho}, \qquad Q_1' = 2\pi\epsilon\rho E = 2\pi\epsilon E_m \exp j(\omega t - kz). \tag{24-2}$$

This charge travels along the guide at the wave speed v. Thus the current that flows along the surface of the inner conductor is

$$I_1 = Q_1' v = 2\pi\epsilon E_m v \exp j(\omega t - kz). \tag{24-3}$$

An equal current flows in the opposite direction on the outer conductor. So $I_1 = I$, the current on the line. This is also a wave: I is a function of both z and t.

The *line voltage* is the potential of the outer conductor with respect to the inner conductor:

$$\mathcal{V} = \int_{\rho_1}^{\rho_2} E\, d\rho = E_m \ln\frac{\rho_2}{\rho_1} \exp j(\omega t - kz). \tag{24-4}$$

Now $\boldsymbol{H}$ is the magnetic field of the axial current flowing in the inner conductor. Then, according to Ampère's law (Sec. 15.4):

$$\boldsymbol{H} = \frac{I}{2\pi\rho}\hat{\boldsymbol{\phi}} = \frac{\epsilon E_m v}{\rho} \exp j(\omega t - kz)\hat{\boldsymbol{\phi}}. \tag{24-5}$$

But what is the value of the wave speed v? It is the speed of a plane wave in the medium of propagation, as in Sec. 22.5:

$$v = 1/(\epsilon\mu_0)^{1/2}. \tag{24-6}$$

We have set $\mu_r = 1$: magnetic materials are normally not used in coaxial lines because they are lossy. Then

$$\boldsymbol{H} = \frac{E_m}{\rho}\left(\frac{\epsilon}{\mu_0}\right)^{1/2} \exp j(\omega t - kz)\hat{\boldsymbol{\phi}}. \tag{24-7}$$

So $E/H = (\mu_0/\epsilon)^{1/2}$, as in Sec. 22.5.

Figure 24-3 shows the relative orientations of $\boldsymbol{E}$ and I over 1.5 wavelengths.

What about the Poynting vector? Return to Fig. 24-2. The Poynting vector $\boldsymbol{E} \times \boldsymbol{H}$ of Sec. 22.4 points to the right, in the direction of propagation of the wave, and

$$\mathcal{S}_{av} = \frac{1}{2}\mathrm{Re}(\boldsymbol{E} \times \boldsymbol{H}^*) = \left(\frac{\epsilon}{\mu_0}\right)^{1/2} \frac{E_m^2}{2\rho^2}\hat{\boldsymbol{z}}. \tag{24-8}$$

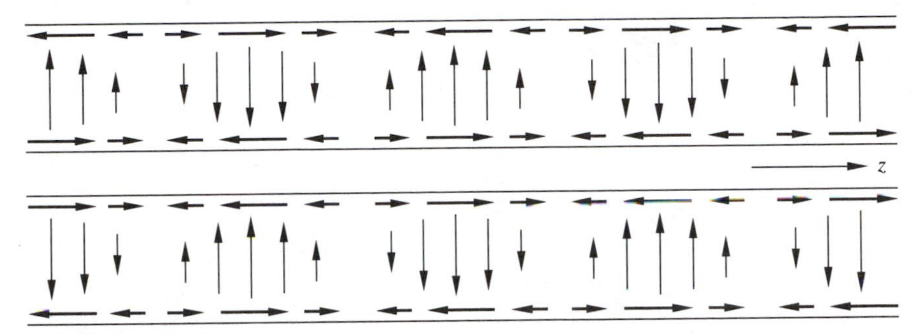

Figure 24-3 Section through a coaxial line showing the electric field strength (vertical arrows) and the current (horizontal arrows) over 1.5 wavelengths at a given instant. The wave travels from left to right. The surface charge density is positive where the current arrows point right, and negative where they point left.

The transmitted power is thus

$$P_T = \int_{\rho_1}^{\rho_2} \mathcal{S}_{av}\, 2\pi\rho\, d\rho = \left(\frac{\epsilon}{\mu_0}\right)^{1/2} \pi E_m^2 \ln\frac{\rho_2}{\rho_1}. \tag{24-9}$$

You can easily check that this is equal to $(1/2)\mathrm{Re}(\mathcal{V}I^*)$.

In practice, there is of course a certain amount of attenuation, because of Joule losses in the conductors, and because of losses in the dielectric.

It is the custom to call the ratio $\mathcal{V}/I$ the characteristic impedance *of the line*:

$$Z_{line} = \frac{\mathcal{V}}{I} = \pi\frac{\ln(\rho_2/\rho_1)}{2\pi(\epsilon/\mu_0)^{1/2}} \approx \frac{60}{\epsilon_r^{1/2}} \ln\frac{\rho_2}{\rho_1} \text{ ohms}. \tag{24-10}$$

If the line is terminated by a resistance of the above value, there is no reflected wave; otherwise there is a reflected wave and the net field is the superposition of a traveling wave and a standing wave.

EXAMPLE The RG-223/U Cable

This cable has silvered copper conductors with diameters of 2.946 and 0.889 millimeters respectively, within a plastic sheath. The outer conductor is double braided and the dielectric is polyethylene, with $\epsilon_r = 2.26$. From Eq. 24-10, $Z = 47.8$ ohms, while the nominal impedance is 50 ohms. The wave speed is

$$v = \frac{1}{(\epsilon_r \epsilon_0 \mu_0)^{1/2}} = \frac{c}{2.26^{1/2}} = 2.00 \times 10^8 \text{ meters/second}. \tag{24-11}$$

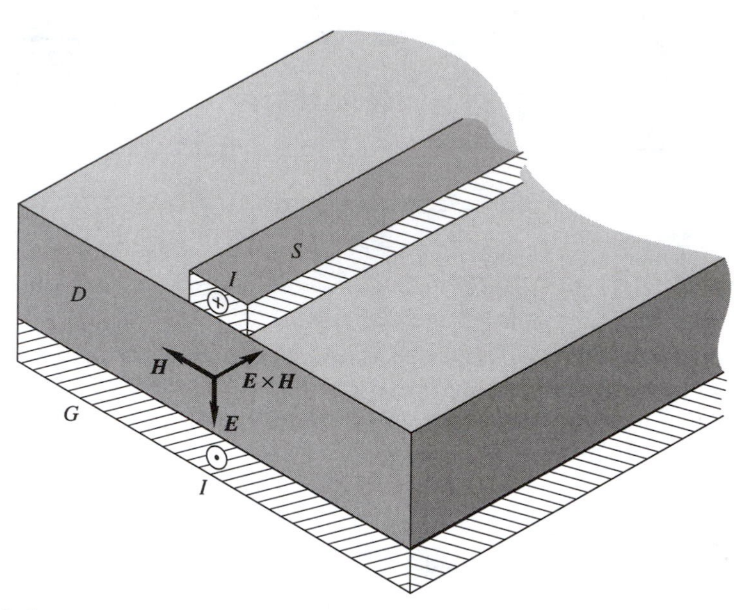

Figure 24-4 Portion of a microstrip line: *G*, ground-plane layer; *D*, insulating sheet; *S*, conducting strip. Here, *S* is positive with respect to *G*, and the currents *I* flow in the directions shown.

24.3 MICROSTRIP LINES

Figure 24-4 shows a portion of a *microstrip line*, consisting of a grounded conducting plane, an insulating sheet, and a conducting strip (Gardiol, 1994). Figure 24-5 shows field lines.

Microstrips serve as transmission lines in printed circuits, or as antennas. They are used in the range of about one to ten gigahertz ($\lambda = 300$ to 30 millimeters). Typically, the insulating sheet has a thickness of 1 millimeter, and the conducting strip a width of a few millimeters. As low-power transmission lines, microstrips have the advantage of being less costly and more compact than coaxial lines. Losses are high, but acceptable, in printed circuits. Microstrips also have the disadvantage that their field is not strictly confined to the region just below the strip. This makes them lossy, and they can interact with other elements in a circuit, unless they are either spaced or shielded properly, say with a conducting plane on top. Microstrips also serve as "patch" antennas that can be applied to curved surfaces, for example on the side or wing of an aircraft.

Roughly speaking, the wave in a microstrip is analogous to that in a coaxial line: it is *TEM* (Transverse Electric and Magnetic), with $\boldsymbol{E}$ and $\boldsymbol{H}$ oriented as in Fig. 24-4. The Poynting vector $\boldsymbol{E} \times \boldsymbol{H}$ points in the direction of propagation of the wave. The wave speed is c/n, where n is the index of refraction of the dielectric, and the wavelength is λ_0/n.

The wave is in fact only quasi-TEM because microstrip lines operate by multiple reflections on the two conducting surfaces, where up-going and down-going waves interfere. As a result, only specific modes are allowed, as in the next section. Also, the field is partly in the dielectric, and partly in air, above the conducting strip.

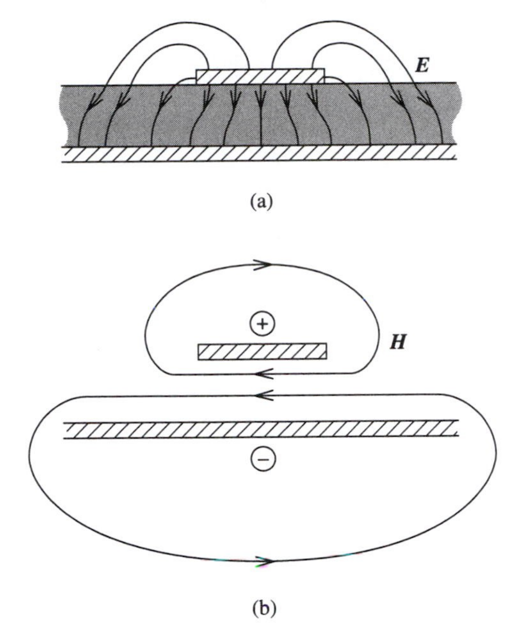

Figure 24-5 (a) Lines of $\boldsymbol{E}$ in a microstrip line. (b) Lines of $\boldsymbol{H}$. Qualitative.

24.4 HOLLOW METALLIC WAVEGUIDES

Hollow metallic waveguides serve at gigahertz frequencies, often at high power over short distances, for example to connect a power amplifier to a radar horn antenna (Sec. 25.2). Propagation occurs through multiple reflections on the guide walls. The most common type is rectangular in cross-section, with one side twice as wide as the other. Standard cross-sections vary from 290×580 millimeters for use at wavelengths of about 750 millimeters at up to 200 megawatts, down to 0.43×0.86 millimeter for use with about 1-millimeter waves. The materials used are either brass or aluminum, and silver for the tiny sizes. The medium of propagation is air.

Figure 24-6 shows how such waveguides can be fed by a coaxial line. The source launches an assortment of modes, but only the allowed mode can propagate. Let us see why.

See Fig. 24-7. The electromagnetic wave propagates down the guide by multiple reflections on the narrow sides, and the angle of incidence is θ. The figure shows a wave whose $\boldsymbol{E}$ vector is perpendicular to the plane of incidence. The net wave that results from the multiple reflections on both sides also has a transverse $\boldsymbol{E}$, but the net $\boldsymbol{H}$ is parallel to the plane of incidence, as we shall see. So this is a Transverse Electric, or *TE* wave, *not* a *TEM* wave.

Refer now to Fig. 24-8. Lines A and C are *fixed* in position, and parallel to wave fronts of the up-going wave. Similarly, line B is fixed and parallel to wave fronts of the down-going wave. Let

$$E_A = E_m \exp j\omega t. \tag{24-12}$$

Then

$$E_C = E_A \exp \frac{2\pi l}{\lambda_0}. \tag{24-13}$$

The up-going wave A results from the metallic reflection of the down-going wave B at D and

$$E_B = E_A \exp j\pi = -E_A. \tag{24-14}$$

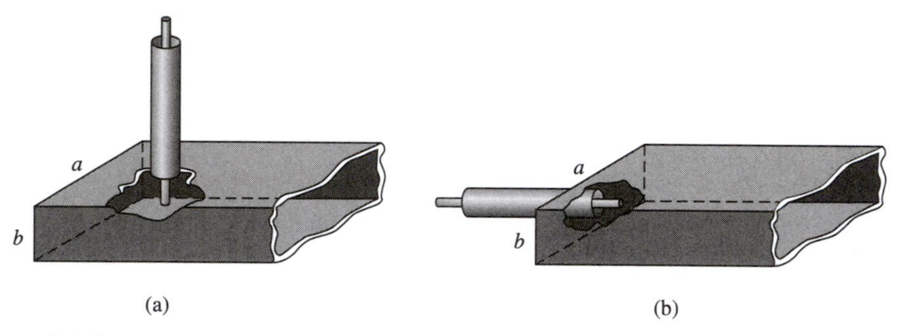

Figure 24-6 Rectangular metallic waveguides fed by a coaxial line.

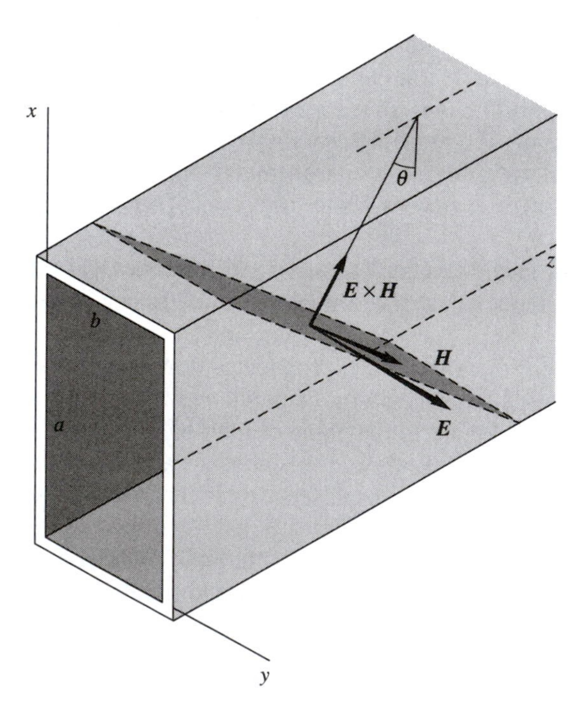

Figure 24-7 Typical wave front inside a metallic rectangular waveguide. Reflection occurs on the narrow faces, and ***E*** is everywhere transverse.

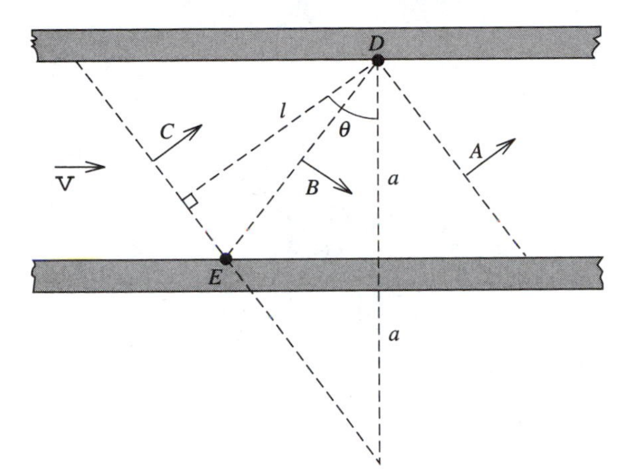

Figure 24-8 Up-going and down-going wave fronts in a hollow metallic rectangular waveguide.

Similarly, the down-going wave B results from the reflection of the up-going wave C at E. So

$$E_C = E_B \exp j\pi = E_A \exp 2j\pi = +E_A. \tag{24-15}$$

Now equating our two values of E_C,

$$\frac{2\pi l}{\lambda_0} = 2\pi. \tag{24-16}$$

But

$$l = 2a\cos\theta. \tag{24-17}$$

Then

$$\cos\theta = \frac{\lambda_0}{2a}. \tag{24-18}$$

This defines the single allowed value of θ. Given a value of a in Fig. 24-8 and a given wavelength, the angle of incidence is fixed. Also, for a given value of a, *increasing* the wavelength *decreases* the angle of incidence. But the limiting value of the wavelength is $2a$; beyond that there is no transmission. So $\lambda_0 = 2a$ is the *cutoff wavelength.*

Figures 24-7 and 24-8 show how a plane wave of speed c zig-zags down the guide. The signal speed v is less than c:

$$v = c\sin\theta = c(1-\cos^2\theta)^{1/2} = c\left(1-\frac{\lambda_0^2}{4a^2}\right)^{1/2} < c. \tag{24-19}$$

Figure 24-9 shows the net $\boldsymbol{E}$ and $\boldsymbol{H}$ vectors.

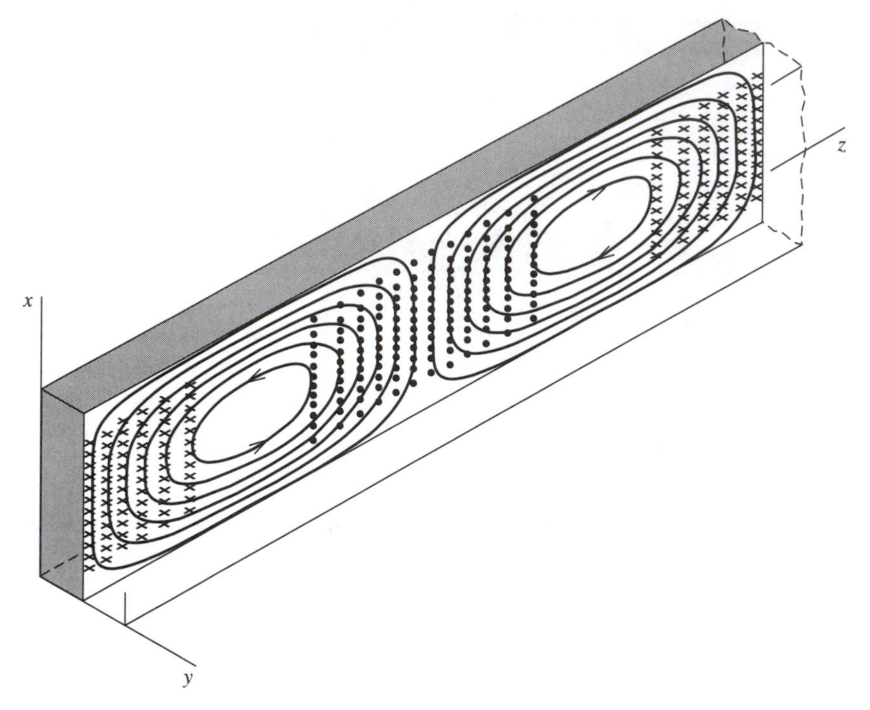

Figure 24-9 Lines of $\boldsymbol{E}$ (dots and crosses) and of $\boldsymbol{H}$ (ovals) in a hollow metallic rectangular waveguide. Reflection takes place on the top and bottom faces.

EXAMPLE The WR(284) Waveguide

This waveguide is designed for the wavelength range 115.3 to 75.9 millimeters, or for the frequency range 2.60 to 3.95 gigahertz. Its inside dimensions are 34.04 by 72.14 millimeters. So $a = 72.14$ millimeters. Say the guide is used at $\lambda_0 = 100$ millimeters. Then

$$\cos\theta = 100/144.28 = 0.6931, \qquad \theta = 0.8052 \text{ rad} = 46.13°, \tag{24-20}$$

and

$$v = c \sin\theta = 3 \times 10^8 \times 0.721 = 2.163 \times 10^8 \text{ meters/second.} \tag{24-21}$$

This particular guide has an attenuation of about 1 decibel over 30.5 meters (100 feet). Then 20 $\log_{10}$ [Amplitude at the output/Amplitude at the input] $= -1$, and the amplitude of the wave decreases by a factor of 0.89 over a distance of 30.5 meters.

Hollow metallic waveguides are not suitable for transmission over long distances.

24.5 OPTICAL WAVEGUIDES

Optical fibers were intensively developed only after it became possible to produce low-loss silica fibers with graded indices of refraction. Several other types of guide had been proposed. For example, a single wire inside a plastic sheath can act as a waveguide. Also, much work was done on the transmission of optical signals over large distances through a series of gas-filled alternately hot and cold cylindrical conductors. The light beam was alternately focused and defocused at the joints, providing a net focusing. That seems ridiculous today, but it was a valiant attempt at solving the problem of long-distance communication.

This is a good illustration of how technology stumbles along, making many false moves before finding an appropriate solution to a practical problem.

Science also stumbles along. See Appendix C, which describes some of Heaviside's work during the late 19th century. While he was forging ahead in electromagnetism and mathematics, his contemporaries were busy arguing about instantaneous action at a distance and about mechanical models of the aether.

Back to optical waveguides. The wave follows a zig-zag path inside a dielectric that is flanked by another dielectric that has a lower index of refraction, and guidance results from total reflection, which has the advantage of being lossless.

Plastic rods, either straight or curved, in air, have been used for many years to convey light from one point to another. *Clad rods*, and *light guides* consisting of a large number of fibers inside a flexible sheath, have the same function. Light guides can be used for concentrating the light from several sources. *Fiber bundles* are light guides with coordinated fibers and lenses at both ends, for transmitting images. *Fiberscopes* comprising a light guide for illumination, and a fiber bundle, serve for medical examinations. Some oscilloscopes have *fiber-optic, or microchannel face plates* consisting of a stack of parallel fibers, with both faces polished, and phosphor on the inner side. The fibers convey the light emitted by the phosphor more efficiently than would a glass plate. See the example in Sec. 23.3.2. The high efficiency of the fibers permits the observation of very fast transients.

24.5.1 Optical Fibers

The use of glass fibers for telecommunications was first proposed in 1966, despite the fact that signal attenuation was, at that time, prohibitively large. Four years later, low-loss fibers had been developed, and now some modern fibers have attenuations as low as 0.5 decibel/kilometer or less.† Optical fibers serve not only for telecommunications, but also in industrial settings and for Local Area Networks (LANs) linking computers. See Waynant and Ediger (2000).

The main advantage of optical fibers for communication is that they operate at frequencies of the order of 10^{14} hertz. Then a tiny frequency shift, in relative terms, corresponds to a huge bandwidth, which permits the simultaneous transmission of a huge amount of information. The transformation of an electrical into an optical signal is done either with light-emission diodes (LEDs) or with laser diodes. The inverse operation at the far end is done with photodiodes.

Optical fiber cables are also much lighter. Thorsen (1998) compares an old-style copper cable with an armored fiber-optic cable. The copper cable, with 900 twisted pairs,

† $20 \log_{10}$ [Amplitude at the output/Amplitude at the input] = –0.5 so that [Amplitude at the output/Amplitude at the input] = 0.944.

can carry 21,000 telephone channels† and weighs 8 tons per kilometer, while an armored fiber-optic cable, with 12 pairs of fibers, can carry 18 times as much, and weighs one hundredth as much.

There are also many other advantages: there is no "cross-talk" between channels, no noise, and the attenuation is so low that repeaters along a line can be spaced by hundreds of kilometers. Also, fibers are immune from electromagnetic interference and from spying.

The fiber itself, which carries the signal, is about the size of a human hair. It is embedded in a cladding having a lower index of refraction, which is itself embedded in a protective coating. For long-distance high-bit-rate service, the core and cladding are made of silica with a graded index of refraction, the index decreasing with increasing radius. The reflection is then gradual as in a mirage.‡ Dispersion $\Delta n/\Delta\lambda$ is minimum at $\lambda = 1320$ nanometers. Low-cost fiber is about 10 times as large in diameter and made of plastic.

24.5.2 Planar Optical Waveguides

Rather than go through the relatively abstruse mathematics of optical fibers, we study the simplest form of optical guide, namely the *planar* optical waveguide, or *slab waveguide,* composed of three layers of dielectric: a *substrate,* a *sheet,* and a *cover.* The Physics of slab waveguides is basically the same as that of optical fibers: the indices of refraction of the substrate and of the cover are slightly lower than that of the sheet, and total reflection occurs at the interfaces. The shift in the index of refraction can be either sudden or gradual.

Slab waveguides serve in *integrated optoelectronic circuits.* Those are analogous to ordinary integrated circuits, except that they operate with optical waves. Slab waveguides have overall thicknesses of the order of 10 micrometers. See Palais (1992) and Waynant and Ediger (2000).

Figure 24-10 shows a portion of a slab waveguide. The wave zig-zags down the sheet in medium 1 between the interfaces 1-2 and 1-3, where there is total reflection. Media 2 and 3 need only be several wavelengths thick. We disregard any wave reflected backward at the far end. We assume that the guide is infinitely wide in the y-direction. We also assume that media 2 and 3 are identical, and that the $\boldsymbol{E}$ vector is normal to the plane of incidence, which is the x-z plane, here.

Propagation can occur only for specific angles of incidence θ as in Fig. 24-8. Waves launched at other angles are strongly attenuated, as for hollow metallic waveguides. The phase speed is $(c/n_2)\sin\theta$.

†Mind you, that is an incredible feat in itself! That is an average of 23 *simultaneous* conversations per pair of wires, each conversation in its own specific frequency band, with negligible cross-talk! This is a good illustration of the nature of technological progress: no matter how marked by genius a device or a process is, it is bound to be replaced by a better one. Moreover, the rate of technological development has been *accelerating* ever since pre-historic times.

‡We have all seen mirages on an asphalt road on a hot summer day. If there is little wind, the air close to the asphalt is relatively hot, and its index of refraction is lower than that of the air higher up. Then a light ray coming from the sky in front gradually bends upward in the hot air, and the road acts as a mirror.

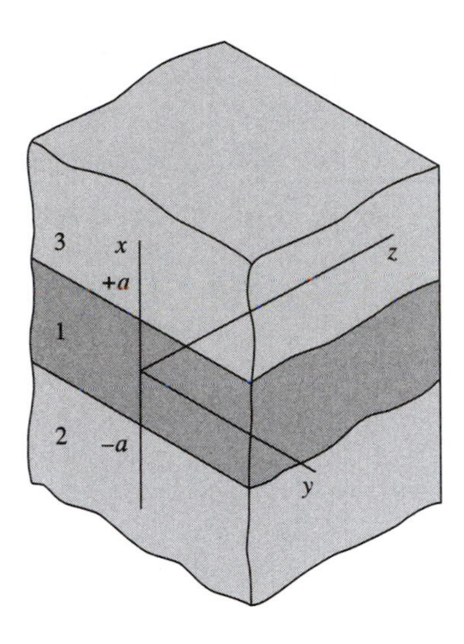

Figure 24-10 Portion of a slab optical waveguide. The three media are infinite in the y- and z-directions. Total reflection occurs at the 1-2 and 1-3 interfaces and the wave zig-zags, in medium 1, in the positive direction of the z axis.

We set

$$E_A = E_m \exp j\omega t, \qquad E_C = E_A \exp j\,\frac{2\pi l}{\lambda_1}, \tag{24-22}$$

with $l = 2a\cos\theta$, but now

$$E_B = E_A \exp j\Phi, \tag{24-23}$$

where Φ is the phase shift that we found in Sec. 23.3.2. Calling θ the angle of incidence,

$$\Phi = 2\arctan\frac{[(n_1/n_2)^2\sin^2\theta - 1]^{1/2}}{(n_1/n_2)\cos\theta} \tag{24-24}$$

and

$$E_C = E_B \exp j\Phi = E_A \exp 2j\Phi. \tag{24-25}$$

Equating the two values of E_C given by Eqs. 24-22 and 24-25, and substituting the above value of l,

$$4\pi\frac{a}{\lambda_1}\cos\theta = 2\Phi + m(2\pi), \tag{24-26}$$

where m is an integer. Of course, $\exp[jm(2\pi)] \equiv 1$. This equation defines the allowed values of the angle of incidence θ.

Now this is an ugly equation! Maybe you can solve it with your math software, but here is a more instructive way. Say $m = 0$. You plot both sides of the equation as functions of θ as in Fig. 24-11. The allowed value of θ is at the intersection of the two

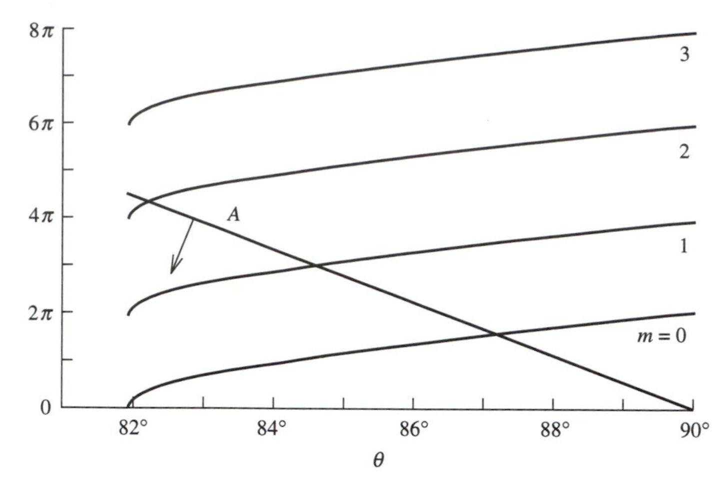

Figure 24-11 Graphical solution of Eq. 24-26 for a symmetric guide with $a = 0.5$ micrometer, $n_1 = 2.0000$, $n_2 = 1.9800$, $n_3 = 1.9800$, $\lambda_0 = 500$ nanometers. Curve A is a plot of the left-hand side of the equation. The other curves are plots of the right-hand side for various values of the mode order m.

curves! Now there is a whole set of curves for the right-hand side, depending on the value of m. So there are several allowed values of θ, depending on the *mode order m*. All the Φ curves start on the left at the critical angle.

Curve A is a cosine curve close to $\pi/2$. *Decreasing* the ratio a/λ_1 sweeps A down in the direction of the arrow, its right-hand end remaining fixed because $\cos(\pi/2) \equiv 0$. This eliminates the higher-order modes $m = 3, 2, 1$, one by one. But the mode $m = 0$ survives, even at long wavelengths. For $a \ll \lambda_1$, θ is approximately equal to the critical angle.

Thus a very thin slab is monomode: it supports only the $m = 0$ mode, and the angle of incidence is only slightly larger than the critical angle.

A thick slab allows a large number of modes, and the angle of incidence can be a few degrees larger than the critical angle. With increasing mode order, the angle of incidence θ decreases and the zig-zagging becomes more pronounced.

For a given half-thickness a, for a given wavelength, and for given dielectrics, a wave can thus propagate down the guide only for specific angles of incidence θ.

Figure 24-12 shows the $\boldsymbol{E}$ and $\boldsymbol{H}$ fields.

EXAMPLE A Monomode Slab Waveguide

For the slab waveguide of Fig. 24-10, from Sec. 23.3.2, with $n_1 = 2.0000$ and $n_2 = 1.9800$,

$$\sin\theta_{Ic} = 1.9800/2.0000, \qquad \theta_{Ic} = 81.89°, \qquad \cos\theta_{Ic} = 0.1411. \tag{24-27}$$

From Fig. 24-11, propagation is monomode if, at θ_{Ic},

$$4\pi(a/\lambda_1)\cos\theta_{Ic} < 2\pi, \qquad 0.2822a < \lambda_1, \qquad 0.5644a < \lambda_0. \tag{24-28}$$

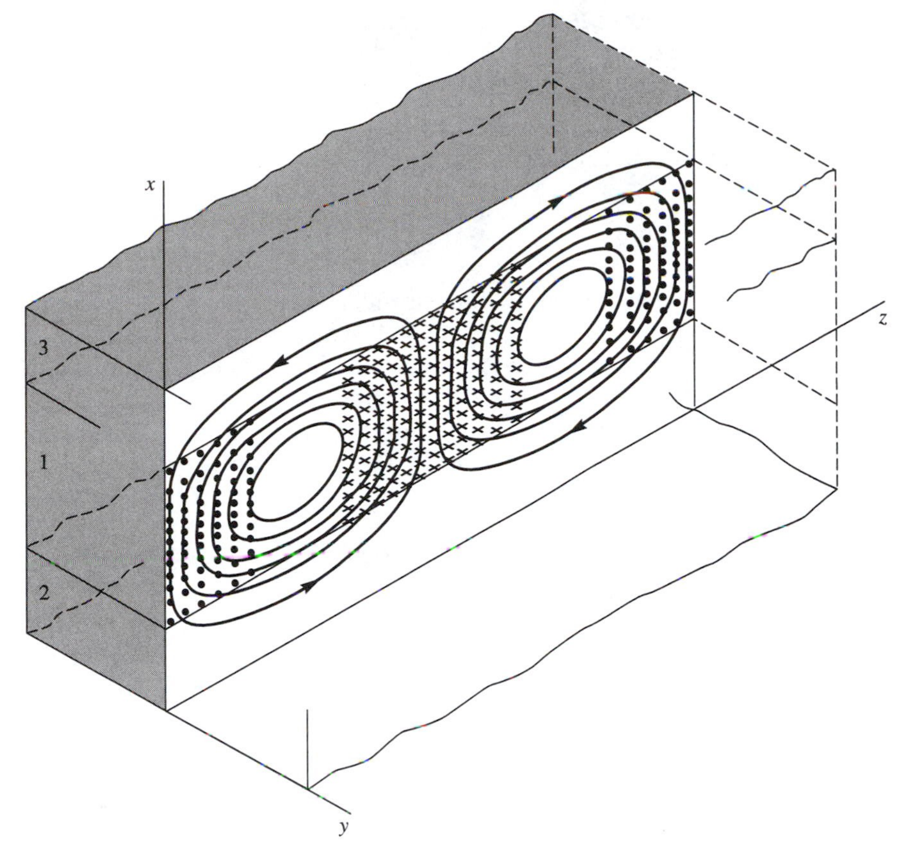

Figure 24-12 Lines of $\boldsymbol{E}$ (dots and crosses) and of $\boldsymbol{H}$ (ovals) in a symmetrical optical waveguide. The wave travels from left to right. Compare with Fig. 24-9.

Suppose you wish to operate at a free-space wavelength λ_0 of 1.3 micrometer. Then you must have that

$$0.5644a < 1.3 \times 10^{-6}, \qquad a < 2.3 \text{ micrometers.} \tag{24-29}$$

In fact, the angle of incidence will be larger than the critical angle and $\cos\theta$ will be less than 0.1411.

24.6 SUMMARY

The most common waveguide is the *parallel-wire line*. Figure 24-1 shows its $\boldsymbol{E}$ and $\boldsymbol{H}$ fields, which are both transverse. The usable bandwidth extends from DC up to frequencies of the order of gigahertz, as long as the wavelength is much larger than the wire spacing.

In the *coaxial line* of Fig. 24-2, the field is again transverse, and the bandwidth is again very large. The wave travels at the speed

$$v = 1/(\epsilon\mu_0)^{1/2} \tag{24-6}$$

as for the parallel-wire line. The transmitted power is

$$P_T = \left(\frac{\epsilon}{\mu_0}\right)^{1/2} \pi E_m^2 \ln \frac{\rho_2}{\rho_1}. \tag{24-9}$$

The advantage of the coaxial line is that its field is strictly confined.

The *microstrip line* of Fig. 24-4 also has a transverse field and a very large bandwidth, and the wavelength is the same as above. Its field is not perfectly confined, but its advantage is that it can be part of an integrated circuit.

The *hollow metallic waveguide* of Fig. 24-6 transmits an electromagnetic wave by internal *metallic* reflection as in Fig. 24-7. Its field is much more complex, as in Fig. 24-9, and only specific wavelengths are allowed. It serves mostly to feed radar antennas, at high power.

Optical waveguides operate by internal *total* reflection and, again, only certain wavelengths are allowed.

The *planar optical waveguide,* as in Fig. 24-10, is basically similar to the optical fiber. Its field, shown in Fig. 24-12, is qualitatively similar to that of the hollow metallic guide.

PROBLEMS

24-1. (*24.2*) The field inside a coaxial line

(a) Sketch a rather large cross-sectional view of a coaxial line in a plane containing the axis. Show lines of $\boldsymbol{E}$ and of $\boldsymbol{H}$ at a given instant over at least one wavelength. The lines should be most closely spaced where the field is strongest. Indicate the directions of the fields by means of arrow heads. The direction of propagation should point to the right.

(b) Add arrows at various points to represent Poynting vectors, using longer arrows where the power flow is larger. Assume that the length of the arrow represents the magnitude of the Poynting vector at its midpoint.

(c) Sketch a cross-sectional view of the coaxial line in a plane perpendicular to the axis, and show lines of $\boldsymbol{E}$ and of $\boldsymbol{H}$ at a particular instant. Relate this plane to the figure you drew in (a).

(d) Add plus and minus signs to both figures to show the surface charges. The spacing between the signs should indicate qualitatively the relative magnitude of the surface charge density.

(e) Now add arrows of various lengths to your first figure to represent surface current densities.

(f) How do the current patterns change with time?

24-2. (*24.2*) The characteristic impedance of a coaxial line

It is known from transmission-line theory that the characteristic impedance of a line is given by $Z_c = (L'/C')^{1/2}$, where L' and C' are, respectively, the inductance and capacitance per meter. Show that this applies to the coaxial line.

24-3. (*24.2*) Eliminating reflection at the end of a coaxial line

An air-insulated coaxial line is terminated by a sheet whose surface resistance is 377 ohms per square (Prob. 4-11).

Show that the resistance of the termination is equal to the characteristic impedance of the line. There is then no reflection at the end of the line.

24-4. (*24.3*) Microstrip line

Figure 24-4 shows a cross-section of a microstrip line.

(a) In practice, the width b of the strip is much larger than its distance h to the ground plane, and edge effects are small. Show that the instantaneous value $I\mathcal{V}$ of the transmitted power is equal to the Poynting vector integrated over the cross-section bh. Assume that there is no reflected wave.

(b) Show that the characteristic impedance $\mathcal{V}/I$ is equal to $(\mu_0/\epsilon)(h/b)$.

(c) Show that the definition of Prob. 24-2 leads to the same result.

(d) Show that the addition of a second ground plane placed symmetrically with the first one reduces the characteristic impedance by a factor of 2.

24-5. (*24.4*) Wavelength in a hollow metallic rectangular waveguide

A wave has a free-space wavelength of 100.0 millimeters. What is its wavelength when traveling inside a WR(284) waveguide?

Justify your answer qualitatively.

24-6. (*24.5.2*) Negative modes in the planar optical waveguide

In Fig. 24-11, curves for $m = -2, -3, -4$, etc. would intersect curve A at angles of incidence larger than $\pi/2$, which is absurd. So these modes are forbidden. However, for mode $m = -1$ the curve would intersect at $\theta = \pi/2$, which is sensible.

Show that the mode $m = -1$ is also forbidden.

24-7. (*24.5.2*) Maximum value of λ_0 as a function of the mode order m.

Find the maximum value of λ_0 as a function of m for a symmetric ($n_2 = n_3$) planar optical waveguide.

24-8. (*24.5.2*) The phase speed v_p in an asymmetric waveguide

Let medium 2 be denser than medium 3 in Fig. 24-10.

Show that, if θ is only slightly larger than the critical angle at the interface 1-2, then the phase speed of the guided wave is approximately equal to that of a uniform plane wave traveling in medium 2.

CHAPTER

25

RADIATION

Electric Dipole Radiation. Antennas

This last chapter concerns the way in which oscillating charges and currents radiate electromagnetic waves.

Say you have, somewhere, stationary and constant electric charges. Then Chapters 3 to 10 apply. For example, Coulomb's law, of Sec. 3.2, defines correctly the force between the charges Q_1 and Q_2. But wait! That can't be right! Suppose that you move Q_1, and that Q_2 is a thousand light-years away. How does Q_2 know about the new position of Q_1? According to Relativity, it takes at least a thousand years for the information to go from Q_1 to Q_2. So Coulomb's law is an approximation. It is fine if the delay is negligible. But it is certainly wrong if the delay is significant.

There is the same problem with magnetic fields. Say you have a coil that carries a current I. You have a $\boldsymbol{B}$-meter a thousand light-years away. You double the current. Surely it takes a long time before the $\boldsymbol{B}$-meter sees a change. So the Physics of magnetic fields that you struggled with in Chapters 14 to 20 is also wrong!

Of course there is the same problem with the law of gravitation. That law says that the gravitational force between two objects is proportional to the product of their masses, and inversely proportional to the square of the distance between them. Period. That law is also wrong! But it is OK if the distance is not too large.

The case of astronomy is striking. We do not, and we cannot, see the stars as they are now, but only as they were millions or billions of years ago. *They are not even where we see them!*

We have here the problem of *instantaneous action at a distance* that physicists struggled with during the nineteenth century. Chapters 3 to 10 and 14 to 20, which assume instantaneous action at a distance, are clearly wrong whenever you have to do with very large distances in Astrophysics or Cosmology. But they are OK most of the time.

Most of the time? Well, instantaneous action at a distance is also wrong in the case of antennas, where the charges and currents change so rapidly that delays can be important, even over short distances.

The radiation fields of antennas. That is the subject of this last chapter.

25.1 ELECTRIC DIPOLE RADIATION

We calculate the far-field of a small oscillating dipole of length s, at a distance

$$r \gg ƛ \gg s. \tag{25-1}$$

This will permit us to calculate the radiation field of the most common types of antenna in Sec. 25.3 below. We calculated the field of a *static* dipole in Sec. 5.1

The above approximation will simplify the calculations, and it will give us the occasion to learn a few tricks. But it has one major disadvantage: the condition $r \gg ƛ$ does not apply to a steady-state dipole because, then, $ƛ$ is infinite! So the calculations that follow will lead us to values of V and of $\boldsymbol{E}$ that disagree with the ones that we found in Sec. 5.1. That is not too serious because we shall give the more general values of V and of $\boldsymbol{E}$, and those values will agree with those of Sec. 5.1.

A steady-state electric dipole has of course no magnetic field because there is no current.

So let us think about the field of an electric dipole at the origin, as in Figs. 25-1 and 25-2, with

$$Q = Q_m \exp j\omega t. \tag{25-2}$$

The charges $-Q$ and Q are separated by a distance $\boldsymbol{s}$, in a vacuum. So the dipole moment is

$$\boldsymbol{p} = (Q_m \exp j\omega t)\boldsymbol{s} = \boldsymbol{p}_m \exp j\omega t, \quad \boldsymbol{p}_m = Q_m \boldsymbol{s}, \tag{25-3}$$

and an alternating current

$$I = \frac{dQ}{dt} = j\omega Q_m \exp j\omega t = I_m \exp j\omega t, \quad I_m = j\omega Q_m \tag{25-4}$$

flows between the charges.

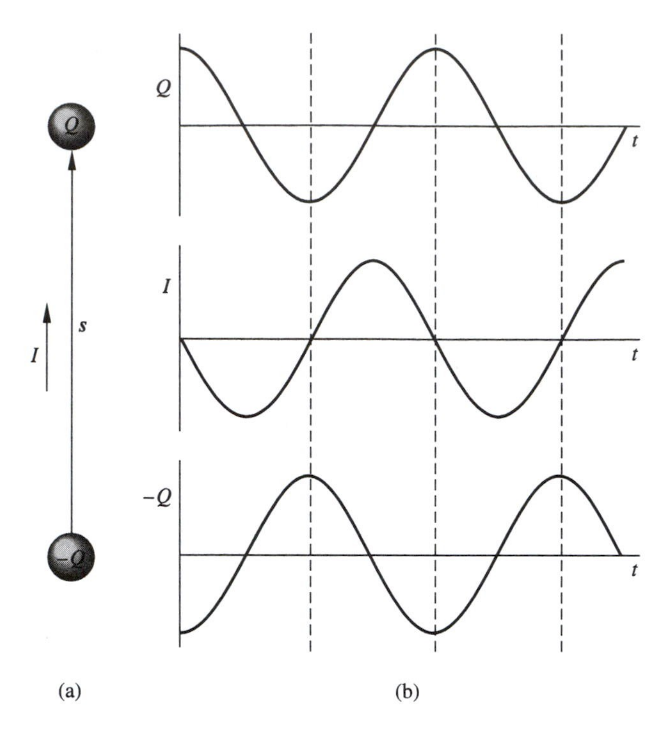

Figure 25-1 An oscillating electric dipole. (a) The vector $\boldsymbol{s}$ points in the direction shown. (b) The charges and the current as functions of the time.

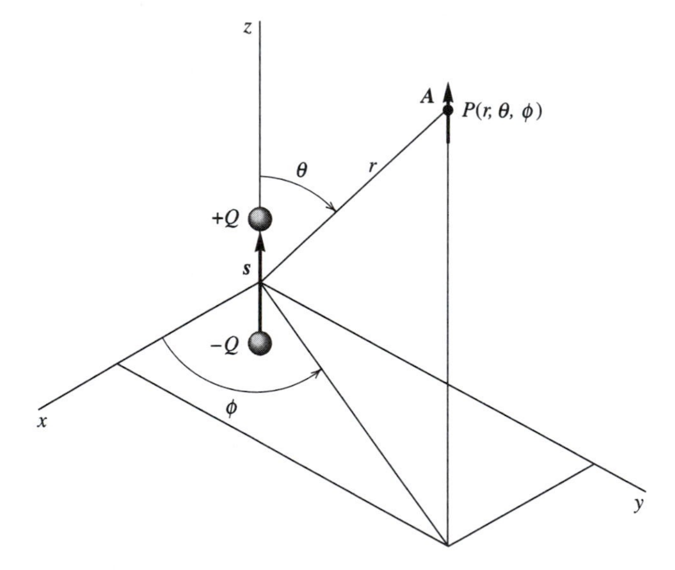

Figure 25-2 Electric dipole of moment $\boldsymbol{p} = Q\boldsymbol{s}$ and the vector potential $\boldsymbol{A}$ at a point P in its field.

We wish to find the *radiation field*, that is $\boldsymbol{E}$ and $\boldsymbol{H}$ at the point (r, θ, ϕ) in Fig. 25-2. We could do that directly, but that would be a bit difficult. See Appendix D. There are cases where brute-force calculations are necessary, but it is usually more effective, more instructive, and more interesting, to be clever. Now we have the Lorentz condition of Sec. 21.7 up our sleeves. Recall that it relates $\boldsymbol{A}$ to V. So, instead of calculating $\boldsymbol{E}$ and $\boldsymbol{H}$ directly, we first calculate $\boldsymbol{A}$, deduce V from the Lorentz condition, and then $\boldsymbol{E}$ and $\boldsymbol{H}$ will follow:

$$\boldsymbol{E} = -\nabla V - \frac{\partial \boldsymbol{A}}{\partial t}, \qquad \boldsymbol{H} = \frac{1}{\mu_0} \nabla \times \boldsymbol{A}. \tag{25-5}$$

Still, calculating V is interesting and simple. So we do it in Sec. 25.1.4 below. There is no harm in skipping that section.

25.1.1 The Retarded Vector Potential $\boldsymbol{A}$

From Sec. 14.4, $\boldsymbol{A}$ is parallel to the z-axis, like the current I, and like $\boldsymbol{s}$:

$$\boldsymbol{A} = \frac{\mu_0 I}{4\pi r}\boldsymbol{s} = \frac{\mu_0 (I_m \exp j\omega t)}{4\pi r}\boldsymbol{s}. \tag{25-6}$$

Right? No, because we have assumed instantaneous action at a distance! The true $\boldsymbol{A}$ corresponds rather to the value of I at the *previous time* $t - (r/c)$, where r/c is the time it takes for the information, or the wave, to travel the distance r. (We assume that the dipole is in free space.) So the vector potential is rather

$$\boldsymbol{A} = \frac{\mu_0 I_m \exp j\omega(t - r/c)}{4\pi r}\boldsymbol{s} \tag{25-7}$$

$$= \frac{\mu_0 I_m \exp j(\omega t - r/\lambda\!\!\!^{-})}{4\pi r}\boldsymbol{s}. \tag{25-8}$$

Recall that

$$\lambda\!\!\!^{-} = \frac{\lambda}{2\pi} = \frac{c/f}{2\pi} = \frac{c}{\omega}. \tag{25-9}$$

This $\boldsymbol{A}$ is the *retarded vector potential* of the oscillating electric dipole. If Q is constant, as in Sec. 5.1, then $I = 0$ and $A = 0$.

From Fig. 25-2,

$$\boldsymbol{s} = s\cos\theta\,\hat{\boldsymbol{r}} - s\sin\theta\,\hat{\boldsymbol{\theta}}. \tag{25-10}$$

Then

$$\boldsymbol{A} = \frac{\mu_0 I_m s \exp j(\omega t - r/\lambda\!\!\!^{-})}{4\pi r}(\cos\theta\,\hat{\boldsymbol{r}} - \sin\theta\,\hat{\boldsymbol{\theta}}) \tag{25-11}$$

$$= \frac{\mu_0 j\omega p_m}{4\pi r}\exp j(\omega t - r/\lambda\!\!\!^{-})(\cos\theta\,\hat{\boldsymbol{r}} - \sin\theta\,\hat{\boldsymbol{\theta}}) \tag{25-12}$$

$$= \frac{\mu_0 j\omega [p]}{4\pi r}(\cos\theta\,\hat{\boldsymbol{r}} - \sin\theta\,\hat{\boldsymbol{\theta}}), \tag{25-13}$$

where $[p]$ *is evaluated at the previous time* $t - r/c$:

$$[p] = p_m \exp j(\omega t - r/\lambda\!\!\!\bar{}). \qquad (25\text{-}14)$$

That is a standard convention, and we shall follow it from now on.

So

$$A_r = \frac{\mu_0 j\omega[p]}{4\pi r}\cos\theta, \quad A_\theta = -\frac{\mu_0 j\omega[p]}{4\pi r}\sin\theta. \qquad (25\text{-}15)$$

25.1.2 The Divergence of A

To use the Lorentz condition of Sec. 21.7 for finding V, we must now calculate $\nabla \cdot \boldsymbol{A}$ in spherical coordinates (Eq. 12 on the back of the front cover), with $\partial/\partial\phi = 0$:

$$\nabla \cdot \boldsymbol{A} = \frac{1}{r^2}\frac{\partial}{\partial r}(r^2 A_r) + \frac{1}{r\sin\theta}\frac{\partial}{\partial\theta}(A_\theta \sin\theta). \qquad (25\text{-}16)$$

A rigorous calculation would be complicated, but we can be more clever than that. What is a divergence? A divergence is a measure of the rates of change of a function with respect to the coordinates. Now $\boldsymbol{A}$ is a function of both r and θ. Let us look at the first term on the right; the second term will be easy to handle. The first term involves the first r-derivative of $r^2 A_r$, and A_r is itself proportional to $1/r$, from Eq. 25-15. So $r^2 A_r$ is proportional to r, and thus increases linearly with r. But A_r is also proportional to $[p]$, and thus to the exponential function, which is a cosine function in disguise, and the cosine function changes rapidly with distance. So it won't do much harm if we remove one of the r's from the derivative and rewrite the first term as

$$\frac{1}{r}\frac{\partial}{\partial r}(rA_r).$$

See the Example below.

Then

$$\nabla \cdot \boldsymbol{A} = \frac{1}{r}\frac{\partial}{\partial r}(rA_r) + \frac{1}{r\sin\theta}\frac{\partial}{\partial\theta}(A_\theta \sin\theta) \qquad (25\text{-}17)$$

$$= \frac{\mu_0 j\omega}{4\pi r}\left\{\left(\frac{\partial}{\partial r}\{[p]\cos\theta\}\right) - \frac{1}{r\sin\theta}\frac{\partial}{\partial\theta}\{[p]\sin^2\theta\}\right\}, \qquad (25\text{-}18)$$

where

$$\frac{\partial}{\partial r}\{[p]\cos\theta\} = \frac{\partial}{\partial r}\{p_m \exp j(\omega t - r/\lambda\!\!\!\bar{})\}\cos\theta = -\frac{j}{\lambda\!\!\!\bar{}}[p]\cos\theta. \qquad (25\text{-}19)$$

Then

$$\nabla \cdot \boldsymbol{A} = \frac{\mu_0 j\omega[p]}{4\pi r}\left(-\frac{j}{\lambda\!\!\!\bar{}}\cos\theta - \frac{2\cos\theta}{r}\right) \qquad (25\text{-}20)$$

$$= \frac{\mu_0 j\omega[p]}{4\pi r}\left(-\frac{j}{\lambda\!\!\!^{-}} - \frac{2}{r}\right)\cos\theta. \tag{25-21}$$

With $r \gg \lambda\!\!\!^{-}$, the last parenthesis simplifies to $-j/\lambda\!\!\!^{-}$ and

$$\nabla \cdot \boldsymbol{A} = -j\frac{\mu_0 j\omega[p]}{4\pi \lambda\!\!\!^{-} r}\cos\theta = \frac{\mu_0 \omega[p]}{4\pi\lambda\!\!\!^{-} r}\cos\theta. \tag{25-22}$$

EXAMPLE The Value of A as a Function of r

The values of E, V, and A vary in the same way with r. Suppose that you measure E one kilometer away from an antenna that radiates a 30-megahertz wave. The wavelength is 10 meters. If you move away one meter, then the *amplitude* of E decreases only by 0.1%. But the *phase* changes by $2\pi/10 = 36°$, which has a large effect on the cosine function and on A.

25.1.3 The Value of V Deduced from the Lorentz Condition

Deducing V from $\nabla \cdot \boldsymbol{A}$ is now trivial. Since

$$\nabla \cdot \boldsymbol{A} + \epsilon_0\mu_0\frac{\partial V}{\partial t} = 0, \quad \nabla \cdot \boldsymbol{A} + j\omega\epsilon_0\mu_0 V = 0, \tag{25-23}$$

and

$$V = -\frac{1}{j\omega\epsilon_0\mu_0}\frac{\mu_0\omega[p]}{4\pi\lambda\!\!\!^{-} r}\cos\theta = \frac{j[p]}{4\pi\epsilon_0\lambda\!\!\!^{-} r}\cos\theta. \tag{25-24}$$

If Q is constant as in Sec. 5.1, then $\boldsymbol{A} = 0$, $dV/dt = 0$, and the Lorentz condition is useless. Also, this value of V does not apply when $\lambda\!\!\!^{-}$ is infinite.

*25.1.4 The Value of V Calculated Directly.†

As a check, we can also calculate V for the oscillating electric dipole directly. From Fig. 25-3,

$$V = \frac{Q_m}{4\pi\epsilon_0}\left\{\frac{\exp j(\omega t - r_b/\lambda\!\!\!^{-})}{r_b} - \frac{\exp j(\omega t - r_a/\lambda\!\!\!^{-})}{r_a}\right\}. \tag{25-25}$$

With $r \gg s$, we can set

$$r_a = r + (s/2)\cos\theta, \qquad r_b = r - (s/2)\cos\theta. \tag{25-26}$$

Then, setting $r_a \approx r_b \approx r$ *in the denominators* of Eq. 25-25,

†Sections marked with asterisk can be skipped without losing continuity.

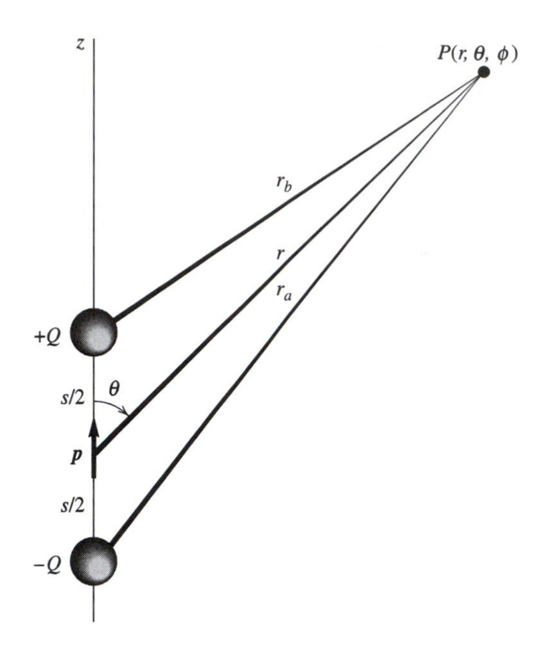

Figure 25-3 Two charges $+Q$ and $-Q$ separated by a distance s and forming a dipole. The dipole moment is $\boldsymbol{p}$. We calculate the potential at point P by summing the retarded potentials of the two charges.

$$V = \frac{Q_m}{4\pi\epsilon_0 r} \exp j(\omega t - r/\lambda\!\!\!^-)\left\{\exp\left(j\frac{s}{2\lambda\!\!\!^-}\cos\theta\right) - \exp\left(-j\frac{s}{2\lambda\!\!\!^-}\cos\theta\right)\right\}, \tag{25-27}$$

where $s \ll \lambda\!\!\!^-$.

Recalling now that $\exp x \approx 1 + x$ if $x \ll 1$, and that $Q_m s = p_m$,

$$V = \frac{Q_m}{4\pi\epsilon_0 r} \exp j\left(\omega t - \frac{r}{\lambda\!\!\!^-}\right)\left(j\frac{s}{\lambda\!\!\!^-}\cos\theta\right) = \frac{j[p]}{4\pi\epsilon_0 \lambda\!\!\!^- r}\cos\theta, \tag{25-28}$$

as above.

25.1.5 The $\boldsymbol{E}$ Field

We calculate

$$\boldsymbol{E} = -\boldsymbol{\nabla} V - \frac{\partial \boldsymbol{A}}{\partial t} \tag{25-29}$$

in the far field of a small electric dipole, assuming again that

$$r \gg \lambda\!\!\!^- \gg s. \tag{25-30}$$

In spherical coordinates, with $\partial/\partial\phi = 0$,

$$\boldsymbol{\nabla} V = \frac{\partial V}{\partial r}\hat{\boldsymbol{r}} + \frac{1}{r}\frac{\partial V}{\partial \theta}\hat{\boldsymbol{\theta}}. \tag{25-31}$$

Within the above approximation,

$$V = \frac{j[p]}{4\pi\epsilon_0 \bar{\lambda} r}\cos\theta = \frac{jp_m}{4\pi\epsilon_0 \bar{\lambda} r}\exp j\left(\omega t - \frac{r}{\bar{\lambda}}\right)\cos\theta, \tag{25-32}$$

$$\frac{\partial V}{\partial r} = \frac{jp_m}{4\pi\epsilon_0 \bar{\lambda}}\frac{\partial}{\partial r}\left\{\frac{\exp j\left(\omega t - \dfrac{r}{\bar{\lambda}}\right)}{r}\cos\theta\right\}. \tag{25-33}$$

Now a gradient is similar to a divergence in that it is a measure of the rates of change of the function with respect to the coordinates. Much as with $\boldsymbol{A}$, we move the $1/r$ term outside the braces and set

$$\frac{\partial V}{\partial r} = \frac{jp_m}{4\pi\epsilon_0 \bar{\lambda} r}\frac{\partial}{\partial r}\left\{\exp j\left(\omega t - \frac{r}{\bar{\lambda}}\right)\cos\theta\right\} \tag{25-34}$$

$$= \frac{j[p]}{4\pi\epsilon_0 \bar{\lambda} r}\left(-\frac{j}{\bar{\lambda}}\right)\cos\theta \tag{25-35}$$

$$= \frac{[p]}{4\pi\epsilon_0 \bar{\lambda}^2 r}\cos\theta. \tag{25-36}$$

Now

$$\frac{1}{r}\frac{\partial V}{\partial \theta} = \frac{1}{r}\frac{j[p]}{4\pi\epsilon_0 \bar{\lambda} r}(-\sin\theta). \tag{25-37}$$

Then

$$\boldsymbol{\nabla} V = \frac{[p]}{4\pi\epsilon_0 \bar{\lambda}^2 r}\left(\cos\theta\,\hat{\boldsymbol{r}} - \frac{j\bar{\lambda}}{r}\sin\theta\,\hat{\boldsymbol{\theta}}\right) \tag{25-38}$$

$$= \frac{[p]}{4\pi\epsilon_0 \bar{\lambda}^2 r}\cos\theta\,\hat{\boldsymbol{r}}. \tag{25-39}$$

We have again set $\bar{\lambda} \ll r$.

Also,

$$\frac{\partial \boldsymbol{A}}{\partial t} = j\omega\boldsymbol{A} = j\omega\frac{\mu_0 j\omega[p]}{4\pi r}(\cos\theta\,\hat{\boldsymbol{r}} - \sin\theta\,\hat{\boldsymbol{\theta}}) \tag{25-40}$$

$$= -\frac{\mu_0\omega^2[p]}{4\pi r}(\cos\theta\,\hat{\boldsymbol{r}} - \sin\theta\,\hat{\boldsymbol{\theta}}). \tag{25-41}$$

But since $\epsilon_0\mu_0 = 1/c^2$, from Sec. 22.3,

$$\mu_0 = \frac{1}{\epsilon_0 c^2} = \frac{1}{\epsilon_0\omega^2\bar{\lambda}^2}, \tag{25-42}$$

and

$$\frac{\partial \boldsymbol{A}}{\partial t} = -\frac{[p]}{4\pi\epsilon_0 \lambda\!\!\!^{-2} r}(\cos\theta\,\hat{\boldsymbol{r}} - \sin\theta\hat{\boldsymbol{\theta}}). \tag{25-43}$$

We now have that

$$\boldsymbol{E} = -\nabla V - \frac{\partial \boldsymbol{A}}{\partial t} \tag{25-44}$$

$$= -\frac{[p]}{4\pi\epsilon_0 \lambda\!\!\!^{-2} r}\cos\theta\,\hat{\boldsymbol{r}} + \frac{[p]}{4\pi\epsilon_0 \lambda\!\!\!^{-2} r}(\cos\theta\,\hat{\boldsymbol{r}} - \sin\theta\,\hat{\boldsymbol{\theta}}). \tag{25-45}$$

(The r-components of ∇V and of $\partial \boldsymbol{A}/\partial t$ cancel, but only in the far field where $r \gg \lambda\!\!\!^{-}$.) Finally,

$$\boldsymbol{E} = -\frac{[p]}{4\pi\epsilon_0 \lambda\!\!\!^{-2} r}\sin\theta\,\hat{\boldsymbol{\theta}}. \tag{25-46}$$

This equation assumes Eq. 25-30. Thus $\boldsymbol{E}$ has only a θ component, as expected. It is inversely proportional to r, but it also depends on both r and t through $[p]$.

25.1.6 The *H* Field

The $\boldsymbol{H}$ field is easier to calculate. In spherical coordinates, from Eq. 13 on the back of the front cover, with $A_\phi = 0$ and $\partial/\partial_\phi = 0$,

$$\boldsymbol{H} = \frac{1}{\mu_0}\nabla \times \boldsymbol{A} = \frac{1}{\mu_0}\frac{1}{r}\left\{\frac{\partial(rA_\theta)}{\partial r} - \frac{\partial A_r}{\partial \theta}\right\}\hat{\boldsymbol{\phi}}. \tag{25-47}$$

From Sec. 25.1.1,

$$\boldsymbol{A} = \frac{\mu_0 j\omega[p]}{4\pi r}(\cos\theta\,\hat{\boldsymbol{r}} - \sin\theta\,\hat{\boldsymbol{\theta}}) \tag{25-48}$$

$$= \frac{\mu_0 j\omega p_m}{4\pi r}\exp j\left(\omega t - \frac{r}{\lambda\!\!\!^{-}}\right)(\cos\theta\,\hat{\boldsymbol{r}} - \sin\theta\,\hat{\boldsymbol{\theta}}). \tag{25-49}$$

Now

$$\frac{\partial(rA_\theta)}{\partial r} = -\left(-\frac{j}{\lambda\!\!\!^{-}}\right)\frac{\mu_0 j\omega[p]}{4\pi}\sin\theta = -\frac{\mu_0\,\omega[p]}{4\pi\lambda\!\!\!^{-}}\sin\theta, \tag{25-50}$$

$$\frac{\partial A_r}{\partial \theta} = -\frac{\mu_0 j\omega[p]}{4\pi r}\sin\theta. \tag{25-51}$$

Since, by hypothesis, $\lambda\!\!\!^{-} \ll r$, $\partial A_r/\partial\theta$ is negligible and

$$\nabla \times \boldsymbol{A} = -\frac{1}{r}\frac{\mu_0 \omega [p]}{4\pi ƛ}\sin\theta\hat{\boldsymbol{\phi}}, \tag{25-52}$$

so that

$$\boldsymbol{H} = -\frac{\omega[p]}{4\pi ƛ r}\sin\theta\hat{\boldsymbol{\phi}} = -\frac{c[p]}{4\pi ƛ^2 r}\sin\theta\hat{\boldsymbol{\phi}}. \tag{25-53}$$

The $\boldsymbol{E}$ and $\boldsymbol{H}$ vectors are orthogonal as in a plane wave and, from Eqs. 25-46 and 25-53,

$$\frac{E}{H} = \frac{1}{\epsilon_0 c} = \left(\frac{\mu_0}{\epsilon_0}\right)^{1/2} = c\mu_0 \approx 377 \quad \text{ohms}. \tag{25-54}$$

What if $s \ll r$ and $s \ll ƛ$, but if r is not necessarily much larger than $ƛ$? In other words, what if we just make the length of the dipole s very small, even with respect to the radian length $ƛ$? Then the expressions for $\boldsymbol{E}$ and $\boldsymbol{H}$ are more complicated (Lorrain, Corson, and Lorrain, 1988). In that case, setting $R = ƛ/r$, in the near field,

$$\boldsymbol{E} = \frac{[p]}{4\pi\epsilon_0 ƛ^2 r}\{2(R^2 + jR)\cos\theta\,\hat{\boldsymbol{r}} + (R^2 - 1 + jR)\sin\theta\,\hat{\boldsymbol{\theta}}\}, \tag{25-55}$$

$$\boldsymbol{H} = \frac{c[p]}{4\pi ƛ^2 r}(-1 + jR)\sin\theta\,\hat{\boldsymbol{\phi}}, \tag{25-56}$$

and an electric field line is obtained by setting

$$\sin^2\theta(R^2+1)^{1/2}\cos\left(\omega t - \frac{1}{R} + \arctan\frac{1}{R}\right)$$

equal to a constant.

Note that R is a *small* quantity. Even at $r = \lambda$, R is only equal to $1/6$.

In the case of a static dipole, $ƛ$ is infinite. Then, in the equation for $\boldsymbol{E}$, only the R^2 terms matter and we are left with the $\boldsymbol{E}$ of Sec. 5.1. Also, for a static dipole, there is no $\boldsymbol{H}$ field.

Figure 25-4 shows lines of $\boldsymbol{E}$ in a plane $\phi = 0$, $\phi = \pi$ as a function of the time for electric dipole radiation, according to Eq. 25-55. The dipole length s is presumed to be much smaller than the radian length $ƛ$, but the ratio $R = ƛ/r$ can have any value.

Figure 25-5 shows the same electric field lines as level lines on the surface defined by the above function. As the time t increases in the above function, the central peaks flip up and down, and the ripples move out.

25.1.7 The Poynting Vector $\boldsymbol{E} \times \boldsymbol{H}$ and the Radiated Power P

From Secs. 25.1.5 and 25.1.6, the Poynting vector $\boldsymbol{E} \times \boldsymbol{H}$ for dipole radiation points in the radial direction because $\hat{\boldsymbol{\theta}} \times \hat{\boldsymbol{\phi}} = \hat{\boldsymbol{r}}$. Let us calculate the *time-averaged*

Poynting vector from the values of $\boldsymbol{E}$ and $\boldsymbol{H}$ that we found there, and which apply to the region $r \gg \lambda\!\!\!^{-}$:

$$\mathcal{S}_{av} = \frac{1}{2}\text{Re}(\boldsymbol{E} \times \boldsymbol{H}^*) = \frac{1}{2}\frac{\omega[p][p^*]}{16\pi^2\epsilon_0 \lambda\!\!\!^{-3} r^2} \sin^2\theta\,\hat{\boldsymbol{r}} \tag{25-57}$$

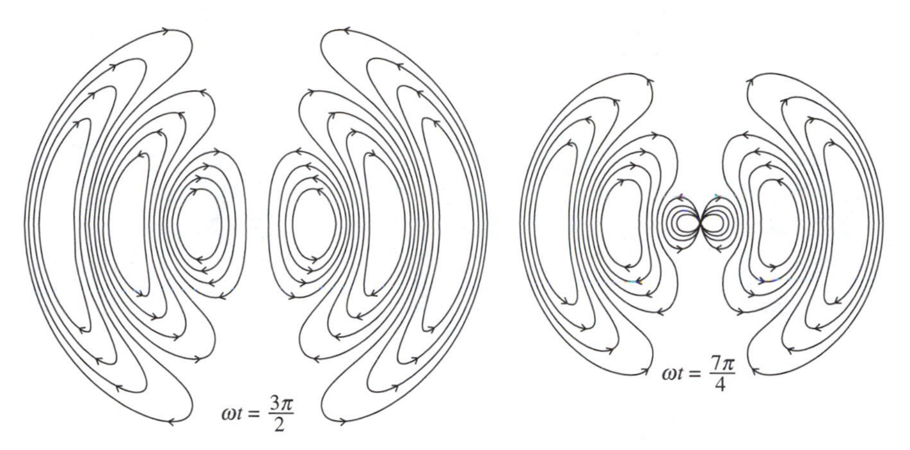

Figure 25-4 Lines of $\boldsymbol{E}$ for an oscillating electric dipole for $\omega t = 0, \pi/4, \pi/2, 3\pi/4, \pi, 5\pi/4, 3\pi/2, 7\pi/4$. The dipole is vertical at the center. Note how the wavelength decreases with distance. The lines of $\boldsymbol{H}$ are circles perpendicular to the paper and centered on the axis of the dipole.

$$= \frac{\mu_0 \pi^2 f^4 p_m^2}{2c} \frac{}{r^2} \sin^2\theta\, \hat{\boldsymbol{r}} \tag{25-58}$$

$$= \frac{\mu_0 \pi^2 f^4 p_{\text{rms}}^2}{c} \frac{}{r^2} \sin^2\theta\, \hat{\boldsymbol{r}}. \tag{25-59}$$

What if we used, instead, the more general expressions for $\boldsymbol{E}$ and $\boldsymbol{H}$ that we gave in Eqs. 25-55 and 25-56? Now this is a surprise: the result is the same!

Figure 25-6 shows the three vectors at (r, θ, ϕ), for $\omega[t] = 0$.

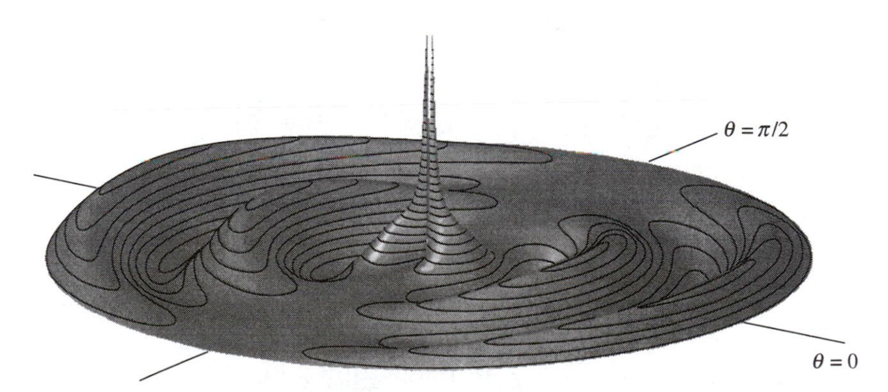

Figure 25-5 The function given at the end of Sec. 25.1.6, plotted as a function of r and θ at $t = 0$. The dipole is at the center, on the $\theta = 0$ axis. As t increases, the twin peaks oscillate in unison between $+\infty$ and $-\infty$, and the ripples move outward radially. The loops are both level lines on that surface, and lines of $\boldsymbol{E}$.

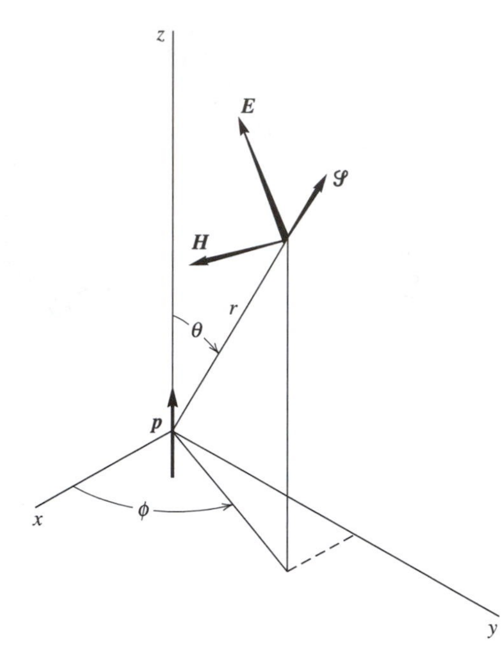

Figure 25-6 The E, H, and $\mathcal{S} = E \times H$ vectors for an oscillating dipole when $\omega[t] = 0$.

Note the following.

1. The power flow is everywhere radial, even close to the dipole, where r is not much larger than λ.
2. The Poynting vector is proportional to $1/r^2$, as one would expect, from the conservation of energy.
3. Because of the $\sin^2\theta$ term, *an electric dipole does not radiate in the direction of its axis.*
4. Because of that same term, the field is maximum in the plane perpendicular to the dipole axis.

Integrating the above Poynting vector over a sphere of radius r yields the radiated power:

$$P = \frac{\mu_0 \pi^2 f^4 p_{\text{rms}}^2}{cr^2} \int_0^{2\pi} \int_0^{\pi} \sin^2\theta (r^2 \sin\theta \, d\theta \, d\phi) \tag{25-60}$$

$$= \frac{8\mu_0 \pi^3}{3c} f^4 p_{\text{rms}}^2 = 3.466 \times 10^{-13} f^4 p_{\text{rms}}^2 \quad \text{watts.} \tag{25-61}$$

The radiated power is proportional to the *fourth* power of the frequency.

EXAMPLES The Colors of the Sky, of the Setting Sun, and of Tobacco Smoke

Dust particles suspended in the atmosphere scatter the light coming from the sun, because the electric field of the incident light wave excites electrons present in the particles. These electrons act as small electric dipoles and reradiate. If we disregard resonances,

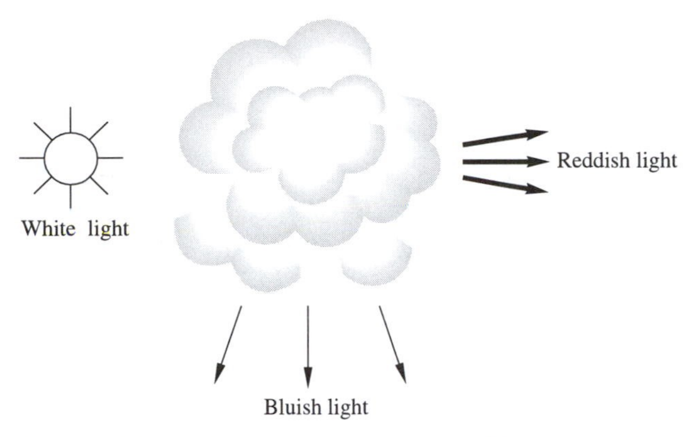

Figure 25-7 Fine particles, as in tobacco smoke, scatter blue light preferentially. The transmitted light is thus reddish.

the p_m of an oscillating electron is proportional to the amplitude of the incident wave. Then the reradiated power is proportional to f^4, and the light scattered by the sky is bluer than sunlight.

If the air were completely dust-free, the sky would still be blue, but darker: atoms and molecules of the air also absorb and reradiate energy, but mostly in the ultraviolet. The light from the sun that reaches the earth, particularly at sunset, is reddish because part of the blue has been diffused out.

It is for the same reason that tobacco smoke is either bluish or reddish, according to the way you look at it, with respect to a source of light. See Fig. 25-7.

25.2 ANTENNAS

Antennas are everywhere: on cordless and cellular telephones, on pagers, on garage-door openers, on TV sets, on vehicles and buildings, on ships and planes, and in the countryside. There exist very many different types. See, for example, Johnson (1993). Most antennas serve for both transmitting and receiving.

The simplest type of antenna is the dipole of Fig. 25-8a. As a *transmitting* antenna, it is fed by an oscillator through a parallel-wire line, and each element of length radiates as a small dipole. When it is used as a *receiving* antenna, the ambient electric field induces a current in the antenna, and a current wave travels back along the line to a receiver. The parallel-wire line can be either open as in the figure, or enclosed in a shield. If the antenna is one-half wavelength long, as in the figure, then there is a standing current wave on it. *Rabbit-ear* antennas on TV sets are of this type.

Possibly the most common type of antenna is the *electric monopole*. It is one-quarter wavelength long and stands on a conducting plane, as in Fig. 25-8b. The antenna has an image in the conducting plane, and it acts as if it were a half-wave antenna fed in the middle. A little thought will show that the current in the image points in the same direction as the current in the antenna: the image of a positive

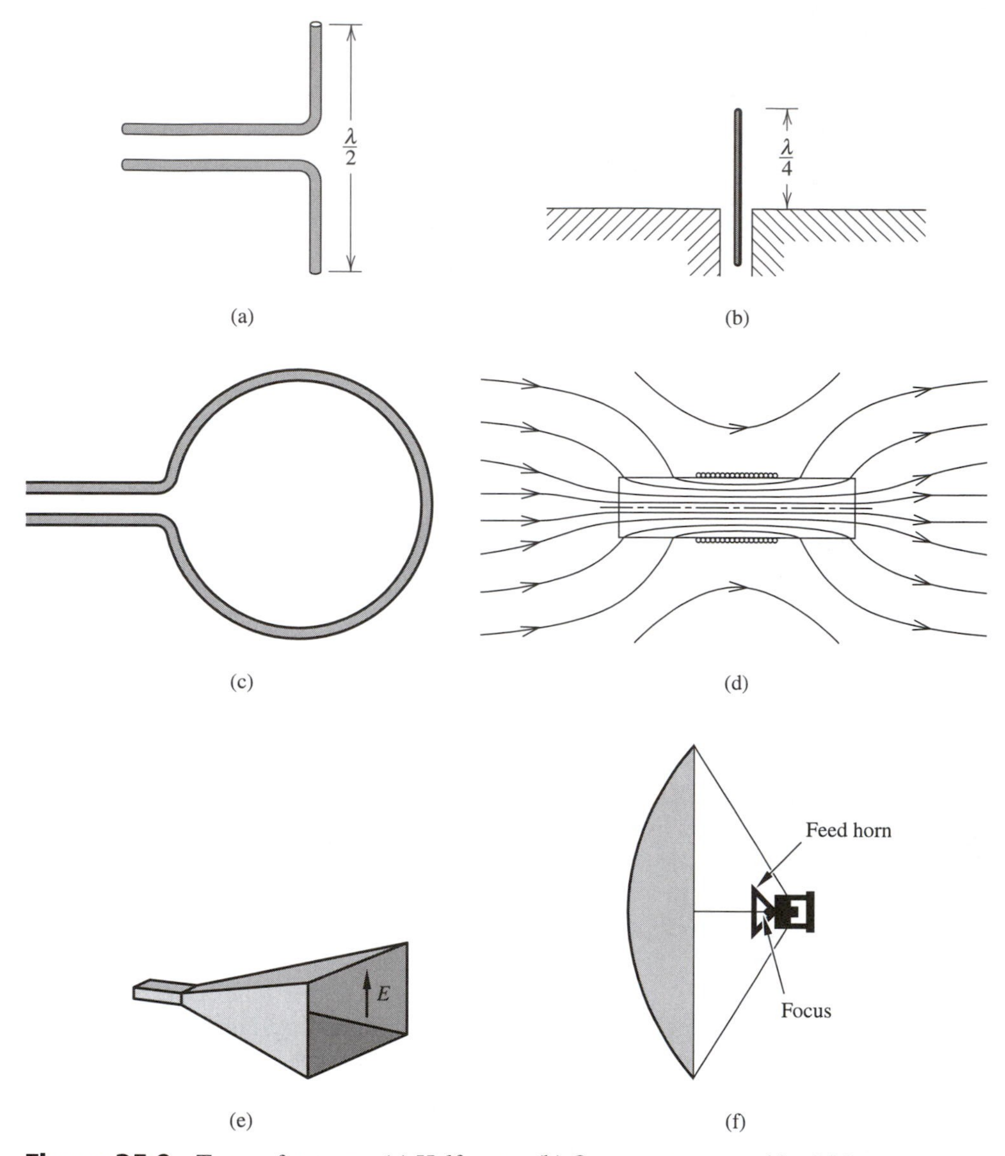

Figure 25-8 Types of antenna. (a) Half-wave. (b) Quarter-wave, or whip. (c) Loop or magnetic dipole. (d) Loop antenna with ferrite core that concentrates the magnetic field. (e) Horn. (f) Paraboloid.

charge that moves up is a negative charge that moves down. *Whips* on vehicles are quarter-wave antennas.

The huge towers, roughly 75 meters high, that one sees in the countryside near cities, are quarter-wave broadcast antennas. They rest on an insulator, and a current is injected at the base. The conducting plane is a radial grid of wires embedded in the ground, and the guy wires are split into sections to avoid parasitic induced currents.

The actual length of a *quarter-wave* antenna is only a rough indication of the operating frequency, because the length of the antenna is chosen to obtain both a radiation pattern and an input impedance that are appropriate for a particular application. For example, the heights of broadcast antennas vary from $\lambda/6$ to $5\lambda/8$, and λ is about 300 meters ($f \approx 1$ megahertz).

Patch antennas are microstrips (Sec. 24.3) applied, for example, to the wing of an aircraft.

Loop, or *magnetic dipole* antennas can be connected to a parallel-wire line, as in Fig. 25-8c. They are best known as indoor antennas for UHF. The AM antennas of small portable radios are usually coils wound on a rod of ferrite (See the Example on eddy currents in Sec. 18.3), as in Fig. 25-8d. The ferrite core amplifies the magnetic flux linking the coil by a factor of about 100.

The FM antennas of small radios are lengths of wire.

Modern cordless and cellular telephones operate at about 900 megahertz, and their antennas are flexible *helical coils* that act more or less as both an electric monopole and a magnetic dipole.

Aperture antennas are of three types. *Horns* are fixed to the end of a hollow metallic waveguide (Sec. 24.4), as in Fig. 25-8e. *Reflectors* are usually paraboloids illuminated by horns, as in Fig. 25-8f. *Lenses* are occasionally used in front of horns or paraboloids. *Microwave relay* antennas, at the top of towers or of tall buildings, are paraboloids, often enclosed in protective *radomes*.

One major use of *Radar* (Radio Direction and Ranging) is the control of air traffic near airports. The antenna is then a paraboloid fed by one or more horns, and the beam is fan-shaped, broad vertically and narrow horizontally. The antenna rotates about a vertical axis. The transmitter emits sharp pulses that are reflected by aircraft in the vicinity, and the time delay of the echo is a measure of the distance to the target. Most radars operate at about 1 to 10 gigahertz.

25.3 THE QUARTER-WAVE AND HALF-WAVE ANTENNAS

The quarter-wave antenna of Fig. 25-8b, together with its image in the conducting plane, acts as if it were one half wavelength long. So we must investigate the radiation of the antenna of Fig. 25-8a, with the length of the antenna equal to $\lambda/2$, where λ is the free-space wavelength. See Fig. 25-9.

The antenna carries a standing wave of current with a maximum at the center and nodes at the ends. Thus

$$I = I_m \cos\frac{l}{\lambda} \exp j\omega t. \tag{25-62}$$

Each element of length dl radiates as an electric dipole.

The radiation is isotropic in a plane perpendicular to the paper.

This description of the half-wave antenna is contradictory because the standing wave along the conductor can be truly sinusoidal only if there is zero energy loss, hence no radiation. In a real antenna the current distribution is not quite sinusoidal, but the distortion hardly affects the field.

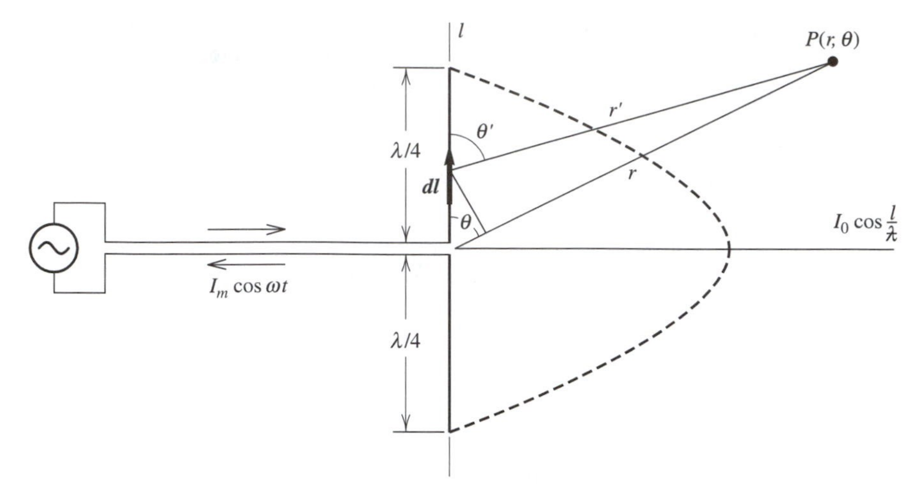

Figure 25-9 Half-wave antenna. The broken line shows the standing wave of current at $\cos \omega t = 1$. The field at $P\,(r, \theta)$ is independent of the ϕ coordinate.

The standing wave of current is the sum of two waves, one in the positive direction of l and the other in the negative direction, each of amplitude $I_m/2$:

$$I = \frac{I_m}{2}\left\{\exp j\left(\omega t - \frac{l}{\lambda\!\!\!^{-}}\right) + \exp j\left(\omega t + \frac{l}{\lambda\!\!\!^{-}}\right)\right\}. \tag{25-63}$$

25.3.1 The E Field

We set $r \gg \lambda\!\!\!^{-}$ and $\theta' \approx \theta$. As usual, we reserve brackets for quantities evaluated at $t - r/c$. From Sec. 25.1.5,

$$d\boldsymbol{E} = -\frac{[dp]}{4\pi\epsilon_0 \lambda\!\!\!^{-2} r}\sin\theta\,\hat{\boldsymbol{\theta}} = -\frac{\omega^2[dp]}{4\pi\epsilon_0 c^2 r}\sin\theta\hat{\boldsymbol{\theta}} \tag{25-64}$$

$$= \frac{\mu_0 j\omega[I]dl}{4\pi r'}\sin\theta\,\hat{\boldsymbol{\theta}}, \tag{25-65}$$

where

$$r' = r - l\cos\theta, \tag{25-66}$$

as in Fig. 25-9. Thus

$$d\boldsymbol{E} = \frac{\mu_0 j\omega I_m}{8\pi r'}\left\{\exp j\left(\omega[t'] - \frac{l}{\lambda\!\!\!^{-}}\right) + \exp j\left(\omega[t'] + \frac{l}{\lambda\!\!\!^{-}}\right)\right\}\sin\theta\, dl\,\hat{\boldsymbol{\theta}}. \tag{25-67}$$

We now integrate over the length of the antenna to find $\boldsymbol{E}$ at the point (r, θ). We can replace the r' in the denominator by r because $r \gg \lambda$, so that $r \gg l$. However, we must *not* replace the r' by r in

$$[t'] = t - \frac{r'}{c}, \tag{25-68}$$

because the phases of the exponential terms vary rapidly with r'. So we set

$$[t'] \approx t - \frac{r - l\cos\theta}{c} = [t] + \frac{l\cos\theta}{c}. \tag{25-69}$$

At a given point in space, the $d\boldsymbol{E}$'s thus all have about the same amplitude and direction, but their phases differ. All these vectors point in the direction of the local unit vector $\hat{\boldsymbol{\theta}}$. Observe that there is no $\hat{\boldsymbol{\phi}}$ component, by symmetry, and that there is also no $\hat{\boldsymbol{r}}$ component. Then

$$\boldsymbol{E} = \frac{\mu_0 j\omega I_m}{8\pi r}\sin\theta \exp j\omega[t] \times \int_{-\lambda/4}^{+\lambda/4}\left\{\exp j\frac{l(\cos\theta - 1)}{\lambda} + \exp j\frac{l(\cos\theta + 1)}{\lambda}\right\}dl\,\hat{\boldsymbol{\theta}}. \tag{25-70}$$

Integrating yields

$$\boldsymbol{E} = \frac{jI_m}{4\pi c\epsilon_0 r}\sin\theta \exp j\omega[t] \times \left\{\frac{\sin\{\pi(\cos\theta - 1)/2\}}{\cos\theta - 1} + \frac{\sin\{\pi(\cos\theta + 1)/2\}}{\cos\theta + 1}\right\}\hat{\boldsymbol{\theta}}, \tag{25-71}$$

where

$$\sin\frac{\pi(\cos\theta - 1)}{2} = -\cos\left(\frac{\pi}{2}\cos\theta\right), \qquad \sin\frac{\pi(\cos\theta + 1)}{2} = +\cos\left(\frac{\pi}{2}\cos\theta\right). \tag{25-72}$$

Thus

$$\boldsymbol{E} = \frac{j}{2\pi c\epsilon_0 r}\frac{\cos\{(\pi/2)\cos\theta\}}{\sin\theta}[I]\,\hat{\boldsymbol{\theta}} \tag{25-73}$$

$$\approx 60.0j\frac{\cos\{(\pi/2)\cos\theta\}}{r\sin\theta}[I]\,\hat{\boldsymbol{\theta}}. \tag{25-74}$$

This expression is indeterminate at $\theta = 0$ and at $\theta = \pi$. But, according to L'Hospital's rule, the limiting value of such a ratio is equal to the limiting value of the ratio of the derivatives. So $\boldsymbol{E}$ is zero on the axis of a half-wave antenna, in agreement with the fact that the elementary dipoles do not radiate along the axis.

The magnitude of $\boldsymbol{E}$ is independent of the frequency! The explanation is that the $\boldsymbol{E}$ of an elementary dipole, for a given current, is proportional to $1/\lambda$, and the antenna is $\lambda/2$ long.

25.3.2 The $\boldsymbol{H}$ Field

The value of $\boldsymbol{H}$ follows immediately. We found in Sec. 25.1.6 that, for an electric dipole, $\boldsymbol{H}$ is azimuthal and that

$$\frac{E}{H} = \left(\frac{\mu_0}{\epsilon_0}\right)^{1/2} = 377 \quad \text{ohms} \qquad (r \gg \lambda). \tag{25-75}$$

Therefore, in the far field of a half-wave antenna,

$$\boldsymbol{H} = \frac{j}{2\pi r}\frac{\cos\{(\pi/2)\cos\theta\}}{\sin\theta}[I]\hat{\boldsymbol{\phi}} \quad (r \gg \lambda). \tag{25-76}$$

25.3.3 The Poynting Vector $\boldsymbol{E} \times \boldsymbol{H}$ and the Radiated Power P

The time-averaged Poynting vector is

$$\mathcal{S}_{av} = \frac{1}{2}\text{Re}(\boldsymbol{E} \times \boldsymbol{H}^*) \tag{25-77}$$

$$= \frac{1}{\pi c\epsilon_0}\frac{\cos^2\{(\pi/2)\cos\theta\}}{\sin^2\theta}\frac{I_{\text{rms}}^2}{4\pi r^2}\hat{\boldsymbol{r}} \tag{25-78}$$

$$= 9.543\frac{\cos^2\{(\pi/2)\cos\theta\}}{\sin^2\theta}\frac{I_{\text{rms}}^2}{r^2}\hat{\boldsymbol{r}} \quad \text{watts/meter}^2. \tag{25-79}$$

To obtain the radiated power, we integrate over a sphere of radius r:

$$P = \frac{I_{\text{rms}}^2}{4\pi^2 c\epsilon_0 r^2}2\pi\int_0^\pi \frac{\cos^2\{(\pi/2)\cos\theta\}}{\sin^2\theta}r^2 \sin\theta \, d\theta. \tag{25-80}$$

The integral is equal to 1.2188267 and

$$P = 73.083 I_{\text{rms}}^2 \approx 73 I_{\text{rms}}^2 \quad \text{watts}. \tag{25-81}$$

The *radiation resistance* of a half-wave antenna is about 73 ohms.

25.4 ANTENNA ARRAYS

Both the electric dipole and the quarter-wave antenna are omnidirectional in the equatorial plane: their radiation fields are independent of the azimuthal angle ϕ.

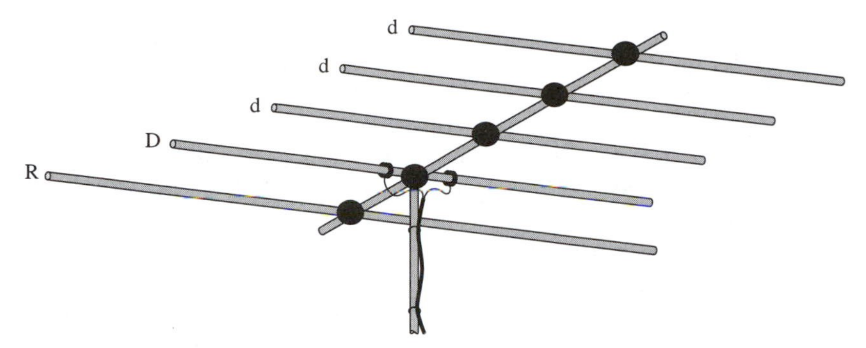

Figure 25-10 Basic Yagi-Uda antenna array, seen from below. The black blobs are insulators. D, half-wave antenna driven at the center; R, reflector; d, director. The elements are all about the same length and are spaced by about one-quarter wavelength. The actual lengths and spacings determine the phases. They are calculated theoretically to achieve an appropriate input impedance and an appropriate directivity.

Omnidirectional antennas have their uses, for example for car radios and for cordless and cellular telephones. But, for many applications, the radiation field, or the sensitivity, should be maximum in a given direction. That is achieved with *arrays* of antennas that are properly spaced and properly phased. The individual antennas can be identical, equally spaced, and oriented similarly, as in the Examples below. The *radiation pattern* of an array typically has one main lobe plus side lobes.

The directivity of an antenna is the same, whether it is used as a transmitter or as a receiver. When thinking about arrays, it is convenient to think of their directivity as *transmitting* antennas.

The *directivity* of an antenna is defined as follows:

$$D = \frac{\Phi_{main\ lobe}}{\Phi_{average}}, \tag{25-82}$$

where Φ is the *radiation intensity* in watts/steradian. You can show, in Prob. 25-12, that the directivity of an electric dipole is 1.5, and that the directivity of a half-wave antenna is slightly larger. Those antennas do not radiate in the axial direction.

The simplest array is a pair of quarter-wave antennas that are spaced and phased so that the interference between the two fields gives an appropriate radiation pattern. See the next Example.

Linear arrays comprise several, often identical, antennas disposed along a straight line. For example, radio telescopes often comprise many parabolic dishes arranged along two perpendicular lines. Changing the phases of the individual dishes rotates the main lobe—or the direction of maximum sensitivity—one way or the other. The antennas of a *broadside* array are all in phase, and the radiation field is maximum in the direction perpendicular to the array. The antennas of an *end-fire* array are spaced and phased so that the radiation is maximum in the direction parallel to the array.

The most common array is the *Yagi-Uda,* or *ladder array.* The Yagi-Uda is an end-fire array. There exists a wide variety of such antennas, and Fig. 25-10 shows the simplest configuration. These arrays are used at frequencies up to 2.5 gigahertz, largely for TV. One rod, which is a half-wave antenna, is driven, while the others are parasitic: the reflector and the directors oscillate in the field of the driver. Directivity is achieved by properly selecting their lengths and spacings. These arrays have directivities of the order of 10.

Some *planar arrays* operate at wavelengths of the order of one centimeter and comprise many antennas, sometimes thousands, arranged over a rectangular or circular plane surface. Accurate beam steering and pattern control are achieved within microseconds by shifting the phases of the individual antennas.

Conformal patch arrays (Sec. 25.2) are part of a given surface, often cylindrical.

Adaptive receiving arrays adjust their patterns automatically to optimize the signal-to-noise ratio in the presence of identifiable noise sources.

EXAMPLE The Radiation Pattern of a Simple End-Fire Array

Figure 25-11 shows a top view of two parallel vertical quarter-wave antennas. Remember, from Sec. 25.3, that the field of one vertical quarter-wave antenna and of its image in a conducting plane is the same as that of the half-wave antenna of Figs. 25-8(a) and 25-9.

We wish to find the three-dimensional radiation pattern of the array, or E in the far field as a function of θ and ϕ, with the origin of the spherical coordinates at the origin O in Fig. 25-11. We first find the ϕ-dependence of E, which results from the interference between the two antennas, and then multiply by the θ-dependence that we found in Sec. 25.3. This simple multiplication is valid *in the far field*, with which we are concerned here.

We wish to illuminate the region to the *right* of the array. Here is how.

Choose a spacing of $\lambda/4$, with the phase of the current in A leading the phase of the current in B by $\pi/2$:

$$I_A = I_B \exp j(\pi/2). \tag{25-83}$$

Then the wave originating at A arrives at B with a phase

$$\frac{\pi}{2} - \frac{2\pi}{\lambda}\frac{\lambda}{4} = 0, \tag{25-84}$$

and the two waves add. But a wave originating at B arrives at A with a phase

$$-\frac{\pi}{2} - \frac{2\pi}{\lambda}\frac{\lambda}{4} = -\pi \tag{25-85}$$

and the two waves cancel. So, on the right, the two waves add, but on the left, the two waves cancel. We are on the right track!

Now recall from Eq. 25-74 that, in the far field of a half-wave antenna, in the plane $\theta = \pi/2$,

$$E = 60.0j\frac{[I]}{r}. \tag{25-86}$$

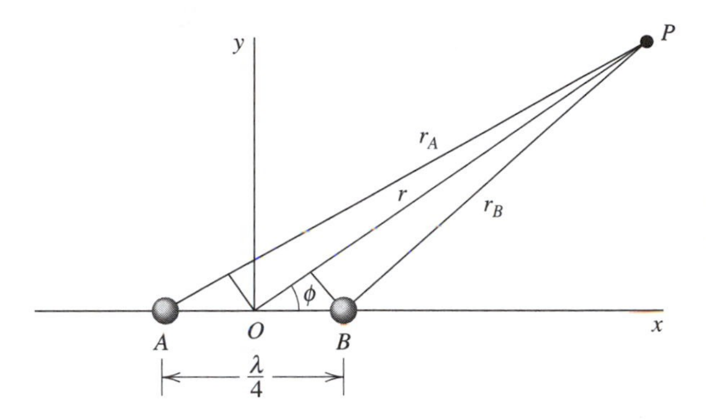

Figure 25-11 End-fire array composed of two parallel, vertical, quarter-wave antennas, seen from above. The phase of A leads the phase of B by $\pi/2$. Figures 25-12 and 25-13 show the radiation pattern in the xy-plane and in three dimensions.

Then at P,

$$E = 60.0j\left(\frac{[I_A]}{r_A} + \frac{[I_B]}{r_B}\right). \tag{25-87}$$

Let us think in terms of an imaginary antenna situated at the origin O and carrying a current I. Then, with respect to that imaginary antenna, the phase of A is $+\pi/4$, and the phase of B is $-\pi/4$. So, at P,

$$[I_A] = I\exp j\left(\frac{\pi}{4} - \frac{r_A}{\lambda}\right), \tag{25-88}$$

$$[I_B] = I\exp j\left(-\frac{\pi}{4} - \frac{r_B}{\lambda}\right). \tag{25-89}$$

The current I, here, is not retarded. It is the current in the imaginary antenna at the origin, at the time t.

Now recall the trick that we used in Sec. 25.1.2: we can set $r_A \approx r_B \approx r$ *in the denominators*. Then

$$E = \frac{60.0j}{r}I\left\{\exp j\left(\frac{\pi}{4} - \frac{r_A}{\lambda}\right) + \exp j\left(-\frac{\pi}{4} - \frac{r_B}{\lambda}\right)\right\}, \tag{25-90}$$

where

$$r_A = r + (\lambda/8)\cos\phi, \qquad r_B = r - (\lambda/8)\cos\phi, \tag{25-91}$$

so that

$$E = \frac{60.0j}{r}I(e_A + e_B), \tag{25-92}$$

with

$$e_A = \exp j\left\{\frac{\pi}{4} - \frac{r + (\lambda/8)\cos\phi}{\lambda}\right\} \tag{25-93}$$

$$= \exp j\left(-\frac{r}{\bar{\lambda}}\right) \exp j\left\{\frac{\pi}{4} - \frac{(\lambda/8)\cos\phi}{\bar{\lambda}}\right\}, \tag{25-94}$$

and

$$e_B = \exp j\left\{-\frac{\pi}{4} - \frac{r - (\lambda/8)\cos\phi}{\bar{\lambda}}\right\} \tag{25-95}$$

$$= \exp j\left(-\frac{r}{\bar{\lambda}}\right) \exp j\left\{-\frac{\pi}{4} + \frac{(\lambda/8)\cos\phi}{\bar{\lambda}}\right\}. \tag{25-96}$$

Since $(\lambda/8)/\bar{\lambda} = \pi/4$,

$$e_A = \exp j\left(-\frac{r}{\bar{\lambda}}\right) \exp j\left\{\frac{\pi}{4}(1 - \cos\phi)\right\}, \tag{25-97}$$

$$e_B = \exp j\left(-\frac{r}{\bar{\lambda}}\right) \exp j\left\{\frac{\pi}{4}(-1 + \cos\phi)\right\}. \tag{25-98}$$

Now†

$$\cos\alpha = \frac{\exp(j\alpha) + \exp(-j\alpha)}{2}. \tag{25-99}$$

Then

$$e_A + e_B = \exp j\left(-\frac{r}{\bar{\lambda}}\right) 2\cos\left\{\frac{\pi}{4}(1 - \cos\phi)\right\}. \tag{25-100}$$

Substituting into Eq. 25-92,

$$E = \frac{60.0j}{r} I \exp j\left(-\frac{r}{\bar{\lambda}}\right) 2\cos\left\{\frac{\pi}{4}(1 - \cos\phi)\right\}. \tag{25-101}$$

This is the radiation pattern in the $\theta = \pi/2$ plane, or in the xy-plane. See Fig. 25-12.

Let us check. At $\phi = 0$, $\cos\phi = 1$ and $2\cos 0 = 2$: the field is twice as large as that of a single antenna. At $\phi = \pi$, $\cos\phi = -1$ and $2\cos(2\pi/4) = 2\cos(\pi/2) = 0$. So E is maximum in the direction $\phi = 0$, and zero in the direction $\phi = \pi$, as required.

To obtain the far-field 3-D radiation pattern of our array of Fig. 25-11, we multiply this function by the θ-dependence that we found in Sec. 25.3.1. See Fig. 25-13. The directivity is adequate for a broadcast array near a city.

The more elements there are in an array, the more directional it is.

---

†If you are not sure, try $\alpha = 0$, $\pi/4$, and $\pi/2$.

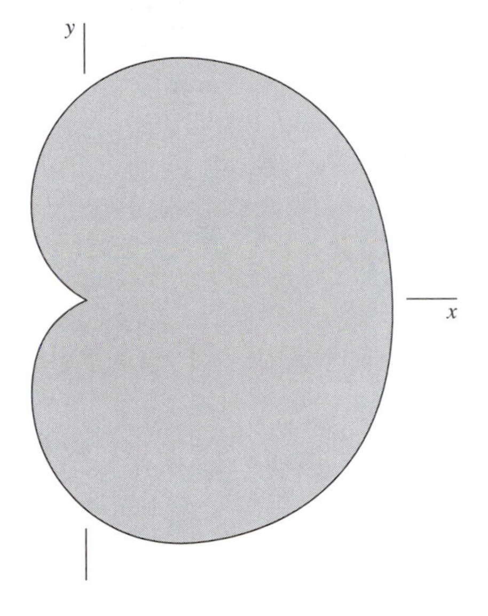

Figure 25-12 The radiation pattern for the end-fire array of Fig. 25-11 in the *xy*-plane. This is a plot of the function $2\cos\{(\pi/4)(1 - \cos\phi)\}$. On the *x*-axis and on the right-hand side, $\phi = 0$ and the net field is twice as large as that of a single antenna. On the left, $\phi = \pi$ and there is zero field.

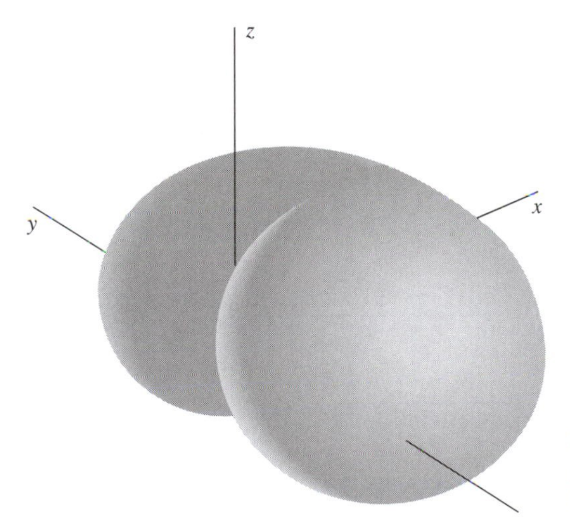

Figure 25-13 The three-dimensional radiation pattern for the end-fire array of Fig. 25-11.

25.5 SUMMARY

In the field of a *small electric dipole* of length **s**, at a point (r,θ), with

$$r \gg ƛ \gg s, \tag{25-1}$$

$$\mathbf{A} = \frac{\mu_0 I_m \exp j(\omega t - r/ƛ)}{4\pi r}\mathbf{s}, \tag{25-8}$$

$$V = \frac{j[p]}{4\pi\epsilon_0 \lambda\!\!\!{}^- r}\cos\theta, \tag{25-24}$$

$$\boldsymbol{E} = -\frac{[p]}{4\pi\epsilon_0 \lambda\!\!\!{}^-{}^2 r}\sin\theta\,\hat{\boldsymbol{\theta}}, \tag{25-46}$$

$$\boldsymbol{H} = -\frac{c[p]}{4\pi \lambda\!\!\!{}^-{}^2 r}\sin\theta\,\hat{\boldsymbol{\phi}}, \tag{25-53}$$

$$\frac{E}{H} = \left(\frac{\mu_0}{\epsilon_0}\right)^{1/2}. \tag{25-54}$$

In the region close to the dipole, where the condition $r \gg \lambda\!\!\!{}^-$ is *not* satisfied, Eqs. 25-55 and 25-56 apply.

Even close to the dipole,

$$\mathcal{S}_{av} = \frac{\mu_0 \pi^2 f^4 p_{\text{rms}}^2}{c r^2}\sin^2\theta\,\hat{\boldsymbol{r}}, \tag{25-59}$$

and the radiated power is given by

$$P = \frac{8\mu_0\pi^3}{3c} f^4 p_{\text{rms}}^2. \tag{25-61}$$

In the field of a *quarter-wave antenna*,

$$\boldsymbol{E} = \frac{j}{2\pi c\epsilon_0 r}\,\frac{\cos\{(\pi/2)\cos\theta\}}{\sin\theta}[I]\,\hat{\boldsymbol{\theta}}, \tag{25-73}$$

$$\boldsymbol{H} = \frac{j}{2\pi r}\,\frac{\cos\{(\pi/2)\cos\theta\}}{\sin\theta}[I]\,\hat{\boldsymbol{\phi}}, \tag{25-76}$$

and E/H has the same value as above. Also,

$$\mathcal{S}_{av} = \frac{1}{\pi c\epsilon_0}\,\frac{\cos^2\{(\pi/2)\cos\theta\}}{\sin^2\theta}\,\frac{I_{\text{rms}}^2}{4\pi r^2}\,\hat{\boldsymbol{r}}, \tag{25-78}$$

and

$$P = 73.083 I_{\text{rms}}^2 \approx 73 I_{\text{rms}}^2 \quad \text{watts}. \tag{25-81}$$

The radiation fields of both electric dipoles and quarter-wave antennas are independent of the azimuthal angle ϕ. But *arrays* of antennas that are properly spaced and properly phased can direct the net radiation field in a given direction. The directivity of an array is the same, whether it is used for transmission or for reception.

PROBLEMS

25-1. (*25.1*) The dipole moment of an oscillating charge

A charge Q oscillates along the z-axis, and $z = z_m \exp j\omega t$.

What is the dipole moment of an equivalent oscillating dipole? You can find the equivalence in the following way. If the currents are the same, then the ***A**'s* are the same. Then the V's are the same, from the Lorentz condition (Sec. 21.7). Then the ***E**'s* and ***H**'s* are the same.

25-2. (*25.1*) The radiation pattern of the electric dipole

What fraction of the total power in the field of an electric dipole is radiated within 45° of the equatorial plane?

25-3. (*25.1*) The electric and magnetic energy densities in the field of an electric dipole

Calculate the ratio of the time-averaged electric energy density to the time-averaged magnetic energy density in the field of an electric dipole (a) for $r \ll \lambda\!\!\!^{-}$, (b) for $r = \lambda\!\!\!^{-}$, and (c) for $r \gg \lambda\!\!\!^{-}$.

25-4. (*25.1*) The Poynting vector and the energy density in the field of an electric dipole

Show that, for $r \gg \lambda\!\!\!^{-}$, the magnitude of the Poynting vector in the field of an electric dipole is equal to the energy density multiplied by c.

25-5. (*25.1*) The light source paradox

All light sources should be black, for the following reason. Take the sun, for example. A cone in the retina of the eye collects radiation emanating from a very large number of atoms. These sources are incoherent.

At any given instant there is a near-infinite number of phasors in the complex plane, all of different magnitudes and different phases, rotating at different velocities. Their vector sum is clearly zero.

The same reasoning applies to any object, say a white wall, illuminated with incoherent light. The radiation that reaches a given cone comes from an area that is a large number of wavelengths in diameter. There again, the net field at the cone should be zero, and the wall should appear black.

To explain this paradox, consider N waves of a single frequency and of a given linear polarization but of random amplitudes and phases. The number N is very large. For the i-th wave, $E_i = E_{mi} \exp j(\omega t - \alpha_i)$ at the cone, and the net E is the sum of the E_i's.

Now the eye is sensitive, not to E but to $\mathcal{S}$, and thus to EE^*. Show that $\mathcal{S}_{av} = \Sigma\, \mathcal{S}_i$. This means that the net energy flux is equal to the sum of the energy fluxes of the individual waves.†

†You have probably noticed that laser light diffused by a wall or a sheet of paper has a granular structure. The pattern moves if one moves one's head from side to side. This phenomenon is called *speckle*. It is used for studying surfaces.

Speckle arises in the following way. Each point on the object, say the sheet of paper, produces on the retina a diffraction pattern whose shape and size depend on the optical characteristics of the eye. Since the radiation is coherent, there is interference between these patterns, and the field varies from point to point on the retina.

25-6. (*25.1*) Cosmological evolution

Consider a particular class of astronomical objects, say quasars. Assume that they are all identical and distributed uniformly in a Euclidean universe.

Show that, if N is the number of objects whose radio-frequency flux is greater than $\mathcal{S}$ at the earth, then a plot of log N against log $\mathcal{S}$ should be a straight line whose slope is −1.5. The slope for quasars is, in fact, larger. This is possibly a measure of cosmological evolution.

25-7. (*25.1*) The polarization of skylight

The atmosphere scatters sunlight. Draw a sketch showing the sun, the earth, and a vector $\boldsymbol{E}$ on a ray of scattered light. Explain why skylight is polarized. The light is only partially polarized because it is scattered many times.

25-8. (*25.3*) The radiation patterns of the electric dipole and of the half-wave antenna

(a) Show that in the far field of an electric dipole $E_{\text{rms}} = 6.71 P^{1/2} \sin\theta / r$.

(b) Show that for the half-wave antenna

$$E_{\text{rms}} = 7.02 \frac{P^{1/2} \cos\{(\pi/2)\cos\theta\}}{r \sin\theta}.$$

(c) Plot E_{rms} as a function of θ for the two fields.

25-9. (*25.3*) The electric field of a radio antenna

Calculate E at a distance of 1 kilometer in the equatorial plane of a half-wave radio antenna radiating 1 kilowatt of power. Set $\lambda \ll 1$ kilometer.

25-10. (*25.3*) The image of a quarter-wave antenna

An antenna is normally situated near a conductor (the earth, an airborne vehicle, a satellite, etc.). Energy radiated toward the conductor is reflected, and the total field is thus the vector sum of the direct wave plus the reflected wave. It is convenient to consider that the latter is generated, not by reflection, but by an image of the antenna located below the surface of the conductor.

(a) Show that the current in the image of a *horizontal* half-wave antenna and the current in the antenna flow in opposite directions.

(b) Show that the current in the image of a *vertical* quarter-wave antenna and that in the antenna flow in the same direction.

Both rules apply to oblique quarter-wave antennas.

(c) We have shown that the radiation resistance of a half-wave antenna is 73.083 ohms. Find the radiation resistance of a quarter-wave antenna perpendicular to a conducting plane.

25-11. (*25.4*) The radiation pattern of a linear array of half-wave antennas

A *linear array* consists of parallel half-wave antennas lying in a plane. Say there are N antennas, uniformly separated by a distance D and excited in phase.

(a) Show that, in the plane perpendicular to the antennas,

$$E \propto \frac{\sin\{(ND/2\lambda\!\!\!^{-})\cos\phi\}}{\sin\{(D/2\lambda\!\!\!^{-})\cos\phi\}},$$

where ϕ is the angle between the direction of observation and the plane of the array. The best approach is to sum the individual $\boldsymbol{E}$ phasors graphically in the complex plane.

(b) Find the angular positions of the minima and maxima of E. Differentiation yields only the maxima.

(c) Show that, for a given spacing D, the main lobe at $\phi = \pi/2$ becomes narrower as N increases.

(d) Plot E as a function of θ between 0 and 360° for an array of 30 parallel half-wave antennas that are in phase and spaced by $\lambda/4$.

(e) Now plot the same function, using Cartesian coordinates, between 0 and 180° with a log scale for the E-axis.

(f) Explain why the main lobe is twice as wide as the two neighboring lobes.

(g) Show that its half-width (the angle between the maximum and the first minimum on one side or the other) is approximately equal to λ/l, where l is the length of the array.

25-12. (*25.4*) The directivity of an antenna

The *directivity* of an antenna can also be defined as the ratio of the Poynting vector at the maximum of the radiation pattern to the Poynting vector averaged over a spherical surface surrounding the antenna:

$$D = \frac{\mathcal{S}_{\max}}{P/(4\pi r^2)} = \frac{\mathcal{S}_{\max}}{\dfrac{1}{4\pi}\displaystyle\int_0^{2\pi}\int_0^{\pi} \mathcal{S} \sin\theta \, d\theta \, d\phi},$$

where P is the radiated power.

(a) Show that the directivity of an electric dipole is 1.5.

(b) Show that the directivity of a half-wave antenna is 1.64.

25-13. (*25.4*) The radiation pattern of a simple array

Two parallel quarter-wave antennas are perpendicular to the ground, in phase, and spaced by $\lambda/2$. Plot the 2-D radiation pattern at ground level, and the 3-D pattern above ground. Show the positions of the antennas on your figures.

APPENDIX

A

SI PREFIXES AND THEIR SYMBOLS

Multiple	Prefix	Symbol	Multiple	Prefix	Symbol
10^{-24}	yocto	y	10	deka†	da
10^{-21}	zepto	z	10^{2}	hecto	h
10^{-18}	atto	a	10^{3}	kilo	k
10^{-15}	femto	f	10^{6}	mega	M
10^{-12}	pico	p	10^{9}	giga	G
10^{-9}	nano	n	10^{12}	tera	T
10^{-6}	micro	μ	10^{15}	peta	P
10^{-3}	milli	m	10^{18}	exa	E
10^{-2}	centi	c	10^{21}	zetta	Z
10^{-1}	deci	d	10^{24}	yotta	Y

†This prefix is written *déca* in French. The English alternate is *deca*.

Caution: the symbol for the prefix is written next to that for the unit *without* a dot. For example, mN stands for millinewton, while $m \cdot N$ is a meter-newton, or a joule.

APPENDIX

B

CONVERSION TABLE: S.I.—c.g.s.

Examples: One meter equals 100 centimeters. One volt = 10^8 electromagnetic units of potential.

		cgs Systems	
Quantity	**SI**	**esu**	**emu**
Length	meter	10^2 centimeters	10^2 centimeters
Mass	kilogram	10^3 grams	10^3 grams
Time	second	1 second	1 second
Force	newton	10^5 dynes	10^5 dynes
Pressure	pascal	10 dynes/centimeter^2	10 dynes/centimeter^2
Energy	joule	10^7 ergs	10^7 ergs
Power	watt	10^7 ergs/ second	10^7 ergs/ second
Charge	coulomb	3×10^9	10^{-1}
Electric potential	volt	$\frac{1}{300}$	10^8
Electric field strength	volt/meter	$1/(3 \times 10^4)$	10^6
Electric flux	coulomb	$12\pi \times 10^9$	$4\pi \times 10^{-1}$
Electric flux density	coulomb/meter^2	$12\pi \times 10^5$	$4\pi \times 10^{-5}$
Polarization	coulomb/meter^2	3×10^5	10^{-5}
Electric current	ampere	3×10^9	10^{-1}
Conductivity	siemens/meter	9×10^9	10^{-11}
Resistance	ohm	$1/(9 \times 10^{11})$	10^9
Conductance	siemens	9×10^{11}	10^{-9}
Capacitance	farad	9×10^{11}	10^{-9}
Magnetic flux	weber	$\frac{1}{300}$	10^8 maxwells
Magnetic flux density	tesla	$1/(3 \times 10^6)$	10^4 gausses
Magnetic field strength	ampere/meter	$12\pi \times 10^7$	$4\pi \times 10^{-3}$ oersted
Magnetomotance	ampere	$12\pi \times 10^9$	$4\pi/10$ gilberts
Magnetization	ampere/meter	$1/(3 \times 10^{13})$	10^{-3}
Inductance	henry	$1/(9 \times 10^{11})$	10^{-9}
Reluctance	ampere/weber	$36\pi \times 10^{11}$	$4\pi \times 10^{-9}$

Note: We have set c = 3×10^8 meters/second.

APPENDIX

C

THE STORY OF THE "MAXWELL" EQUATIONS

The story of the Maxwell equations is fascinating in many ways.†

First, as we shall see, the equations should be called, more appropriately, the Heaviside equations.

Second, the story covers less than the fourth quarter of the nineteenth century.

Third, there were few *Maxwellians*, the main actors being Oliver Heaviside (1850–1925), who was self-taught, George Francis FitzGerald (1851–1901), who worked at Trinity College in Dublin, Oliver Lodge (1851–1940), who was at University College in Liverpool, and Heinrich Hertz (1857–1894), who was an experimental physicist at the Karlsruhe Physical Institute. John-Henry Poynting (1852–1914) contributed a major concept when he proposed his theorem in 1884, but Heaviside discovered it practically at the same time. James Clerk Maxwell (1831–1879) was an actor through the publication in 1873 of his *Treatise on Electricity and Magnetism*. At the time of his death, he was working on a second edition of the *Treatise*.

Fourth, all the actors were British, except for Hertz, who was German. Maxwell, who was Scottish, became Professor of Experimental Physics at Cambridge but, after his death, there were no Maxwellians at Cambridge, and none at Oxford.

Fifth, the main actor was Heaviside, who had to leave school when he was sixteen. He started life as a telegrapher in Newcastle, where he stayed for eight years, but had no real job for the rest of his life. He was a prolific writer, despite the fact that he had poor health and lived as a recluse, partly by temperament, partly because of his deafness, and partly because of his great poverty. For many years he published his theories in *The Electrician*, which was a trade journal. For this he was paid 40 pounds per year, which was less than the salary of a laborer. Near the end of his life he received a modest government pension of 120 pounds per year, but remained exceedingly poor and miserable, despite the many honors that were bestowed upon him. Although he was self-taught, he had become one of the best, if not *the* best, physicist-mathematician of his time.

Maxwell's *Treatise* is unreadable. Hunt (1991), who clearly regrets having spent so much time on it, says that it is rambling, obscure, inconsistent, contradictory, awkward, confusing, and on some points simply wrong; that it has no clear focus, no or-

†The best reference is Hunt (1991), but see also Bork (1963), Heaviside (1950 and 1971), Gillispie (1980), and Harman (1998).

derly presentation, and that the mathematics is clumsy and involved! It is a wonder that the Maxwellians managed to master it. Maxwell had great admiration for Faraday, and he was attempting to put Faraday's ideas in mathematical form.

The *Treatise* does *not* state the Maxwell equations because Maxwell reasoned mostly in terms of the potentials V (then called ψ) and $\boldsymbol{A}$. Shifting from V and $\boldsymbol{A}$ to $\boldsymbol{E}$ and $\boldsymbol{B}$ caused much debate about "the murder of ψ." The potentials then came to be considered as mathematical fictions.

Heaviside focused on the electromagnetic *field*, like Faraday and Maxwell. His main motivation was to improve signaling on submarine cables. He finally recast the Maxwell theory in the form of the four vector "Maxwell" equations in 1884, in the same year that Hertz stated them in Cartesian form. Hertz admitted that Heaviside had the priority, in view of the fact that his own "proof" was "very shaky." At the time, vector notation seemed to everyone to be highly esoteric.

Modern Physics owes very much to Heaviside. He was the first to apply vector analysis as we know it today to electromagnetic fields, and the first to use *rationalized* units. (With unrationalized units, factors of 4π appear in most of the important equations.) He discovered the *Poynting* theorem the same year as Poynting. He proposed the existence of the Heaviside layer† and developed Operational Calculus. He also discovered the "Lorentz" force $Q\boldsymbol{v} \times \boldsymbol{B}$ fifteen years before Hendrik Lorentz (1853–1928). As early as 1888 he calculated the $\boldsymbol{E}$ and $\boldsymbol{B}$ fields of a moving charge (Sec. 13.2). That was a most important result because it showed how the field contracts lengthwise and that *the contraction involves the factor γ of relativity!* This phenomenon was later called the *FitzGerald contraction* and served to explain the negative result of the Michelson-Morley experiment.

†This is a region in the upper atmosphere where the degree of ionization varies with altitude in such a way that low-frequency radio waves are reflected downward. At TV frequencies, waves bend slightly and escape.

APPENDIX

THE RETARDED POTENTIALS AND FIELDS

More advanced books, for example Lorrain, Corson, and Lorrain (1988) deduce the following equations.

D.1 THE RETARDED ELECTRIC POTENTIAL V

The *wave equation for V* is

$$\nabla^2 V - \epsilon_0 \mu_0 \frac{\partial^2 V}{\partial t^2} = -\frac{\rho}{\epsilon_0}. \tag{1}$$

This equation applies at a point where the electric potential is V, and the electric charge density is ρ. We have assumed that the charges are situated in a vacuum. At points where $\rho = 0$ the right-hand side is zero and, if $\partial/\partial t = 0$, we have Poisson's equation of Sec. 4.1.

If ρ is not constant, then, at x, y, z at the time t,

$$V(x,y,z,t) = \frac{1}{4\pi\epsilon_0}\int_{v'} \frac{\rho(x',y',z',t - r/c)}{r} dv', \tag{2}$$

where the primed coordinates refer to the charge distribution, r is the distance between the source point x', y', z' and the field point x, y, z and $dv' = dx'dy'dz'$. This is the *retarded electric potential.*

D.2 THE RETARDED VECTOR POTENTIAL A

The *wave equation for* $\boldsymbol{A}$ is similar to that for V:

$$\nabla^2 \boldsymbol{A} - \epsilon_0 \mu_0 \frac{\partial^2 \boldsymbol{A}}{\partial t^2} = -\mu_0 \boldsymbol{J}, \tag{3}$$

and its solution is similar: if $\boldsymbol{J}$ is not constant, then the *retarded vector potential* is

$$\boldsymbol{A}(x,y,z,t) = \frac{\mu_0}{4\pi}\int_{v'} \frac{\boldsymbol{J}(x',y',z',t-r/c)}{r} dv'. \tag{4}$$

D.3 THE RETARDED ELECTRIC FIELD STRENGTH $\boldsymbol{E}$

The *wave equation for* $\boldsymbol{E}$, again in a vacuum, is more complicated:

$$\nabla^2 \boldsymbol{E} - \epsilon_0\mu_0 \frac{\partial^2 \boldsymbol{E}}{\partial t^2} = \frac{\nabla\rho}{\epsilon_0} + \mu_0 \frac{\partial \boldsymbol{J}}{\partial t}. \tag{5}$$

The first term on the right comes from charge accumulations, and the second term from changing magnetic fields. The *retarded electric field* is

$$\boldsymbol{E}(x,y,z,t) = -\frac{1}{4\pi\epsilon_0}\int_{v'} \frac{[\nabla'\rho + \epsilon_0\mu_0(\partial \boldsymbol{J}/\partial t)]}{r} dv'. \tag{6}$$

The *brackets* here have a special meaning: they designate quantities that must be evaluated at the *previous time* $t - r/c$. The primed gradient is evaluated with respect to the coordinates x', y', z' of the source point.

D.4 THE RETARDED MAGNETIC FIELD STRENGTH $\boldsymbol{H}$

The *wave equation for* $\boldsymbol{H}$, in a vacuum, is

$$\nabla^2 \boldsymbol{H} - \epsilon_0\mu_0 \frac{\partial^2 \boldsymbol{H}}{\partial t^2} = -\mu_0 \nabla \times \boldsymbol{J}, \tag{7}$$

and the *retarded magnetic field strength* is

$$\boldsymbol{H}(x,y,z,t) = \frac{1}{4\pi}\int_{v'} \frac{[\nabla' \times \boldsymbol{J}]}{r} dv'. \tag{8}$$

Again, the primed curl concerns the source variables x', y', z', and the brackets mean that the curl is evaluated at the time $t - r/c$.†

†In the case of the electric dipole, in cylindrical coordinates, $\nabla \times \boldsymbol{J}$ reduces to $-(\partial J_z/\partial \rho)\,\hat{\boldsymbol{\phi}}$, and the derivative is undefined.

ANSWERS

Problems often require the demonstration of a given result; the list below provides about half of the remaining answers, to two significant figures.

1-5. $4\pi a^3$.

2-7. 85 watts.

3-1. $E = -[Q/(2^{5/2}\pi\epsilon_0 a^2)]\hat{x}$.

3-6. 89 meters/second.

3-9. 440 millimeters.

3-11. (b) 3.3×10^{-3} newton, (d) 6 microseconds.

3-14. (a) (i) 10^{19} atoms/meter^2, (ii) 1.6 coulombs/meter^2, (iii) 1.8×10^{11} volts/meter, (b) 240 atomic diameters, or 0.07 micrometer.

3-17. (a) $Q\lambda/(2\pi\epsilon_0 r)$.

4-2. (a) *B, E, D* are at the same potential; *F, C, H* are at another potential.

4-3. (a) 1.5×10^{20} newtons. (b) 1.5×10^{20} joules (c) 4.1×10^{13} \$. (d) 5000 years.

4-13. $E_p s - Is/(\mathcal{A}\sigma)$.

4-16. 1.9 newtons.

5-3. $(4/3)\pi R^3 \sigma_0 \hat{z}$.

5-7. $$\frac{Q}{4\pi\epsilon_0 r}, \qquad \frac{Qs\cos\theta}{4\pi\epsilon_0 r^2}, \qquad \frac{Qs^2}{4\pi\epsilon_0 r^3}\left(\frac{3\cos^2\theta}{2} - \frac{1}{2}\right).$$

5-9. $\rho a^3/(4\pi\epsilon_0 r)$, 0, 0.

5-11. 0, $[a^4 Q/(16\pi\epsilon_0 r^5)][35(l^4 + m^4 + n^4) - 21]$.

6-2. (a) $0.15Q^2/(\pi\epsilon_0 R)$. (b) $3GM^2/(5R')$. (c) 1.2×10^{29} joules. (d) 0.17 meter. (e) 1.0×10^{20}volts.

6-10. (a) $W_1/W_2 = C_2/C_1$. (b) $W_1/W_2 = C_1/C_2$.

6-13. (b) 56 picofarads/meter.

6-20. (a) 150 kilovolts. (b) 4×10^{-4} atmosphere. (c) 4 kilograms/meter^2.

7-1. (a) 5.7×10^{-37} coulomb-meter. (b) 5.9×10^{-19} meter.

7-12. (a) $[2\pi\epsilon_r\epsilon_0/\ln(R_2/R_1)]V$.

8-2. $$\text{(a)}\ \frac{R_1/(j\omega C_1)}{R_1 + 1/(j\omega C_1)} + \frac{R_2/(j\omega C_2)}{R_2 + 1/(j\omega C_2)}, \qquad \text{(b)}\ \frac{(\epsilon_{r2}\sigma_{co1} - \epsilon_{r1}\sigma_{co2})\epsilon_0}{s_1\sigma_{co2} + s_2\sigma_{co1}}V.$$

8-4. 4000.

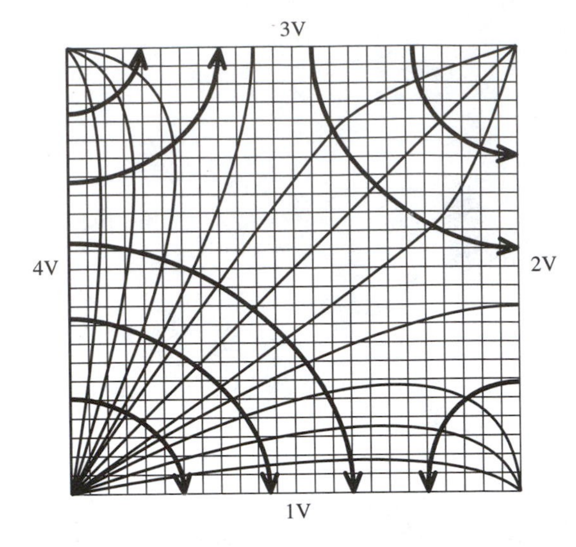

Problem 9-10 Equipotentials (every 0.25 V) and lines of $\boldsymbol{E}$.

8-9. (b) 10 meters^3, 10 tons.

8-14. 25 micrometers.

8-18. 2.5 atmospheres.

9-2. $-2QD/[4\pi(D^2+r^2)^{3/2}]$.

9-10. 25×25 cells, 400 iterations.

10-6. $\rho_0(a^2+b^2)(a+b)/(12\epsilon_0 r^2)$.

11-2. According to O', the signals were emitted simultaneously. According to O, B emitted her signal first.

11-3. (a) $\tan\alpha = \gamma \tan\alpha'$. (b) $\pi/2$.

12-4. 2.6×10^{-27} kilogram, 2.27×10^8 meters/second.

12-6. (a) 7.1×10^{-26} kilogram. (b) 38 millimeters. (c) 10^{-5} second, 1.28×10^{-10} second.

12-7. (a) Negative. The light is redder. (b) For the sun, -2.1×10^{-6}. For the earth, -7.0×10^{-10}. (c) -2.1×10^{-6}. (d) 5.2×10^7 times the density of water. (e) $\Delta\nu/\nu = 6.9 \times 10^{-6}$, which is about three times the red shift. (f) As the proton flies away, its increase of potential energy just compensates for its decrease in kinetic energy.

13-3. (a) 2.3×10^{-22} newton. (b) 7.7×10^{-23} newton.

14-2. $\dfrac{\mu_0 I}{2\pi}\left(\dfrac{1}{D-x}+\dfrac{1}{D+x}\right)$.

14-5. (a) $erv/2$. (b) 9.3×10^{-24} ampere-meter^2.

14-9. $\mu_0 N I a^2 \left\{\dfrac{1}{2[(z+a/2)^2+a^2]^{3/2}}+\dfrac{1}{2[(z-a/2)^2+a^2]^{3/2}}\right\}$.

15-2. (a) 330 microamperes. (b) 4.2×10^{-2} tesla.

15-4. $\mu_0\alpha/2$.

15-8. $L \ln \ldots /(R_2 - R_1)$.

16-1. 2.

16-4. $\theta = \pi/4$.

16-11. $B, B/\mu_0$.

17-1. (a) 3×10^{13} meters. (b) 7.3 days.

17-2. (b) $2.2 \times 10^{18}\,\hat{\boldsymbol{y}}$ meters/second^2.

17-5. 1.2, 2.7.

17-7. (c) 1.4×10^{-4} volt.

17-10. (b) For H_1^+, 0.077 tesla; for H_2^+, 0.11 tesla; for H_3^+, 0.13 tesla.

17-13. (a) 8.7×10^8 amperes/meter^2. (b) Very hot. (c) East.

17-17. (a) $AV\omega/R - A^2\omega^2/R$.

18-3. (a) Charge flows to cancel the $\boldsymbol{v} \times \boldsymbol{B}$ field. (b) 31 amperes. (c) 1.5×10^{-2} newton. (d) For a diameter of 10 meters, $I = 30$ microamperes, $P \approx 1$ microwatt, $F \approx 2 \times 10^{-10}$ newton.

18-4. (a) 1.2 millivolts. (b) Zero.

18-11. $P_{\text{max}} = \pi\sigma\omega^2 B_{\text{rms}}^2 sa^4/8$.

18-13. (a) $B_0\omega^2 \cos(2\omega t - \phi)$. (b) 2 kilohertz.

19-2. $3.4 \times 10^{-7}\,R$ henry.

19-5. (b) The mutal inductance varies as the cosine of the angular rotation. (c) No.

19-7. (a) $R' = \dfrac{2\pi a}{\sigma b},\ L' = \mu_0 \pi^2$, (b) $\left(\dfrac{4 + x^2}{4 + 4x^2}\right)^{1/2}$, $x = \omega\mu_0\sigma ab$.

19-8. $L = R_b R_d C,\ R = R_b R_d/R_c$.

20-3. (a) 4.4 joules/meter^3. (b) 4.0×10^5 joules/meter^3.

20-6. 2.9×10^4 ampere-turns.

20-9. (a) 22 kilowatt-hours/meter^3. (b) 780 atmospheres. (c) 0.083 kilowatt-hour/meter^3.

20-14. (a) 4 atmospheres. (b) The same.

20-18. (c) $-(R/2)(\partial B_z/\partial z),\ m(\partial B_z/\partial z)$. (d) $I\pi R^2(\partial B_z/\partial z)$.

21-2. (a) $\mu_0 Q\mathcal{V} \sin\theta/(4\pi r^2)$. (b) Set $\gamma = 1$, $\mathcal{V}/c = 0$. The two values agree. (d) Q.

21-8. (a) $\dfrac{\epsilon_0 kb}{s}(a - x)$. (b) $\boldsymbol{B} = \dfrac{\epsilon_0\mu_0 k}{s}(a - x)\hat{\boldsymbol{y}}$, $\boldsymbol{A} = \dfrac{\epsilon_0\mu_0 k}{s}(a - x)z\hat{\boldsymbol{x}}$.

(c) 0. Above the upper plate, $(\epsilon_0\mu_0 k/s)(a - x)\hat{\boldsymbol{x}}/2$. Below the upper plate, $\boldsymbol{A}$ has the same magnitude and the opposite sign.

21-9. (e) 1.2×10^{-8} meter.

21-10. (a) $-\sigma_{co}RC/\epsilon$.

22-6. (a) $E = iR'$, $H = I/(2\pi a)$, $|\mathbf{E} \times \mathbf{H}| = I^2R'/(2\pi a)$, pointing inward. $2\pi aEH = I^2R'$ is the power dissipation per meter. Power flows radially from the field into the wire. (b) Outside, $2\pi rEH = I^2R'(r^2/a^2)$ is the power flowing into the region inside r.

22-11. (a) (i) $\dfrac{I^2}{16\pi\epsilon_0 v^2}\left(1 + 4\ln\dfrac{R_2}{R_1}\right)$. (ii) $\mathscr{E}'_M = \dfrac{v^2}{c^2}\mathscr{E}'_E$. (iii) $2\mathscr{E}'_E v$. (iv) $\dfrac{Imv^2}{2e}$, or IV.

(b) (i) 2.0×10^{-10} joule/meter. (ii) 4.2×10^{-13} joule/meter. (iii) 5.5×10^{-3} watt. (iv) 10^3 watts.

23-7. $F = [1 - (1 - n_2^2/n_1^2)^{1/2}]/2$.

23-8. (a) 17°. (e) $F \approx 0.020$, $T \approx 0.69$, $FT \approx 1.4\%$.

24-7. $\lambda_0 = 4a(n_2^2 - n_3^2)^{1/2}$ meter.

25-2. 88%.

25-9. 0.22 volt/meter.

REFERENCES

J. van Bladel, 1984, *Relativity and Engineering*, Springer-Verlag, Berlin.

J. van Bladel, 1985, *Electromagnetic Fields*, Hemisphere Publishing, Washington.

H.G. Booker, 1982, *Energy in Electromagnetism*, Institution of Electrical Engineers, London.

A.M. Bork, 1963, "Maxwell, Displacement Current, and Symmetry," *American Journal of Physics* **31**, 854–859.

L.G. Chambers, 1963, *Journal of Mathematical Physics* **4**, p. 1373.

D.R. Corson, 1956, "Electromagnetic induction in moving systems," *American Journal of Physics* **24**, 126–130.

H. Erlichson, 1999, "Ampère was not the author of Ampère's circuital law," *American Journal of Physics* **67**, 448.

R.P. Feynman, R.B. Leighton, and M. Sands, 1964, *Lectures on Physics*, Addison-Wesley, Reading, Mass.

A.P. French, 1994, "Question #6. Faraday's law," *American Journal of Physics* **62**, 972.

F. Friedman, D. Frisch, and V. Smith, 1963, *Time Dilation: An Experiment in Mu-Mesons* [Motion Picture], Educational Development Center, Newton, Mass.

F. Gardiol, 1994, *Microstrip Circuits*, John Wiley, New York, N.Y.

C.C. Gillispie, 1980, editor in chief, *Dictionary of Scientific Biography*, Charles Scribner's Sons, New York.

L. Golub and J. Pasachoff, 1997, *The Solar Corona*, Cambridge University Press, Cambridge, U.K.

P. Harman, 1998, *The Natural Philosophy of James Clerk Maxwell*, Cambridge University Press, Cambridge, U.K.

O. Heaviside, 1950, *Electromagnetic Theory*, Volumes I, II, and III, originally published in 1893, 1899, and 1912, with a Critical and Historical Introduction by Ernst Weber, with references, Dover Publications, New York.

O. Heaviside, 1971, *Electromagnetic Theory*, Volumes I, II, and III as above, with a Foreword by Sir Edmund Whittaker and references, Chelsea Publishing Company, New York.

B.J. Hunt, 1991, *The Maxwellians*, Cornell University Press, Ithaca, N.Y.

J.D. Jackson, 1999, *Classical Electrodynamics*, John Wiley, New York.

B. Jakosky, 1998, *The Search for Life on Other Planets*, Cambridge University Press, Cambridge, U.K.

R.C. Johnson, 1993, *Antenna Engineering Handbook*, 3rd ed, McGraw-Hill, New York.

C. Kittel, 1976, *Introduction to Solid State Physics*, 5th ed, John Wiley, New York, p. 166.

P. Lorrain, 1967, "The Lorentz Condition," *Transactions of the Royal Society of Canada* V, 233–240.

P. Lorrain, 1978, "Motional E.M.F.," *Physics Education* **13**, 203–204.

P. Lorrain, 1982, "Alternative Choice for the Energy Flow Vector of the Electromagnetic Field," *American Journal of Physics* **50**, 492.

P. Lorrain, 1984, "The Poynting Vector in a Transformer," *American Journal of Physics* **52**, 987–988.

P. Lorrain, 1985, "Clarification" (in relation with the Collett method for the measurement of an index of refraction), *Laser Focus/Electro-Optics* November 1985, p. 20.

P. Lorrain, 1987, "Comment on the Poynting Vector Distribution in a Transformer," *American Journal of Physics* **55**, 474.

P. Lorrain, 1990, "Electrostatic Charges in $\boldsymbol{v} \times \boldsymbol{B}$ Fields: the Faraday Disk and the Rotating Sphere," *European Journal of Physics* **11**, 94–98.

P. Lorrain, 1993, "Azimuthal Magnetic Fields in the Earth's Core," *Physica Scripta* **47**, 461–468.

P. Lorrain, 1995, "Self-Excited Dynamos in Nature: The Disk Dynamo Model," *Physica Scripta* **52**, 349–352.

P. Lorrain, D.R. Corson, F. Lorrain, *Electromagnetic Fields and Waves*, 3rd ed, 1988, Freeman, NewYork.

P. Lorrain and O. Koutchmy, 1998, "The Sunspot as a Self-Excited Dynamo," *Astronomy and Astrophysics* **339**, 610–614.

P. Lorrain and S. Koutchmy, 1993, "Photospheric Electric Currents in Solar Magnetic Elements," *Astronomy and Astrophysics* **269**, 518–526.

P. Lorrain and S. Koutchmy, 1996, "Two Dynamic Models for Solar Spicules," *Solar Physics* **165**, 115-137.

P. Lorrain and S. Koutchmy, 1998, "Chromospheric Heating by Electric Currents Induced by Fluctuating Magnetic Elements," *Solar Physics* **178**, 39–42.

P. Lorrain, J. McTavish, and F. Lorrain, 1998, "Magnetic Fields in Moving Conductors: Four Simple Examples," *European Journal of Physics* **19**, 451–457.

P. Lorrain and N. Salingaros, 1993, "Local Currents in Magnetic Flux Tubes and Flux Ropes," *American Journal of Physics* **61**, 811–817.

P. Lorrain, R. Tyler, and F. Lorrain, probably 2001, *Magneto-Fluid-Dynamics, with Applications to Natural Phenomena*, Springer, New York.

V.K. McElheny, 1998, *Insisting on the Impossible: The Life of Edwin Lamb*, The Sloan Technology Series, Perseus Books, Reading, Massachusetts.

R.T. Merrill, M.W. McElhinny, and P.L. McFadden, 1996, *The Magnetic Field of the Earth*, International Geophysics Series, Volume 63, Academic Press.

H.K. Messerle, 1995, *Magnetohydrodynamic Power Generation*, John Wiley and Sons, New York.

M.S. Mirotznik and D. Prather, 1997, "How to Choose Electromagnetic Software," *Spectrum*, Institute of Electrical and Electronic Engineers, December, 53–58.

J.C. Palais, 1992, *Fiber Optic Communications*, 3rd ed, Prentice Hall, Upper Saddle River, N.J.

P. Penfield and H. Haus, 1967, *Electrodynamics of Moving Media*, Research Monograph 40, M.I.T. Press, Cambridge, Mass.

J.C. Sabonadiere and A. Konrad, 1992, "Computing Electromagnetic Fields," *Spectrum*, Institute of Electrical and Electronic Engineers, November, 52–56.

R.V. Southwell, 1940, *Relaxation Methods in Engineering Science; a Treatise on Approximate Computation,* Clarendon Press, Oxford, U.K.

R.V. Southwell, 1946, *Relaxation Methods in Theoretical Physics; a continuation of the treatise Relaxation Methods in Engineering Science,* Oxford University Press, Oxford, U.K.

F.D. Stacey, 1992, *Physics of the Earth*, 3rd ed, Brookfield Press, Brisbane, Australia.

E.F. Taylor and J.A. Wheeler, 1992, *Spacetime Physics*, 2nd ed, W.H. Freeman, New York.

R.S. Tebble, 1969, *Magnetic Domains*, Methuen, London.

J.J. Thomson, 1893, *Notes on Recent Researches on Electricity and Magnetism*, Clarendon Press, Oxford, p. 534.

N. Thorsen, 1998, *Fiber Optics and the Telecommunications Explosion*, Prentice Hall, Upper Saddle River, N.J.

W.H. Watson, 1950, *Proceedings of the 2nd symposium on Applied Mathematics*, p. 49.

R. Waynant and M. Ediger, 2000, *Electro-Optics Handbook*, McGraw-Hill Inc., New York.

Y. Xu, 1991, *Ferroelectric Materials and Their Applications*, North Holland, New York, 1991.

H. Zirin, 1988, *Astrophysics of the Sun*, Cambridge University Press, Cambridge, U.K.

INDEX